Evaluation and Utilization of Bioethanol Fuels. II.

The sixth volume of ***Handbook of Bioethanol Fuels*** provides an overview of the research on the country-based experience of bioethanol fuels at large, Chinese, US, and European experience of bioethanol fuels in particular, production of bioethanol fuel-based biohydrogen fuels for fuel cells, bioethanol fuel cells, and bioethanol fuel-based biochemicals with a collection of 17 chapters. Thus, it complements the fifth volume of this handbook.

Hence, this book indicates that the research on the evaluation and utilization of bioethanol fuels has intensified in recent years to become a major part of the bioenergy and biofuels research together primarily with biodiesel, biohydrogen, and biogas research as a sustainable alternative to crude oil-based gasoline and petrodiesel fuels as well as natural gas and syngas.

This book is a valuable resource for the stakeholders primarily in the research fields of energy and fuels, chemical engineering, environmental science and engineering, biotechnology, microbiology, chemistry, physics, mechanical engineering, agricultural sciences, food science and engineering, materials science, biochemistry, genetics, molecular biology, plant sciences, water resources, economics, business, management, transportation science and technology, ecology, public, environmental and occupational health, social sciences, toxicology, multidisciplinary sciences, and humanities among others.

Evaluation and Utilization of Bioethanol Fuels. II.

Biohydrogen Fuels, Fuel Cells, Biochemicals, and Country Experiences

Edited by
Ozcan Konur

CRC Press
Taylor & Francis Group
Boca Raton London New York

CRC Press is an imprint of the
Taylor & Francis Group, an **informa** business

Designed cover image: © Shutterstock

First edition published 2024
by CRC Press
2385 NW Executive Center Drive, Suite 320, Boca Raton FL 33431

and by CRC Press
4 Park Square, Milton Park, Abingdon, Oxon, OX14 4RN

CRC Press is an imprint of Taylor & Francis Group, LLC

ISBN: 978-1-032-12761-3 (hbk)
ISBN: 978-1-032-12868-9 (pbk)
ISBN: 978-1-003-22657-4 (ebk)

DOI: 10.1201/9781003226574

Typeset in Times
by codeMantra

Contents

PART 29 Country-based Experience of Bioethanol Fuels

PART 30 Bioethanol Fuel-based Biohydrogen Fuels

PART 31 Bioethanol Fuel Cells

Preface

Recent supply shocks caused first by the COVID-19 pandemic and later by the Ukrainian war have shown that biofuels such as bioethanol, biohydrogen, biogas, biosyngas, and biodiesel fuels could play a vital role in maintaining the energy security and indirectly food security at the global scale. These shocks have also resulted in the need for further setup of incentive structures for the production and consumption of bioethanol fuels in blends with crude oil-based gasoline, petrodiesel, or liquefied natural gas (LNG) in gasoline and diesel engines, for their direct utilization in direct ethanol fuel cells (DEFCs), and for the production of biohydrogen fuels for fuel cells and valuable biochemicals from bioethanol fuels.

Thus, it is essential to assess the research on the production, evaluation, and utilization of bioethanol fuels from a wide range of biomass including first generation starch and sugar feedstocks, wood, grass, second generation lignocellulosic biomass including waste biomass and agricultural residues such as starch feedstock residues and sugar feedstock residues, and the third generation algal biomass.

Thus, this six-volume *Handbook of Bioethanol Fuels* assesses the research on the production, evaluation, and utilization of bioethanol fuels and presents a representative sample of this interdisciplinary research population with a collection of 110 chapters (Table 1.1).

The first two volumes provide an overview of the research on fundamental processes for bioethanol fuel production with a collection of 39 chapters: Pretreatments of biomass, hydrolysis of the pretreated biomass, microbial fermentation of hydrolysates with yeasts, and separation and distillation of bioethanol fuels from the fermentation broth. They also provide an overview of the research on bioethanol fuels and production processes for bioethanol fuels (Tables 1.2 and 1.3).

The third and fourth volumes provide an overview of research on the production of bioethanol fuels from non-waste and waste biomass, respectively, with a collection of 36 chapters. In this context, the third volume covers the production of bioethanol fuels from first generation starch feedstocks and sugar feedstocks, grass biomass, wood biomass, cellulose, biosyngas, and third generation algae (Table 1.4) while the fourth volume covers the production of second generation bioethanol fuels from residual sugar feedstocks, residual starch feedstocks, food waste, industrial waste, urban waste, forestry waste, and lignocellulosic biomass at large (Table 1.5). They also provide an overview of the research on feedstock-based bioethanol fuels, non-waste feedstock-based bioethanol fuels, and second generation waste biomass-based bioethanol fuels (Tables 1.4 and 1.5).

Finally, the fifth and sixth volumes provide an overview of the research on the evaluation and utilization of bioethanol fuels with a collection of 37 chapters. In this context, the fifth volume covers the evaluation and utilization of bioethanol fuels in general, gasoline fuels, nanotechnology applications in bioethanol fuels, utilization of bioethanol fuels in transport engines, evaluation of bioethanol fuels, utilization of bioethanol fuels, and development and utilization of bioethanol fuel sensors (Table 1.6).

Thus, the sixth volume of this handbook provides an overview of the research on the country-based experience of bioethanol fuels at large, Chinese, US, and European experience of bioethanol fuels, production of bioethanol fuel-based biohydrogen fuels for fuel cells, bioethanol fuel cells, and bioethanol fuel-based biochemicals with a collection of 17 chapters (Table 1.7).

Hence, the sixth volume indicates that the research on the evaluation and utilization of bioethanol fuels has intensified in recent years to become a major part of the bioenergy and biofuels research together primarily with biodiesel, biohydrogen, and biogas research as a sustainable alternative to crude oil-based gasoline and petrodiesel fuels as well as natural gas and syngas.

The sixth volume also indicates that the production of bioethanol fuels and their derivatives and coproducts in a biorefinery context increases the utility of bioethanol fuels and reduces their production cost in relation to the crude oil-, natural gas-, syngas-, and coal-based fuels as well as other biofuels such as biodiesel fuels.

The sixth volume also indicates that bioethanol fuels are primarily used directly in the DEFCs as an alternative to crude oil- and natural gas-based fuels, biohydrogen fuels used in fuel cells as an alternative to their direct use in DEFCs, and biochemicals used as an alternative to crude oil-based chemicals. It also shows that the USA, Brazil, Europe, and China have been the major producers and consumers of the bioethanol fuels and their derivatives, setting up national policies for developing incentive structures for the production and consumption of bioethanol fuels and their derivatives and coproducts.

The sixth volume also indicates that a small number of documents, authors, institutions, publication years, source titles, countries, Scopus subject categories, Scopus keywords, and research fronts have shaped the research on the evaluation and utilization of bioethanol fuels.

The sixth volume also indicates that the level of funding for the research on evaluation and utilization of bioethanol fuels has not been sufficient with the resulting loss of momentum in the research output in recent years. Thus, there is a crucial need to improve the incentive structures for the major stakeholders such as researchers and their institutions as well as source titles and academic databases to improve the volume and quality of the research output in these fields. This is a crucial need to maintain the energy security and food security indirectly at a global scale in light of the recent supply shocks caused by the COVID-19 pandemics and the Ukrainian war.

The sixth volume also indicates that the contribution of the social sciences and humanities to the research in these fields has been minimal, due to in part by the restrictive editorial policies of the source titles in these fields toward social science- and humanities-based interdisciplinary studies. Thus, there is ample room to improve incentive structures for the inclusion of social sciences and humanities into these fields.

The sixth volume also indicates that China, Europe as a whole, and the USA have been major producers of research in these research fields and there has been heavy competition among them in terms of both volume and citation impact of the research output. The USA and Europe as a whole have had a higher citation impact in relation to China benefiting from their first-mover advantage starting their research in these fields in the 1970s. China as a late mover has had more intensive research funding initiatives in relation to the USA and Europe, improving its both research output and citation impact through the provision of efficient incentive structures for its major stakeholders in the last two decades. In this way, China might also overtake both the USA and Europe in terms of citation impact of the research output in addition to the volume of the research output in the future.

This handbook at large and sixth volume are a valuable resource for the stakeholders primarily in the research fields of energy and fuels, chemical engineering, environmental science and engineering, biotechnology, microbiology, chemistry, physics, mechanical engineering, agricultural sciences, food science and engineering, materials science, biochemistry, genetics, molecular biology, plant sciences, water resources, economics, business, management, transportation science and technology, ecology, public, environmental and occupational health, social sciences, toxicology, multidisciplinary sciences, and humanities among others.

Ozcan Konur

Acknowledgments

This handbook has been a multi-stakeholder project from its conception to its publication. CRC Press and Taylor and Francis Group have been the major stakeholders in financing and executing it. Marc Gutierrez has been the executive director of the project. A large number of teams from the Publisher have contributed immensely to the production of the handbook. Only a limited number of authors have participated in this project due to the low level of incentives, compared to journals. A small number of highly cited scholars have shaped the research on bioethanol fuels. The contribution of all these and other stakeholders to this handbook has been greatly acknowledged.

Editor

The Editor has interdisciplinary research interests and has published primarily in the areas of bioenergy and biofuels, algal bioenergy and biofuels, nanoenergy and nanofuels, nanobiomedicine, algal biomedicine, disability studies, higher education, biodiesel fuels, algal biomass, lignocellulosic biomass, scientometrics, and bioethanol fuels. He has edited a book titled *Bioenergy and Biofuels* (CRC Press, 2018), a handbook titled *Handbook of Algal Science, Technology, and Medicine* (Elsevier, 2020), and a handbook titled *Handbook of Biodiesel and Petrodiesel Fuels: Science, Technology, Health, and Environment* (CRC Press, 2021) in three volumes.

Contributors

Angelo Basile
Hydrogenia S.R.L.
Genova, Italy

Vannessa Caballero
Gene and Linda Voiland School of Chemical Engineering and Bioengineering, Washington State University
Pullman, WA

Francesco Dalena
Department of Environmental and Chemical Engineering, University of Calabria
Rende, Italy

Ozcan Konur
(Formerly) Department of Materials Engineering, Ankara Yildirim Beyazit University
Ankara, Turkey

Thanh Khoa Phung
Department of Chemical Engineering, School of Biotechnology, International University
Ho Chi Minh City, Vietnam
and
Vietnam National University
Ho Chi Minh City, Vietnam

Anthony William Savoy
Gene and Linda Voiland School of Chemical Engineering and Bioengineering, Washington State University
Pullman, WA

Alessandro Senatore
Department of Chemistry and Chemical Technologies, University of Calabria
Rende, Italy

Junming Sun
Gene and Linda Voiland School of Chemical Engineering and Bioengineering, Washington State University
Pullman, WA

Khanh B. Vu
Department of Chemical Engineering, School of Biotechnology, International University;
Ho Chi Minh City, Vietnam
Vietnam National University, Ho Chi Minh City, Vietnam

Yong Wang
Pacific Northwest National Laboratory, Institute for Integrated Catalysis
Richland, WA
and
Gene and Linda Voiland School of Chemical Engineering and Bioengineering, Washington State University
Pullman, WA

Part 29

Country-based Experience of Bioethanol Fuels

94 Country-based Experience of Bioethanol Fuels
Scientometric Study

Ozcan Konur
(Formerly) Ankara Yildirim Beyazit University

94.1 INTRODUCTION

Crude oil-based gasoline fuels (Ma et al., 2002; Newman and Kenworthy, 1989) have been widely used in the transportation sector since the 1920s. However, there have been great public concerns over the adverse environmental and human impact of these fuels (Hill et al., 2006, 2009). Hence, biomass-based bioethanol fuels (Hill et al., 2006; Konur, 2012e, 2015, 2019, 2020a) have increasingly been used in blending gasoline fuels (Hsieh et al., 2002; Najafi et al., 2009) and in fuel cells (Antolini, 2007, 2009). Additionally, bioethanol fuels have been used to produce valuable biochemicals (Angenent et al., 2004; Nikolau et al., 2008) in a biorefinery (Cherubini et al., 2009; Maity, 2015) context.

The research in the field of experiences of countries has also intensified in recent years (Amorim et al., 2011; Lopes et al., 2016). The primary focus of research on these country-based experiences has been the optimization of bioethanol and biomass production and the development of policies to promote the production and use of bioethanol fuels worldwide. In this context, there has been a significant focus on evaluative studies in bioethanol fuels (Farrell et al., 2006; Hill et al., 2006).

However, it is essential to develop efficient incentive structures (North, 1991) for the primary stakeholders to enhance research in this field (Konur, 2000, 2002a,b,c, 2006a,b, 2007a,b). The scientometric analysis has been used in this context to inform the primary stakeholders about the current state of research in a selected research field (Garfield, 1955; Konur, 2011, 2012a,b,c,d,e,f,g,h,i, 2015, 2018b, 2019, 2020a).

As there have been no scientometric studies on the country-based experiences of bioethanol fuels, this book chapter presents a scientometric study of research in the country-based experience of bioethanol fuels. It examines the scientometric characteristics of both the sample and population data presenting the scientometric characteristics of these both datasets in the order of documents, authors, publication years, institutions, funding bodies, source titles, countries, Scopus subject categories, keywords, and research fronts.

94.2 MATERIALS AND METHODS

The search for this study was carried out using the Scopus database (Burnham, 2006) in October 2021.

As a first step for the search of the relevant literature, the keywords were selected using the first 200 most-cited papers. The selected keyword list was optimized to obtain a representative sample of papers for the searched research field. This keyword list was provided in the appendix for future replication studies. Additionally, the information about the most-used keywords was given in Section 94.3.9 to highlight the key research fronts in Section 94.3.10.

DOI: 10.1201/9781003226574-127

As a second step, two sets of data were used for this study. First, a population sample of over 1,300 papers was used to examine the scientometric characteristics of the population data. Second, a sample of 100 most-cited papers was used to examine the scientometric characteristics of these citation classics with over 59 citations each.

The scientometric characteristics of these both sample and population datasets were presented in the order of documents, authors, publication years, institutions, funding bodies, source titles, countries, Scopus subject categories, keywords, and research fronts.

Lastly, the key scientometric findings for both datasets were discussed to highlight the research landscape for the country-based experience of bioethanol fuels. Additionally, several brief conclusions were drawn and many relevant recommendations were made to enhance the future research landscape.

94.3 RESULTS

94.3.1 The Most Prolific Documents in the Country-based Experience of Bioethanol Fuels

The information on the types of documents for both datasets is given in Table 94.1. The articles and review papers dominate both the sample and population datasets, while the review papers and book chapters have a surplus and deficit, respectively.

It is also interesting to note that all of the papers in the sample dataset were published in journals, while only 92.1% of the papers were published in journals for the population dataset. Furthermore, 6.1% and 1.8% of the population papers were published in books and book series, respectively.

94.3.2 The Most Prolific Authors in the Country-based Experience of Bioethanol Fuels

The information about the 29 most prolific authors with at least two sample papers and three population papers each is given in Table 94.2.

The most prolific authors are Antonio Bonomi, Shabbir H. Gheewala, Joaquim E. A. Sabra, and Arnaldo Walter with at least five sample papers each. Andre P. C. Faaij, Otavio Cavalett, Thu L. T. Nguyen, Isaisas de Carvalho Macedo, and H. Martin Junginger follow these top authors with four sample papers each.

TABLE 94.1
Documents in the Country-based Experience of Bioethanol Fuels

Documents	Sample Dataset (%)	Population Dataset (%)	Surplus (%)
Article	74	76.7	–2.7
Review	21	8.8	12.2
Conference paper	3	4.0	–1
Short survey	2	0.4	1.6
Book chapter	0	6.0	–6
Note	0	2.5	–2.5
Letter	0	0.9	–0.9
Book	0	0.4	–0.4
Editorial	0	0.2	–0.2
Sample size	100	1,335	

TABLE 94.2
Most Prolific Authors in the Country-based Experience of Bioethanol Fuels

No.	Author Name	Author Code	Sample Papers	Population Papers	Institutions	Country	Res. Front
1	Bonomi, Antonio	7004767629	6	14	Univ. Campinas	Brazil	Brazil
2	Gheewala, Shabbir H.	6602264724	5	26	King Mongkut's Univ. Technol.	Thailand	Thailand
3	Seabra, Joaquim E. A.	23390441500	5	10	Univ. Campinas	Brazil	Brazil
4	Walter, Arnaldo	7102205549	5	10	Univ. Campinas	Brazil	Brazil
5	Faaij, Andre P. C. (A)	6701681600	4	11	Utrecht Univ.	Netherlands	Brazil
6	Cavalett, Otavio	9846419000	4	10	Natl. Bioethanol Sci. Technol. Lab.	Brazil	Brazil
7	Nguyen, Thu L. T.	16480815100	4	9	King Mongkut's Univ. Technol.	Thailand	Thailand
8	De Carvalho Macedo, Isaias	24324505900 6603142465	4	8	Interdiscip. Ctr. Ener. Plan.	Brazil	Brazil
9	Junginger, H. Martin (M)	57195636600	4	6	Utrecht Univ.	Netherlands	Brazil, US
10	Lopes, Mario L.	22986007500	3	6	Fermentec	Brazil	Brazil
11	Pu, Gengqiang	7003913249	3	6	Shanghai Jiao Tong Univ.	China	China
12	Wang, Chengtao	35231632300	3	5	Shanghai Jiao Tong Univ.	China	China
13	Garivait, Savitri	16479645800	3	4	King Mongkut's Univ. Technol.	Thailand	Thailand
14	Buckeridge, Marcos S.	6701624952	2	8	Sao Paulo Univ.	Brazil	Brazil
15	Cardona, Carlos A.	57214443163	2	7	Natl. Colombia Univ.	Colombia	Colombia, Africa
16	Griffin, W. Michael	7201673419	2	7	Carnegie Mellon Univ.	USA	USA, Brazil
17	Leal, Manoel R. L. V.	25925560500	2	7	Univ. Campinas	Brazil	Brazil
18	De Amorim, Henrique V	6602097475	2	7	Fermentec	Brazil	Brazil
19	Martinelli, Luiz A.	7102366222	2	6	Sao Paulo Univ.	Brazil	Brazil
20	Chagas, Mateus F.	55266291800	2	5	Natl. Bioethanol Sci. Technol. Lab.	Brazil	Brazil
21	Basso, Luiz C.	8352847700	2	4	Sao Paulo Univ.	Brazil	Brazil
22	Hu, Zhiyuan	57208640287	2	4	Tongji Univ.	China	China
23	Lee, Keat T.	8675851300	2	4	Univ. Sains Malaysia	Malaysia	Malaysia
24	Filho, Rubens M.	7003732915 57132293600	2	4	Univ. Campinas	Brazil	Brazil
25	Quintero, Julian A.	23482772800	2	4	Natl. Colombia Univ.	Colombia	Colombia, Africa
26	Galdos Marcelo V. (M)	26534213300	2	4	Natl. Bioethanol Sci. Technol. Lab.	Brazil	Brazil
27	Filoso, Solange	6603438972	2	3	Univ. Maryland	USA	Brazil
28	Gnansounou, Edgard	6508334495	2	3	Swiss Fed. Inst. Technol.	Switzerland	Europe, China
29	Rossell, Carlos E. V.	6505763454 12780694800	2	3	Natl. Bioethanol Sci. Technol. Lab.	Brazil	Brazil

Author Code, the unique code given by Scopus to the authors; Population Papers, the number of papers authored in the population dataset; Sample Papers: the number of papers authored in the sample dataset.

The most prolific institution for the sample dataset is the University of Campinas with five authors. The National Bioethanol Science and Technology Laboratory, King Mongkut's University of Technology, and Sao Paulo University follow these top institutions with at least three authors each. Additionally, Fermentec, National Colombia University, Sao Paulo University, Shanghai Jiaotong University, and Utrecht University published two papers each. In total, 13 institutions house these prolific authors.

The most prolific country for the sample dataset is Brazil with 15 authors. China and Thailand follow Brazil with three authors each. The other prolific countries are Colombia, the Netherlands, and the USA. In total, eight countries house these authors.

The most prolific research front is the Brazilian experience of bioethanol with 18 authors. The other prolific research fronts are the Chinese, Colombian, African, and US experiences of bioethanol fuels with at least two papers each. There is also one paper each for the European and Malaysian experiences.

However, there is a significant gender deficit (Beaudry and Lariviere, 2016) for the sample dataset as surprisingly nearly all of these top researchers are male.

94.3.3 The Most Prolific Research Output by Years in Country-based Experience of Bioethanol Fuels

Information about papers published between 1970 and 2021 is given in Figure 94.1. This figure clearly shows that the bulk of research papers in the population dataset were published primarily in the 2010s with 64.4% of the population datasets. The publication rates for the 2020s, 2000s, 1990s, 1980s, 1970s, and pre-1970s were 9.6%, 20.2%, 2.5%, 2.3%, 0.3%, and 0.2%, respectively. There was a rising trend for the population papers between 2006 and 2011, and after 2011, it became steady at around 6% of the population papers for each year, losing its momentum.

Similarly, the bulk of research papers in the sample dataset were published in the 2000s and 2010s with 40% and 46% of the sample datasets, respectively. The publication rates for the 1990s, 1980s, and 1970s were 4%, 4%, and 2% of the sample papers, respectively.

The most prolific publication years for the population dataset were after 2007 with at least 4.1% each year of the dataset. Similarly, 73% of the sample papers were published between 2007 and 2013.

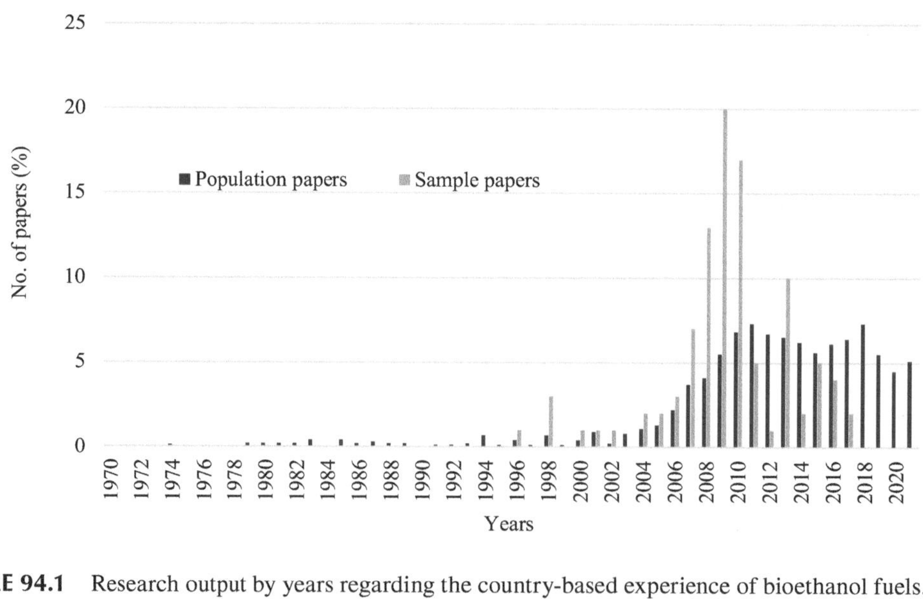

FIGURE 94.1 Research output by years regarding the country-based experience of bioethanol fuels.

94.3.4 The Most Prolific Institutions in the Country-based Experience of Bioethanol Fuels

Information about the 22 most prolific institutions publishing papers on the country-based experiences of bioethanol fuels with at least two sample papers and 0.4% of the population papers each is given in Table 94.3.

The most prolific institutions are the State University of Campinas, University of Sao Paulo, and National Biorenewables Laboratory with 15, 11, and 10 papers, respectively. The King Mongkut's University of Technology, Chinese Academy of Sciences, University of Utrecht, and National Research Center of Energy and Materials follow these top institutions with at least four sample papers each.

The top country for these most prolific institutions is Brazil with ten institutions. Next, the USA and China follow Brazil with at least two institutions each. In total, ten countries house these top institutions.

However, the institutions with the most citation impact are State University of Campinas and National Biorenewables Laboratory with 11.2% and 8.7% surplus, respectively. Similarly, the institutions with the least impact are Federal University Rio de Janeiro and Iowa State University with 0.5% surplus each.

94.3.5 The Most Prolific Funding Bodies in the Country-based Experience of Bioethanol Fuels

Information about the 13 most prolific funding bodies funding at least two sample papers and 0.4% of the population papers each is given in Table 94.4.

TABLE 94.3
Most Prolific Institutions in Country-based Experience of Bioethanol Fuels

No.	Institutions	Country	Sample Papers (%)	Population Papers (%)	Surplus (%)
1	State Univ. Campinas	Brazil	15	3.8	11.2
2	Univ. Sao Paulo	Brazil	11	7.9	3.1
3	Natl. Biorenewables Lab.	Brazil	10	1.9	8.1
4	King Mongkut's Univ. Technol.	Thailand	5	2.2	2.8
5	Chinese Acad. Sci.	China	4	1.4	2.6
6	Univ. Utrecht	Netherlands	4	1.0	3
7	Natl. Res. Ctr. Energ. Mater.	Brazil	4	0.9	3.1
8	Fed. Univ. Rio de Janeiro	Brazil	3	2.5	0.5
9	Shanghai Jiao Tong Univ.	China	3	1.0	2
10	Fermentec	Brazil	3	0.4	2.6
11	Iowa State Univ.	USA	2	1.5	0.5
12	Imperial Coll.	UK	2	1.2	0.8
13	Univ. Illinois	USA	2	1.0	1
14	Fed. Univ. Parana	Brazil	2	0.7	1.3
15	Univ. Brasilia	Brazil	2	0.7	1.3
16	Carnegie Mellon Univ.	Brazil	2	0.7	1.3
17	Purdue Univ.	USA	2	0.6	1.4
18	Univ. British Columbia	Canada	2	0.6	1.4
19	Fed. Univ. Sao Carlos	Brazil	2	0.6	1.4
20	Natl. Univ. Colombia	Colombia	2	0.5	1.5
21	Ind. Agric. Res. Inst.	India	2	0.4	1.6
22	Int. Inst. Appl. Syst. Analy.	Austria	2	0.4	1.6

TABLE 94.4
Most Prolific Funding Bodies in Country-based Experience of Bioethanol Fuels

No.	Funding Bodies	Country	Sample Paper No. (%)	Population Paper No. (%)	Surplus (%)
1	Sao Paulo State Res. Found	Brazil	9	3.3	5.7
2	European Commission	EU	7	1.3	5.7
3	Natl. Counc. Sci. Technol. Dev.	Brazil	5	4.8	0.2
4	National Natural Science Foundation of China	China	5	3.1	1.9
5	Minist. Sci. Technol. Innov.	Brazil	4	4.1	–0.1
6	Coord. Impov. Higher Educ. Person.	Brazil	4	3.3	0.7
7	National Science Foundation	USA	3	1.8	1.2
8	Funding Stud. Proj.	Brazil	3	0.7	2.3
9	Minist. Sci. Technol.	India	3	0.6	2.4
10	King Mongkut's Univ. Technol.	Thailand	2	1.0	1.0
11	Swedish Ener. Agcy.	Sweden	2	0.6	1.4
12	Govnt. Canada	Canada	2	0.6	1.4
13	Natrl. Sci. Eng. Res. Counc.	Canada	2	0.4	1.6

The most prolific funding bodies are the Sao Paulo State Research Foundation and the European Commission with nine and seven sample papers, respectively. The National Council for Science and Technology Development follows these top funding bodies with five sample papers each.

It is notable that 48% and 34.2% of the sample and population papers are funded, respectively.

The most prolific countries for these top funding bodies are Brazil and Canada with five and two funding bodies, respectively. In total, seven countries and the European Union house these top funding bodies.

The funding bodies with the most citation impact are Sao Paulo State Research Foundation and the European Commission with 5.7% surplus each. Similarly, the funding bodies with the least citation impact are Ministry of Science, Technology, and Innovation and National Council for Science and Technology Development with at least 0.1% deficit each.

94.3.6 The Most Prolific Source Titles in the Country-based Experience of Bioethanol Fuels

Information about the 14 most prolific source titles publishing at least two sample papers and 0.3% of the population papers each in the country-based experience of bioethanol fuels is given in Table 94.5.

The most prolific source titles are 'Renewable and Sustainable Energy Reviews', 'Applied Energy', and 'Bioresource Technology' with 10 sample papers each. The 'Energy Policy' and 'Biomass and Bioenergy' follow these top titles with nine and eight sample papers, respectively.

However, the source titles with the most citation impact are 'Bioresource Technology' and 'Applied Energy' with at least 8.1% surplus each. The 'Renewable and Sustainable Energy Reviews', Energy Policy', and 'Biomass and Bioenergy' follow these top titles with at least 4.4% surplus each. Similarly, the source titles with the least impact are the 'Journal of Cleaner Production', 'Renewable Energy', and 'Energy for Sustainable Development' with at least 1.1% deficit each.

94.3.7 The Most Prolific Countries in the Country-based Experience of Bioethanol Fuels

Information about the 16 most prolific countries publishing at least two sample papers and 0.7% of the population papers each in the country-based experience of bioethanol fuels is given in Table 94.6.

TABLE 94.5
Most Prolific Source Titles in Country-based Experience of Bioethanol Fuels

No.	Source Titles	Sample Papers (%)	Population Papers (%)	Surplus (%)
1	Renewable and Sustainable Energy Reviews	10	3.9	6.1
2	Applied Energy	10	1.9	8.1
3	Bioresource Technology	10	1.6	8.4
4	Energy Policy	9	4.3	4.7
5	Biomass and Bioenergy	8	3.6	4.4
6	Environmental Science and Technology	5	1.5	3.5
7	International Journal of Life Cycle Assessment	5	1.4	3.6
8	Energy	4	2.2	1.8
9	American Journal of Agricultural Economics	3	0.4	2.6
10	Applied Microbiology and Biotechnology	3	0.2	2.8
11	Journal of Cleaner Production	2	3.1	−1.1
12	Renewable Energy	2	1.6	0.4
13	Energy for Sustainable Development	2	1.1	0.9
14	Bioscience	2	0.3	1.7

TABLE 94.6
Most Prolific Countries in the Country-based Experience of Bioethanol Fuels

No.	Countries	Sample Papers (%)	Population Papers (%)	Surplus (%)
1	Brazil	35	23.7	11.3
2	USA	27	21.3	5.7
3	China	11	8.9	2.1
4	Netherlands	8	2.8	5.2
5	UK	7	5.9	1.1
6	Thailand	6	4.6	1.4
7	Canada	5	4.1	0.9
8	Sweden	4	2.8	1.2
9	India	3	4.6	−1.6
10	Germany	3	2.8	0.2
11	Australia	3	1.6	1.4
12	Malaysia	3	1.5	1.5
13	Spain	2	1.6	0.4
14	Colombia	2	1.6	0.4
15	Austria	2	0.8	1.2
16	Switzerland	2	0.7	1.3

The most prolific countries are Brazil and the USA with 35 and 27 sample papers, respectively. China, the Netherlands, the UK, and Thailand are the other prolific countries with at least six sample papers each. Furthermore, seven European countries produce 28% and 17% of the sample and population papers, respectively.

However, the country with the most citation impact is Brazil with 11.3% surplus. The USA and the Netherlands follow Brazil with at least 5.1% surplus each. Similarly, the countries with the least citation impact are India, Germany, Spain, and Colombia with at least 1.6% deficit each.

94.3.8 The Most Prolific Scopus Subject Categories in the Country-based Experience of Bioethanol Fuels

Information about the 11 most prolific Scopus subject categories indexing at least 6% and 6.1% of the sample and population papers, respectively, is given in Table 94.7.

The most prolific Scopus subject categories in the country-based experiences of bioethanol fuels are 'Environmental Science' and 'Energy' with 67 and 61 sample papers, respectively. 'Agricultural and Biological Sciences', 'Engineering', and 'Chemical Engineering' follow these top categories with at least 14 papers each.

However, the Scopus subject categories with the most citation impact are 'Environmental Science' and 'Energy' with 24.8% and 19.1% surplus, respectively. Similarly, the Scopus subject categories with the least citation impact are 'Agricultural and Biological Sciences', 'Business, Management and Accounting', and 'Social Sciences' with at least 6.4% deficit each.

94.3.9 The Most Prolific Keywords in the Country-based Experience of Bioethanol Fuels

Information about the keywords used in at least 5% of the sample or population papers each is given in Table 94.8. For this purpose, keywords related to the keyword set given in the appendix are selected from a list of the most prolific keyword set provided by the Scopus database.

These keywords are grouped under five headings: bioethanol fuels, biomass, issues, and countries and continents related to the countries' experiences of bioethanol fuels.

There are eight keywords used related to bioethanol fuels. The prolific keywords are ethanol, bioethanol, ethanol fuels, and bio-ethanol production with at least 14 sample papers each. It is notable that bioethanol keyword appears in the sample paper keyword list with around 29% and 5% of the sample papers for the keywords of bioethanol and bio-ethanol, respectively.

The prolific keywords related to biomass are sugarcane, biomass, zea, energy crops, and agricultural wastes with at least 13 papers each, while the prolific keywords related to the issues are greenhouse gases (GHGs), life cycle assessment, fermentation, economics, costs, and energy policy with at least 11 sample papers each.

The prolific keywords related to the countries and continents are Brazil, China, the USA, Eurasia, and South America with at least 11 sample papers each.

TABLE 94.7
Most Prolific Scopus Subject Categories in the Country-based Experience of Bioethanol Fuels

No.	Scopus Subject Categories	Sample Papers (%)	Population Papers (%)	Surplus (%)
1	Environmental Science	67	42.2	24.8
2	Energy	61	41.9	19.1
3	Agricultural and Biological Sciences	16	22.4	–6.4
4	Engineering	16	16.3	–0.3
5	Chemical Engineering	14	14.8	–0.8
6	Immunology and Microbiology	8	5.5	2.5
7	Biochemistry, Genetics and Molecular Biology	7	9.6	–2.6
8	Social Sciences	6	9.7	–3.7
9	Chemistry	6	7.3	–1.3
10	Economics, Econometrics and Finance	5	7.9	–2.9
11	Business, Management and Accounting	2	6.1	–4.1

TABLE 94.8
Most Prolific Keywords in Country-based Experience of Bioethanol Fuels

No.	Keywords	Sample Papers (%)	Population Papers (%)	Surplus (%)
1.	Bioethanol fuels			
	Ethanol	90	69.6	20.4
	Bioethanol	29	26.8	2.2
	Ethanol production	23	11.6	11.4
	Ethanol fuels	14	6.8	7.2
	Bio-ethanol production	14	9.9	4.1
	Fuel ethanol	9	3.3	5.7
	Ethanol yield	6	0.0	6
	Bio-ethanol	5	2.1	2.9
2.	Biomass			
	Sugarcane	44	19.3	24.7
	Biomass	23	11.0	12
	Zea	14	7.3	6.7
	Energy crops	13	2.8	10.2
	Agricultural wastes	13	3.1	9.9
	Feedstocks	12	7.2	4.8
	Cellulose	11	6.3	4.7
	Bagasse	10	3.7	6.3
	Lignocellulose	9	2.5	6.5
	Lignin	9	4.0	5
	Saccharum	8	2.8	5.2
	Cassava	8	2.5	5.5
	Biomass production	8	0.0	8
	Molasses	7	3.5	3.5
	Manihot	7	3.0	4
	Corn	7	1.9	5.1
	Maize	6	3.7	2.3
3.	Issues			
	Greenhouse gases	26	17.0	9
	Life cycle assessment	21	8.5	12.5
	Fermentation	14	10.3	3.7
	Economics	13	5.8	7.2
	Costs	12	9.1	2.9
	Energy policy	12	7.5	4.5
	Environmental impact	11	7.3	3.7
	Global warming	11	3.9	7.1
	Land use	11	5.2	5.8
	Sustainable development	11	6.4	4.6
	Biorefineries	11	3.8	7.2
	Carbon dioxide	9	7.2	1.8
	Hydrolysis	9	4.3	4.7
	Sustainability	9	4.8	4.2

(Continued)

TABLE 94.8 (*Continued*)
Most Prolific Keywords in Country-based Experience of Bioethanol Fuels

No.	Keywords	Sample Papers (%)	Population Papers (%)	Surplus (%)
	Energy balance	8	2.5	5.5
	Energy efficiency	8	3.8	4.2
	Gas emissions	8	6.3	1.7
	Air pollution	7	2.4	4.6
	Electricity	7	0.0	7
	International trade	7	2.3	4.7
	Life cycle analysis	6	4.4	1.6
	Biodiversity	6	1.4	4.6
	Competition	6	0.0	6
	Conservation of natural resources	6	0.0	6
	Cost-benefit analysis	6	1.7	4.3
	Economic analysis	6	4.7	1.3
	Energy conservation	6	1.6	4.4
	Energy yield	6	0.0	6
	Ozone	6	0.0	6
	Air quality	5	0.0	5
	Atmospheric pollution	5	0.0	5
	Carbon emission	5	1.7	3.3
	Competitiveness	5	0.0	5
	Electricity generation	5	0.0	5
	Environmental impact assessment	5	2.0	3
	Environmental protection	5	0.0	5
	Environmental sustainability	5	0.0	5
	Greenhouse effect	5	1.9	3.1
	Yeast	5	4.3	0.7
	Life cycle	0	7.2	–7.2
4.	Countries and continents			
	Brazil	31	16.2	14.8
	China	12	5.0	7
	USA	11	6.3	4.7
	Eurasia	11	2.1	8.9
	South America	11	0.0	11
	Asia	9	1.5	7.5
	North America	7	0	7
	Far East	6	0	6
	Thailand	6	0	6

However, the keywords with the most citation impact are sugarcane, ethanol, Brazil, life cycle assessment, biomass, ethanol production, and South America with at least 11% surplus each. Similarly, the keywords with the least citation impact are life cycle, yeast, economic analysis, life cycle analysis, gas emissions, and carbon dioxide.

94.3.10 The Most Prolific Research Fronts in Country-based Experience of Bioethanol Fuels

Information about the most prolific research fronts for the sample papers in the country-based experience of bioethanol fuels is given in Table 94.9.

As Table 94.9 shows, there are four primary research fronts for these 25 most-cited papers: Brazil, the USA, China, and other countries.

94.3.10.1 The Brazilian Experience of Bioethanol Fuels

Brazil has been one of the major producers, users, and exporters of bioethanol fuels as it has started a sugarcane-based bioethanol program in the 1970s. Over 40% of the 100 most-cited papers were related to Brazil's experiences with bioethanol fuels (Table 94.5). All these reviewed papers focused on sugarcane and sugarcane bagasse.

A number of research issues emerge from these reviewed studies. Technoeconomics of bioethanol production was a primary topic considered in these papers (Dias et al., 2011; Luo et al., 2009). A number of ways were advanced to increase the net energy value and economic competitiveness of bioethanol fuels produced in Brazil in these papers.

The GHG emissions from bioethanol fuels during both the production and utilization stages were another major issue in these reviewed papers (Luo et al., 2009; Macedo et al., 2008). A number of ways were advanced to decrease the GHG emissions during both production and utilization of bioethanol fuels produced in Brazil in these papers.

As the amount of bioethanol production has increased since the 1970s, the size of the sugarcane plantations has also increased putting pressure on the lands available for sugarcane production (Martinelli and Filoso, 2008; Rudorff et al., 2010).

These findings are thought-provoking in seeking ways to optimize bioethanol production and utilization in Brazil. The technoeconomics, environmental and societal impacts, and land use change for bioethanol production and utilization emerge as the key issues for the Brazilian experience of bioethanol fuels.

TABLE 94.9
Most Prolific Research Fronts in Country-based Experience of Bioethanol Fuels

No.	Countries	Sample Papers (%)
1	Brazil	41
2	USA	15
3	China	12
4	Other countries	26
	Thailand	6
	Europe	4
	India	4
	Malaysia	3
	Sweden	3
	Colombia	2
	Canada	2
	UK	2
	Sample size	100

94.3.10.2 The US Experience of Bioethanol Fuels

The USA has been one of the major producers, users, and exporters of bioethanol fuels as it has started a corn grain and corn stover-based bioethanol program in the 1970s. Over 15% of the 100 most-cited papers, respectively, were related to the USA's experiences of bioethanol fuels (Table 94.5). All these reviewed papers focused primarily on corn grains.

A number of research issues emerge from these reviewed studies. Technoeconomics of bioethanol production was a primary issue considered in these papers (Hettinga et al., 2009; Tyner, 2008). A number of ways were advanced to increase the net energy value and economic competitiveness of bioethanol fuels produced in the USA in these papers.

The GHG emissions from bioethanol fuels during both the production and utilization stages were another major issue in these reviewed papers (Hertel et al., 2010; Jacobson, 2007). A number of ways were advanced to decrease the GHG emissions during both production and utilization of bioethanol fuels produced in the USA in these papers. Jacobson (2007) also studied the effect of ozone emissions from bioethanol fuels on human health for the E85 bioethanol–gasoline fuel blend. It has to be noted that the traditional bioethanol–gasoline blends do not exceed 20% bioethanol additives (E20). Hence, the importance of this study should be devalued.

As the amount of bioethanol production has increased, the size of the corn plantations has also increased putting pressure on the lands available for corn production (Hertel et al., 2010). Additionally, Chiu et al. (2009) studied the impact of irrigation practices on the water requirement for bioethanol production. This study highlighted the importance of water resources for bioethanol production.

These findings are thought-provoking in seeking ways to optimize bioethanol production and utilization in the USA. The technoeconomics, environmental and societal impact, land use change, and water requirements emerge as the key issues for the US experience of bioethanol fuels.

94.3.10.3 The Chinese Experience of Bioethanol Fuels

China has been one of the major producers and users of bioethanol fuels as it has intensified various biomass-based bioethanol programs in recent years. Around 12% of the 100 most-cited papers were related to China's experiences of bioethanol fuels (Table 94.5). All these reviewed papers focused on cassava and sweet sorghum.

A number of research issues emerge from these reviewed studies. Technoeconomics of bioethanol production was a primary issue considered in these papers (Dai et al., 2006). A number of ways were advanced to increase the net energy value and economic competitiveness of bioethanol fuels produced in China in these papers.

The GHG emissions from bioethanol fuels during both the production and utilization stages were another major issue in these reviewed papers (Ren et al., 2015). A number of ways were advanced to decrease the GHG emissions during both production and utilization of bioethanol fuels produced in China in these papers.

These findings are thought-provoking in seeking ways to optimize bioethanol production and utilization in China. The technoeconomics and environmental and societal impact of bioethanol production and utilization emerge as the key issues for the Chinese experience of bioethanol fuels.

94.3.10.4 The Other Countries' Experience of Bioethanol Fuels

Europe, Colombia, Malaysia, and India have been among the major producers and users of bioethanol fuels as they have started a various biomass-based bioethanol programs in recent years. However, around 26% of the 100 most-cited papers were related to these countries' experiences of bioethanol fuels (Table 94.5). All these reviewed papers focused on several feedstocks such as sugarcane, corn, wastes, sweet sorghum, and lignocellulosic biomass. It should be noted that there were three and two papers on the experiences of bioethanol fuels for Sweden and the UK in the 100 most-cited paper sets.

A number of research issues emerge from these review studies. Technoeconomics of bioethanol production was a primary use considered in these papers (Quintero et al., 2008). A number of ways were advanced to increase the net energy value and economic competitiveness of bioethanol fuels produced in these countries in these papers.

The GHG emissions from bioethanol fuels during both the production and utilization stages were another major issue in these reviewed papers (Quintero et al., 2008). A number of ways were advanced to decrease the GHG emissions during both production and utilization of bioethanol fuels produced in these countries in these papers.

These findings are thought-provoking in seeking ways to optimize bioethanol production and utilization in these countries. The technoeconomics and environmental and societal impact of bioethanol production and utilization emerge as the key issues for these countries' experience of bioethanol fuels.

94.4 DISCUSSION

94.4.1 INTRODUCTION

Crude oil-based gasoline fuels have been widely used in the transportation sector since the 1920s. However, there have been great public concerns over the adverse environmental and human impact of these fuels. Hence, biomass-based bioethanol fuels have increasingly been used in blending gasoline fuels. Additionally, bioethanol fuels have been used to produce valuable biochemicals in a biorefinery context.

The research in the field of experiences of countries has also intensified in recent years. The primary focus of research on these country-based experiences has been the optimization of bioethanol and biomass production and the development of policies to promote the production and use of bioethanol fuels worldwide. In this context, there has been a significant focus on evaluative studies in bioethanol fuels.

However, it is essential to develop efficient incentive structures for the primary stakeholders to enhance research in this field. The scientometric analysis has been used in this context to inform the primary stakeholders about the current state of research in a selected research field.

As there have been no scientometric studies on the country-based experience of bioethanol fuels, this book chapter presents a scientometric study of research in this field. It examines the scientometric characteristics of both the sample and population data presenting the scientometric characteristics of these both datasets in the order of documents, authors, publication years, institutions, funding bodies, source titles, countries, Scopus subject categories, keywords, and research fronts.

As a first step for the search of the relevant literature, the keywords were selected using the first 200 most-cited papers. The selected keyword list was optimized to obtain a representative sample of papers for the searched research field. This keyword list was provided in the appendix for future replication studies. Additionally, the information about the most-used keywords is given in Section 94.3.9 to highlight the key research fronts in Section 94.3.10.

As a second step, two sets of data were used for this study. First, a population sample of over 1,300 papers was used to examine the scientometric characteristics of the population data. Second, a sample of 100 most-cited papers was used to examine the scientometric characteristics of these citation classics with over 59 citations each.

The scientometric characteristics of these both sample and population datasets were presented in the order of documents, authors, publication years, institutions, funding bodies, source titles, countries, Scopus subject categories, keywords, and research fronts.

Lastly, the key scientometric findings for both datasets were discussed to highlight the research landscape for the country-based experience of bioethanol fuels. Additionally, several brief conclusions were drawn and many relevant recommendations were made to enhance the future research landscape.

94.4.2 The Most Prolific Documents in the Country-based Experience of Bioethanol Fuels

The articles and review papers dominate both the sample and population datasets, while the review papers and book chapters have a surplus and deficit, respectively.

Scopus differs from the Web of Science database in differentiating and showing articles and conference papers published in journals separately. Similarly, Scopus differs from the Web of Science database in introducing short surveys. Hence, the total number of articles and review papers in the sample dataset is 77% and 23%, respectively.

It is observed during the search process that there has been inconsistency in the classification of the documents in Scopus and in other databases such as Web of Science. This is especially relevant for the classification of papers as reviews or articles as the papers not involving a literature review may be erroneously classified as a review paper. There is also a case of review papers being classified as articles. For example, although there are 23 review papers and short surveys as classified by the Scopus database, only ten of the sample papers are review papers based on the literature reviews.

In this context, it would be helpful to provide a classification note for the published papers in the books and journals at the first instance. It would also be helpful to use the document types listed in Table 94.1 for this purpose. Book chapters may also be classified as articles or reviews as an additional classification to differentiate review chapters from experimental chapters as is done by the Web of Science. It would be further helpful to additionally classify the conference papers as articles or review papers and it is done in the Web of Science database.

94.4.3 The Most Prolific Authors in the Country-based Experience of Bioethanol Fuels

There have been 29 most prolific authors with at least two sample papers and three population papers each as given in Table 94.2. These authors have shaped the development of research in this field.

The most prolific authors are Antonio Bonomi, Shabbir H. Gheewala, Joaquim E. A. Sabra, Arnaldo Walter, and to a lesser extent Andre P. C. Faaij, Otavio Cavalett, Thu L. T. Nguyen, Isaisas de Carvalho Macedo, and H. Martin Junginger.

It is important to note the inconsistencies in the indexing of author names in Scopus and other databases. It is especially an issue for names with more than two components such as 'Judge Alex de Camp Sirous'. The probable outcomes are 'Sirous, J.A.D.C.', 'de Camp Sirous, J.A.', or 'Camp Sirous, J.A.D.'. The first choice is the gold standard of the publishing sector as the last word in the name is taken as the last name. In most of academic databases such as PubMed and EBSCO databases, this version is used predominantly. The second choice is a strong alternative, while the last choice is an undesired outcome as two last words are taken as the last name.

For example, in this study 'Isaisas de Carvalho Macedo' was abbreviated as 'Macedo, I.C.' and 'de Carvalho Macedo, I.'. It is good practice to combine the words of the last name by a hyphen: 'Camp-Sirous, J.A.D.'. It is notable that inconsistent indexing of author names may cause substantial inefficiencies in the search process for the papers and allocating credit to the authors as there are different author entries for each outcome in the databases.

There is also a case of shortening Chinese names. For example, 'Yuoyang Wang' is often shortened as 'Wang, Y.', 'Wang, Y.-Y.', and 'Wang Y.Y.' as it is done in the Web of Science database as well. However, the gold standard in this case is 'Wang Y' where the last word is taken as the last name and the first word is taken as a single forename. In most of academic databases such as PubMed and EBSCO, this first version is used predominantly. However, it makes sense to use the third option to differentiate Chinese names efficiently: 'Wang Y.Y.'. Therefore, there have been difficulties to locate papers for Chinese authors. In such cases, the use of the unique author codes provided for each author by the Scopus database has been helpful.

There is also a difficulty in allowing credit for the authors, especially for the authors with common names such as 'Wang, Y.', or 'Huang, Y.', or 'Zhu, Y.' in conducting scientometric studies. These difficulties strongly influence the efficiency of the scientometric studies and allocating credit to the authors as there are the same author entries for different authors with the same name, e.g., 'Wang Y.' in the databases.

In this context, the coding of authors in the Scopus database is a welcome innovation compared with other databases such as Web of Science. In this process, Scopus allocates a unique number to each author in the database (Aman, 2018). However, there might still be substantial inefficiencies in this coding system, especially for common names. For example, some of the papers for a certain author may be allocated to another researcher with a different author code. It is possible that Scopus uses many software programs to differentiate the author names and the program may not be false-proof (Shin et al., 2014).

In this context, it does not help that author names are not given in full in some journals and books. This makes it difficult to differentiate authors with common names and makes scientometric studies further difficult in the author domain. Therefore, the author names should be given in all books and journals at the first instance. There is also a cultural issue where some authors do not use their full names in their papers. Instead, they use initials for their forenames: 'Coutancy, A.P.' or just 'Coutancy' instead of 'Coutancy, Alas Padras'.

There are also inconsistencies in the naming of authors with more than two components by the authors themselves in journal papers and book chapters. For example, 'Alaspanda, A.P.C.', 'Sakoura, C.E.', and 'Mentaslo, S.J.' might be given as 'Alaspanda, A.', 'Sakoura, C.', or 'Mentaslo, S.', respectively, in journals and books. In this study, there were inconsistencies in the naming of 'Faaij, A.P.C.', 'Junginger, H.M.', and 'Galdos, M.V.'. This also makes scientometric studies difficult in the author domain. Hence, contributing authors should use their name consistently in their publications.

The other critical issue regarding the author names is the spelling of author names in the national spellings (e.g., Gonçalves, Übeiro) rather than in the English spellings (e.g., Goncalves, Ubeiro) in the Scopus database. Scopus differs from the Web of Science database and many other databases in this respect where the author names are given only in English spellings. It is observed that the national spellings of the author names do not help in conducting scientometric studies and in allocating credits to the authors as sometimes there are different author entries for the English and national spellings in the Scopus database.

The most prolific institution for the sample dataset is the University of Campinas, National Bioethanol Science and Technology Laboratory, and to a lesser extent King Mongkut's University of Technology, Sao Paulo University, Fermentec, National Colombia University, Sao Paulo University, Shanghai Jiao Tong University, and Utrecht University. In total, 13 institutions house these prolific authors.

The most prolific countries for the sample dataset are Brazil, China, Thailand, and to a lesser extent Colombia, the Netherlands, and the USA. In total, eight countries house these authors.

It is also notable that there is significant gender deficit for the sample dataset as surprisingly nearly all of these top researchers are male. This finding is the most thought-provoking with strong public policy implications. Hence, institutions, funding bodies, and policymakers should take efficient measures to reduce the gender deficit in this field and other scientific fields with strong gender deficit. In this context, it is worth to note the level of representation of researchers from minority groups in science based on race, sexuality, age, and disability, besides gender (Blankenship, 1993; Dirth and Branscombe, 2017; Konur, 2000, 2002a,b,c, 2006a,b, 2007a,b).

94.4.4 The Most Prolific Research Output by Years in the Country-based Experience of Bioethanol Fuels

The research output observed between 1970 and 2021 is illustrated in Figure 94.1. This figure clearly shows that the bulk of research papers in the population dataset were published primarily in the 2010s and to a lesser extent in the 2000s. Similarly, the bulk of research papers in the sample dataset were published in the 2000s and 2010s.

These data suggest that the most-cited sample and population papers were primarily published in the 2000s and 2010s. These are thought-provoking findings as there has been no significant research in this field in the pre-2000s, but there has been a significant research boom in the last two decades. In this context, the increasing public concerns about climate change (Change, 2007), GHG emissions (Carlson et al., 2017), and global warming (Kerr, 2007) have been certainly behind the boom in the research in this field in the last two decades.

Based on these findings, the size of the population papers is likely to more than double in the current decade, provided that the public concerns about climate change, GHG emissions, and global warming are translated efficiently to the research funding in this field. However, it should be noted the research output for the population papers stagnated after 2011, losing its momentum.

94.4.5 The Most Prolific Institutions in the Country-based Experience of Bioethanol Fuels

The 22 most prolific institutions publishing papers on the country-based experiences of bioethanol fuels with at least two sample papers and 0.4% of the population papers each given in Table 94.3 have shaped the development of research in this field.

The most prolific institutions are the State University of Campinas, University of Sao Paulo, National Biorenewables Laboratory, and to a lesser extent King Mongkut's University of Technology, Chinese Academy of Sciences, University of Utrecht, and National Research Center of Energy and Materials.

The top countries for these most prolific institutions are Brazil and to a lesser extent the USA and China. In total, ten countries house these top institutions.

However, the institutions with the most citation impact are State University of Campinas and National Biorenewables Laboratory, respectively. Similarly, the institutions with the least impact are Federal University Rio de Janeiro and Iowa State University. It appears that the Brazilian institutions dominate the top institution list as it has been a major producer, consumer, and exporter of first generation sugarcane-based bioethanol fuels since the 1970s, following the oil crisis in the early 1970s.

94.4.6 The Most Prolific Funding Bodies in the Country-based Experience of Bioethanol Fuels

The 12 most prolific funding bodies funding at least two sample papers and 0.4% of the population papers each are given in Table 94.4.

The most prolific funding bodies are the Sao Paulo State Research Foundation, the European Commission, and to a lesser extent National Council for Science and Technology Development. It is notable that 48% and 34% of the sample and population papers are funded, respectively. The most prolific countries for these top funding bodies are Brazil and Canada. In total, seven countries and the European Union house these top funding bodies. The heavy funding by the Brazilian and Canadian funding bodies is notable as these countries are all major producers of research in this field.

These findings on the funding of research in this field suggest that the level of funding, mostly in the last two decades, has been largely instrumental in enhancing the research in this field (Ebadi and Schiffauerova, 2016) in light of North's institutional framework (North, 1991).

94.4.7 The Most Prolific Source Titles in Country-based Experience of Bioethanol Fuels

The 14 most prolific source titles publishing at least two sample papers and 0.3% of the population papers each in the country-based experience of bioethanol fuels have shaped the development of research in this field (Table 94.5).

The most prolific source titles are 'Renewable and Sustainable Energy Reviews', 'Applied Energy', 'Bioresource Technology', and to a lesser extent 'Energy Policy' and 'Biomass and Bioenergy'. However, the source titles with the most citation impact are 'Bioresource Technology', 'Applied Energy', and to a lesser extent 'Renewable and Sustainable Energy Reviews', Energy Policy', and 'Biomass and Bioenergy'. Similarly, the source titles with the least impact are the 'Journal of Cleaner Production', 'Renewable Energy', and 'Energy for Sustainable Development'.

It is notable that these top source titles are related to energy. This finding suggests that the journals in these fields have significantly shaped the development of research in this field as they focus on the evaluative studies for the country-based experiences of bioethanol fuels.

94.4.8 The Most Prolific Countries in Country-based Experience of Bioethanol Fuels

The 16 most prolific countries publishing at least two papers and 0.7% of the population papers each have significantly shaped the development of research in this field (Table 94.6).

The most prolific countries are Brazil, the USA, and to a lesser extent China, the Netherlands, the UK, and Thailand. However, the countries with the most citation impact are Brazil and to a lesser extent USA and the Netherlands. Similarly, the countries with the least citation impact are India, Germany, Spain, and Colombia.

A close examination of these findings suggests that Brazil, the USA, China, and Europe, and to a lesser extent Thailand, Canada, Sweden, India, Australia, Malaysia, and Colombia are the major producers of research in this field.

It is a fact that the USA has been a major player in science (Leydesdorff and Wagner, 2009; Leydesdorff et al., 2014). The USA has further developed a strong research infrastructure to support its corn- and grass-based bioethanol industry (Vadas et al., 2008).

However, China has been a rising star in scientific research in competition with the USA and Europe (Leydesdorff and Zhou, 2005). China is also a major player in this field as a major producer of bioethanol (Li and Chan-Halbrendt, 2009).

Next, Europe has been a persistent player in scientific research in competition with both the USA and China (Leydesdorff, 2000). Europe has also been a persistent producer of bioethanol along with the USA and Brazil (Gnansounou, 2010).

Additionally, Brazil has also been a persistent player in scientific research at a moderate level (Glanzel et al., 2006). Brazil has also developed a strong research infrastructure to support its biomass-based bioethanol industry (Soccol et al., 2010).

94.4.9 The Most Prolific Scopus Subject Categories in Country-based Experience of Bioethanol Fuels

The eight most prolific Scopus subject categories indexing at least 11% and 6.1% of the sample and population papers, respectively, given in Table 94.7 have shaped the development of research in this field.

The most prolific Scopus subject categories in the country-based experience of bioethanol fuels are 'Environmental Science' and 'Energy'. 'Agricultural and Biological Sciences', 'Engineering', and 'Chemical Engineering' follow these top categories. On the other hand, the Scopus subject categories with the most citation impact are 'Environmental Science' and 'Energy'. Similarly, the Scopus subject categories with the least citation impact are 'Agricultural and Biological Sciences', 'Business, Management and Accounting', and 'Social Sciences'.

These findings are thought-provoking, suggesting that the primary subject categories are 'Environmental Science' and 'Energy'. The other key finding is that social sciences are relatively well represented in both the sample and population papers, unlike most fields in bioethanol fuels. These findings are not surprising as the key research fronts in this field are the evaluative studies on bioethanol fuels focusing on the technoeconomics and life cycle assessment of bioethanol fuels.

94.4.10 The Most Prolific Keywords in Country-based Experience of Bioethanol Fuels

A limited number of keywords have shaped the development of research in this field as shown in Table 94.8 and the Appendix.

These keywords are grouped under five headings: bioethanol fuels, biomass, issues, and countries and continents related to the countries' experiences of bioethanol fuels.

The prolific keywords related to the countries and continents are Brazil, China, the USA, Eurasia, and South America. Furthermore, the keywords with the most citation impact are sugarcane, ethanol, Brazil, life cycle assessment, biomass, ethanol production, and South America. These prolific keywords highlight the key research fronts in this field and reflect well the keywords used in the sample papers.

94.4.11 The Most Prolific Research Fronts in Country-based Experience of Bioethanol Fuels

As Table 94.9 shows there are four primary research fronts for these 25 most-cited papers: Brazil, the USA, China, and other countries.

94.4.11.1 The Brazilian Experience of Bioethanol Fuels

Brazil has been one of the major producers, users, and exporters of bioethanol fuels as it has started a sugarcane-based bioethanol program in the 1970s. Over 40% of the 100 most-cited papers were related to Brazil's experiences of bioethanol fuels (Table 94.5). All these reviewed papers focused on first generation sugarcane and second generation sugarcane bagasse. The technoeconomics, environmental and societal impacts, and land use change for bioethanol production and utilization emerge as the key issues for the Brazilian experience of bioethanol fuels.

94.3.11.2 The US Experience of Bioethanol Fuels

The USA has been one of the major producers, users, and exporters of bioethanol fuels as it has started a corn grain- and corn bagasse-based bioethanol program in the 1970s. Over 15% of the 100 most-cited papers were related to the USA's experiences of bioethanol fuels (Table 94.5). All these reviewed papers focused primarily on corn grains. The technoeconomics, environmental and societal impact, land use change, water requirements, and blends with high octane numbers for bioethanol production and utilization emerge as the key issues for the US experience of bioethanol fuels.

94.3.11.3 The Chinese Experience of Bioethanol Fuels

China has been one of the major producers and users of bioethanol fuels as it has intensified various biomass-based bioethanol program in recent years. Around 12% of the 100 most-cited papers were related to China's experiences of bioethanol fuels (Table 94.5). All these reviewed papers focused on cassava and sweet sorghum. The technoeconomics and environmental and societal impact of bioethanol production and utilization emerge as the key issues for the Chinese experience of bioethanol fuels.

94.3.11.4 The Other Countries' Experience of Bioethanol Fuels

Europe, Colombia, Malaysia, and India have been among the major producers and users of bioethanol fuels as they have started a various biomass-based bioethanol programs in recent years. However, around 26% of the 100 most-cited papers were related to these countries' experiences of bioethanol fuels (Table 94.5). All these reviewed papers focused on several feedstocks such as sugarcane, corn, wastes, sweet sorghum, and lignocellulosic biomass. It should be noted that there were three and two papers on the experiences of bioethanol fuels for Sweden and the UK in the 100 most-cited paper sets. The technoeconomics and environmental and societal impact of bioethanol production and utilization emerge as the key issues for these countries' experience of bioethanol fuels.

In the end, these most-cited papers in this field hint that the efficiency of bioethanol fuels could be optimized using the structure, processing, and property relationships of bioethanol fuels (Formela et al., 2016; Konur, 2018a, 2020b, 2021a,b,c,d; Konur and Matthews, 1989).

94.5 CONCLUSION AND FUTURE RESEARCH

The research on the country-based experience of bioethanol fuels has been mapped through a scientometric study of both sample and population datasets.

The critical issue in this study has been to obtain a representative sample of research as in any other scientometric study. Therefore, the keyword set has been carefully devised and optimized after several runs in the Scopus database. It should be noted that the focus in this book chapter has been on four major producers, users, and exporters of bioethanol fuels: Brazil, the USA, China, and Europe besides other countries. It is a representative sample of the wider population studies, which include large numbers of countries in this field.

The other issue has been the selection of a multidisciplinary database to carry out the scientometric study of research in this field. For this purpose, the Scopus database has been selected. The journal coverage of this database has been wider than that of the Web of Science.

The key scientometric properties of research in this field have been determined and discussed in this book chapter. It is evident that a limited number of documents, authors, institutions, publication periods, institutions, funding bodies, source titles, countries, Scopus subject categories, keywords, and research fronts have shaped the development of research in this field.

There is ample scope to increase the efficiency of the scientometric studies in this field in the author and document domains by developing consistent policies and practices in both domains across all academic databases. In this respect, authors, journals, and academic databases have a lot to do. Furthermore, the significant gender deficit as in most scientific fields emerges as a public policy issue. The potential deficits based on age, race, disability, and sexuality need also to be explored in this field as in other scientific fields.

The research in this field has boomed in the 2000s and 2010s possibly promoted by the public concerns on global warming, GHG emissions, and climate change. The institutions from Brazil, China, and the USA have mostly shaped the research in this field.

The relatively low funding rate of 48% and 34% for sample and population papers, respectively, suggests that funding in this field significantly enhanced the research in this field primarily in the 2010s, possibly more than doubling in the current decade. However, there is ample room for more funding regarding the stagnation of research output after 2011 and the recent supply shocks caused by the coronavirus disease 2019 (COVID-19) pandemic and the Ukrainian war.

The most prolific journals have been mostly indexed by the subject categories of energy as the focus of the sample papers has been on the evaluative studies of bioethanol fuels with a focus on the net energy and environmental impact of bioethanol fuels.

Brazil, the USA, China, and Europe have been the major producers of research in this field as the major producers and users of bioethanol fuels from different types of biomass such as corn, sugarcane, and grass as well as other types of biomass. These countries have well-developed research infrastructure in bioethanol fuels and have been the major producers and consumers of bioethanol fuels.

The primary subject categories have been 'Environmental Science' and 'Energy'. Furthermore, social sciences are relatively well represented in both the sample and population papers, unlike most fields in bioethanol fuels. These findings are not surprising as the focus of the sample papers has been on the evaluative studies of bioethanol fuels with a focus on the net energy and environmental impact of bioethanol fuels.

Ethanol is more popular than bioethanol as a keyword with strong implications for the search strategy. In other words, the search strategy using only bioethanol keyword would not be much helpful.

These keywords are grouped under the five headings: bioethanol fuels, biomass, issues, and countries and continents related to the countries' experiences of bioethanol fuels. These groups of keywords highlight the potential primary research fronts for these fields: evaluative studies based on the countries. These findings are thought-provoking. The focus of these 100 most-cited papers is the evaluative studies for bioethanol fuels highlighting strong structure–processing–property relationships for bioethanol fuels.

Thus, the scientometric analysis has a great potential to gain valuable insights into the evolution of research in this field as in other scientific fields.

It is recommended that further scientometric studies are carried out about the other aspects of both production and utilization of bioethanol fuels. It is further recommended that reviews of the most-cited papers are carried out for each research front to complement these scientometric studies. Next, the scientometric studies of the hot papers in these primary fields are carried out.

ACKNOWLEDGMENTS

The contribution of the highly cited researchers in the field of the country-based experience of bioethanol fuels has been gratefully acknowledged.

APPENDIX: THE KEYWORD SET FOR COUNTRY-BASED EXPERIENCE OF BIOETHANOL FUELS

((TITLE (ethanol OR bioethanol OR {ethyl alcohol*} OR c2h5oh OR e85 OR e10 OR defc*) OR SRCTITLE (ethanol OR bioethanol)) AND (TITLE (brazil* OR us OR usa OR "united states" OR america OR american* OR europ* OR eu OR german* OR france OR french OR ireland OR irish OR italy OR italian OR uk OR netherlands OR dutch OR denmark OR danish OR sweden OR swedish OR "united kingdom" OR finland OR finnish OR poland OR polish OR ukrain* OR turkey OR turkish OR switzerland OR swiss OR greece OR greek OR spain* OR spanish OR portug* OR serbia* OR chinese OR china OR asia OR japan* OR korea* OR russia* OR colombia* OR bangladesh* OR argentin* OR mexic* OR india* OR canadian OR canada OR iran* OR thailand OR thai OR malaysia* OR pakistan* OR africa* OR indonesia* OR asian OR belgium OR africa* OR australia* OR taiwan* OR colombia* OR egypt* OR czech* OR romania* OR hungar* OR austria* OR israel* OR chile* OR nigeria* OR arabia* OR norway* OR cameroon OR ethiopia OR ghana* OR kenya* OR sudan* OR "hong kong" OR zealand OR singapor* OR slovenia* OR tunisia* OR brunei* OR lebanon OR libya* OR lithuania OR luxembourg* OR peru OR philippines OR zambia OR belarus))) AND NOT (SUBJAREA (dent OR heal OR medi OR neur OR nurs OR phar OR psyc OR vete) OR TITLE (injection* OR solvent* OR extract* OR anti* OR vitro OR beetle* OR carcinoma OR persimm* OR lipid* OR disease* OR drug* OR wine* OR injury OR vivo OR brain OR oxidative OR hepat* OR apopt* OR cells OR leaf* OR diester* OR bats OR curd* OR tumor* OR ablation OR gene OR postharvest OR blood OR rats OR mice OR hamster* OR larva*)) AND (LIMIT-TO (SRCTYPE, "j") OR LIMIT-TO (SRCTYPE, "b") OR LIMIT-TO (SRCTYPE, "k")) AND (LIMIT-TO (DOCTYPE, "ar") OR LIMIT-TO (DOCTYPE, "cp") OR LIMIT-TO (DOCTYPE, "re") OR LIMIT-TO (DOCTYPE, "ch") OR LIMIT-TO (DOCTYPE, "no") OR LIMIT-TO (DOCTYPE, "le") OR LIMIT-TO (DOCTYPE, "sh") OR LIMIT-TO (DOCTYPE, "bk") OR LIMIT-TO (DOCTYPE, "ed")) AND (LIMIT-TO (LANGUAGE, "English"))

REFERENCES

Aman, V. 2018. Does the Scopus author ID suffice to track scientific international mobility? A case study based on Leibniz laureates. *Scientometrics* 117:705–720.

Amorim, H. V., M. L. Lopes, J. V. de Castro Oliveira, M. S. Buckeridge and G. H. Goldman. 2011. Scientific challenges of bioethanol production in Brazil. *Applied Microbiology and Biotechnology* 91:1267–1275.

Angenent, L. T., K. Karim, M. H. Al-Dahhan, B. A. Wrenn and R. Domiguez-Espinosa. 2004. Production of bioenergy and biochemicals from industrial and agricultural wastewater. *Trends in Biotechnology* 22:477–485.

Antolini, E. 2007. Catalysts for direct ethanol fuel cells. *Journal of Power Sources* 170:1–12.

Antolini, E. 2009. Palladium in fuel cell catalysis. *Energy and Environmental Science* 2:915–931.

Beaudry, C. and V. Lariviere. 2016. Which gender gap? Factors affecting researchers' scientific impact in science and medicine. *Research Policy* 45:1790–1817.

Blankenship, K. M. 1993. Bringing gender and race in: US employment discrimination policy. *Gender & Society* 7:204–226.

Burnham, J. F. 2006. Scopus database: A review. *Biomedical Digital Libraries* 3:1–8.

Carlson, K. M., J. S. Gerber and D. Mueller, et al. 2017. Greenhouse gas emissions intensity of global croplands. *Nature Climate Change* 7:63–68.

Change, C. 2007. Climate change impacts, adaptation and vulnerability. *Science of the Total Environment* 326:95–112.

Cherubini, F., G. Jungmeier and M. Wellisch, et al. 2009. Toward a common classification approach for biorefinery systems. *Biofuels, Bioproducts and Biorefining* 3:534–546.

Chiu, Y. W., B. Walseth and S. Suh. 2009. Water embodied in bioethanol in the United States. *Environmental Science and Technology* 43:2688–2692.

Dai, D., Z. Hu, G. Pu, H. Li and C. Wang. 2006. Energy efficiency and potentials of cassava fuel ethanol in Guangxi region of China. *Energy Conversion and Management* 47:1686–1699.

Dias, M. O. S., M. P. Cunha and C. D. F. Jesus, et al. 2011. Second generation ethanol in Brazil: Can it compete with electricity production? *Bioresource Technology* 102:8964–8971.

Dirth, T. P. and N. R. Branscombe. 2017. Disability models affect disability policy support through awareness of structural discrimination. *Journal of Social Issues* 73:413–442.

Ebadi, A. and A. Schiffauerova. 2016. How to boost scientific production? A statistical analysis of research funding and other influencing factors. *Scientometrics* 106:1093–1116.

Farrell, A. E., R. J. Plevin and B. T. Turner, et al. 2006. Ethanol can contribute to energy and environmental goals. *Science* 311:506–508.

Formela, K., A. Hejna, L. Piszczyk, M. R. Saeb and X. Colom. 2016. Processing and structure-property relationships of natural rubber/wheat bran biocomposites. *Cellulose* 23:3157–3175.

Garfield, E. 1955. Citation indexes for science. *Science* 122:108–111.

Glanzel, W., J. Leta and B. Thijs. 2006. Science in Brazil. Part 1: A macro-level comparative study. *Scientometrics* 67:67–86.

Gnansounou, E. 2010. Production and use of lignocellulosic bioethanol in Europe: Current situation and perspectives. *Bioresource Technology* 101:4842–4850.

Hertel, T. W., A. A. Golub and A. D. Jones, et al. 2010. Effects of US Maize ethanol on global land use and greenhouse gas emissions: Estimating market-mediated responses. *BioScience* 60:223–231.

Hettinga, W. G., H. M. Junginger and S. C. Dekker, et al. 2009. Understanding the reductions in US corn ethanol production costs: An experience curve approach. *Energy Policy* 37:190–203.

Hill, J., E. Nelson, D. Tilman, S. Polasky and D. Tiffany. 2006. Environmental, economic, and energetic costs and benefits of biodiesel and ethanol biofuels. *Proceedings of the National Academy of Sciences of the United States of America* 103:11206–11210.

Hill, J., S. Polasky and E. Nelson, et al. 2009. Climate change and health costs of air emissions from biofuels and gasoline. *Proceedings of the National Academy of Sciences of the United States of America* 106:2077–2082.

Hsieh, W. D., R. H. Chen, T. L. Wu and T. H. Lin. 2002. Engine performance and pollutant emission of an SI engine using ethanol-gasoline blended fuels. *Atmospheric Environment* 36:403–410.

Jacobson, M. Z. 2007. Effects of ethanol (E85) versus gasoline vehicles on cancer and mortality in the United States. *Environmental Science and Technology* 41:4150–4157.

Kerr, R. A. 2007. Global warming is changing the world. *Science* 316:188–190.

Konur, O. 2000. Creating enforceable civil rights for disabled students in higher education: An institutional theory perspective. *Disability & Society* 15:1041–1063.

Konur, O. 2002a. Access to nursing education by disabled students: Rights and duties of nursing programs. *Nurse Education Today* 22:364–374.

Konur, O. 2002b. Assessment of disabled students in higher education: Current public policy issues. *Assessment and Evaluation in Higher Education* 27:131–152.

Konur, O. 2002c. Access to employment by disabled people in the UK: Is the Disability Discrimination Act working? *International Journal of Discrimination and the Law* 5:247–279.

Konur, O. 2006a. Participation of children with dyslexia in compulsory education: Current public policy issues. *Dyslexia* 12:51–67.

Konur, O. 2006b. Teaching disabled students in higher education. *Teaching in Higher Education* 11:351–363.

Konur, O. 2007a. A judicial outcome analysis of the *Disability Discrimination Act*: A windfall for the employers? *Disability & Society* 22:187–204.

Konur, O. 2007b. Computer-assisted teaching and assessment of disabled students in higher education: The interface between academic standards and disability rights. *Journal of Computer Assisted Learning* 23:207–219.

Konur, O. 2011. The scientometric evaluation of the research on the algae and bio-energy. *Applied Energy* 88:3532–3540.

Konur, O. 2012a. Prof. Dr. Ayhan Demirbas' scientometric biography. *Energy Education Science and Technology Part A: Energy Science and Research* 28:727–738.

Konur, O. 2012b. The evaluation of the biogas research: A scientometric approach. *Energy Education Science and Technology Part A: Energy Science and Research* 29:1277–1292.

Konur, O. 2012c. The evaluation of the global energy and fuels research: A scientometric approach. *Energy Education Science and Technology Part A: Energy Science and Research* 30:613–628.

Konur, O. 2012d. The evaluation of the research on the biodiesel: A scientometric approach. *Energy Education Science and Technology Part A: Energy Science and Research* 28:1003–1014.

Konur, O. 2012e. The evaluation of the research on the bioethanol: A scientometric approach. *Energy Education Science and Technology Part A: Energy Science and Research* 28:1051–1064.

Konur, O. 2012f. The evaluation of the research on the biofuels: A scientometric approach. *Energy Education Science and Technology Part A: Energy Science and Research* 28:903–916.

Konur, O. 2012g. The evaluation of the research on the biohydrogen: A scientometric approach. *Energy Education Science and Technology Part A: Energy Science and Research* 29:323–338.

Konur, O. 2012h. The evaluation of the research on the microbial fuel cells: A scientometric approach. *Energy Education Science and Technology Part A: Energy Science and Research* 29:309–322.

Konur, O. 2012i. The scientometric evaluation of the research on the production of bioenergy from biomass. *Biomass and Bioenergy* 47:504–515.

Konur, O. 2015. Current state of research on algal bioethanol. In *Marine Bioenergy: Trends and Developments*, Ed. S. K. Kim and C. G. Lee, pp. 217–244. Boca Raton, FL: CRC Press.

Konur, O., Ed. 2018a. *Bioenergy and Biofuels*. Boca Raton, FL: CRC Press.

Konur, O. 2018b. Bioenergy and biofuels science and technology: Scientometric overview and citation classics. In *Bioenergy and Biofuels*, Ed. O. Konur, pp. 3–63. Boca Raton: CRC Press.

Konur, O. 2019. Cyanobacterial bioenergy and biofuels science and technology: A scientometric overview. In *Cyanobacteria: From Basic Science to Applications*, Ed. A. K. Mishra, D. N. Tiwari and A. N. Rai, pp. 419–442. Amsterdam: Elsevier.

Konur, O. 2020a. The scientometric analysis of the research on the bioethanol production from green macroalgae. In *Handbook of Algal Science, Technology and Medicine*, Ed. O. Konur, pp. 385–401. London: Academic Press.

Konur, O., Ed. 2020b. *Handbook of Algal Science, Technology and Medicine.* London: Academic Press.

Konur, O., Ed. 2021a. *Handbook of Biodiesel and Petrodiesel Fuels: Science, Technology, Health, and Environment*. Boca Raton, FL: CRC Press.

Konur, O., Ed. 2021b. *Handbook of Biodiesel and Petrodiesel Fuels: Science, Technology, Health, and Environment. Volume 1. Biodiesel Fuels: Science, Technology, Health, and Environment*. Boca Raton, FL: CRC Press.

Konur, O., Ed. 2021c. *Handbook of Biodiesel and Petrodiesel Fuels: Science, Technology, Health, and Environment. Volume 2. Biodiesel Fuels based on the Edible and Nonedible Feedstocks, Wastes, and Algae: Science, Technology, Health, and Environment.* Boca Raton, FL: CRC Press.

Konur, O., Ed. 2021d. *Handbook of Biodiesel and Petrodiesel Fuels: Science, Technology, Health, and Environment. Volume 3. Petrodiesel Fuels: Science, Technology, Health, and Environment*. Boca Raton, FL: CRC Press.

Konur, O. and F. L. Matthews. 1989. Effect of the properties of the constituents on the fatigue performance of composites: A review. *Composites* 20:317–328.

Leydesdorff, L. 2000. Is the European Union becoming a single publication system? *Scientometrics* 47:265–280.

Leydesdorff, L. and C. Wagner. 2009. Is the United States losing ground in science? A global perspective on the world science system. *Scientometrics* 78:23–36.

Leydesdorff, L., C. S. Wagner and L. Bornmann. 2014. The European Union, China, and the United States in the top-1% and top-10% layers of most-frequently cited publications: Competition and collaborations. *Journal of Informetrics* 8:606–617.

Leydesdorff, L. and P. Zhou. 2005. Are the contributions of China and Korea upsetting the world system of science? *Scientometrics* 63:617–630.

Li, S. Z. and C. Chan-Halbrendt. 2009. Ethanol production in (the) People's Republic of China: Potential and technologies. *Applied Energy* 86:S162–S169.

Lopes, M. L., S. C. D. L. Paulillo and A. Godoy, et al. 2016. Ethanol production in Brazil: A bridge between science and industry. *Brazilian Journal of Microbiology* 47:64–76.

Luo, L., E. van der Voet and G. Huppes. 2009. Life cycle assessment and life cycle costing of bioethanol from sugarcane in Brazil. *Renewable and Sustainable Energy Reviews* 13:1613–1619.

Ma, X., L. Sun and C. Song. 2002. A new approach to deep desulfurization of gasoline, diesel fuel and jet fuel by selective adsorption for ultra-clean fuels and for fuel cell applications. *Catalysis Today* 77:107–116.

Macedo, I. C., J. E. A. Seabra and J. E. A. R. Silva. 2008. Green house gases emissions in the production and use of ethanol from sugarcane in Brazil: The 2005/2006 averages and a prediction for 2020. *Biomass and Bioenergy* 32:582–595.

Maity, S. K. 2015. Opportunities, recent trends and challenges of integrated biorefinery: Part I. *Renewable and Sustainable Energy Reviews* 43:1427–1445.

Martinelli, L. A. and S. Filoso. 2008. Expansion of sugarcane ethanol production in Brazil: Environmental and social challenges. *Ecological Applications* 18:885–898.

Najafi, G., B. Ghobadian and T. Tavakoli, et al. 2009. Performance and exhaust emissions of a gasoline engine with ethanol blended gasoline fuels using artificial neural network. *Applied Energy* 86:630–639.

Newman, P. W. G. and J. R. Kenworthy. 1989. Gasoline consumption and cities: A comparison of U.S. cities with a global survey. *Journal of the American Planning Association* 55:24–37.

Nikolau, B. J., M. A. D. Perera, L. Brachova and B. Shanks. 2008. Platform biochemicals for a biorenewable chemical industry. *Plant Journal* 54:536–545.

North, D. C. 1991. Institutions. *Journal of Economic Perspectives* 5:97–112.

Quintero, J. A., M. I. Montoya, O. J. Sanchez, O. H. Giraldo and C. A. Cardona. 2008. Fuel ethanol production from sugarcane and corn: Comparative analysis for a Colombian case. *Energy* 33:385–399.

Ren, J., A. Manzardo, A. Mazzi, F. Zuliani and A. Scipioni. 2015. Prioritization of bioethanol production pathways in China based on life cycle sustainability assessment and multicriteria decision-making. *International Journal of Life Cycle Assessment* 20:842–853.

Rudorff, B. F. T., D. A. de Aguiar and W. F. da Silva, et al. 2010. Studies on the rapid expansion of sugarcane for ethanol production in Sao Paulo state (Brazil) using Landsat data. *Remote Sensing* 2:1057–1076.

Shin, D., T. Kim, J. Choi and J. Kim. 2014. Author name disambiguation using a graph model with node splitting and merging based on bibliographic information. *Scientometrics* 100:15–50.

Soccol, C. R., L. P. de Souza Vandenberghe and A. B. P. Medeiros, et al. 2010. Bioethanol from lignocelluloses: Status and perspectives in Brazil. *Bioresource Technology* 101:4820–4825.

Tyner, W.E. 2008. The US ethanol and biofuels boom: Its origins, current status, and future prospects. *BioScience* 58:646–653.

Vadas, P. A., K. H. Barnett and D. J. Undersander 2008. Economics and energy of ethanol production from alfalfa, corn, and switchgrass in the Upper Midwest, USA. *Bioenergy Research* 1:44–55.

95 Country-based Experience of Bioethanol Fuels

Review

Ozcan Konur
(Formerly) Ankara Yildirim Beyazit University

95.1 INTRODUCTION

Crude oil-based gasoline fuels (Ma et al., 2002; Newman and Kenworthy, 1989) have been widely used in transportation sector since the 1920s. However, there have been great public concerns over adverse environmental and human impact of these fuels (Hill et al., 2006, 2009). Hence, biomass-based bioethanol fuels (Hill et al., 2006; Konur, 2012, 2015, 2019, 2020) have increasingly been used in blending gasoline fuels (Hsieh et al., 2002; Najafi et al., 2009) and in fuel cells (Antolini, 2007, 2009).

Research in the field of experiences of countries has also intensified in recent years (Amorim et al., 2011; Lopes et al., 2016). The primary focus of research on these country-based experiences has been optimization of bioethanol and biomass production and development of the policies to promote production and use of bioethanol fuels worldwide. In this context, there has been a significant focus on evaluative studies in bioethanol fuels (Farrell et al., 2006; Hill et al., 2006).

However, it is essential to develop efficient incentive structures (North, 1991) for the primary stakeholders to enhance research in this field (Konur, 2000, 2002a,b,c, 2006a,b, 2007a,b).

Although there has been a number of review papers on country experiences of bioethanol fuels (Amorim et al., 2011; Lopes et al., 2016), there has been no review of the 25 most-cited articles in this field.

This book chapter presents a review of the 25 most-cited articles in the field of country-based experiences of bioethanol fuels. Then, it discusses the key findings of these highly influential papers and comments on future research priorities in this field.

95.2 MATERIALS AND METHODS

Search for this study was carried out using Scopus database (Burnham, 2006) in October 2021.

As the first step for search of relevant literature, keywords were selected using the first 200 most-cited papers. The selected keyword list was optimized to obtain a representative sample of papers for the searched research field. This keyword list was provided in appendix of Konur (2023) for future replication studies.

As the second step, a sample data set was used for this study. The first 25 articles in the sample of 100 most-cited papers with at least 126 citations each were selected for review study. Key findings from each paper were taken from abstracts of these papers and were discussed. Additionally, a number of brief conclusions were drawn and a number of relevant recommendations were made to enhance future research landscape.

95.3 RESULTS

Brief information about 25 most-cited papers with at least 126 citations each on country-based experiences of bioethanol fuels is given below.

DOI: 10.1201/9781003226574-128

95.3.1 Brazilian Experience of Bioethanol Fuels

Brief information about 11 prolific studies with at least 142 citations each on Brazilian experience of bioethanol fuels is given in Table 95.1. Furthermore, brief notes on contents of these studies are also given below.

Macedo et al. (2008) evaluated energy balance and greenhouse gas (GHG) emission balance in both production and use of bioethanol fuels from sugarcane in Brazil in a paper with 607 citations. They used a sample of sugar mills with a maximum capacity of 100 million tons of sugarcane per year. They found that net energy ratio was 9.3 for 2005/2006 and would be 11.6 in 2020. The total GHG emissions were 436 kg CO_2 eq/m^3 bioethanol for 2005/2006 and would decrease to 345 kg CO_2 eq/m^3 in 2020. Further, avoided GHG emissions for E100 use in Brazil were 2,181 kg CO_2 eq/m^3 bioethanol for 2005/2006, while for E25, they were 2,323 kg CO_2 eq/m^3 bioethanol. Both values would increase about 26% in 2020 mostly due to large increase in sales of electricity surpluses. There was high impact of sugarcane productivity and bioethanol yield variation on energy and emission balances as well as impacts of average sugarcane transportation distances and level of soil cultivation. There was also impact of sugarcane bagasse and bioelectricity surpluses on GHG emissions avoidance.

Basso et al. (2008) selected *Saccharomyces cerevisiae* strains for bioethanol fuel production in Brazil in a paper with 297 citations. They used them for fermentation of sugarcane juice and molasses. They noted a positive impact of these selected yeast strains in increasing bioethanol yield and reducing production costs due to their higher fermentation performance. These strains resulted in high bioethanol yield, reduced glycerol and foam formation, and maintained high viability during recycling and very high implantation capability into industrial fermenters. They also noted that the

TABLE 95.1
Brazilian Experience of Bioethanol Fuels

No.	Papers	Biomass	Parameters	Cits
1	Macedo et al. (2008)	Sugarcane	Energy balance, GHG emission balance, avoided GHG emissions	607
2	Basso et al. (2008)	Sugarcane juice and molasses	Yeasts, bioethanol productivity	297
3	Soccol et al. (2010)	Sugarcane bagasse	Bioethanol productivity, sugarcane plantation, sugarcane bagasse use	289
4	Rudorff et al. (2010)	Sugarcane	Sugarcane plantation expansion, land use change, sugarcane burning	271
5	Martinelli and Filoso (2008)	Sugarcane	Environmental sustainability, economic competitiveness	256
6	Goldemberg et al. (2004)	Sugarcane	Economic competitiveness, economic of scale, technological advances	242
7	Luo et al. (2009)	Sugarcane and sugarcane bagasse	LCA, LCC, E0, E10, E86, E100, GHG emissions, economic competitiveness	229
8	Smeets et al. (2008)	Sugarcane and sugarcane bagasse	Environmental and socioeconomic impacts of bioethanol production	163
9	Dias et al. (2011)	Sugarcane and sugarcane bagasse	Bagasse-based bioethanol production, techno-economic analysis, bagasse-based power production	160
10	De Cerqueira Leite et al. (2009)	Sugarcane	Land use change, bioethanol production	144
11	van den Wall Bake et al. (2009)	Sugarcane	Feedstock costs and industrial production costs	142

Cits., the number of the citations received by each paper.

great yeast biodiversity could be an important source of strains. During yeast cell recycling, selective pressure was imposed on cells, leading to strains with higher tolerance to stressful conditions of industrial fermentation.

Soccol et al. (2010) evaluated production of bioethanol from sugarcane bagasse in Brazil in a paper with 289 citations. They noted that addition of 25% bioethanol to gasoline (E25) reduced import of 550 million barrels of crude oil and also reduced CO_2 emissions by 110 million tons. Sugarcane planted in 0.9% of the land available for agriculture contributed to the 13.5% of the energy supply in 2010. Hence, there was ample room to expand sugarcane plantation. Furthermore, bioethanol yield per hectare of sugarcane could rise from 6,000 L/ha to 10,000 L/ha, if 50% of the produced bagasse would be converted to bioethanol.

Rudorff et al. (2010) evaluated the areal extent and characteristics of the rapid sugarcane expansion and land use change for bioethanol production in Brazil using Landsat data in a paper with 271 citations. They found that sugarcane plantations expanded from 2.57 million ha in 2003 to 4.45 million ha in 2008. They noted that almost all the land use change, for sugarcane expansion of crop year 2008/09, took place on pasture and annual crop land. Further, during 2008 harvest season, burned sugarcane area was reduced to 50% of the total harvested area due to stopping of sugarcane straw burning practice by 2014 for mechanized areas.

Martinelli and Filoso (2008) evaluated the expansion of sugarcane-based bioethanol production to be environmentally sustainable and economically competitive in Brazil in a paper with 256 citations. They recommended proper planning and environmental risk assessments for expansion of sugarcane to new regions, improvement of land use practices to reduce soil erosion and N pollution, proper protection of streams and riparian ecosystems, banning of sugarcane burning practices, and fair working conditions for sugarcane cutters. They also supported creation of a more constructive approach for international stakeholders and trade organizations to promote sustainable development for bioethanol production in developing countries. Finally, they supported inclusion of environmental values in price of bioethanol fuels in order to discourage excessive replacement of natural ecosystems such as forests, wetlands, and pasture by bioethanol crops.

Goldemberg et al. (2004) evaluated the Brazilian experience for bioethanol production from sugarcane in a paper with 242 citations. They showed that economies of scale and technological advances led to increased competitiveness of sugarcane-based bioethanol fuels, thus reducing the gap with crude oil-based gasoline fuels.

Luo et al. (2009) carried out life cycle assessment (LCA) and life cycle costing (LCC) of bioethanol from sugarcane compared to gasoline fuels in Brazil in a paper with 229 citations. They considered bioethanol production from sugarcane and power generation from sugarcane bagasse as a base case, while bioethanol production from both sugarcane and bagasse and electricity generation from wastes was a future case, while in both cases, sugar was co-produced. Life cycles of fuels included gasoline production; agricultural production of sugarcane; bioethanol production; sugar and power coproduction; blending ethanol with gasoline to produce E10 and E85; and finally, use of E0, E10, E85, and E100. They found that in base case, less GHG was emitted, while overall evaluation of these fuel options depended on importance attached to different impacts. Future case was certainly more economically attractive. Nevertheless, outcomes depended very much on the assumed price for crude oil as in the real market, prices of fuels were very much dependent on taxes and subsidies. They asserted that technological developments could help in lowering both environmental impact and prices of bioethanol fuels.

Smeets et al. (2008) evaluated the environmental and socioeconomic impacts of bioethanol production to determine major bottlenecks for production and for the development of a practically applicable certification system in Brazil in a paper with 163 citations. Due to higher yields and overlapping costs, they found that the total additional production costs of compliance with various environmental and socioeconomic criteria were about +36%. They asserted that the net energy balance could be increased and GHG emissions reduced by increasing bioethanol production per ton sugarcane and by increasing use of sugarcane bagasse for power production. A major bottleneck for a sustainable and certified production was the increase in sugarcane production and possible impacts on biodiversity and competition with food production. Finally, both ban on and allowance of the use of genetically modified sugarcane could also become a major bottleneck.

Dias et al. (2011) evaluated second generation sugarcane bagasse-based bioethanol production in relation to the power production in Brazil in a paper with 160 citations. They noted that bagasse-based bioethanol production might be primarily in competition with power production from sugarcane bagasse. They carried out a techno-economic analysis of integrated production of bioethanol from sugarcane and sugarcane bagasse evaluating different technological scenarios. They showed the importance of the integrated use of sugarcane for bioethanol production. They asserted that bagasse-based bioethanol might favorably compete with power production when low-cost enzyme and improved technologies became commercially available.

De Cerqueira Leite et al. (2009) evaluated production of bioethanol fuels in Brazil in a paper with 144 citations. They considered the expansion of sugarcane-derived bioethanol fuels to displace 5% of projected global gasoline use in 2025. With existing technology, they estimated that 21 million ha of land would be required to produce the necessary bioethanol, less than 7% of current Brazilian agricultural land and equivalent to current soybean land use. New production lands could come from pasture made available through improving pasture management in cattle industry. They asserted that with continued introduction of new sugarcane species and new bioethanol production technologies, this could reduce these modest land requirements by 29%–38%.

Van den Wall Bake et al. (2009) evaluated cost reductions of bioethanol production from sugarcane to describe development of feedstock costs and industrial production costs in Brazil in a paper with 142 citations. They found that progress ratio (PR) for feedstock costs and industrial costs were 0.68 and 0.81, respectively, while the experience curve of total production costs resulted in a PR of 0.80. All production subprocesses of sugarcane production contributed to total costs, while increasing yields were the main driving force. Industrial costs mainly decreased due to increasing scales of bioethanol plants. Total production costs in 2009 were approximately 340 US\$/$m_{ethanol}^3$ (16 US\$/giga joules-GJ). Based on the experience curves for feedstock and industrial costs, they estimated total bioethanol production costs in 2020 between US\$ 200 and 260/m^3 (9.4–12.2 US\$/GJ).

95.3.2 The US Experience of Bioethanol Fuels

Brief information about seven prolific studies with at least 127 citations each on the US experience of bioethanol fuels is given in Table 95.2. Furthermore, brief notes on contents of these studies are also given below.

Hertel et al. (2010) evaluated effects of the US first generation corn-based bioethanol on global indirect land use and GHG emissions in a paper with 355 citations. They found that factoring market-mediated responses and byproduct use into analysis reduced cropland conversion by 72% from the land used for bioethanol feedstock. Consequently, the associated GHG release estimated in the framework was 800 g of CO_2 per megajoule (MJ) and 27 g/MJ/year, over 30 years of ethanol production. This was roughly a quarter of only other published estimate of GHG releases attributable to changes in indirect land use. Nonetheless, they argued that 800 g of CO_2 were enough to cancel out benefits that corn-based bioethanol had on global warming.

Jacobson (2007) evaluated effects of bioethanol-gasoline fuel blends (E85) on cancer, mortality, and hospitalization in the USA in a paper with 242 citations. He found that E85 fuels might increase ozone-related mortality, hospitalization, and asthma by about 9% in Los Angeles and 4% in the USA as a whole relative to 100% gasoline. Ozone increases in Los Angeles and the northeast were partially offset by decreases in the southeast. Further, E85 also increased peroxyacetyl nitrate (PAN) in the USA but caused little change in cancer risk. They asserted that due to its ozone effects, future E85 might be a greater overall public health risk than gasoline fuels and E85 would unlikely improve air quality over future gasoline vehicles. Further, unburned bioethanol emissions from E85 might result in a global-scale source of acetaldehyde larger than that of direct emissions.

TABLE 95.2
The US Experience of Bioethanol Fuels

No.	Papers	Biomass	Parameters	Cits
1	Hertel et al. (2010)	Corn grains	GHG emissions, indirect land use, global warming	355
2	Jacobson (2007)	Corn grains	Ozone-related mortality, hospitalization, and asthma, ozone emissions, pan and acetaldehyde emissions	242
3	Anderson et al. (2012)	Corn grains	Bioethanol-gasoline fuel blends, high octane numbers	171
4	Tyner (2008)	Corn grains	Bioethanol fuel production and price, crude oil price	141
5	Roberts and Schlenker (2013)	Corn, rice, soybeans, wheat	Supply and demand elasticities of agricultural commodities	133
6	Hettinga et al. (2009)	Corn grains	Corn-based bioethanol production cost reductions	131
7	Chiu et al. (2009)	Corn grains	Bioethanol water requirement, irrigation practices	127

Anderson et al. (2012) evaluated high-octane number bioethanol-gasoline fuel blends to increase minimum octane number (Research Octane Number, RON) of regular-grade gasoline in a paper with 171 citations. They noted that higher RON would enable greater thermal efficiency in future engines through higher compression ratio (CR) and/or more aggressive turbocharging and downsizing, and in current engines, through more aggressive spark timing under some driving conditions. For bioethanol and blendstock RON scenarios considered, they estimated CR increases on the order of 1–3 CR-units for port fuel injection engines as well as for direct injection engines. They asserted that substantial societal benefits might be associated with capitalizing on inherent high-octane rating of bioethanol in future higher octane number bioethanol-gasoline blends.

Tyner (2008) evaluated bioethanol boom in the USA in a paper with 141 citations. He reasoned that this boom was an unintended consequence of a fixed bioethanol subsidy that was keyed to $ 20 per barrel crude oil, combined with a surge in crude oil prices from $60 per barrel to $120 per barrel. Thus, future prospects for first generation corn-based bioethanol would depend on crude oil price, price of corn and distillers' grains, market value of bioethanol, plant capital and operating costs, and federal bioethanol policies. He asserted that policy choices would be absolutely critical in determining the extent to which bioethanol targets were achieved and at what cost. He predicted that if price of crude oil remains above $100 per barrel, bioethanol fuels would continue to be produced even without government interventions and subsidies.

Roberts and Schlenker (2013) evaluated supply and demand elasticities of agricultural commodities in the USA with implications for bioethanol mandate in a paper with 133 citations. They developed a new framework to identify supply elasticities of storable commodities where past shocks were used as exogenous price shifters. In agricultural context, they found that past yield shocks changed inventory levels and futures prices of agricultural commodities. They used the estimated elasticities to evaluate impact of the 2009 Renewable Fuel Standard on commodity prices, quantities, and food consumers' surplus for four basic staples of corn, rice, soybeans, and wheat. They found that prices increased 20% if one-third of commodities used to produce bioethanol were recycled as feedstock.

Hettinga et al. (2009) evaluated reductions in the first generation corn-based bioethanol production costs by using experience curve approach in the USA between 1980 and 2005 in a paper with 131 citations. They found that corn production costs declined by 62% over 30 years, down to $100\$_{2005}$/ton in 2005, while corn production volumes almost doubled since 1975. They calculated a progress ratio (PR) of 0.55 indicating a 45% cost decline over each doubling in cumulative production where higher corn yields and increasing farm sizes were the most important drivers behind this cost decline. Industrial processing costs of bioethanol declined by 45% since 1983, to below

$130\$_{2005}/m^3$ in 2005 (excluding costs for corn and capital), equivalent to a PR of 0.87. Similarly, total bioethanol production costs (including capital and net corn costs) declined approximately 60% from $800\$_{2005}/m^3$ in the early 1980s, to $300\$_{2005}/m^3$ in 2005. They reasoned that higher bioethanol yields, lower energy use, and the replacement of beverage alcohol-based production technologies mostly contributed to this substantial cost decline. In addition, average size of dry grind bioethanol plants increased by 235% since 1990. They further estimated that solely due to technological learning, production costs of bioethanol might decline 28%–44% in future, though this excluded effects of the current rising corn and fossil fuel costs.

Chiu et al. (2009) evaluated water use in corn-based bioethanol production in the USA in a paper with 127 citations. They noted a previous estimate of a liter of bioethanol required 263–784 L of water from corn farm to fuel pump. They estimated state-level field-to-pump water requirement of bioethanol across the nation. They found that bioethanol's water requirements could range from 5 to 2,138 L/L of bioethanol depending on regional irrigation practices. Finally, they found that as the bioethanol industry expanded to areas that apply more irrigated water than others, consumptive water appropriation by bioethanol increased 246% from 1.9 to 6.1 trillion L between 2005 and 2008, whereas U.S. bioethanol production increased only 133% from 15 to 34 billion L during the same period.

95.3.3 The Chinese Experience of Bioethanol Fuels

Brief information about three prolific studies with at least 126 citations each on the Chinese experience of bioethanol fuels is given in Table 95.3. Furthermore, brief notes on contents of these studies are also given below.

Gnansounou et al. (2005) evaluated production of bioethanol fuels from sweet sorghum juice and bagasse in China in a paper with 287 citations. Alternative options were selling sugar from juice or burning bagasse for power production. They noted that bioethanol production from bagasse was more favorable than burning it to produce power, while relative merits of making bioethanol or sugar from the juice was very sensitive to sugar price. Thus, they recommended a flexible plant capable of making both sugar and bioethanol fuel from the juice. Overall, bioethanol production from sorghum bagasse was very favorable. However, they emphasized that corn stover and rice hulls would be more attractive feedstocks for producing bioethanol in the long run due to their extensive availability as well as their independence from other markets. Furthermore, new technologies could enhance competitiveness of bioethanol production.

Ren et al. (2015) evaluated first generation bioethanol production pathways in China based on life cycle sustainability assessment (LCSA) and multicriteria decision-making (MCDM) in a paper with 132 citations. They considered an illustrative case about three alternative bioethanol production scenarios based on wheat, corn, and cassava feedstocks. They found that the prior sequence based on sustainability performances in descending order was cassava-based, corn-based, and wheat-based bioethanol fuel production. This methodology allowed decision-makers/stakeholders to select the most sustainable scenario among many alternatives.

TABLE 95.3
Chinese Experience of Bioethanol Fuels

No.	Papers	Biomass	Parameters	Cits
1	Gnansounou et al. (2005)	Sweet sorghum juice and bagasse	Bioethanol productivity, sugar production, power production	287
2	Ren et al. (2015)	Wheat, corn, cassava	Alternative bioethanol production scenarios, LCSA, MCDM, sustainability decision-making	132
3	Dai et al. (2006)	Cassava	Energy efficiency and potentials of cassava-based bioethanol fuel production	126

Dai et al. (2006) evaluated energy efficiency and potentials of first generation cassava-based bioethanol fuels in Guangxi region of China in a paper with 126 citations. They used net energy value (NEV) and net renewable energy value (NREV) to assess the energy and renewable energy efficiency of this bioethanol system during its life cycle. They divided this system into five subsystems including cassava plantation/treatment, bioethanol conversion, denaturing, refueling, and transportation. They found that this system was energy and renewable energy efficient as indicated by positive NEV and NREV values of 7.475 and 7.881 MJ/L, respectively. Through this system, one Joule of crude oil-based fuel could produce 9.8 J of fuel bioethanol. With cassava output in 2003, it could substitute for 166.107 million L of gasoline, while with cassava output potential, it could substitute for 618.162 million L of gasoline. They asserted that cassava-based bioethanol was more energy efficient than gasoline, diesel fuel, and corn-based bioethanol fuels but less efficient than biodiesel.

95.3.4 Other Countries' Experience of Bioethanol Fuels

Brief information about four prolific studies with at least 134 citations each on other countries' experience of bioethanol fuels is given in Table 95.4. Furthermore, brief notes on contents of these studies are also given below.

Quintero et al. (2008) evaluated economic and environmental performance of first generation bioethanol production from sugarcane and corn in Colombia in a paper with 240 citations. They used net present value (NPV) and total output rate of potential environmental impact as economic and environmental indicators, respectively. They found sugarcane-based bioethanol process as the best choice for bioethanol production facilities. However, they asserted that starch feedstocks like corn, cassava, or potatoes could potentially cause a higher impact on rural communities and boost their economies if social matters are considered.

Goh et al. (2010) evaluated second generation bioethanol production from waste feedstocks in Malaysia in a paper with 228 citations. They estimated amount of waste feedstocks as 47,402 dry kton/year and the total capacity and demand of waste-based bioethanol production as 26,161 and 6,677 ton/day, respectively. Thus, 19% of the total CO_2 emissions could be avoided. They proposed an integrated national supply network together with collection, storage, and transportation of raw materials and products.

Prasad et al. (2007) evaluated first generation bioethanol production from sweet sorghum syrup as an additive for gasoline and petrodiesel fuels in India in a paper with 169 citations. They noted that sweet sorghum had potential as a raw material for bioethanol production due to its rapid growth rate and early maturity, greater water use efficiency, limited fertilizer requirement, high total value, and wide adoptability. They asserted that the major stakeholders such as bioethanol producers, research institutions, and governments could coordinate with farmers to strategically develop value-added utilization of sweet sorghum and bioethanol production from this syrup and could

TABLE 95.4
Other Countries' Experience of Bioethanol Fuels

No.	Papers	Biomass	Parameters	Country	Cits
1	Quintero et al. (2008)	Sugarcane, corn	Economic and environmental performance of bioethanol production	Colombia	240
2	Goh et al. (2010)	Wastes	Bioethanol production capacity and demand, CO_2 emissions	Malaysia	228
3	Prasad et al. (2007)	Sweet sorghum syrup	Bioethanol production from sweet sorghum syrup	India	169
4	Gnansounou (2010)	Lignocellulosic biomass	Lignocellulosic biomass and bioethanol production	Europe	134

significantly reduce India's dependence on the imported crude oil and minimize the environmental threats caused by the crude oil-based gasoline fuels.

Gnansounou (2010) evaluated production and use of lignocellulosic bioethanol in Europe in a paper with 134 citations. He noted that in most of the European countries, sustainable lignocellulosic resources might not be widely available in future for bioethanol production due to possible competition between several potential usages. Thus, actual deployment of lignocellulosic bioethanol in Europe would depend on opportunity costs of biomass on one side and on prices of bioethanol and gasoline fuels on the other side. They recommended that as security of crude oil supply would be lower in long term, policy instruments should explicitly reward the higher value of lignocellulosic bioethanol compared to first generation bioethanol and gasoline fuels in future.

95.4 DISCUSSION

95.4.1 Introduction

Crude oil-based gasoline fuels have been widely used in transportation sector since the 1920s. However, there have been great public concerns over adverse environmental and human impact of these fuels. Hence, biomass-based bioethanol fuels have increasingly been used in blending gasoline fuels and in fuel cells.

Research in the field of experiences of countries has also intensified in recent years. The primary focus of research on these country-based experiences has been optimization of bioethanol and biomass production and development of policies to promote production and use of bioethanol fuels worldwide.

However, it is essential to develop efficient incentive structures for the primary stakeholders to enhance research in this field. Although there has been a number of review papers on country experiences of bioethanol fuels, there has been no review of the 25 most-cited articles in this field. Hence, this book chapter presents a review of the 25 most-cited articles in the field of country-based experiences of bioethanol fuels. Then, it discusses key findings of these highly influential papers and comments on future research priorities in this field.

As the first step for search of relevant literature, keywords were selected using the first 200 most-cited papers. The selected keyword list was optimized to obtain a representative sample of papers for the searched research field. This keyword list was provided in appendix of Konur (2023) for future replication studies.

As the second step, a sample dataset was used for this study. The first 25 articles in the sample of 100 most-cited papers with at least 126 citations each were selected for review study. Key findings from each paper were taken from abstracts of these papers and were discussed. Additionally, a number of brief conclusions were drawn and a number of relevant recommendations were made to enhance future research landscape.

As Table 95.5 shows, there are four primary research fronts for these 25 most-cited papers: Brazil, the USA, China, and other countries.

95.4.2 Brazilian Experience of Bioethanol Fuels

Brazil has been one of the major producers, users, and exporters of bioethanol fuels since it has started a sugarcane-based bioethanol program in the 1970s. Over 40% of these reviewed papers and the 100 most-cited papers were related to Brazil's experiences with bioethanol fuels (Table 95.5). All these reviewed papers focused on sugarcane and sugarcane bagasse.

A number of research issues emerge from these reviewed studies. Techno-economics of bioethanol production was a primary topic considered in these papers (Dias et al., 2011; Goldemberg et al., 2004; Luo et al., 2009, Macedo et al., 2008; Martinelli and Filoso, 2008; Smeets et al., 2008;

TABLE 95.5
The Country-based Experiences of Bioethanol Fuels: Research Fronts

No.	Countries	Sample Papers (%)	Reviewed Papers (%)
1	Brazil	41	44
2	USA	15	28
3	China	12	12
4	Other countries	26	16
	Europe	4	4
	India	4	4
	Malaysia	3	4
	Colombia	2	4
	Thailand	6	0
	Sweden	3	0
	Canada	2	0
	UK	2	0
	Sample size	100	25

van den Wall Bake et al., 2009). A number of ways were advanced to increase net energy value and economic competitiveness of bioethanol fuels produced in Brazil in these papers.

GHG emissions from bioethanol fuels during both production and utilization stages were another major issue in these reviewed papers (Luo et al., 2009; Macedo et al., 2008; Martinelli and Filoso, 2008; Smeets et al., 2008; Soccol et al., 2010). A number of ways were advanced to decrease GHG emissions during both production and utilization of bioethanol fuels produced in Brazil in these papers.

As the amount of bioethanol production has increased since the 1970s, the size of sugarcane plantations has also increased putting pressure on lands available for sugarcane production (De Cerqueira Leite et al., 2009; Martinelli and Filoso, 2008; Rudorff et al., 2010; Smeets et al., 2008; Soccol et al., 2010). Additionally, Basso et al. (2008) studied the impact of yeast strains on bioethanol productivity.

These findings are thought-provoking in seeking ways to optimize bioethanol production and utilization in Brazil. The techno-economics, environmental and societal impacts, and land use change for bioethanol production and utilization emerge as key issues for the Brazilian experience of bioethanol fuels.

95.4.3 The US Experience of Bioethanol Fuels

The USA has been one of the major producers, users, and exporters of bioethanol fuels as it has started a corn grain and corn bagasse-based bioethanol program in the 1970s. Over 25% and 15% of these reviewed papers and the 100 most-cited papers, respectively, were related to the USA's experiences with bioethanol fuels (Table 95.5). All these reviewed papers focused primarily on corn grains.

A number of research issues emerge from these reviewed studies. Techno-economics of bioethanol production was a primary issue considered in these papers (Hettinga et al., 2009; Roberts and Schlenker, 2013; Tyner, 2008). A number of ways were advanced to increase net energy value and economic competitiveness of bioethanol fuels produced in the USA in these papers.

GHG emissions from bioethanol fuels during both production and utilization stages were another major issues in these reviewed papers (Hertel et al., 2010; Jacobson et al., 2007). A number of ways were advanced to decrease GHG emissions during both production and utilization of bioethanol fuels produced in the USA in these papers. Jacobson et al. (2007) also studied effect of ozone emissions from bioethanol fuels on human health for the E85 bioethanol-gasoline fuel blend. It has to be noted that traditional bioethanol fuel-gasoline blends do not exceed 20% bioethanol additives. Hence, the importance of this study should be devalued.

As the amount of bioethanol production has increased, the size of the corn plantations has also increased putting pressure on lands available for corn production (Hertel et al., 2010). Additionally, Chiu et al. (2009) studied the impact of irrigation practices on water requirement for bioethanol production. This study highlighted the importance of water resources for bioethanol production. Similarly, Anderson et al. (2012) studied bioethanol-gasoline fuel blends with high octane numbers as a way to increase the engine performance.

These findings are thought-provoking in seeking ways to optimize bioethanol production and utilization in the USA. The techno-economics, environmental and societal impact, land use change, water requirements, and blends with high octane numbers for bioethanol production and utilization emerge as key issues for the US experience of bioethanol fuels.

95.4.4 The Chinese Experience of Bioethanol Fuels

China has been one of the major producers and users of bioethanol fuels as it has intensified a various biomass-based bioethanol program in recent years. Around 12% of these reviewed papers and the 100 most-cited papers were related to China's experiences with bioethanol fuels (Table 95.5). All these reviewed papers focused on cassava and sweet sorghum.

A number of research issues emerge from these reviewed studies. Techno-economics of bioethanol production was a primary issue considered in these papers (Dai et al., 2006; Gnansounou et al., 2005). A number of ways were advanced to increase net energy value and economic competitiveness of bioethanol fuels produced in China in these papers.

GHG emissions from bioethanol fuels during both production and utilization stages were another major issues in these reviewed papers (Ren et al., 2015). A number of ways were advanced to decrease GHG emissions during both production and utilization of bioethanol fuels produced in China in these papers.

These findings are thought-provoking in seeking ways to optimize bioethanol production and utilization in China. Techno-economics and environmental and societal impact of bioethanol production and utilization emerge as key issues for the Chinese experience of bioethanol fuels.

95.4.5 Other Countries' Experience of Bioethanol Fuels

Europe, Colombia, Malaysia, and India have been among the major producers and users of bioethanol fuels as they have had started various biomass-based bioethanol programs in recent years. However, around 12% of these reviewed papers and the 100 most-cited papers were related to these countries' experiences with bioethanol fuels (Table 95.5).

All these reviewed papers focused on a number of feedstocks such as sugarcane, corn, wastes, sweet sorghum, and lignocellulosic biomass. It should be noted that there were three and two papers on the experiences of bioethanol fuels for Sweden and the UK in the 100 most-cited 100 papers set although there were no papers in the reviewed paper set.

A number of research issues emerge from these review studies. Techno-economics of bioethanol production was a primary use considered in these papers (Gnansounou, 2010; Goh et al., 2010; Prasad et al., 2007; Quintero et al., 2008). A number of ways were advanced to increase net energy value and economic competitiveness of bioethanol fuels produced in these countries in these papers.

GHG emissions from bioethanol fuels during both production and utilization stages were another major issues in these reviewed papers (Goh et al., 2010; Quintero et al., 2008). A number of ways were advanced to decrease GHG emissions during both production and utilization of bioethanol fuels produced in these countries in these papers.

These findings are thought-provoking in seeking ways to optimize bioethanol production and utilization in these countries. Techno-economics and environmental and societal impact of bioethanol production and utilization emerge as key issues for these countries' experience of bioethanol fuels.

95.4.6 The Overall Remarks

It is not surprising that Brazil dominated the research on country-based experiences of bioethanol fuels with over 40% of both reviewed paper and 100 most-cited paper sets. The USA, China, and Europe were other prolific countries in this field. All these countries were major producers, users, and exporters of bioethanol fuels in recent years.

As expected, emissions of bioethanol fuels are less than those of gasoline or petrodiesel fuels in general. Furthermore, operating conditions of gasoline or diesel engines as well as proportions of bioethanol in these blends strongly impact emissions of these bioethanol fuel blends. With the optimization of operating conditions of engines as well as the proportions of bioethanol, bioethanol blends might have better emission performance compared to both gasoline and petrodiesel fuels.

Emissions (Lapuerta et al., 2008) of these fuels are also of the most importance in the light of public concerns about climate change (Change, 2007), GHG emissions (Carlson et al., 2017), and global warming (Kerr, 2007). As these concerns have been certainly behind the boom in research in this field in the last two decades, it is not surprising that the bulk of majority of the most-cited papers sought to evaluate the environmental impact of bioethanol fuels during both its production and utilization.

Emissions of CO_2 (Ang, 2007) are especially important and any potential of reduction of these emissions by bioethanol blends compared to gasoline or petrodiesel fuels would have significant implications for environmental protection. Emissions of particulate matter (PM) are also important for the protection of environment (Abdullahi et al., 2013).

Key findings from these studies also hint that bioethanol fuels have comparable net energy gain, better emissions, and comparable economic competitiveness following the list of issues given by Hill et al. (2006). As expected, emissions, net energy, and economic competitiveness of second generation bioethanol fuels are better than those of first generation bioethanol fuels and much better than those of gasoline or petrodiesel fuels in general.

The studies also evaluate the impact of production and utilization of bioethanol fuels on biodiversity, deforestation, soil biodegradation, soil decarbonization, water resources contamination and depletion, competition between food and fuels, water pollution, air pollution, land use, global warming, and ecological footprints. It is essential that these issues should be incorporated into the life cycle and techno-economic assessments of bioethanol fuels.

As expected, the impact of second generation bioethanol fuels on these issues is better than those of the first generation bioethanol fuels and much better than those of gasoline or petrodiesel fuels in general.

The change in land use (Foley et al., 2005) due to production of bioethanol fuels emerges as a critical public policy issue to be considered by the major stakeholders. Such changes in land use potentially undermine the capacity of ecosystems to sustain food production, maintain freshwater and forest resources, regulate climate and air quality, and ameliorate infectious diseases (Foley et al., 2005).

Next, with increasing change in land use due to bioethanol production, change in biodiversity (Jenkins, 2003), deforestation (Achard et al., 2002), and water pollution (Wang and Yang, 2016) also emerge as critical public policy issues.

One of the major problems with the use of first generation food crop-based feedstocks for bioethanol production is potential threat to food security (Maxwell, 1996) for global society (Pimentel et al., 2009). For example, growing crops for bioethanol fuels squanders land, water, and energy resources vital for production of food for human consumption. Further, using corn for ethanol increases the price of US foods and causes food shortages for the poor of the world (Pimentel et al., 2009).

Key findings from these studies hint that as expected second generation bioethanol fuels have better net energy gain and higher economic competitiveness than those of first generation bioethanol fuels and much better than those of gasoline or petrodiesel fuels in general. These studies also highlight the strong impact of methodological issues as well as the efficiency of the conversion technologies on the outcome of the techno-economic and life cycle assessments as these issues account for different assessment outcomes.

In the end, these most-cited papers in this field hint that the efficiency of bioethanol production could be optimized using structure, processing, and property relationships of bioethanol fuels and their feedstocks, derivatives, and coproducts (Formela et al., 2016; Konur, 2018, 2020b, 2021a,b, c, d; Konur and Matthews, 1989).

These reviewed studies also show the importance of incentive structures for development of bioethanol industry and research at a global scale in the light of North's institutional framework (North, 1991). In this context, the major producers and users of bioethanol fuels such as Brazil, the USA, China, and Europe have developed strong incentive structures for effective development of bioethanol industry and research.

95.5 CONCLUSION AND FUTURE RESEARCH

Brief information about key research fronts covered by the 25 most-cited papers with at least 126 citations each is given under four headings: Brazil, the USA, China, and other countries (Table 95.5).

These papers can be further grouped under headings of emissions during production of bioethanol fuels, emissions from utilization of bioethanol fuels in gasoline and diesel engines, life cycle assessment of bioethanol fuels, environmental issues other than emissions, and techno-economics for bioethanol fuels.

Key findings on these research fronts should be read in the light of increasing public concerns about climate change, GHG emissions, and global warming as these concerns have been certainly behind the boom in research in this field in the last two decades. It is therefore not surprising that the bulk of the majority of the most-cited papers sought to evaluate environmental impact of bioethanol fuels both during its production and utilization in gasoline and diesel engines.

These findings confirm that bioethanol fuels are a viable alternative to crude oil-based gasoline and petrodiesel fuels, have a net energy gain, have environmental benefits impacting favorably global warming, GHG emissions, and climate change, are economically competitive, and are producible in large quantities without reducing food supplies especially using second generation bioethanol feedstocks such as lignocellulosic feedstock using criteria introduced by Hill et al. (2006).

Fuel properties are important as they determine emission, combustion, and engine performance of these fuels. These studies hint that bioethanol fuels have better fuel properties in general compared to gasoline and petrodiesel fuels and there is a room to optimize these properties by adjusting composition and blend ratios of these fuels.

There is a persistent trend for bioethanol research to focus more on second and third generation feedstocks such as sugarcane bagasse and algae and to move away from first generation bioethanol feedstocks such as corn grains and sugarcane. These feedstocks would certainly impact properties, engine performance, emissions as well as net energy gain, economic competitiveness, and emissions of resulting bioethanol fuels and their blends.

This trend should be followed closely by the major stakeholders in future. For this reason, the impact of source of biomass on properties, engine performance as well as net energy gain, economic competitiveness, and emissions of bioethanol fuels and their gasoline or petrodiesel blends in diesel or gasoline engines should be focused on to establish structure-property-processing relationships for these bioethanol fuels in future studies as it has been done in other research streams of bioethanol production and utilization.

Performance of these fuels in gasoline or diesel engines is of the utmost importance as a function of the properties of these fuels as well as proportions of bioethanol in these blends. These studies hint that bioethanol fuels have better fuel performance compared to both gasoline and petrodiesel fuels in both gasoline and diesel engines in general.

The emissions of CO_2 are especially important and any potential of reduction of these emissions by bioethanol blends compared to gasoline or petrodiesel fuels would have significant implications for environmental protection. Emissions of PM are also important for the protection of environment.

As expected, emissions of second generation bioethanol fuels are less than those of first generation bioethanol fuels and much less than those of gasoline or petrodiesel fuels in general. Furthermore, besides the type of biomass, biomass productivity, bioethanol yield, and efficiency of conversion technologies strongly impact emissions of resulting bioethanol fuels and their blends.

Key findings from life cycle assessment studies hint that methodological issues for life cycle assessment strongly impact the outcome of these assessments. As expected, emissions, net energy, and economic competitiveness of second generation bioethanol fuels are better than those of first generation bioethanol fuels and much better than those of gasoline or petrodiesel fuels in general.

Other environmental studies evaluate the impact of production and utilization of bioethanol fuels on biodiversity, deforestation, soil biodegradation, soil decarbonization, water resources contamination and depletion, competition between food and fuels, water pollution, air pollution, land use, global warming, and ecological footprints. It is essential that these issues should be incorporated into life cycle and techno-economic assessments of bioethanol fuels. As expected, the impact of second generation bioethanol fuels on these issues is better than those of first generation bioethanol fuels and much better than those of gasoline or petrodiesel fuels in general.

Change in the land use due to production of bioethanol fuels emerges as a critical public policy issue to be considered by the major stakeholders. Such changes in land use potentially undermine the capacity of ecosystems to sustain food production, maintain freshwater and forest resources, regulate climate and air quality, and ameliorate infectious diseases (Foley et al., 2005). Next, with the increasing change in the land use due to bioethanol production, change in biodiversity, deforestation, and water pollution also emerge as a critical public policy issues.

One of the major problems with the use of first generation food crop-based feedstocks for bioethanol production is the potential threat to food security for global society. For example, growing crops for bioethanol fuels squanders land, water. and energy resources vital for production of food for human consumption. Furthermore, using corn for bioethanol increases the price of US foods and causes food shortages for the poor of the world (Pimentel et al., 2009).

In the end, these most-cited papers in the field of production and utilization of bioethanol fuels and their blends hint that benefits sought from these fuels could be maximized using structure, processing, and property relationships of these fuels as in materials science and engineering. These findings confirm that bioethanol fuels are a viable alternative to crude oil-based gasoline and petrodiesel fuels and have environmental and techno-economic benefits impacting favorably global warming, GHG emissions, and climate change.

These reviewed studies also show the importance of incentive structures for the development of bioethanol industry and research at a global scale in the light of North's institutional framework (North, 1991). In this context, major producers and users of bioethanol fuels such as Brazil, the USA, China, and Europe have developed strong incentive structures for the effective development of bioethanol industry and research.

It is recommended that such review studies should be performed for other research fronts on both production and utilization of bioethanol fuels complementing the corresponding scientometric studies.

ACKNOWLEDGMENTS

Contribution of the highly cited researchers in the field of country-based experiences of bioethanol fuels has been gratefully acknowledged.

REFERENCES

Abdullahi, K. L., J. M. Delgado-Saborit and R. M. Harrison. 2013. Emissions and indoor concentrations of particulate matter and its specific chemical components from cooking: A review. *Atmospheric Environment* 71:260–294.

Achard, F., H. D. Eva and H. J. Stibig, et al. 2002. Determination of deforestation rates of the world's humid tropical forests. *Science* 297:999–1002.

Amorim, H.V., M. L. Lopes, J. V. de Castro Oliveira, M. S. Buckeridge and G. H. Goldman. 2011. Scientific challenges of bioethanol production in Brazil. *Applied Microbiology and Biotechnology* 91:1267–1275.

Anderson, J. E., D. M. Dicicco and J. M. Ginder, et al. 2012. High octane number ethanol-gasoline blends: Quantifying the potential benefits in the United States. *Fuel* 97:585–594.

Ang, J. B. 2007. CO_2 emissions, energy consumption, and output in France. *Energy Policy* 35:4772–4778.

Antolini, E. 2007. Catalysts for direct ethanol fuel cells. *Journal of Power Sources* 170:1–12.

Antolini, E. 2009. Palladium in fuel cell catalysis. *Energy and Environmental Science* 2:915–931.

Basso, L. C., H. V. de Amorim, A. J. de Oliveira and M. L. Lopes. 2008. Yeast selection for fuel ethanol production in Brazil. *FEMS Yeast Research* 8:1155–1163.

Burnham, J. F. 2006. Scopus database: A review. *Biomedical Digital Libraries* 3:1–8.

Carlson, K. M., J. S. Gerber and N. D. Mueller, et al. 2017. Greenhouse gas emissions intensity of global croplands. *Nature Climate Change* 7:63–68.

Change, C. 2007. Climate change impacts, adaptation and vulnerability. *Science of the Total Environment* 326:95–112.

Chiu, Y. W., B. Walseth and S. Suh. 2009. Water embodied in bioethanol in the United States. *Environmental Science and Technology* 43:2688–2692.

Dai, D., Z. Hu, G. Pu, H. Li and C. Wang. 2006. Energy efficiency and potentials of cassava fuel ethanol in Guangxi region of China. *Energy Conversion and Management* 47:1686–1699.

De Cerqueira Leite, R. C., M. R. L. V. Leal, L. A. B. Cortez, W. M. Griffin and M. I. G. Scandiffio. 2009. Can Brazil replace 5% of the 2025 gasoline world demand with ethanol? *Energy* 34:655–661.

Dias, M. O. S., M. P. Cunha and C. D. F. Jesus, et al. 2011. Second generation ethanol in Brazil: Can it compete with electricity production? *Bioresource Technology* 102:8964–8971.

Farrell, A. E., R. J. Plevin and B. T. Turner, et al. 2006. Ethanol can contribute to energy and environmental goals. *Science* 311:506–508.

Foley, J. A., R. deFries and G. P. Asner, et al. 2005. Global consequences of land use. *Science* 309:570–574.

Formela, K., A. Hejna, L. Piszczyk, M. R. Saeb and X. Colom. 2016. Processing and structure-property relationships of natural rubber/wheat bran biocomposites. *Cellulose* 23:3157–3175.

Gnansounou, E. 2010. Production and use of lignocellulosic bioethanol in Europe: Current situation and perspectives. *Bioresource Technology* 101:4842–4850.

Gnansounou, E., A. Dauriat and C. E. Wyman. 2005. Refining sweet sorghum to ethanol and sugar: Economic trade-offs in the context of North China. *Bioresource Technology* 96:985–1002.

Goh, C.S., K. T. Tan, K. T. Lee and S. Bhatia. 2010. Bio-ethanol from lignocellulose: Status, perspectives and challenges in Malaysia. *Bioresource Technology* 101:4834–4841.

Goldemberg, J., S. T. Coelho, P. M. Nastari and O. Lucon. 2004. Ethanol learning curve - The Brazilian experience. *Biomass and Bioenergy* 26:301–304.

Hertel, T. W., A. A. Golub and A. D. Jones, et al. 2010. Effects of US Maize ethanol on global land use and greenhouse gas emissions: Estimating market-mediated responses. *BioScience* 60:223–231.

Hettinga, W. G., H. M. Junginger and S. C. Dekker, et al. 2009. Understanding the reductions in US corn ethanol production costs: An experience curve approach. *Energy Policy* 37:190–203.

Hill, J., E. Nelson, D. Tilman, S. Polasky and D. Tiffany. 2006. Environmental, economic, and energetic costs and benefits of biodiesel and ethanol biofuels. *Proceedings of the National Academy of Sciences of the United States of America* 103:11206–11210.

Hill, J., S. Polasky and E. Nelson, et al. 2009. Climate change and health costs of air emissions from biofuels and gasoline. *Proceedings of the National Academy of Sciences of the United States of America* 106:2077–2082.

Hsieh, W. D., R. H. Chen, T. L. Wu and T. H. Lin. 2002. Engine performance and pollutant emission of an SI engine using ethanol-gasoline blended fuels. *Atmospheric Environment* 36:403–410.

Jacobson, M. Z. 2007. Effects of ethanol (E85) versus gasoline vehicles on cancer and mortality in the United States. *Environmental Science and Technology* 41:4150–4157.

Jenkins, M. 2003. Prospects for biodiversity. *Science* 302:1175–1177.

Kerr, R. A. 2007. Global warming is changing the world. *Science* 316:188–190.

Konur, O. 2000. Creating enforceable civil rights for disabled students in higher education: An institutional theory perspective. *Disability & Society* 15:1041–1063.

Konur, O. 2002a. Access to nursing education by disabled students: Rights and duties of nursing programs. *Nurse Education Today* 22:364–374.

Konur, O. 2002b. Assessment of disabled students in higher education: Current public policy issues. *Assessment and Evaluation in Higher Education* 27:131–152.

Konur, O. 2002c. Access to employment by disabled people in the UK: Is the Disability Discrimination Act working? *International Journal of Discrimination and the Law* 5:247–279.

Konur, O. 2006a. Participation of children with dyslexia in compulsory education: Current public policy issues. *Dyslexia* 12:51–67.

Konur, O. 2006b. Teaching disabled students in higher education. *Teaching in Higher Education* 11:351–363.

Konur, O. 2007a. A judicial outcome analysis of the *Disability Discrimination Act*: A windfall for the employers? *Disability & Society* 22:187–204.

Konur, O. 2007b. Computer-assisted teaching and assessment of disabled students in higher education: The interface between academic standards and disability rights. *Journal of Computer Assisted Learning* 23:207–219.

Konur, O. 2012. The evaluation of the research on the bioethanol: A scientometric approach. *Energy Education Science and Technology Part A: Energy Science and Research* 28:1051–1064.

Konur, O. 2015. Current state of research on algal bioethanol. In *Marine Bioenergy: Trends and Developments*, Ed. S. K. Kim and C. G. Lee, pp. 217–244. Boca Raton, FL: CRC Press.

Konur, O., Ed. 2018. *Bioenergy and Biofuels*. Boca Raton, FL: CRC Press.

Konur, O. 2019. Cyanobacterial bioenergy and biofuels science and technology: A scientometric overview. In *Cyanobacteria: From Basic Science to Applications*, Ed. A. K. Mishra, D. N. Tiwari and A. N. Rai, pp. 419–442. Amsterdam: Elsevier.

Konur, O. 2020. The scientometric analysis of the research on the bioethanol production from green macroalgae. In *Handbook of Algal Science, Technology and Medicine*, Ed. O. Konur, pp. 385–401. London: Academic Press.

Konur, O., Ed. 2020b. *Handbook of Algal Science, Technology and Medicine*. London: Academic Press.

Konur, O., Ed. 2021a. *Handbook of Biodiesel and Petrodiesel Fuels: Science, Technology, Health, and Environment*. Boca Raton, FL: CRC Press.

Konur, O., Ed. 2021b. *Handbook of Biodiesel and Petrodiesel Fuels: Science, Technology, Health, and Environment. Volume 1. Biodiesel Fuels: Science, Technology, Health, and Environment*. Boca Raton, FL: CRC Press.

Konur, O., Ed. 2021c. *Handbook of Biodiesel and Petrodiesel Fuels: Science, Technology, Health, and Environment. Volume 2. Biodiesel Fuels based on the Edible and Nonedible Feedstocks, Wastes, and Algae: Science, Technology, Health, and Environment*. Boca Raton, FL: CRC Press.

Konur, O., Ed. 2021d. *Handbook of Biodiesel and Petrodiesel Fuels: Science, Technology, Health, and Environment. Volume 3. Petrodiesel Fuels: Science, Technology, Health, and Environment*. Boca Raton, FL: CRC Press.

Konur, O. 2023. Country-based experience of bioethanol fuels: Scientometric study. In *Evaluation and Utilization of Bioethanol Fuels. II.: Biohydrogen Fuels, Fuel Cells, Biochemicals, and Country Experiences. Handbook of Bioethanol Fuels Volume 6*, Ed. O. Konur, pp. 3–25. Boca Raton, FL: CRC Press.

Konur, O. and F. L. Matthews. 1989. Effect of the properties of the constituents on the fatigue performance of composites: A review. *Composites* 20:317–328.

Lapuerta, M., O. Armas and J. M. Herreros. 2008. Emissions from a diesel-bioethanol blend in an automotive diesel engine. *Fuel* 87:25–31.

Lopes, M. L., S. C. D. L. Paulillo and A. Godoy, et al. 2016. Ethanol production in Brazil: A bridge between science and industry. *Brazilian Journal of Microbiology* 47:64–76.

Luo, L., E. van der Voet and G. Huppes. 2009. Life cycle assessment and life cycle costing of bioethanol from sugarcane in Brazil. *Renewable and Sustainable Energy Reviews* 13:1613–1619.

Ma, X., L. Sun and C. Song. 2002. A new approach to deep desulfurization of gasoline, diesel fuel and jet fuel by selective adsorption for ultra-clean fuels and for fuel cell applications. *Catalysis Today* 77:107–116.

Macedo, I. C., J. E. A. Seabra and J. E. A. R. Silva. 2008. Green house gases emissions in the production and use of ethanol from sugarcane in Brazil: The 2005/2006 averages and a prediction for 2020. *Biomass and Bioenergy* 32:582–595.

Martinelli, L. A. and S. Filoso. 2008. Expansion of sugarcane ethanol production in Brazil: Environmental and social challenges. *Ecological Applications* 18:885–898.

Maxwell, S. 1996. Food security: A post-modern perspective. *Food Policy* 21:155–170.

Najafi, G., B. Ghobadian and T. Tavakoli, et al. 2009. Performance and exhaust emissions of a gasoline engine with ethanol blended gasoline fuels using artificial neural network. *Applied Energy* 86:630–639.

Newman, P. W. G. and J. R. Kenworthy. 1989. Gasoline consumption and cities: A comparison of U.S. cities with a global survey. *Journal of the American Planning Association* 55:24–37.

North, D. C. 1991. Institutions. *Journal of Economic Perspectives* 5:97–112.

Pimentel, D., A. Marklein and M. A. Toth, et al. 2009. Food versus biofuels: Environmental and economic costs. *Human Ecology* 37:1–12.

Prasad, S., A. Singh and H. C. Joshi. 2007. Ethanol production from sweet sorghum syrup for utilization as automotive fuel in India. *Energy and Fuels* 21:415–2420.

Quintero, J. A., M. I. Montoya, O. J. Sanchez, O. H. Giraldo and C. A. Cardona. 2008. Fuel ethanol production from sugarcane and corn: Comparative analysis for a Colombian case. *Energy* 33:385–399.

Ren, J., A. Manzardo, A. Mazzi, F. Zuliani and A. Scipioni. 2015. Prioritization of bioethanol production pathways in China based on life cycle sustainability assessment and multicriteria decision-making. *International Journal of Life Cycle Assessment* 20:842–853.

Roberts, M. J. and W. Schlenker. 2013. Identifying supply and demand elasticities of agricultural commodities: Implications for the US ethanol mandate. *American Economic Review* 103:2265–2295.

Rudorff, B. F. T., D. A. de Aguiar and W. F. da Silva, et al. 2010. Studies on the rapid expansion of sugarcane for ethanol production in Sao Paulo state (Brazil) using Landsat data. *Remote Sensing* 2:1057–1076.

Smeets, E., M. Junginger and A. Faaij, et al. 2008. The sustainability of Brazilian ethanol-An assessment of the possibilities of certified production. *Biomass and Bioenergy* 32:781–813.

Soccol, C. R., L. P. de Souza Vandenberghe and A. B. P. Medeiros, et al. 2010. Bioethanol from lignocelluloses: Status and perspectives in Brazil. *Bioresource Technology* 101:4820–4825.

Tyner, W.E. 2008. The US ethanol and biofuels boom: Its origins, current status, and future prospects. *BioScience* 58:646–653.

Van den Wall Bake, J. D., M. Junginger, A. Faaij, T. Poot and A. Walter. 2009. Explaining the experience curve: Cost reductions of Brazilian ethanol from sugarcane. *Biomass and Bioenergy* 33:644–658.

Wang, Q. and Z. Yang. 2016. Industrial water pollution, water environment treatment, and health risks in China. *Environmental pollution* 218: 358–365.

96 Ethanol

A Brief Review of the Major Production Methods in the World

Alessandro Senatore and Francesco Dalena
University of Calabria

Angelo Basile
Hydrogenia S.R.L.

96.1 INTRODUCTION

Ethanol, an alkyl alcohol with the aspect of a clear colorless liquid with a density of 789 kg/m³, a flash point of 13°C, and a boiling point of 78.5°C, has today (based on its degree of purity) a wide range of applications, from a polar solvent, an antiseptic drug, a central nervous system depressant, and other pharmacological applications, up to an interesting biofuel (Pubchem, 2021). In 1925, probably for the first time, Henry Ford cited this molecule as 'the fuel of the future' (Chandel et al., 2007). Due to the rapid industrialization and urbanization of the world, the first 15 years of the 21st century have known the necessity of addressing the new climate changes and improving the world's energy supply (Castaneda-Ayarza and Cortez, 2017).

In this context, ethanol has been viewed with respect. Due to its low cetane number of 8, an octane number of about 107, and a high heat of vaporization of about 840 kJ/kg, 1 L of ethanol provides about 66% of the energy of the same amount of fossil fuel (Dalena et al., 2019; Senatore et al., 2019). Thanks to its properties, ethanol can be burned at a higher compression ratio and shorter burning time than fossil fuel; it can be mixed with the current gasoline or petrodiesel fuels; and it can be used with the current engines on the market, making it a good candidate as a replacement for fossil fuels (Nigam and Singh, 2011; Senatore et al., 2019). An ethanol-gasoline blend is usually indicated with the letter 'E' followed by a number, indicating the percentage of ethanol in the blend (Dalena et al., 2019). The most common is the E10 blend, which contains 10% of ethanol that reduces up to 6% of petroleum use, 2% of greenhouse gas (GHG) emissions, and 3% of fossil energy use (Ho et al., 2014). From another point of view, the growth of photosynthetic plants, which need their conversion into ethanol and which produce O_2 by sequestering CO_2 from the environment, reduces the environmental impact and the related GHGs emitted during the subsequential conversion process for obtaining ethanol (Naik et al., 2010).

Based on the raw material used, three generations of biofuels have been characterized: (i) a first generation feedstock (FGF), which includes seeds used in the production of biodiesel or simple sugars that are fermented by yeast to produce ethanol; (ii) a second generation feedstock (SGF), which includes lignocellulosic materials to produce bioethanol and biobutanol (via enzymatic hydrolysis) or methanol and biodiesel (by thermochemical processes); and (iii) a third generation feedstock (TGF), which exploits algal biomass for the production of bioethanol, biodiesel, and biohydrogen (Dalena et al., 2019). The advantages of the FGF discussed in this chapter derive mainly from their simplicity of processing and low cost of production, as well as the good ratio between cultivation costs and ethanol yield (Bringezu et al., 2007). The good temperatures and the crop choice with a low lignin content and high easy fermentable C_6 carbohydrates in the form of starch or saccharide (65%–76% in corn and 67% in sugarcane) are critical for the success of this conversion process

 DOI: 10.1201/9781003226574-129

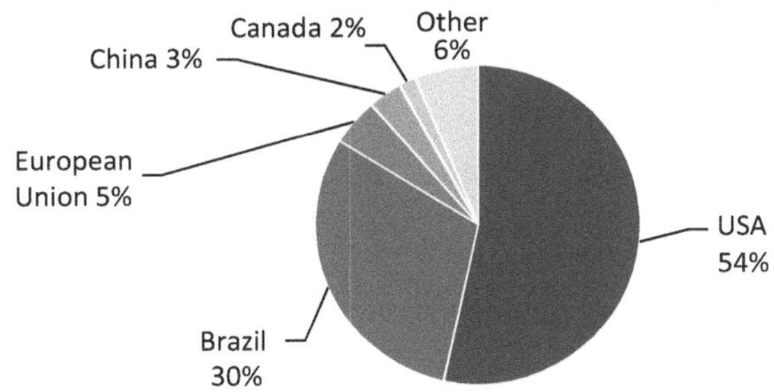

FIGURE 96.1 Worldwide Ethanol production in 2020 (AFDC, 2020a).

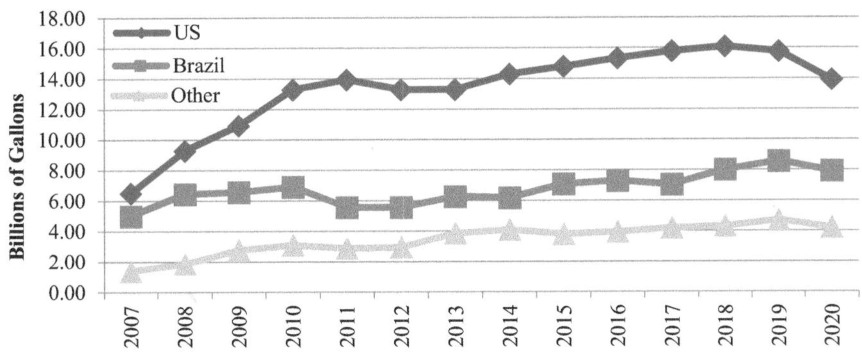

FIGURE 96.2 Billions of gallons in ethanol production from 2007 up to 2020.

(Bothast and Schlicher, 2005; Dalena et al., 2019, Naik et al., 2010). Despite these advantages, its use as a fuel has raised controversy due to its competition with agricultural lands for food purposes (Castaneda-Ayarza and Cortez, 2017; Singh, 2012). Some recent data indicate that this phenomenon cannot be a real problem, and the research world continues to consider the ethanol solution (Shrestha et al., 2019). In 2020 (Figure 96.1), the major producers of ethanol as a biofuel were the USA from corn with a production of 13.93 billion gallons and Brazil from sugarcane with a production of 7.93 billion gallons (AFDC, 2020a). Despite global ethanol production declining in recent years due to the pandemic, according to the World Economic Outlook, researchers are confident that it will increase as previously predicted (Chum et al., 2013).

The production trend from 2007 to 2020 for the major producers is summarized in Figure 96.2., in which there is a decrease since the beginning of the pandemic in 2019 (AFDC, 2020a).

In this chapter, the costs and methods of production are briefly reviewed for the two most important industrial producers: On one hand, the USA converts corn into ethanol; on the other hand, Brazil produces ethanol from sugarcane (Goldemberg, 2008; Senatore et al., 2019; Shapouri and Gallagher, 2005; Shikida et al., 2014).

96.2 THE US AND THE BRAZILIAN ETHANOL PRODUCTION

The main worldwide producer of ethanol from starch, the USA, converts maize (corn) and other starchy plants or waste streams into ethanol through two well-known processes: the dry grind (DG) and the wet milling (WM) processes (Senatore et al., 2019). Until today, in the USA, there are 201 functional plants that, in 90% of the cases, taking advantage of the DG process for economic reasons, have the ability to produce up to 13.38 mmgal of ethanol per year (EIA, 2020a; RFA, 2017). According to FAOSTAT data (Figure 96.3), in 2019, the US production of maize (corn) reached a value of 347 Mt (FAOSTAT, 2021a).

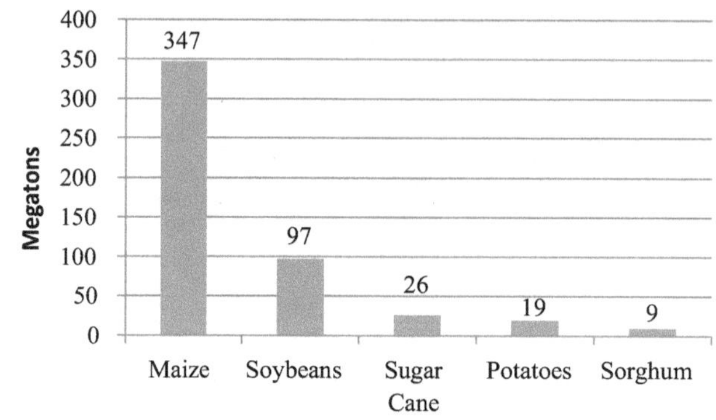

FIGURE 96.3 US Production of maize and other starchy plants in 2019.

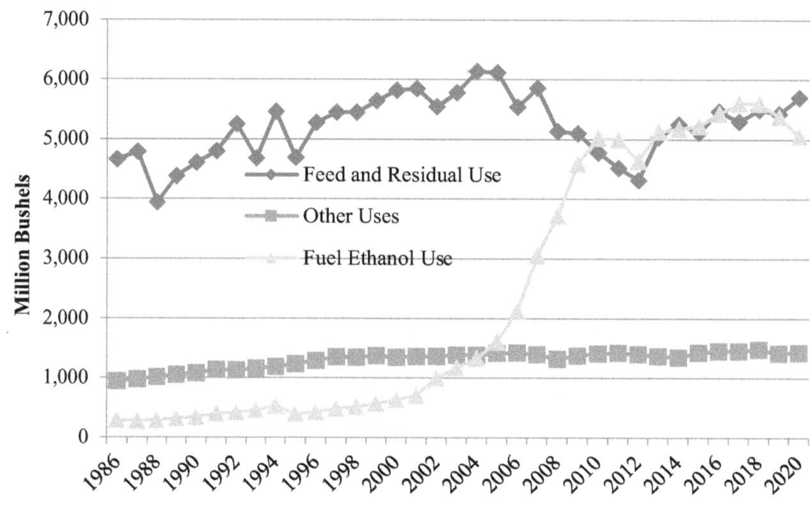

FIGURE 96.4 The US corn production divided according to its use: Feed and residual, ethanol and other uses, from 1986 up to 2020 (AFDC, 2020b).

The data on production, however, gives us little information on the distinction of its final products. Figure 96.4 shows the main uses of corn from 1986 to 2020, with an exponential increase in ethanol production from 2005 to 2010.

From the beginning of the pandemic in 2019, however, some fuel ethanol plants began to produce industrial alcohol to meet the increased demand for hand sanitizers, so there has been a reduction in ethanol production as a fuel; a trend that, despite moving away from that predicted some years ago, will tend to increase again in the coming years (EIA, 2020b). On the other hand, Brazil's large production of ethanol as a biofuel, favored by government economic subsidies, takes advantage at first of the well-known MB and MMB processes to convert sugarcane into biofuel (Senatore et al., 2019). Until today, there have been about 350 sugarcane ethanol plants in Brazil and a large production of sugarcane (USDA, 2020) of about 753 Mt (Figure 96.5).

In 2019, Brazil produced about 37.38 billion L of ethanol (Figure 96.6), 96% of which came from sugarcane (USDA, 2020). However, in recent years, a new ethanol production market has caught on starting with corn. 16 corn ethanol plants, four of which use only corn while the others use a mixture of sugarcane and ethanol, are already working in Brazil (USDA, 2020).

The FAOSTAT data (2021a) show the increase in sugarcane production and the doubling in corn production, which goes from approximately 58.93 Mt in 2008 to 101.14 Mt in 2019 (Figure 96.7).

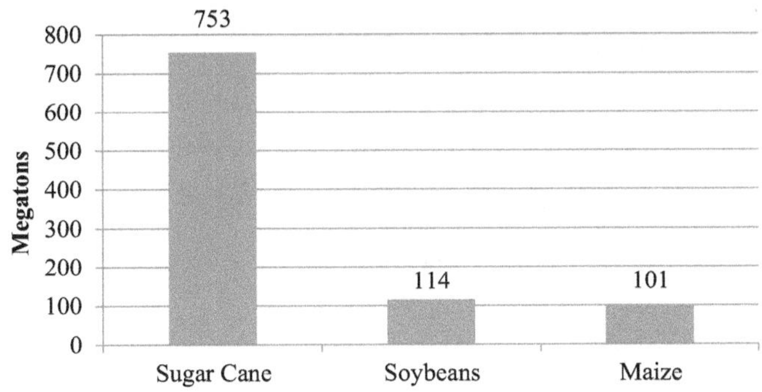

FIGURE 96.5 Brazilian production of sugarcane, soybeans, and maize in 2019.

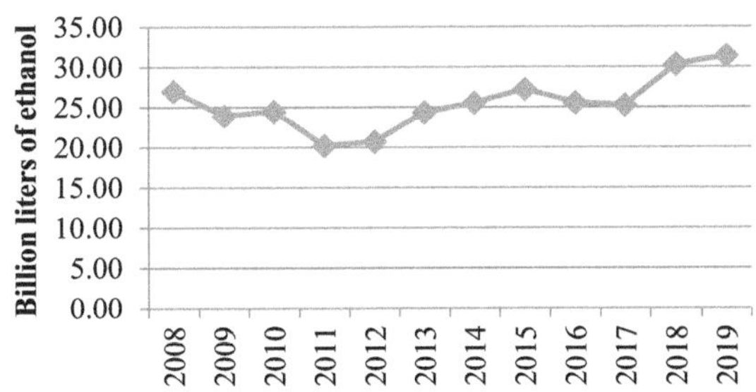

FIGURE 96.6 Ethanol production in Brazil from 2008 to 2019.

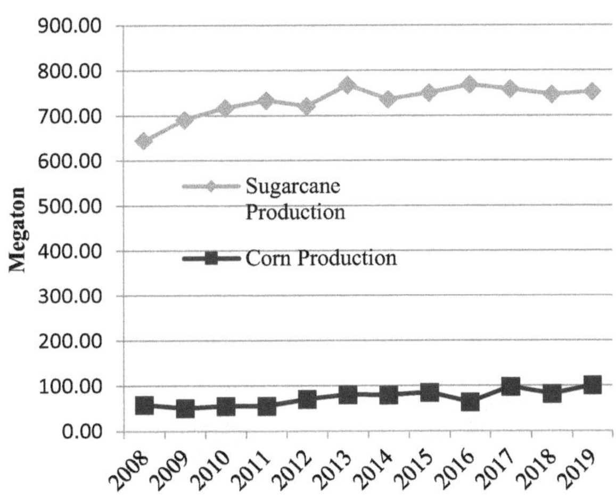

FIGURE 96.7 Brazilian production of corn and sugarcane from 2008 to 2019.

From the data shown below, which includes the period before the pandemic and the pandemic (Figure 96.8), it can be seen the slight increase in ethanol production from corn but the general decrease in production due to the pandemic (Statista, 2021a).

96.3 ETHANOL PLANTS: THE DG, THE WM, AND THE MB PROCESSES

There are several ways to obtain ethanol from plants; it depends on the feedstocks and the pretreatments they need (Dalena et al., 2019). From a general point of view, we can schematize the whole process of converting a FGF as follows (Figure 96.9):

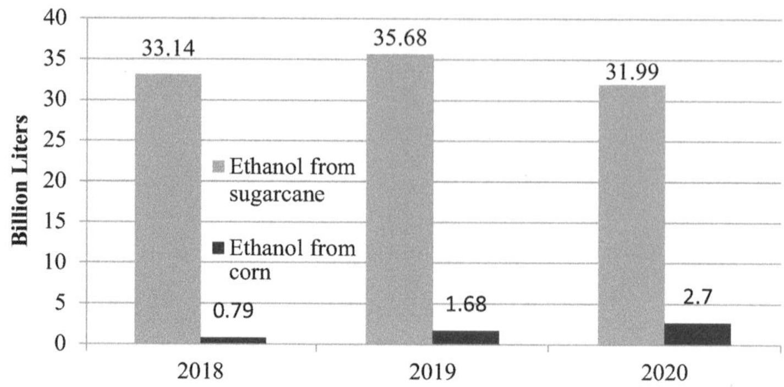

FIGURE 96.8 Brazilian ethanol production from sugarcane and corn, in the 3-year period, from 2018 to 2020 (during COVID19 pandemic).

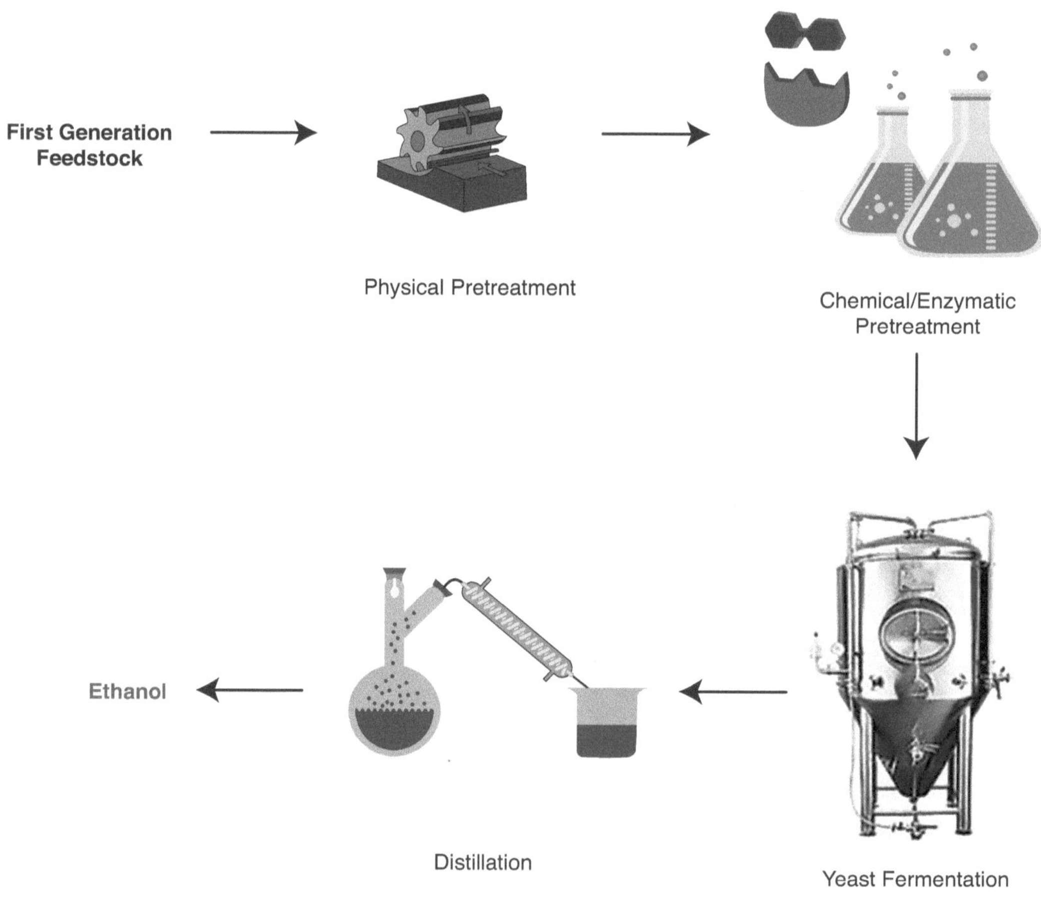

FIGURE 96.9 General view of a FGF treatment to obtain ethanol.

Although the conversion methods used in the US and Brazilian processes are quite different, the first major difference is whether or not enzymes are used in the process. In the conversion of Brazilian sugarcane, for example, a plant rich in simple sugars is used; it can be easily extracted through physical and chemical treatments before the fermentation process (Senatore et al., 2019). On the other hand, the use of starchy materials such as corn requires an enzymatic treatment before the fermentation process (Senatore et al., 2019).

96.4 PLANT COMPOSITION AND ECONOMICS

Among the billions of plants available in nature, those that had already been cultivated for centuries in the same area (exploiting the climate of the region) were chosen for their transformation into ethanol. The production of corn and sugarcane, respectively, in the United States and Brazil represents a milestone that has very ancient commercial roots. While from 1 ha of cultivation it is possible to produce up to 12.3 t of corn (Sandhu et al., 2020), in the case of sugarcane it is possible to produce about 72.4 t (Upreti and Singh, 2017). Figure 96.10 shows the prices (in USD) per ton of raw material for corn and sugarcane, respectively. The data comes from two different sources. In the case of corn (Macrotrends, 2021), the data originally reported in bushels has been converted to tons, and the 1-year average has been done on a random value in the middle of each month. In the case of sugarcane instead (Statista, 2021b), the values refer to the average of the price in four different cities of the United States, which are, respectively, Hawaii, Florida, Louisiana, and Texas.

Since 1960, when the price of corn was around USD 1.20 per bushel, which corresponds to 0.0254 mt, it has more than tripled, reaching probably the highest historical price of USD 6.80 between 2011 and 2012. *Saccharum officinarum* (sugarcane), on the other hand, kept costs down by about USD 29 per ton, with a peak of about USD 47 reached in 2011. Both sugarcane and *Zea mays* (corn) are C photosynthetic plants, in which, on the one hand, the stalk rich in sucrose, an easily fermentable sugar, is used for the process; on the other hand, in the case of corn, there is greater processing due to the composition of the plant, which consists of an endosperm as a source of proteins and starch, which is the main element converted into ethanol; a germ with enzymes and micronutrients required for plant germination and growth; and a pericarp, the fibrous part of the corn. Other sugars we find are glucose and other polysaccharides, such as galactose, arabinose, and xylose (Lopes et al., 2016; Nogueira et al., 2008; Senatore et al., 2019; Singh et al., 2001). In the following table, the percentage composition by weight is reported for sugarcane and corn (Table 96.1).

Despite the high percentage of carbohydrates in corn compared to sugarcane, a difference must be made on quality and therefore on processing costs: On the one hand, corn is mainly constituted by starch, a polysaccharide made up of glucose monomers joined together by α-glycosidic bonds (Figure 96.11). Based on the bond, two types of molecules are distinguished in the structure: The linear and helical amylose (20%–30% by weight) with α1–4 glycosidic linkages and the branched amylopectin (70%–80% by weight) with α1–4 as well as α1–6 glycosidic linkages (Nawaz et al., 2020; Swinkels, 1985).

Due to its polymeric and branching nature, starch shows poor solubility in water and oil and relatively low enzymatic digestibility (Nawaz et al., 2020). Both fractions of amylose and amylopectin,

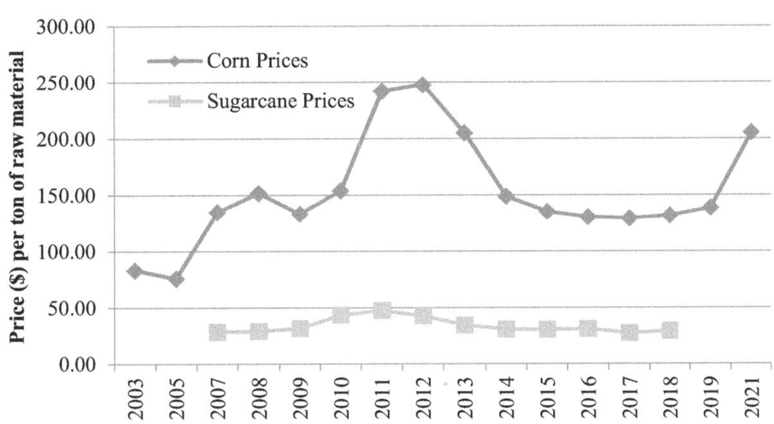

FIGURE 96.10 Price of corn from 2003 to August 2021 (Macrotrends, 2021) and price of sugarcane from 2007 to 2018 (Statista, 2021b).

TABLE 96.1
Corn and Sugarcane Plant Composition Divided by Colors

	Corn Composition (%)		Sugarcane Composition (%)
Starch	60.6	Sugars	14
Cellulose	3.46	Fiber	13.5
Hemicellulose	4.6	Proteins	0.4
Glucose	8.7	Acids and Fats	0.6
Lignin	0.4	Ash	1.5
Proteins	2.2	Moisture	70
Fatty Acids	3.64		
Ash	1.17		
Moisture	15.5		

are hydrolyzed by enzymes (Figure 96.12), and, in particular, the α-1–4-linkages are broken by amylases, while the α-1–6-linkages are broken by glucosidases (Araujo et al., 2004).

On the other hand, sucrose from sugarcane, a water soluble disaccharide molecule (Queneau et al., 2007), is composed of glucose and fructose (Figure 96.13). Its interglycosidic bond is easily hydrolyzed at pH 4, making any acid-catalyzed conversion difficult (Queneau et al., 2007).

On the contrary, its simple hydrolysis (Figure 96.14) does not necessarily require enzymes and is sensitive to heat under a wide range of pH conditions, such as 0.1 HCl in MeOH for 30 min, providing glucose and methyl fructoside (Queneau et al., 2007).

The concentration of sugar in the dry stem of the plant varies according to the type of sugarcane, the type of cultivation, and the climate. According to Muchow et al. (1996), from Q117 sugarcane production, it is possible to extract up to 0.48 t of sucrose from 1 t of dry stem. Similarly, according to studies by Shao et al. (2020), it is possible to extract up to 0.54 t of glucose from 1 t of corn stover; regarding the yield in ethanol, it has been estimated that from 1 t of sugarcane, it is possible to obtain up to 19.5 gallons of ethanol, and from 1 bushel of corn, it is possible to obtain 2.65 gallons of ethanol through the wet milling process and 2,75 gallons of ethanol through the dry grind process, respectively (Pimentel and Patzec, 2005; USDA, 2006). Considering that 1 gallon corresponds to 3,785411784 L and 1 bushel of corn equals 0.0254 mt, from 1 t of sugarcane it is possible to obtain up to 73.81 L of ethanol, while from 1 mt of corn it is possible to obtain, respectively, 394.88 L of ethanol through the WM process and 409.84 L through the DG process. Starting from the data reported in the chapter and found from different sources, it has been possible to draw the following summary theoretical table (Table 96.2).

Making an estimate on production costs is very difficult, as each cost can vary a lot from area to area and from period to period. In any case, starting from the literature data, it has been possible to realize the following Table 96.3.

The cultivation costs include the use of fertilizers, fungicides, pesticides, insecticides, labor, irrigation, energy, and much more, making this calculation quite complex. According to these data, the production costs of sugarcane are much higher than those of corn (Table 96.3), but according to the data in Table 96.2, the yield of a sugarcane crop per hectare of land is higher than that of corn (Table 96.2), with a ratio of 3:1.

96.5 THE PROCESSING TECHNIQUES AND COSTS

Based on the characteristics of the plant, one type of procedure rather than another is required in order to extract as much fermentable sugar as possible. Concerning corn, two industrial transformation/extraction processes for sugars have been developed: The dry grind and the wet mill process

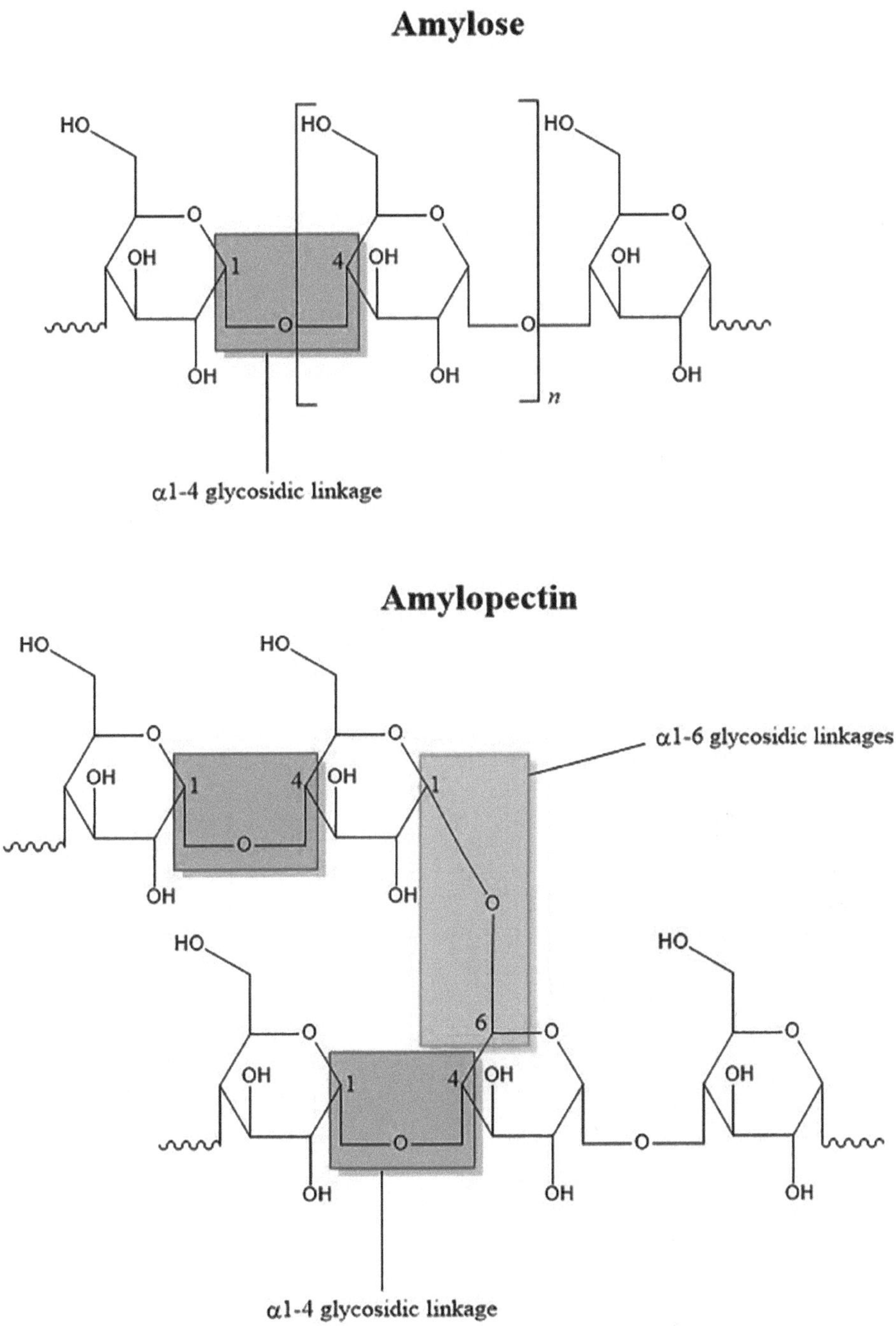

FIGURE 96.11 Starch components: Amylase and amylopectin structures.

(Hettinga et al., 2009; Senatore et al., 2019; Shapouri and Gallagher, 2005). In the DG process (the best known and used process for converting corn into ethanol in the U.S.), the whole kernel is grinded and water is added before saccharification and fermentation (Kim et al., 2008; Senatore et al., 2019; Singh et al., 2001); while in the WM variant, a more sophisticated process, the grain components such as starch, fiber, germ, gluten, and steep liquor are separated, in order to obtain better quality products (Bothast and Schlicher, 2005; Ramirez et al., 2009; Senatore et al., 2019).

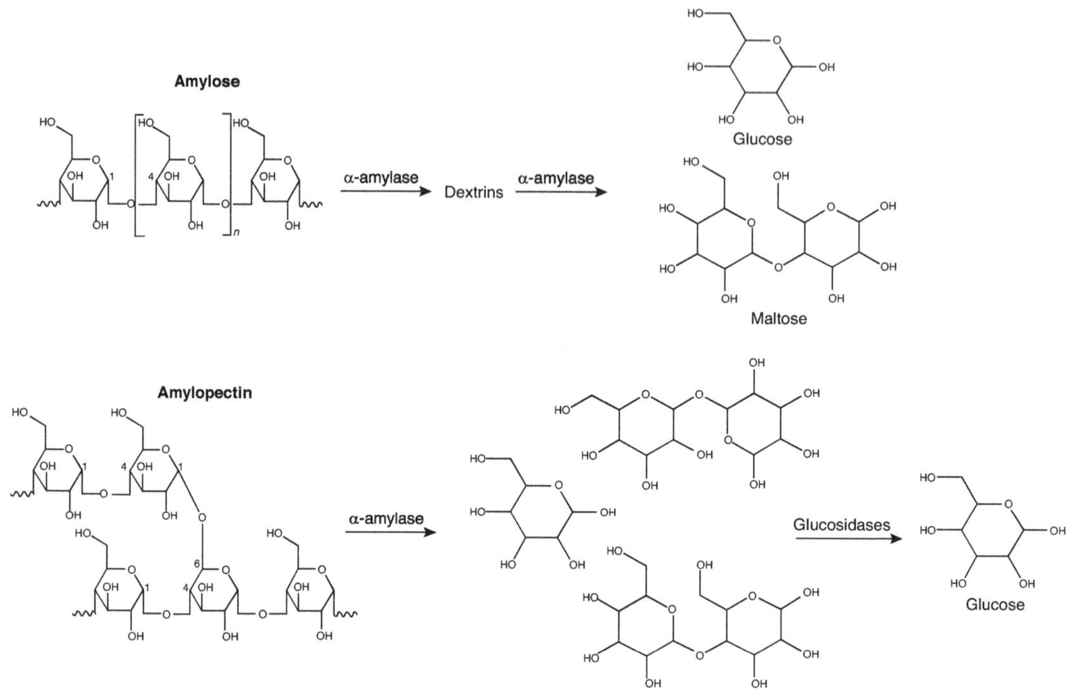

FIGURE 96.12 Starch enzymatic hydrolysis by α-amylases and glucosidases.

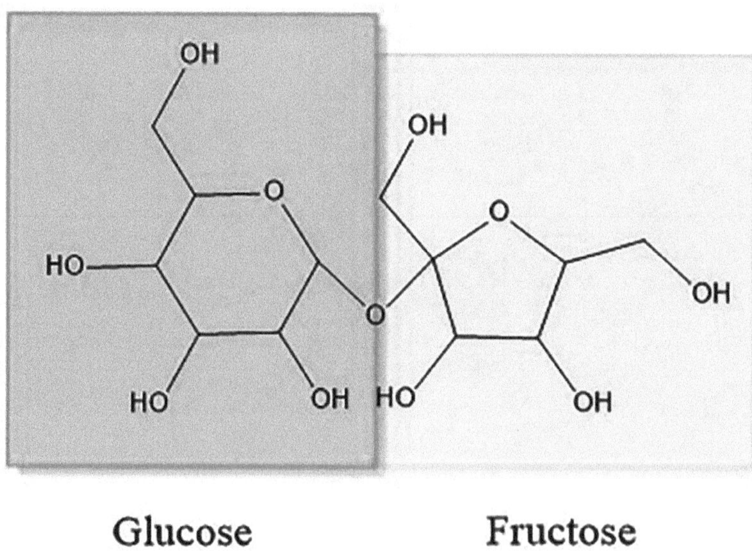

FIGURE 96.13 Sucrose disaccharide, composed by glucose and fructose.

On the other side, two processes are used in Brazil to convert sugarcane into ethanol: the Melle-Boinot(MB), or batch fermentation process (which includes about 75% of the distilleries), and the modified Melle-Boinot (MMB), also known as the fed-batch fermentation process (Basso et al., 2011; Kelsall and Lyons, 2003; Soccol et al., 2005). These processes will be described in detail in the following paragraphs.

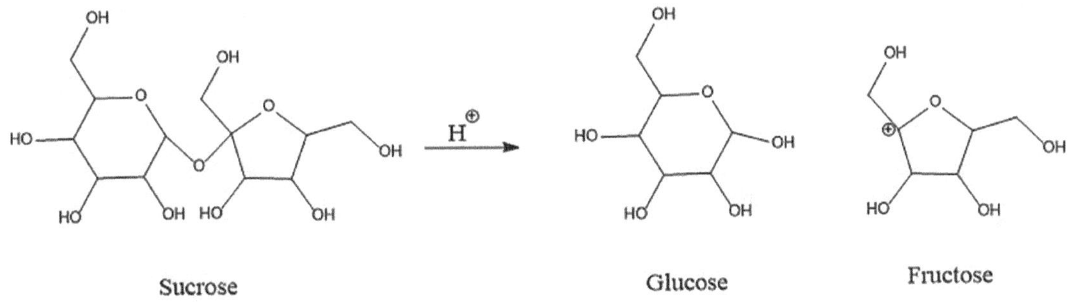

FIGURE 96.14 Sucrose hydrolysis at acid pH.

TABLE 96.2
Theoretical Table on Yields per One Hectare of Cultivated Land

	Corn Production	Brazilian Production	References
t/ha	12.3	72.4	Sandhu et al. (2020); Upreti and Singh (2017)
Dry matter (t/ha)[a]	10.4	21.7	-
Fermentable sugars (t/ha)[b]	5.6	10.4	-
Theoretical Ethanol yield[b] (L/ha)	3,867	11,582	-

[a] Data obtained knowing the moisture content reported in Table 96.1.
[b] Data obtained from the above information and contained in the respective references USDA (2006) and Pimentel and Patzec (2005).

TABLE 96.3
Costs on Cultivation and Harvest

	Year	Total Operating Costs ($/ha)	References
Corn	Av. 1983–1985[a]	789–914	Hettinga et al. (2009)
	1996	508.07	USDA (2006)
	2002	470.25	
	2003	506.32	
	2004	543.12	
	2005	589.87	
	1999–2003	891.76–916.93	Pimentel and Patzec (2005); Pimentel et al. (2007)
	Av. 2003–2005[a]	309–324	Hettinga et al. (2009)
	2010	1,310–1,516	Sandhu et al. (2020)
	2015–2016	1,672.72	De Moraes Rego et al. (2017)
Sugarcane	2007–2008	1,278.29 (Sand soil) (cost/acre)	Roka et al. (2009)
	2008–2009	2,419 (Muck soil)-3,572 (Sand soil)	Roka et al. (2010)
	Hypothetical scenarios	1,937.93–2,689.32	Cardoso et al. (2017)

[a] Some expenses, such as labor, are not added

96.5.1 The DG Process and Economics

The DG process (Figure 96.15) is the most widely used method in the U.S. for producing fuel ethanol by fermentation of corn grain. The first phase of the process involves the accumulation of corn in silos (sized to hold sufficient corn for 12 days of plant operation) and the cleaning of non-workable residues using a blower and screens (Kwiatkowski et al., 2006). The beginning of the corn transformation process and the extraction of substances useful for the production of ethanol occur when this is ground and mixed with tap water at 35°C in order to obtain a mash (Kwiatkowski et al., 2006; Senatore et al., 2019; Singh et al., 2001). Ground corn size seems to be very important for the process in order to improve ethanol yield (up to 7.5%); it should be constituted by particles as small as possible (1–8 mm) to be easily recovered during the process (Dalena et al., 2019; Kelsall and Lyons, 2003; Rausch et al., 2005). The temperatures are then increased to 85°C in order to obtain a slurry, and α-amylase (0.082% db), ammonia (90 kg/h), and lime (54 kg/h) are added to the solution in order to break down the α-1–4-linkages of amylase and amylopectin of the starch (Figures 96.11 and 96.12) and release as much dextrins and glucose as possible (Kwiatkowski et al., 2006; Senatore et al., 2019). During this process, the liquefaction step takes place and continues at pH 6.5 and 880°C for 1 h under continuous stirring (Kwiatkowski et al., 2006). The output of this step is mixed with 'backset', a recycled stream taken from the stillage obtained later in the process and rich in nutrients useful for the fermentation process (Kwiatkowski et al., 2006).

The temperature is subsequently decreased to 60°C and the pH adjusted to 4.1–4.2, by the addition of 1N H_2SO_4 solution in order to proceed with the next step: Saccharification (Senatore et al., 2019). To start the saccharification process, the enzyme glucoamylase (0.11% db) is added to the solution to break down the α-1–6-linkages (Figures 96.11 and 96.12) and finally to release glucose, which will be fermented (Kwiatkowski et al., 2006; Sauer et al., 2000; Senatore et al., 2019). This reaction proceeds for 2 h at 60°C under continuous stirring before the beginning of the fermentation process (Senatore et al., 2019). To start one of the last important steps, it is necessary to lower the temperature of the container to 30°C–32°C; at this temperature, the fermentation can begin without problems, so the *Saccharomyces cerevisiae* is added to the solution, and the following (Eq. 96.1)

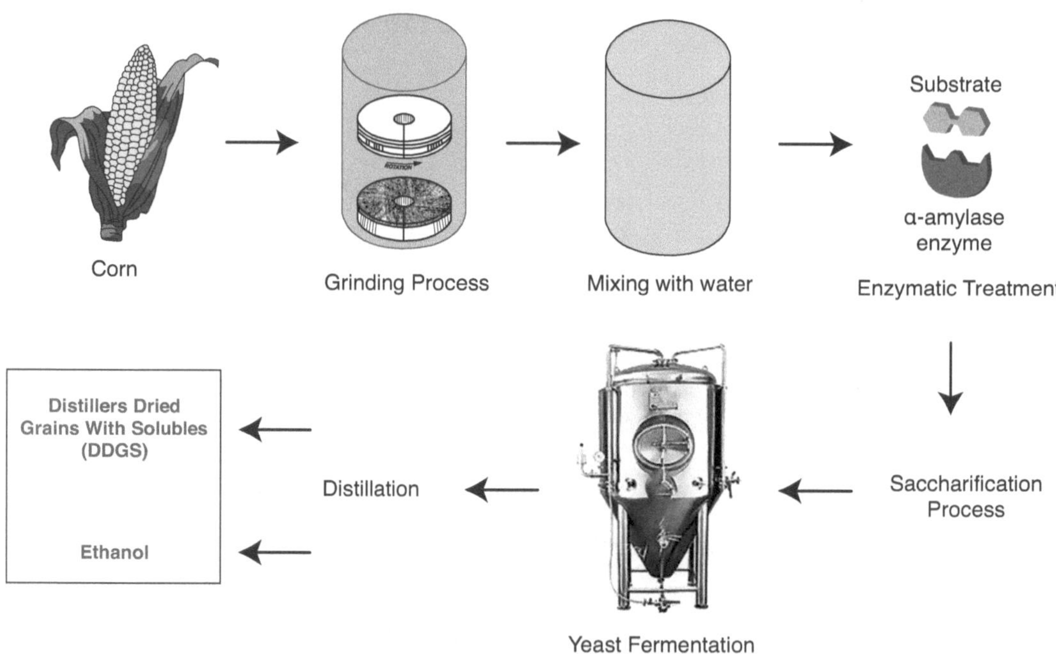

FIGURE 96.15 DG process, a schematic representation.

reaction can occur for the subsequent 72 h (Bothast and Schlicher, 2005; Senatore et al., 2019). This yeast, with the addition of nutrients such as urea or ammonium sulfate, through the Embden-Meyerhof-Parnas pathway, is able to convert glucose to ethanol (Bothast and Schlicher, 2005; Lin and Tanaka, 2006).

$$C_6H_{12}O_6 + H_2O + \text{Yeast} \rightarrow 2\ CO_2 + 2\ C_2H_5OH + H_2O + \text{Heat} \tag{96.1}$$

To optimize both the costs of production and ethanol yield/concentration, saccharification and fermentation can be carried out simultaneously, in a process known as simultaneous saccharification and fermentation (Dalena et al., 2019). In this case, the enzymatic hydrolysis is performed by α-glucosidases cellulases (endoglucanase and exoglucanase) that break down cellulose into glucose molecules (Dalena et al., 2019; Olofsson et al., 2008). The last steps of the DG process involve the separation of ethanol from water and include distillation and/or rectification/stripping to obtain the desired degree of purity of the final product (Kwiatkowski et al., 2006).

Since ethyl alcohol evaporates at a lower temperature than water (78.4°C compared to 100°C in water), it can be separated by distillation, a process that exploits these differences in evaporation (Pimentel et al., 2007). During the process, however, a certain amount of water ends up in the ethanol portion, which reaches a purity of about 95% w/w (Pimentel et al., 2007; Singh et al., 2001). In order to be mixed with gasoline, this fraction must be distilled again or passed through a molecular sieve that retains a certain fraction of water to reach a higher degree of purity of about 99.5% w/w (Kwiatkowski et al., 2006; Pimentel et al., 2007; Singh et al., 2001). From the remaining not distilled part that contains a solid and liquid fraction made up of oil, proteins, yeast cells, fibers, and unfermented starch, can be recovered the distillers' dried grains with solubles (DDGS) and the wet distillers' grains (WDG) (Kwiatkowski et al., 2006; Senatore et al., 2019). The two parts can be recovered through centrifugation: The WDG, also known as 'backset' is used back in the process while the DDGS is sold as animal feed and is really important for the economics of the whole DG process (Kwiatkowski et al., 2006; Rausch et al., 2005; Senatore et al., 2019; Singh et al., 2005).

In order to calculate the costs related to the operation of such a plant, it is necessary to take into account the raw materials (corn, denaturant, enzymes, yeast, and others), the utilities (electricity, steam, natural gas, cooling water), the workforce, the maintenance work, as well as other expenses (Kwiatkowski et al., 2006). Except for the cost of feedstock, the costs of enzymes, steam, and natural gas are the ones that have the greatest influence on the process and correspond respectively to about USD 2,016,000, 5,054,000, and 3,222,000 per year (Kwiatkowski et al., 2006). The following summary table (Table 96.4) proposes the costs studied by Kwiatkowski et al. (2006) by changing the original price of corn that the author had taken into account; which was USD 0.0866/kg (USD 2,20/bushel) to USD 0.205/kg in 2021.

In this case, the price for corn for a year at that studied industrial plant would be USD 73,440,185, and the final costs have almost doubled compared to the USD 53,050,000/year proposed by the author. In the final phase of sale and profit, the DDGS are of fundamental importance. Thanks to

TABLE 96.4
DG Process Plant Expenses (Kwiatkowski et al., 2006)

Expenses	USD/year
Raw material	77,467,185
Utilities	10,261,000
Labor and maintenance	2,352,000
Insurance and administration	722,000
Depreciation (10 years straight line)	4,664,000
TOTAL	9,546,6185

the increase or decrease of these products, the price of ethanol can still vary within certain ranges. According to the online data (U.S. Grains Council, 2021a), the price for DDGS reported from 33 locations on July 22 was USD 177/t. According to Kwiatkowski et al. (2006) data, corn is produced at about 119,000,000 kg/year (119,000 t/year); assuming that the average annual price of DDGS is about USD 117/t, at the end of the year, such a plant would make about USD 13,923,000, and about 15% of the total costs would be cut.

96.5.2 The WM Process and Economics

The WM process (Figure 96.16), the second most popular type of plant in the United States, allows for the production of a whole series of high value products thanks to the separation of all the corn components: Germ, gluten, fiber, and steep liquor (Ramirez et al., 2008; Senatore et al., 2019). According to Lyons (2003), a WM facility in 2002 produced about 450,000 US t of corn gluten meal, 2,500,000t of corn gluten feed and germ meal, and 530,000,000 pounds of corn oil. As for the DG process, a first preliminary phase foresees the accumulation of corn in silos (sized to hold sufficient corn for the facility) and the cleaning of unwanted residues (Ramirez et al., 2008). The first step of the whole process, also named the steeping process, begins when the corn is immersed in an aqueous solution of SO_2 and lactic acid for about 24–48 h at 50°C in order to release the starch granules (Jackson and Shandera, 1995; Ramirez et al., 2008; Senatore et al., 2019). During the subsequent degermination process, in a first step, coarse grinding is needed to separate the oil-rich corn germ from starchy kernels, while in a second step of the process, the remaining germ is recovered (Ramirez et al., 2008; Senatore et al., 2019). The germ-free corn is passed on a grit screen to separate water, mill starch (starch and gluten), and fiber; gluten is separated from starch by a centrifuge, and the remaining fiber and mill starch are finally ground to release the starch into the medium (Ramirez et al., 2008). In the DG process, the starch has to be treated with enzymes to extract the fermentable sugar, which will be converted into ethanol (Senatore et al., 2019). Through a series of

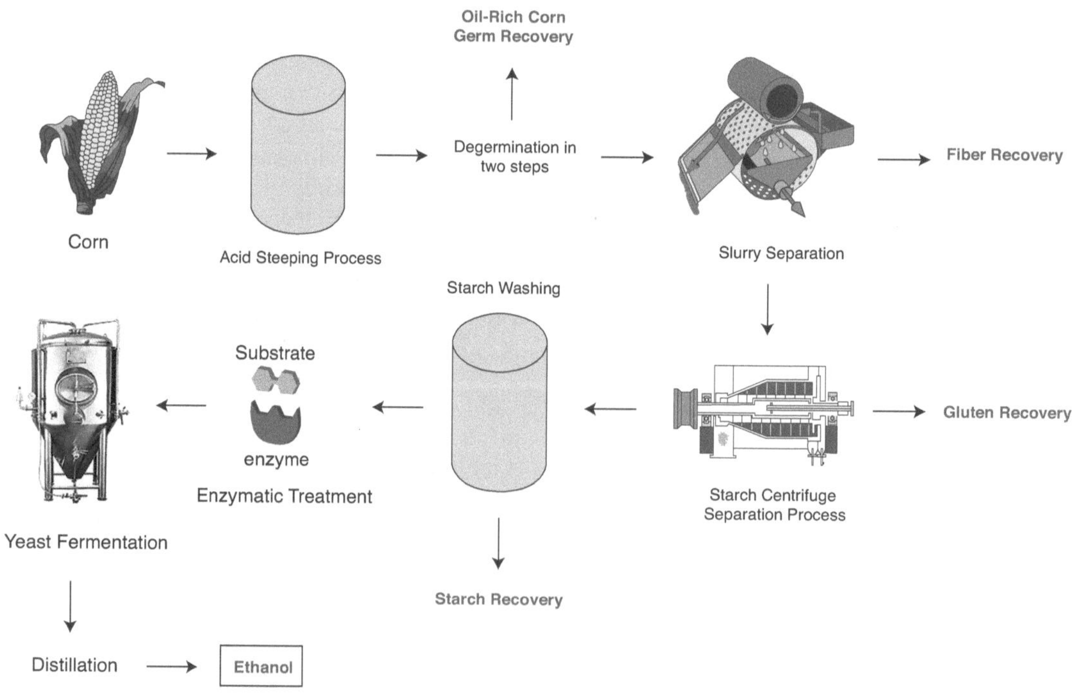

FIGURE 96.16 WM process, a schematic representation.

cleaning steps, the final extracted sugar can be used for ethanol production due to its low content of impurities (about 1%) (Ramirez et al., 2008; Senatore et al., 2019).

The evaluation of the costs of such a complex plant becomes even more difficult. According to Ramirez et al. (2008), who used Superpro Designer® software to simulate the costs of a generic non-existent plant, taking into account real costs, raw materials (corn, sulfur, and water), utilities (natural gas, steam, and electricity), and other expenses, a table has been created (Table 96.5).

Also in this case, the costs of corn have been varied, trying to adapt them to the prices of 2021 of about USD 205/t instead of the 2007 prices mentioned by the author, which were around USD 135/t. As in the case of the DG process, the feedstock price also has the greatest weight. About 822,356 t of raw material per year have been calculated for this WM process, which for USD 205/t becomes about USD 168,582,888. In this case, the price of enzymes and yeasts does not seem to have been taken into account. For a WM plant, however, in addition to ethanol, there are more quality products that are extracted and that can be sold, lowering the high production costs; those are, respectively, corn gluten meal, corn gluten feed, and corn germ (Johnston et al., 2005; Ramirez et al., 2008; RFA, 2008). In 2019, more than 1.86 million t of corn gluten meal and corn gluten feed left the U.S. for the export market, where Ireland (23%), Israel (12%), and Indonesia (9%) were the main buyers (U.S. Grains Council, 2021b). Not being able to obtain the quantities of products (corn gluten meal, corn gluten feed, and corn germ) produced by the hypothetical plant mentioned above, the appropriate conversions with the prices currently on the market have not been made, and the prices of the Ramirez et al. (2008) article have been maintained for the latest considerations. According to this article, in fact, the hypothetical prices for corn gluten meal, corn gluten feed, and corn germ are USD 19,255,000, 12,071,000, and 16,482,000, respectively, so from these products there would be a profit of USD 47,808,000, and the final price of plant management and production would drop to USD 147,100,888/year, resulting in a reduction in production costs of about 24.5%.

96.5.3 The MB Process and Economics

Until the 1990s, batch processes, also known as the Melle-Boinot process (Figure 96.17) and the modified Melle-Boinot process, were used as the main fermentation processes in Brazil (Goldemberg, 1994; Senatore et al., 2019). In 2006, approximately 325 plants crushed 212,500,000 t of sugarcane every year to produce ethanol and about the same quantity to produce sugar (Goldemberg, 2008). In the same years, 147 new distilleries were in the planning stages, 86 of which were due to be completed by 2015 (Goldemberg, 2008). A large facility today can process about 2,000,000 t of sugarcane stalks per year (Cardoso et al., 2017) and can produce about 250,000,000–300,000,000 L of ethanol per year (Pimentel et al., 2007). Despite this being the main Brazilian market, in recent years this nation has exploited not only the production of ethanol from sugarcane but also from corn, as in the classic U.S. process, to increase the total ethanol production (Da Silva and Castaneda-Ayarza, 2021; Eckert et al., 2018). The main difference from the classic U.S. DG process is the recovery of yeasts during the fermentation step (Senatore et al., 2019). At the beginning of the MB

TABLE 96.5
WM Process Plant Expenses (Ramirez et al., 2008)

Expenses	US$/year
Raw material	168,978,888
Utilities	12,550,000
Labor and maintenance	5,447,000
Separation operations	79,300,000
Depreciation	7,933,000
TOTAL	274,208,888

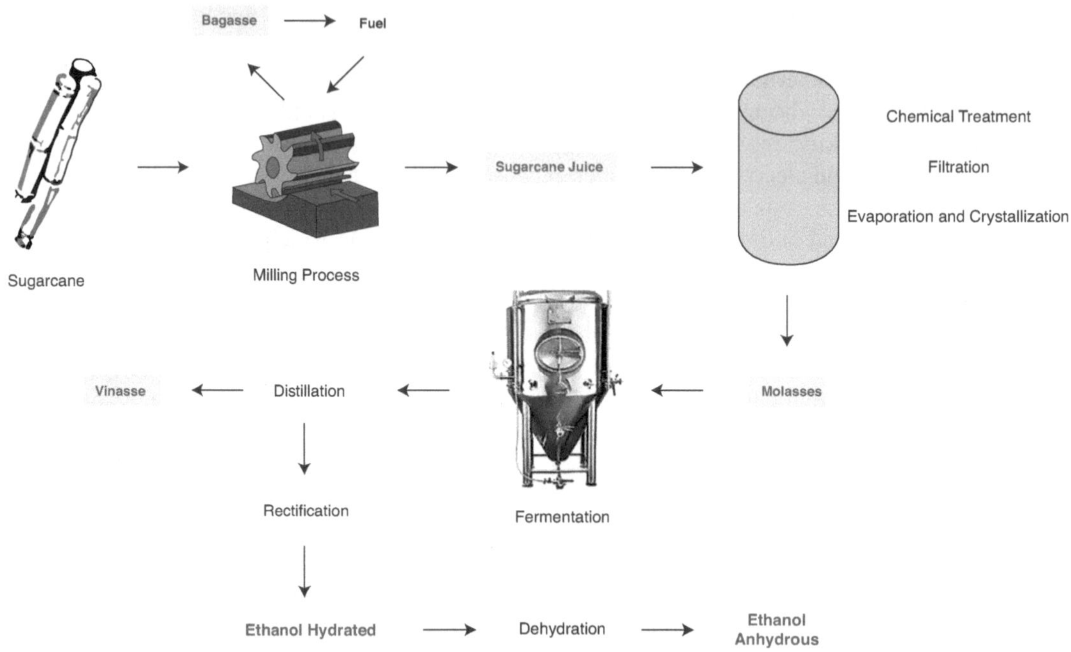

FIGURE 96.17 MB process, a schematic representation.

process, the stalk is washed and shredded into pieces of 20–25 cm in length (Senatore et al., 2019; Van den Wall Bake et al., 2009). The pieces are sent to a 4–7 mills extraction unit to separate a juice rich in sugars from the so-called bagasse, a fibrous part burned as a fuel in the same MB process (Chum et al., 2013; Senatore et al., 2019; Van den Wall Bake et al., 2009). About 1.5 L of water are added to about 1 kg of juice for the fermentation and distillation processes (Pimentel et al., 2007). The juice is filtered in order to eliminate the larger particles, warmed up (about 95°C), and treated with lime and phosphate to neutralize some organic acid; a then is pasteurized (Chum et al., 2013; EPA, 1997; Senatore et al., 2019). Through a centrifuge, it is possible to separate the juice from an insoluble particle mass called mud (EPA, 1997).

In the next evaporation step, the water is evaporated to concentrate the sugars. This step can be divided into two parts: A first step to concentrate the juice and a second step through a vacuum pump to crystallize the sugars (EPA, 1997). During the first step, a highly concentrated sugar solution is formed (Chum et al., 2013; Senatore et al., 2019). Subsequent chemical treatments, based mainly on coagulation and flocculation, are needed to precipitate pollutants, while crystallization is carried out to concentrate the fermentable sugar (Nogueira et al., 2008; Senatore et al., 2019). A remaining part called molasses, consisting mainly of glucose and fructose, is sterilized and used with saccharide in the fermentation process (Chum et al., 2013; Nogueira et al., 2008; Senatore et al., 2019).

Fermentation can finally take place through the addition of the well-known *Saccharomyces cerevisiae* yeast (Senatore et al., 2019; Van den Wall Bake et al., 2009). This step lasts from 4 to 12 h with a conversion yield of about 80%–90% and an alcohol content of about 7%–10% by volume (Chum et al., 2013). Through distillation, alcohol can be separated from the remaining aqueous solution containing yeasts, minerals, non-fermented sugars, CO_2, and SO_2 (Chum et al., 2013). As long as they remain vital, the yeast cells are recovered and reused for the next process, both in the MB and MMB processes (Lopes et al., 2016; Senatore et al., 2019).

From 11 L of ethanol/water mixture, up to 1 L of 95% pure ethanol can be obtained, which must be further processed in order to obtain 99.5% pure ethanol suitable for gasoline blends (Pimentel et al., 2007). Due to a lack of information, the following production costs refer to those from U.S.

sugarcane production reported in the article by Pimentel et al. (2007) and not those from Brazil, which probably have to be similar. As done for the previous tables (Tables 96.4 and 96.5), the costs will be divided into raw materials (sugarcane, water, and cement), utilities (steam and electricity), and other expenses (Table 96.6).

To obtain the results shown in Table 96.6 and relating to a hypothetical facility that treats 2,000,000t of sugarcane per year, as reported by Cardoso et al. (2017), some conversion has been necessary: First of all, the price of sugarcane has been converted to that of 2018, which was about USD 37/t (FAOSTAT, 2021b). According to this data, the price of 2 million t of sugarcane per year has been calculated, equivalent to USD 74 million. Always referring to the same annual quantity of sugarcane, all the quantities reported in the article by Pimentel et al. (2007) have been re-proportioned and the prices recalculated for one year of production.

96.6 CONCLUSIONS

Among the processes currently most commonly used to convert first generation feedstocks into ethanol, the DG and WM processes for corn and the MB process for sugarcane are among those with the best yields. The data already seen in the previous sections have been equated for a million t of material (Table 96.7) for easier reading. From this data, it can be seen that the WM process is about 25% more expensive than the MB process, while the sugarcane conversion process is considerably lower than the others. Despite this, calculating exactly the cost of the process becomes very difficult, as not all costs are taken into account by different authors. Despite the low price of sugarcane and its conversion to ethanol, the yield is very low compared to the DG and WM processes, so it would have to be almost quadrupled to achieve ethanol yields similar to corn conversion processes. In fact, by multiplying the values by 4.7, we would obtain an ethanol yield of 347.8 million L, but we would have to process 4.7 million t of sugarcane, which would have a cost of USD 108 million

TABLE 96.6
Expenses for a 2,000,000t of Sugarcane Treated through a MB Process Plant (Pimentel et al., 2007)

Expenses	US$/year
Raw material	75,767,999
Utilities	3,533,334
Labor and maintenance	5,447,000
Other	4,433,333
TOTAL	89,181,666

TABLE 96.7
Comparison among the Summary Data of the DG, WM and MB Production Processes Respectively

Process	Feedstock Quantity (million t)	Feedstock Price (million $)	Total Process Costs (million $)	Ethanol Theoretical Yield (million L/million t of feedstock)	Raw Material to Ethanol Theoretical Conversion Yield (%)
DG	1	243[a]	266	410	29
WM	1	210[a]	333	395	29
MB	1	23	44	74	42

[a] Prices for two different years, respectively those of 2021 and 2018.

and a processing price of at least USD 207 million. This is probably also among the reasons why Brazil has started producing corn in recent years and is considering the production processes of the U.S. to increase its ethanol production. As regards the DG and WM processes, on the other hand, it is necessary to consider the greater expense of a WM plant compared to a DG plant that divides and processes corn in a more meticulous way than the DG process. In order to make the most of a WM plant, it is necessary to consider the sale of high quality byproducts such as corn gluten meal, corn gluten feed, germ meal, and corn oil. The DG process, on the other hand, remains a milestone as although it treats corn roughly and the processing costs are not so low, it manages to extract good yields of ethanol (410 L from 1 mt of corn) and mainly considers DDGS as products to be sold to bring back the costs of the process.

REFERENCES

AFDC. 2020a. *Global Ethanol Production by Country*. Washington, DC: Alternative Fuels Data Center. https://afdc.energy.gov/data/10331. Accessed 23.08.2021.

AFDC. 2020b. *U.S. Total Corn Production and Corn Used for Fuel Ethanol Production*. Washington, DC: Alternative Fuels Data Center. https://www.afdc.energy.gov/data/10339. Accessed 26.08.2021.

Araujo, M., A. M. Cunha and M. Mota. 2004. Enzymatic degradation of starch-based thermoplastic compounds used in protheses: Identification of the degradation products in solution. *Biomaterials* 25:2687–2693.

Basso, L. C., T. O. Basso and S. N. Rocha. 2011. Ethanol production in Brazil: The industrial process and its impact on yeast fermentation. In *Biofuel Production, Recent Developments and Prospects*, Ed. by M. A. dos Santos Bernardes, pp. 85–100. London: IntechOpen.

Bothast, R. J. and M. A. Schlicher. 2005. Biotechnological processes for conversion of corn into ethanol. *Applied Microbiology and Biotechnology* 67:19–25.

Bringezu, S., S. Ramesohl and K. Arnold, et al. 2007. Toward a *Sustainable Biomass Strategy: What We Know and What We Should Know*. Wuppertal: Wuppertal Institute for Climate, Environment and Energy.

Cardoso, T. F., M. D. B. Watanabe and A. Souza, et al. 2017. Economic, environmental, and social impacts of different sugarcane production systems. *Biofuels, Bioproducts and Biorefining* 12:68–82.

Castaneda-Ayarza, J. A. and L. A. B. Cortez. 2017. Final and B molasses for fuel ethanol production and some market implications. *Renewable and Sustainable Energy Reviews* 70:1059–1065.

Chandel, A. K., E. S. Chan and R. Rudravaram, et al. 2007. Economics and environmental impact of bioethanol production technologies: An appraisal. *Biotechnology and Molecular Biology Reviews* 2:14–32.

Chum, H. L., E. Warner, J. E. A. Seabra and I. C. Macedo. 2013. A comparison of commercial ethanol production systems from Brazilian sugarcane and US corn. *Biofuels, Bioproducts and Biorefining* 8:205–223.

Da Silva, A. L. and J. A. Castaneda-Ayarza. 2021. Macro-environment analysis of the corn ethanol fuel development in Brazil. *Renewable and Sustainable Energy Reviews* 135:110387.

Dalena, F., A. Senatore and A. Iulianelli, et al. 2019. Ethanol from biomass: Future and perspectives. In *Ethanol: Science and Engineering*, Ed. F. Dalena, A. Basile, T. N. Veziroglu and A. Iulianelli, pp. 25–59. London: Elsevier.

De Moraes Rego, C. R., V. R. Reis and A. Wander, et al. 2017. Cost analysis of corn cultivation in the setup of the crop-livestock-forest integration system to recover degraded pastures. *Journal of Agricultural Science* 9:168.

Eckert, C. T., E. P. Frigo and L. P. Albrecht, et al. 2018. Maize ethanol production in Brazil: Characteristics and perspectives. *Renewable and Sustainable Energy Reviews* 82:3907–3912.

EIA. 2020a. *U.S. Fuel Ethanol Plant Production Capacity*. Washington, DC: Energy Information Administration. https://www.eia.gov/petroleum/ethanolcapacity/. Accessed 24.08.2021.

EIA. 2020b. *U.S. Fuel Ethanol Production Capacity Increased by 3% in 2019*. Washington, DC: Energy Information Administration. https://www.eia.gov/todayinenergy/detail.php?id=45316. Accessed 26.08.2021.

EPA. 1997. *Sugarcane Processing*. Washington, DC: U.S. Environmental Protection Agency. https://www3.epa.gov/ttn/chief/ap42/ch09/final/c9s10-1a.pdf. Accessed 18.09.2021.

FAOSTAT. 2021a. *Crops and Livestock Products*. Rome: Food and Agriculture Organization. https://www.fao.org/faostat/en/#data/QCL. Accessed 25.08.21.

FAOSTAT. 2021b. *Producer Prices*. Rome: Food and Agriculture Organization. https://www.fao.org/faostat/en/#data/PP. Accessed 18.09.2021.

Goldemberg, J. 2008. The Brazilian biofuels industry. *Biotechnology for Biofuels* 1:6.

Goldemberg, J. and I. C. Macedo. 1994. Brazilian alcohol program: An overview. *Energy for Sustainable Development* 1:17–22.

Hettinga, W. G., H. M. Junginger and S. C. Dekker, et al. 2009. Understanding the reductions in US corn ethanol production costs: An experience curve approach. *Energy Policy* 37:190–203.

Ho, D. P., H. H. Ngo and W. Guo. 2014. A mini review on renewable sources for biofuel. *Bioresource Technology* 169:742–749.

Jackson, D. S. and D. L. Shandera. 1995. Corn wet milling: Separation chemistry and technology. *Advances in Food and Nutrition Research* 38:271–300.

Johnston, D. B., A. J. McAloon, R. A. Moreau, K. B. Hicks and V. Singh. 2005. Composition and economic comparison of germ fractions from modified corn processing technologies. *Journal of the American Oil Chemists' Society* 82:603–608.

Kelsall, D. R. and T. P. Lyons. 2003. Grain dry milling and cooking for alcohol production. In *The Alcohol Textbook*, Ed. K. A. Jacques, T. P. Lyons and D. R. Kelsall. Nottingham: Nottingham University Press.

Kim, Y., N. S. Mosier and R. Hendrickson, et al. 2008. Composition of corn dry-grind ethanol by-products: DDGS, wet cake, and thin stillage. *Bioresource Technology* 99:5165–5176.

Kwiatkowski, J. R., A. J. McAloon, F. Taylor and D. B. Johnston. 2006. Modeling the process and costs of fuel ethanol production by the corn dry-grind process. *Industrial Crops and Products* 23:288–296.

Lin, Y. and S. Tanaka. 2006. Ethanol fermentation from biomass resources: Current state and prospects. *Applied Microbiology and Biotechnology* 69:627–642.

Lopes, M. L., S. C. De Lima Paulillo and A. Godoy, et al. 2016. Ethanol production in Brazil: A bridge between science and industry. *Brazilian Journal of Microbiology* 47:64–76.

Lyons, T. P. 2003. Ethanol around the world: Rapid growth in policies, technology and production. In *The Alcohol Textbook*, Ed. K. A. Jacques, T. P. Lyons and D. R. Kelsall. Nottingham: Nottingham University Press.

Macrotrends. 2021. *Corn Prices -59 Year Historical Chart.* https://www.macrotrends.net/2532/corn-prices-historical-chart-data. Accessed 01.09.2021.

Muchow, R. C., M. J. Robertson and A. W. Wood. 1996. Growth of sugarcane under high input conditions in tropical Australia. II. Sucrose accumulation and commercial yield. *Field Crops Research* 48:27–36.

Naik, S. N., V. V. Goud, P. K. Rout and A. K. Dalai. 2010. Production of first and second generation biofuels: A comprehensive review. *Renewable and Sustainable Energy Reviews* 14:578–597.

Nawaz, H., R. Waheed, M. Nawaz and D. Shahwar. 2020. Physical and chemical modifications in starch structure and reactivity. In *Chemical Properties of Starch*, Ed. M. Emeje. London: IntechOpen.

Nigam, P. S. and A. Singh. 2011. Production of liquid biofuels from renewable resources. *Progress in Energy and Combustion Science* 37:52–68.

Nogueira, L. A. A., J. E. A. Seabra, G. Best, M. R. L. V. Leal and M. K. Roppe. 2008. Bioethanol production. In *Sugarcane-Based Bioethanol Energy for Sustainable Development.* Rio de Janeiro: National Development Bank of Brazil.

Olofsson, K., M. Bertilsson and G. Liden. 2008. A short review on SSF - an interesting process option for ethanol production from lignocellulosic feedstocks. *Biotechnology for Biofuels* 1:7.

Pimentel, D. and T. W. Patzek. 2005. Ethanol production using corn, switchgrass, and wood; biodiesel production using soybean and sunflower. *Natural Resources Research* 14:65–76.

Pimentel, D., T. Patzek and G. Cecil. 2007. Ethanol production: Energy, economic, and environmental losses. *Reviews of Environmental Contamination and Toxicology* 189:25–41.

Pubchem. 2021. *Ethanol.* https://pubchem.ncbi.nlm.nih.gov/compound/Ethanol. Accessed 01.04.2023.

Queneau, Y., S. Jarosz, B. Lewandowski and J. Fitremann. 2007. Sucrose chemistry and applications of sucrochemicals. *Advances in Carbohydrate Chemistry and Biochemistry* 61:217–292.

Ramirez, E. C., D. B. Johnston, A. J. McAloon and V. Singh. 2009. Enzymatic corn wet milling: Engineering process and cost model. *Biotechnology for Biofuel* 2:2.

Ramirez, E. C., D. B. Johnston, A. J. McAloon, W. Yee and V. Singh. 2008. Engineering process and cost model for a conventional corn wet milling facility. *Industrial Crops and Products* 27:91–97.

Rausch, K. D., R. L. Belyea and M. R. Ellersieck, et al. 2005. Particle size distributions of ground corn and DDGS from dry grind processing. *Transactions of the ASAE* 48:273–277.

RFA. 2008. *Feed Co-Products.* Washington, DC: Renewable Fuel Association. https://ethanolrfa.org/wp-content/uploads/2015/09/feed_co-products.pdf. Accessed 13.09.2021.

RFA. 2017. *Building Partnerships: Growing Markets 2017 Ethanol Industry Outlook.* Washington, DC: Renewable Fuel Association. https://ethanolrfa.org/wp-content/uploads/2017/02/Ethanol-Industry-Outlook-2017.pdf. Accessed 24.08.2021.

Roka, F. M., J. Alvarez and L. E. Baucum. 2009. *Projected Costs and Returns for Sugarcane Production on Mineral Soils of South Florida, 2007-2008.* EDIS 2009:SC087.

Roka, F. M., L. F. Baucum, R. W. Rice and J. Alvarez. 2010. Comparing costs and returns for sugarcane production on sand and muck soils of Southern Florida, 2008-2009. *Journal American Society of Sugar Cane Technologists* 30:50–66.

Sandhu, H., N. E. H. Scialabba and C. Warner, et al. 2020. Evaluating the holistic costs and benefits of corn production systems in Minnesota. *Scientific Reports* 10:3922.

Sauer, J., B. W. Sigurskjold and U. Christensen, et al. 2000. Glucoamylase: Structure/function relationships, and protein engineering. *Biochimica et Biophysica Acta (BBA) - Protein Structure and Molecular Enzymology* 1543:275–293.

Senatore, A., F. Dalena and A. Sola, et al. 2019. First-generation feedstock for bioenergy production. In *Second and Third Generation of Feedstocks*, Ed. A. Basile and F. Dalena, pp. 35–57. London: Elsevier.

Shao, L., H. Chen and Y. Li, et al. 2020. Pretreatment of corn stover via sodium hydroxide-urea solutions to improve the glucose yield. *Bioresource Technology* 307:123191.

Shapouri, H. and P. Gallagher. 2005. *USDA's 2002 Ethanol Cost of Production Survey*. Economics Technical Reports and White Papers. Washington, DC: US Department of Agriculture.

Shikida, P. F. A., A. Finco and B. F. Cardoso, et al. 2014. A comparison between ethanol and biodiesel production: The Brazilian and European experiences. *Lecture Notes in Energy* 27:25–53.

Shrestha, D. S., B. D. Staab and J. A. Duffield. 2019. Biofuel impact on food prices index and land use change. *Biomass and Bioenergy* 124:43–53.

Singh, A. 2012. Biofuels and their impact on food prices and food security. *Presented at National Conference on Emerging Trends in Renewable Energy Technology-2012*, Bhilwara, India.

Singh, V., D. B. Johnston and K. Naidu, et al. 2005. Comparison of modified dry-grind corn processes for fermentation characteristics and DDGS composition. *Cereal Chemistry* 82:187–190.

Singh, V., K. D. Rausch and P. Yang, et al. 2001. *Modified Dry Grind Ethanol Process*. Urban, IL: University of Illinois at Urbana-Champaign.

Soccol, C. R., L.P. S. Vandenberghe and B. Costa. 2005. Brazilian biofuel program: An overview. *Journal of Scientific and Industrial Research* 64:897–904.

Statista. 2021a. *Fuel Ethanol Production in Brazil From Crop Year 2018/19 to 2021/22, by Feedstock (in Billion Liters)*. New York: Statista. https://www.statista.com/statistics/1177543/fuel-ethanol-production-brazil-feedstock/. Accessed 29.08.2021.

Statista. 2021b. *Price per Ton of Sugar Cane in the U.S. from 2007 to 2018, by State (in U.S.* Dollars). New York: Statista. https://www.statista.com/statistics/191995/price-per-ton-of-sugarcane-in-the-us-by-state/. Accessed 01.09.2021.

Swinkels, J. J. M. 1985. Composition and properties of commercial native starches. *Starch* 37:1–5.

U.S. Grain Council. 2021a. *DDGS Reports*. https://grains.org/ddgs_report/. Accessed 10.09.2021.

U.S. Grain Council. 2021b. *Corn Gluten*. https://grains.org/buying-selling/corn-gluten/. Accessed 13.09.2021.

Upreti, P. and A. Singh. 2017. An economic analysis of sugarcane cultivation and its productivity in major sugar producing states of Uttar Pradesh and Maharashtra. *Economic Affairs* 62:711.

USDA. 2006. *The Economic Feasibility of Ethanol Production from Sugar in the United States*. Washington, DC: United States Department of Agriculture. https://www.fsa.usda.gov/Internet/FSA_File/ethanol_fromsugar_july06.pdf. Accessed 04.09.2021.

USDA. 2020. *Brazil: Corn Ethanol Production Booms in Brazil.* Washington, DC: United States Department of Agriculture.

Van den Wall Bake, J. D., M. Junginger, A. Faaij, T. Poot and A. Walter. 2009. Explaining the experience curve: Cost reductions of Brazilian ethanol from sugarcane. *Biomass and Bioenergy* 33:644–658.

97 The Chinese Experience of Bioethanol Fuels

Scientometric Study

Ozcan Konur
(Formerly) Ankara Yildirim Beyazit University

97.1 INTRODUCTION

The crude oil-based gasoline fuels (Ma et al., 2002; Newman and Kenworthy, 1989) have been widely used in the transportation sector since the 1920s. However, there have been great public concerns over the adverse environmental and human impact of these fuels (Hill et al., 2006, 2009). Hence, biomass-based bioethanol fuels (Hill et al., 2006; Konur, 2012e, 2015, 2019, 2020a) have increasingly been used in blending gasoline fuels (Hsieh et al., 2002; Najafi et al., 2009), in the fuel cells (Antolini, 2007, 2009), and in the biochemical production (Angelici et al., 2013; Morschbacker, 2009) in a biorefinery context (Fernando et al., 2006; Huang et al., 2008).

Bioethanol fuels also play a critical role in maintaining the energy security (Kruyt et al., 2009; Winzer, 2012) in the supply shocks (Kilian, 2008, 2009) related to oil price shocks (Hamilton, 1983, 2003), COVID-19 pandemics (Fauci et al., 2020; Li et al., 2020), or wars (Hamilton, 1983; Jones, 2012) in the aftermath of the Russian invasion of Ukraine (Reeves, 2014).

However, it is necessary to pretreat the biomass (Taherzadeh and Karimi, 2008; Yang and Wyman, 2008) to enhance the yield of the bioethanol (Hahn-Hagerdal et al., 2006; Sanchez and Cardona, 2008) prior to the bioethanol production through the hydrolysis (Sun and Cheng, 2002; Taherzadeh and Karimi, 2007) and fermentation (Lin and Tanaka, 2006; Olsson and Hahn-Hagerdal, 1996) of the biomass and the resulting hydrolysates, respectively.

China has been one of the most-prolific countries engaged in the bioethanol fuel research (Konur, 2023a,b). The Chinese research in the field of the bioethanol fuels has intensified in this context in the key research fronts of the production of bioethanol fuels, utilization of bioethanol fuels (Antolini, 2007; Hansen et al., 2005), and to a lesser extent evaluation (Hamelinck et al., 2005; Pimentel and Patzek, 2005) of bioethanol fuels. For the first research front, pretreatment (Alvira et al., 2010; Hendriks and Zeeman, 2009) and hydrolysis (Alvira et al., 2010; Sun and Cheng, 2002) of the feedstocks, fermentation of the hydrolysates (Jonsson and Martin, 2016; Lin and Tanaka, 2006), production of bioethanol fuels (Limayem and Ricke, 2012, Lin and Tanaka, 2006), and to a lesser extent separation and distillation of bioethanol fuels (Sano et al., 1994; Vane, 2005) are the key research areas, while for the second research front, utilization of bioethanol fuels in fuel cells and transport engines, bioethanol-based biohydrogen fuels, bioethanol sensors, and to a lesser extent bioethanol-based biochemicals are the key research areas.

The research in this field has also intensified for the feedstocks of lignocellulosic biomass at large (Hendriks and Zeeman, 2009; Sun and Cheng, 2002), bioethanol fuels for their utilization (Antolini, 2007; Hansen et al., 2005), cellulose (Pinkert et al., 2009; Zhang and Lynd, 2004), starch feedstock residues (Binod et al., 2010; Talebnia et al., 2010), wood biomass (Galbe and Zacchi, 2002; Zhu and Pan, 2010), and to a lesser extent industrial waste (Cardona et al., 2010), lignin (Bourbonnais and Paice, 1990), grass biomass (Pimentel and Patzek, 2005), hydrolysates (Palmqvist and Hahn-Hagerdal, 2000), biomass at large (Lin and Tanaka, 2006), sugar feedstocks (Bai et al., 2008), sugar

DOI: 10.1201/9781003226574-130

feedstock residues (Cardona et al., 2010), algal biomass (John et al., 2011), lignocellulosic wastes (Sanchez, 2009), starch feedstocks (Bai et al., 2008), urban wastes (Ravindran and Jaiswal, 2016), forestry wastes (Duff and Murray, 1996), food wastes (Ravindran and Jaiswal, 2016), biosyngas (Henstra et al., 2007), and plants (Sukumaran et al., 2009).

However, it is essential to develop efficient incentive structures (North, 1991) for the primary stakeholders to enhance the research in this field (Konur, 2000, 2002a,b,c, 2006a,b, 2007a,b). The scientometric analysis has been used in this context to inform the primary stakeholders about the current state of the research in a selected research field (Garfield, 1955; Konur, 2011, 2012a,b,c,d,e,f,g,h,i, 2015, 2018b, 2019, 2020a).

As there have been no published current scientometric studies in this field, this book chapter presents a scientometric study of the Chinese research in the bioethanol fuels. It examines the scientometric characteristics of both the sample and population data presenting scientometric characteristics of these both datasets in the order of documents, authors, publication years, institutions, funding bodies, source titles, countries, Scopus subject categories, Scopus keywords, and research fronts.

97.2 MATERIALS AND METHODS

The search for this study was carried out using Scopus database (Burnham, 2006) in November 2022.

As a first step for the search of the relevant literature, the keywords were selected using the most-cited first 300 population papers for each research front. The selected keyword list was then optimized to obtain a representative sample of papers for the each research field. These keyword lists were then integrated to obtain the keyword list for this research field (Konur, 2023a,b).

As a second step, two sets of data were used for this study. First, a population sample of 15,913 papers was used to examine the scientometric characteristics of the population data. Second, a sample of 318 most-cited papers, corresponding to 2% of the population papers, was used to examine the scientometric characteristics of these citation classics.

The scientometric characteristics of these both sample and population datasets were presented in the order of documents, authors, publication years, institutions, funding bodies, source titles, countries, Scopus subject categories, Scopus keywords, and research fronts.

Lastly, the key scientometric findings for both datasets were discussed to highlight the research landscape for the bioethanol fuels. Additionally, a number of brief conclusions were drawn, and a number of relevant recommendations were made to enhance the future research landscape.

97.3 RESULTS

97.3.1 The Most-Prolific Documents in the Chinese Experience of Bioethanol Fuels

The information on the types of documents for both datasets is given in Table 97.1. The articles and conference papers, published in journals, dominate both the sample (87%) and population (97%) papers with 10% deficit. Further, review papers and short surveys have 11% surplus as they are over-represented in the sample papers as they constitute 13% and 2% of the sample and population papers, respectively. Additionally, 1% of the sample papers were published as book chapters.

It is further notable that 96%, 3%, and 1% of the population papers were published in journals, book series, and books, respectively. Similarly, 100% of the sample papers were published solely in the journals.

97.3.2 The Most-Prolific Authors in the Chinese Experience of Bioethanol Fuels

The information about the most-prolific 28 authors with at least 1.3% of sample papers each is given in Table 97.2. The most-prolific authors are Runcang Sun, Gongquan Sun, Quin Xin, and Peikang Shen with 4.4%, 3.8%, 2.8% and 2.5% of the sample papers, respectively. The other prolific authors

TABLE 97.1
Documents in the Chinese Experience of Bioethanol Fuels

Documents	Sample Dataset (%)	Population Dataset (%)	Surplus (%)
Article	85.2	92.9	–7.7
Review	11.9	2.2	9.7
Conference paper	1.9	3.9	–2.0
Short survey	0.9	0.0	0.9
Book chapter	0.0	0.6	–0.6
Note	0.0	0.2	–0.2
Editorial	0.0	0.1	–0.1
Letter	0.0	0.0	0.0
Book	0.0	0.0	0.0
Sample size	318	15,913	

Population dataset (%), the number of papers (%) in the set of the 15,913 population papers; sample dataset (%), the number of papers (%) in the set of 318 highly cited papers.

TABLE 97.2
The Most-Prolific Authors in the Chinese Experience of Bioethanol Fuels

No.	Author Name	Author Code	Sample Papers (%)	Population Papers (%)	Surplus	Institution	Country	HI	N	Res. Front
1	Sun, Runcang	55661525600	4.4	1.5	2.9	Dalian Polytech. Univ.	China	119	1083	P, H
2	Sun, Gongquan	7402760735	3.8	0.3	3.5	Chinese Acad. Sci.	China	66	283	A
3	Xin, Quin	28167866000	2.8	0.2	2.6	Chinese Acad. Sci.	China	68	325	A
4	Shen, Peikang	7201767641	2.5	0.2	2.3	Guangxi Univ.	China	81	441	A
5	Xu, Feng*	56420960200	2.2	0.6	1.6	Beijing Forestry Univ.	China	59	270	P
6	Jiang, Luhua*	57209054216	2.2	0.2	2.0	Qingdao Univ. Sci. Technol.	China	48	160	A
7	Chen, Hongzhang	7501614171	1.9	0.4	1.5	Chinese Acad. Sci.	China	46	234	P, H, F, R
8	Xu, Changwei	9248835900	1.9	0.2	1.7	Quangdong Ocean Univ.	China	35	72	A
9	Song, Shuqin	7403349881	1.9	0.2	1.7	Sun Yat-Sen Univ.	China	53	156	A
10	Shuai, Shijin	6603356005	1.9	0.0	1.9	Tsinghua Univ.	China	47	145	I
11	Zhao, Xuebing	8961267200	1.6	0.3	1.3	Tsinghua Univ.	China	41	122	P, H
12	Liu, Dehua	35233867100	1.6	0.3	1.3	Tsinghua Univ.	China	58	310	P, H
13	Wang, Jianji	55904673200	1.6	0.2	1.4	Henan Normal Univ.	China	69	628	P
14	Zhao, Tianshou	13004121800	1.6	0.2	1.4	Southern Univ. Sci. Technol.	China	90	494	A

(Continued)

TABLE 97.2 (*Continued*)
The Most-Prolific Authors in the Chinese Experience of Bioethanol Fuels

No.	Author Name	Author Code	Sample Papers (%)	Population Papers (%)	Surplus	Institution	Country	HI	N	Res. Front
15	Chen, Yujin	7601437135	1.6	0.2	1.4	Chinese Acad. Sci.	China	56	245	S
16	Wang, Jianxin	35254238800	1.6	0.2	1.4	Tsinghua Univ.	China	50	348	I
17	Bao, Jie	57189034954	1.3	0.6	0.7	East China Univ. Sci. Technol.	China	35	163	P, H, F, R
18	Tan, Tian-Wei	57020620400	1.3	0.2	1.1	Beijing Univ. Chem. Technol.	China	80	720	P, H, F, R
19	Yu, Ziniu	7404346720	1.3	0.2	1.1	Huazhong Agr. Univ.	China	53	297	P, H
20	Zhang, Lina*	55917992100	1.3	0.2	1.1	Wuhan Univ.	China	100	712	P
21	Liang, Zhen Xing	7402178316	1.3	0.2	1.1	South China Univ. Technol.	China	40	112	A
22	Sun, Shaoni	55533508500	1.3	0.2	1.1	Beijing Forestry Univ.	China	33	89	P, H
23	Sun, Shi-Gang	7404510197	1.3	0.2	1.1	Univ. Sci. Technol. China	China	86	758	A
24	Zhao, Zongbao K.	56972812400 35197609300	1.3	0.2	1.1	Chinese Acad. Sci.	China	68	228	P
25	Ren, Nanqi	7004987612	1.3	0.1	1.2	Harbin Inst. Technol.	China	100	1250	B
26	Shen, Wenjie		1.3	0.1	1.2	Chinese Acad. Sci.	China	60	235	B
27	Xia, Liming	7201955949	1.3	0.1	1.2	Zhejiang Univ.	China	22	100	P, H
28	Zhu, Chunling	7403439505	1.3	0.1	1.2	Harbin Eng. Univ.	China	57	120	S

*, Female; A, utilization of bioethanol fuels in fuel cells; Author code, the unique code given by Scopus to the authors; B, utilization of bioethanol fuels for biohydrogen fuel production; E, evaluation of bioethanol fuels; F, fermentation of the feedstock-based hydrolysates; H, hydrolysis of the feedstock; HI, H-index; I, utilization of bioethanol fuels in the transport engines; N, number of papers published by each author; P, pretreatment of the feedstock; population papers, the number of papers authored in the population dataset; R, bioethanol fuel production; S, bioethanol sensors; sample papers, the number of papers authored in the sample dataset.

are Feng Xu, Luhua Jiang, Hongzhang Chen, Changwei Xu, Shuqin Song, and Shijin Shuai with 1.9%–2.2% of the sample papers each.

On the other hand, the most influential authors are Gongquan Sun, Runcang Sun, and Quin Xin with 3.5%, 2.9%, and 2.6% surplus, respectively. The other influential authors are Peikang Shen, Luhua Jiang, Shijin Shuai, Shuqin Song, Changwei Xu, Feng Xu, and Hongzhang Chen with 1.5%–2.3% surplus each.

The most-prolific institutions for the sample dataset are the Chinese Academy of Sciences, Tsinghua University, and Beijing Forestry University with six, four, and two institutions, respectively. In total, only 19 institutions house these top authors.

The most-prolific research front for these top authors is the pretreatments of the feedstocks with 13 authors followed by the hydrolysis of the feedstocks and utilization of bioethanol fuels in fuel cells with nine authors each. The other prolific research fronts are the fermentation of the feedstock-based hydrolysates, bioethanol production, utilization of bioethanol fuels for biohydrogen fuel production, utilization of bioethanol fuels in transport engines, and bioethanol sensors with two to three authors each.

On the other hand, there is a significant gender deficit (Beaudry and Lariviere, 2016) for the sample dataset as surprisingly only three of these top researchers are female with a representation rate of 8%.

On the other hand, the most-prolific collaborating authors are Junyong Zhu, Bruce E. Dale, Venkatesh Balan, Panagiotis Tsiakaras, Jinlong Gong, Rolf J. Behm, Shishir P. S. Chundawat, Roland Gleisner, Xuejun Pan, Jeremy Tomkinson, Nirmal Uppugundhla, Charles E. Wyman, Weijiang Zhou, and Zhenhua Zhou.

Additionally, there are other prolific China-based authors with the relatively low citation impact contributing to the research in this area: Qiang Yong, Yide Xu, Jian Zhang, Yu Zhang, Xinqing Zhao, Jian Zhao, Xin Li, Fei Shen, Yue-Qin Tang, Tong-Qi Yuan, Caoxing Huang, Fengwu Bai, Chenhuan Lai, Liangcai Peng, Tao Xia, Jianxin Jiang, Di Cai, Jinguang Hu, Qiong Wang, Jia-Long Wen, Junhua Li, Peiyong Qin, Zhihao Dong, Bing-Zhi Li, He Huang, Yanting Wang, Jun Zhang, Yimin Zhang, and Jing Zhao.

97.3.3 The Most-Prolific Research Output by Years in the Chinese Experience of Bioethanol Fuels

Information about papers published between 1970 and 2022 is given in Figure 97.1. This figure clearly shows that the bulk of the research papers in the population dataset were published primarily in the 2010s and the early 2020s with 58% and 32% of the population dataset, respectively. Similarly, the publication rates for the 2000s, 1990s, 1980s, and 1970s were 9%, 1%, 0%, and 0% respectively. Further, the rate for the pre-1970s was 0.1%.

Similarly, the bulk of the research papers in the sample dataset were published in the 2010s and 2000s with 47% and 39% of the sample dataset, respectively. Similarly, the publication rates for the 1990s, 1980s, and 1970s were 2%, 0%, and 0% of the sample papers, respectively. Further, the rate for the pre-1970s was 0%.

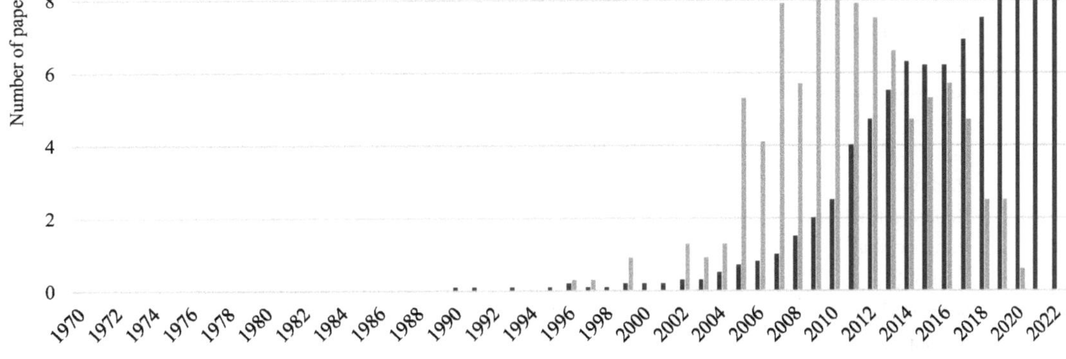

FIGURE 97.1 The research output by years regarding the bioethanol fuels.

The most-prolific publication years for the population dataset were 2019, 2020, 2021, and 2022 with 8.6%, 9.4%, 10.4%, and 12% of the dataset, respectively, while 94% of the population papers were published between 2008 and 2022. Further, 40% of the population papers were published in the last 4 years. Similarly, 94% of the sample papers were published between 2005 and 2019, while the most-prolific publication years were 2009 and 2010 with 12.6% and 11.3% of the sample papers, respectively.

97.3.4 The Most-Prolific Institutions in the Chinese Experience of Bioethanol Fuels

Information about the most-prolific 31 institutions publishing papers on the Chinese experience of bioethanol fuels with at least 1.3% of the sample papers each is given in Table 97.3.

The most-prolific institutions are the Chinese Academy of Sciences with 23% of the sample papers. The other prolific institutions are Tsinghua University, South China University of Technology, Tianjin University, Sun Yat-Sen University, Beijing Forestry University, Beijing University of Chemical Technology, Shanghai Jiao Tong University, and Xiamen University with 3%–8% of the sample papers.

TABLE 97.3
The Most-Prolific Institutions in the Bioethanol Fuels

No.	Institutions	Sample Papers (%)	Population Papers (%)	Surplus (%)
1	Chinese Acad. Sci.	22.6	11.1	11.5
2	Tsinghua Univ.	7.9	2.8	5.1
3	South China Univ. Technol.	6.3	5.1	1.2
4	Tianjin Univ.	4.1	3.5	0.6
5	Sun Yat-Sen Univ.	4.1	0.6	3.5
6	Beijing Forestry Univ.	3.5	3.5	0.0
7	Beijing Univ. Chem. Technol.	2.8	2.2	0.6
8	Shanghai Jiao Tong Univ.	2.8	1.9	0.9
9	Xiamen Univ.	2.5	1.2	1.3
10	Zhejiang Univ.	1.9	2.3	–0.4
11	East China Univ. Sci. Technol.	1.9	1.8	0.1
12	Shandong Univ.	1.9	1.8	0.1
14	Harbin Inst. Technol.	1.9	1.5	0.4
15	Beijing Inst. Technol.	1.9	0.7	1.2
16	Nanjing Forestry Univ.	1.6	3.8	–2.2
17	China Agr. Univ.	1.6	2.2	–0.6
18	Nanjing Tech Univ.	1.6	1.7	–0.1
19	Northwest A&F Univ.	1.6	1.6	0.0
20	Jilin Univ.	1.6	1.3	0.3
21	Huazhong Agr. Univ.	1.6	1.2	0.4
22	Univ. Sci. Technol. China	1.6	1.2	0.4
23	Wuhan Univ.	1.6	0.6	1.0
24	Jinan Univ.	1.6	0.5	1.1
25	Henan Normal Univ.	1.6	0.3	1.3
26	Jiangnan Univ.	1.3	2.1	–0.8
27	Tianjin Univ. Sci. Technol.	1.3	2.0	–0.7
28	Soochow Univ.	1.3	0.6	0.7
29	Nankai Univ.	1.3	0.5	0.8
30	Fudan Univ.	1.3	0.5	0.8
31	Harbin Eng. Univ.	1.3	0.3	1.0

On the other hand, the institution with the most citation impact is the Chinese Academy of Sciences with 12% surplus. The other influential institutions are Tsinghua University, Sun Yat-Sen University, Xiamen University, Henan Normal University, South China University of Technology, and Beijing Institute of Technology with 1%–5% surplus each.

Further, the most-prolific collaborating overseas institutions are USDA Forest Service, University of Wisconsin-Madison, National University of Singapore, Oak Ridge National Laboratory, Michigan State University, Bangor University, University of Thessaly, NC State University, Nanyang Technological University, University of Illinois Urbana-Champaign, NC State University, University of New Brunswick, University of Tennessee Knoxville, SUNY, University of British Columbia, and University of Minnesota Twin Cities.

Additionally, there are other Chinese institutions with the relatively low citation impact: Qilu University of Technology, Sichuan University, Guangxi University, Dalian University of Technology, Chinese Academy of Agricultural Sciences, Jiangsu University, Nanjing Agricultural University, Chinese Academy of Forestry, Northeast Forestry University, Huazhong University of Science and Technology, South China Agricultural University, Dalian Polytechnic University, Hongqing University, Xi'an Jiaotong University, Qingdao Institute of Bioenergy and Bioprocess Technology, Hunan University, Tongji University, Central South University, Nanjing University, Henan Agricultural University, University of Science and Technology Beijing, Fujian Agriculture and Forestry University, Zhengzhou University, Northeast Agricultural University, Taiyuan University of Technology, Zhejiang University of Technology, Kunming University of Science and Technology, and Nanjing University of Science and Technology.

97.3.5 The Most-Prolific Funding Bodies in the Chinese Experience of Bioethanol Fuels

Information about the most-prolific 17 funding bodies funding at least 1.3% of the sample papers each is given in Table 97.4. Further, only 55% and 70% of the sample and population papers each were funded, respectively.

TABLE 97.4
The Most-Prolific Funding Bodies in the Chinese Experience of Bioethanol Fuels

No.	Funding Bodies	Sample Paper No. (%)	Population Paper No. (%)	Surplus (%)
1	National Natural Science Foundation of China	33.0	46.4	–13.4
2	National Key Research and Development Program of China	6.0	9.1	–3.1
3	Chinese Academy of Sciences	5.3	2.8	2.5
4	National High-tech Research and Development Program	3.8	2.1	1.7
5	China Postdoctoral Science Foundation	2.8	3.3	–0.5
6	Fundamental Research Funds for the Central Universities	2.5	6.7	–4.2
7	Ministry of Education of China	2.5	1.9	0.6
8	Natural Science Foundation of Jiangsu Province	2.2	2.5	–0.3
9	National Basic Research Program of China (973 Program)	2.2	2.5	–0.3
10	Ministry of Science and Technology of China	2.2	1.3	0.9

(Continued)

TABLE 97.4 (*Continued*)
The Most-Prolific Funding Bodies in the Chinese Experience of Bioethanol Fuels

No.	Funding Bodies	Sample Paper No. (%)	Population Paper No. (%)	Surplus (%)
11	Natural Science Foundation of Guangdong Province	2.2	1.2	1.0
12	Priority Academic Program Development of Jiangsu Higher Education Institutions	1.3	2.8	−1.5
13	Higher Education Discipline Innovation Project	1.3	1.2	0.1
14	Program for New Century Excellent Talents in University	1.3	1.1	0.2
15	Science and Technology Planning Project of Guangdong Province	1.3	0.9	0.4
16	Project 211	1.3	0.4	0.9
17	Scientific Research Foundation for Returned Scholars of Ministry of Education	1.3	0.4	0.9

The most-prolific funding body is the National Natural Science Foundation of China with 33% of the sample papers. The other prolific funding bodies are National Key Research and Development Program of China, Chinese Academy of Sciences, National High-Tech Research and Development Program, China Postdoctoral Science Foundation, Fundamental Research Funds for the Central Universities, and Ministry of Education of China with 3%–6% of the sample papers each. It is notable that the National Natural Science Foundation of China also funds 46.1% of the population papers.

The funding bodies with the most citation impact are the Chinese Academy of Sciences, National High-tech Research and Development Program, Natural Science Foundation of Guangdong Province, Ministry of Science and Technology of China, Project 211, and Scientific Research Foundation for Returned Scholars of Ministry of Education with 1%–3% surplus each. Further, the funding body with the least citation impact is the National Science Foundation and Natural Sciences with 13% deficit. The other funding bodies with the least impact are Fundamental Research Funds for the Central Universities, National Key Research and Development Program of China, and Priority Academic Program Development of Jiangsu Higher Education Institutions with 2%–4% deficit each.

On the other hand, the prolific overseas collaborating funding bodies are the Office of Science, U.S. Department of Energy, Great Lakes Bioenergy Research Center, U.S. Forest Service, Basic Energy Sciences, Engineering and Physical Sciences Research Council, European Commission, USDA Forest Service, National Science Foundation, Natural Sciences and Engineering Research Council of Canada, U.S. Department of Agriculture, Alexander von Humboldt Foundation, Army Research Office, Ministry of Science and Technology Taiwan, Oak Ridge National Laboratory, and Research Grants Council.

The other Chinese funding bodies with the relatively low citation impact are the China Scholarship Council, Natural Science Foundation of Shandong Province, Natural Science Foundation of Zhejiang Province, Specialized Research Fund for the Doctoral Program of Higher Education of China, Nanjing Forestry University, State Key Laboratory of Pulp and Paper Engineering, Natural Science Foundation of Beijing Municipality, Science and Technology Commission of Shanghai Municipality, Agriculture Research System of China, Natural Science Foundation of Guangxi Province, Natural Science Foundation of Fujian Province, Youth Innovation Promotion Association of the Chinese Academy of Sciences, Natural Science Foundation of Tianjin City, Special Fund for Agro-scientific Research in the Public Interest, Key Technology Research and Development Program of Shandong, Natural Science Foundation of Heilongjiang Province, Natural Science Foundation of Anhui Province, and Guangzhou Municipal Science and Technology Project.

97.3.6 The Most-Prolific Source Titles in the Chinese Experience of Bioethanol Fuels

Information about the most-prolific 23 source titles publishing at least 1.3% of the sample papers each in the bioethanol fuels is given in Table 97.5.

The most-prolific source title is the Bioresource Technology with 13.5% of the sample papers. The other prolific journals are Green Chemistry, Journal of Power Sources, Sensors and Actuators B Chemical, and Electrochimica Acta with 3%–5% of the sample papers each.

On the other hand, the source titles with the most citation impact are Green Chemistry and Journal of Power Sources with 4.1% surplus each, followed by Bioresource Technology with 3.9% surplus. The other influential titles are the Sensors and Actuators B Chemical, Energy and Environmental Science, Electrochimica Acta, Applied Catalysis B Environmental, ACS Applied Materials and Interfaces, and Advanced Materials with 1.5%–2.6% surplus each.

The other source titles with the relatively low citation impact are Bioresources, Industrial Crops and Products, Advanced Materials Research, RSC Advances, ACS Sustainable Chemistry and Engineering, Carbohydrate Polymers, International Journal of Biological Macromolecules, Energy and Fuels, Process Biochemistry, Renewable Energy, Science of the Total Environment, Journal of Cleaner Production, Chemical Engineering Journal, Journal of Agricultural and Food Chemistry,

TABLE 97.5
The Most-Prolific Source Titles in the Chinese Experience of Bioethanol Fuels

No.	Source Titles	Sample Papers (%)	Population Papers (%)	Surplus (%)
1	Bioresource Technology	13.5	9.6	3.9
2	Green Chemistry	5.0	0.9	4.1
3	Journal of Power Sources	4.7	0.6	4.1
4	Sensors and Actuators B Chemical	3.8	1.2	2.6
5	Electrochimica Acta	2.5	0.6	1.9
6	Energy and Environmental Science	2.2	0.1	2.1
7	Biotechnology for Biofuels	1.9	1.6	0.3
8	International Journal of Hydrogen Energy	1.9	1.4	0.5
9	Applied Microbiology and Biotechnology	1.9	0.9	1.0
10	ACS Applied Materials and Interfaces	1.9	0.4	1.5
11	Applied Catalysis B Environmental	1.9	0.3	1.6
12	Fuel	1.6	1.7	–0.1
13	Chemsuschem	1.6	0.2	1.4
14	Renewable and Sustainable Energy Reviews	1.6	0.2	1.4
15	Advanced Materials	1.6	0.1	1.5
16	Cellulose	1.3	1.3	0.0
17	Applied Biochemistry and Biotechnology	1.3	1.2	0.1
18	Biomass and Bioenergy	1.3	0.7	0.6
19	Industrial and Engineering Chemistry Research	1.3	0.7	0.6
20	Energy	1.3	0.6	0.7
21	Electrochemistry Communications	1.3	0.1	1.2
22	Angewandte Chemie International Edition	1.3	0.1	1.2
23	Journal of the American Chemical Society	1.3	0.1	1.2

Separation and Purification Technology, Food Chemistry, Scientific Reports, Biochemical Engineering Journal, Biomass Conversion and Biorefinery, Journal of Chemical Technology and Biotechnology, Biotechnology Letters, Applied Surface Science, Bioprocess and Biosystems Engineering, Chinese Journal of Chemical Engineering, Plos One, Applied Energy, and Applied Mechanics and Materials.

97.3.7 The Most-Prolific Collaborating Countries in the Chinese Experience of Bioethanol Fuels

Information about the most-prolific 13 collaborating countries publishing at least 0.6% of sample papers each in the bioethanol fuels is given in Table 97.6.

The most-prolific collaborating country is the USA with 17% of the sample papers. The other prolific countries are the UK, Japan, Greece, Canada, Singapore, and Germany with 1.6%–4.1% of the sample papers each. Nine European countries listed in Table 97.6 as a whole produce 12.5% and 4.5% of the sample and population papers, respectively, with 8% surplus.

On the other hand, the collaborating country with the most citation impact is the USA with 8.3% surplus. The other influential countries are the UK, Japan, and Greece with 1.7%–2.4% surplus each.

Additionally, there are other collaborating countries with relatively low citation impact: Australia, Taiwan, Pakistan, India, France, Malaysia, Egypt, Saudi Arabia, Thailand, Netherlands, Iran, New Zealand, Mexico, South Africa, Brazil, Italy, Russia, Bangladesh, Ireland, Nigeria, Switzerland, Austria, and Sudan.

97.3.8 The Most-Prolific Scopus Subject Categories in the Chinese Experience of Bioethanol Fuels

Information about the most-prolific 11 Scopus subject categories indexing at least 2.2% of the sample papers each is given in Table 97.7.

The most-prolific Scopus subject categories in the bioethanol fuels are the Chemical Engineering, Energy, Environmental Science, and Chemistry with 31%–44% of the sample papers each. The other

TABLE 97.6
The Most-Prolific Collaborating Countries in the Chinese Experience of Bioethanol Fuels

No.	Countries	Sample Papers (%)	Population Papers (%)	Surplus (%)
	USA	17.0	8.7	8.3
1	UK	4.1	1.7	2.4
2	Japan	3.8	2.1	1.7
3	Greece	1.9	0.2	1.7
4	Canada	1.6	2.4	−0.8
5	Singapore	1.6	0.7	0.9
6	Germany	1.6	0.7	0.9
7	Sweden	1.3	0.4	0.9
8	S. Korea	0.9	0.8	0.1
9	Finland	0.9	0.5	0.4
10	Spain	0.9	0.2	0.7
11	Denmark	0.6	0.4	0.2
12	Belgium	0.6	0.2	0.4
13	Italy	0.6	0.2	0.4

TABLE 97.7
The Most-Prolific Scopus Subject Categories in the Chinese Experience Bioethanol Fuels

No.	Scopus Subject Categories	Sample Papers (%)	Population Papers (%)	Surplus (%)
1	Chemical Engineering	43.7	41.1	2.6
2	Energy	38.1	29.3	8.8
3	Environmental Science	36.8	30.6	6.2
4	Chemistry	31.8	27.4	4.4
5	Engineering	18.9	17.4	1.5
6	Materials Science	18.9	16.6	2.3
7	Biochemistry, Genetics, and Molecular Biology	18.2	20.4	–2.2
8	Immunology and Microbiology	12.3	12.6	–0.3
9	Physics & Astronomy	9.7	8.9	0.8
10	Agricultural and Biological Sciences	6.9	14.4	–7.5
11	Multidisciplinary	2.2	1.5	0.7

prolific categories are Engineering, Materials Science, Biochemistry, Genetics and Molecular Biology, and Immunology and Microbiology with 12%–19% of the sample papers each.

It is notable that Social Sciences including Economics and Business account for only 0% and 2% of the sample and population studies, respectively, mostly published outside the energy and fuel journals.

On the other hand, the Scopus subject categories with the most citation impact are the Energy, Environmental Science, and Chemistry with 4%–9% surplus each. Similarly, the least influential subject categories are the Agricultural and Biological Sciences and Biochemistry, Genetics and Molecular Biology with 8% and 2% deficit, separately.

97.3.9 The Most-Prolific Keywords in the Chinese Experience of Bioethanol Fuels

Information about the Scopus keywords used with at least 5.3% or 3.8% of the sample or population papers, respectively, is given in Table 97.8. For this purpose, keywords related to the keyword set given in the appendix of the related papers are selected from a list of the most-prolific keyword set provided by Scopus database.

These keywords are grouped under the six headings: Feedstocks, pretreatments of the biomass, fermentation of the hydrolysates, hydrolysis of the biomass and hydrolysates, utilization of bioethanol fuels, and products.

The most-prolific keyword related to the feedstocks is cellulose with 37% of the sample papers, followed by lignin, biomass, lignocellulose, and lignocellulosic biomass with 11%–29% of the sample papers each. Further, the most-prolific keyword related to the pretreatments is ionic liquids (ILs) with 15% of the sample papers, followed by pretreatment, solvents, pretreatment, and cellulases with 9%–12% of the sample papers each, respectively.

The most-prolific keyword related to the fermentation is fermentation with 15% of the sample papers, followed by bacteria and saccharomyces with 9% and 8% of the sample papers, respectively. Further, the most-prolific keyword related to the hydrolysis and hydrolysates is hydrolysis with 26% of the sample papers, followed by sugar, enzymatic hydrolysis, enzyme activity, and glucose with 12%–19% of the sample papers each.

Finally, the most-prolific keyword related to the utilization of bioethanol fuels is catalysts with 12% of the sample papers, followed by oxidation, catalysis, and electrocatalysts with 9%–11% of the sample papers each. Further, the most-prolific keywords related to products are ethanol with 52% of the sample papers, followed by biofuels and bioethanol with 19% and 7% of the sample papers, respectively.

TABLE 97.8
The Most-Prolific Keywords in the Chinese Experience of Bioethanol Fuels

No.	Keywords	Sample Papers (%)	Population Papers (%)	Surplus (%)
1	Feedstocks			
	Cellulose	37.1	23.7	13.4
	Lignin	28.9	18.2	10.7
	Biomass	22.6	13.5	9.1
	Lignocellulose	13.5	7.4	6.1
	Lignocellulosic biomass	11.3	4.4	6.9
	Hemicellulose	9.4	4.5	4.9
	Straw	8.2	4.1	4.1
	Zea	7.5	5.6	1.9
	Maize	4.4	4.5	–0.1
2	Pretreatments			
	ILs	15.4	6.5	8.9
	Pretreatment	11.9	5.5	6.4
	Solvents	11.6	5.3	6.3
	Pretreatment	9.7	6.1	3.6
	Cellulases	9.1	6.5	2.6
	Enzymes	8.2	8.7	–0.5
	Temperature	7.9	7.0	0.9
	pH	4.7	6.9	–2.2
	Water	4.7	3.8	0.9
3	Fermentation			
	Fermentation	14.8	22.6	–7.8
	Bacteria	9.4	11.6	–2.2
	Saccharomyces	8.2	4.8	3.4
	Yeast	6.0	5.8	0.2
	Acetic acid	4.1	4.1	0.0
4	Hydrolysis			
	Hydrolysis	25.5	19.1	6.4
	Sugar	18.6	7.6	11.0
	Enzymatic hydrolysis	16.0	12.0	4.0
	Enzyme activity	13.5	11.9	1.6
	Glucose	12.3	11.3	1.0
	Saccharification	9.1	7.7	1.4
	Xylose	5.3	4.5	0.8
5.	Utilization			
	Catalysts	11.6	7.4	4.2
	Oxidation	11.0	5.3	5.7
	Catalysis	9.7	4.5	5.2
	Electrocatalysts	8.8	2.9	5.9
	Fuel cells	8.2	1.9	6.3
	Catalyst activity	6.9	4.4	2.5
	Electrooxidation	6.9	2.7	4.2
	Palladium	6.3	1.8	4.5
	Ethanol oxidation	5.7	2.5	3.2
	Carbon	5.0	4.8	0.2
6	Products			
	Ethanol	51.9	35.8	16.1
	Biofuels	19.2	9.1	10.1
	Bioethanol	6.9	6.0	0.9

On the other hand, the most-prolific keywords across all of the research fronts are ethanol, cellulose, lignin, hydrolysis, biomass, biofuels, sugar, enzymatic hydrolysis, ILs, fermentation, enzyme activity, lignocellulose, glucose, pretreatment, catalysts, and solvents with 12%–52% of the sample papers each.

Similarly, the most influential keywords are ethanol, cellulose, sugar, lignin, biofuels, biomass, ILs, lignocellulosic biomass, hydrolysis, pretreatment, solvents, fuel cells, lignocellulose, and electrocatalysts with 6%–16% surplus each.

97.3.10 The Most-Prolific Research Fronts in the Chinese Experience of Bioethanol Fuels

Information about the research fronts for feedstocks used for both sample and population papers in the Chinese experience of bioethanol fuels is given in Table 97.9.

As this table shows, the most-prolific research front for this field is the bioethanol fuels as feedstocks for the utilization of bioethanol fuels with 39% of the sample papers, followed by starch feedstock residues and cellulose with 15% and 13% of the sample papers, respectively. The other prolific research fronts are the lignocellulosic biomass at large, wood biomass, industrial waste, lignin, sugar feedstock residues, biomass at large, algal biomass, grass biomass, and urban wastes

TABLE 97.9
The Most-Prolific Research Fronts for the Chinese Experience of Bioethanol Fuels

No.	Research Fronts	N Paper (%) Sample I	N Paper (%) Sample II	Surplus (%)
1	Bioethanol fuels	39.3	19.6	19.7
2	Starch feedstock residues	15.1	9.7	5.4
3	Cellulose	12.9	15.1	–2.2
4	Lignocellulosic biomass at large	8.8	23.0	–14.2
5	Wood biomass	7.9	9.0	–1.1
6	Industrial waste	5.3	6.5	–1.2
7	Lignin	3.8	6.2	–2.4
8	Sugar feedstock residues	2.8	2.2	0.6
9	Biomass	2.2	2.6	–0.4
10	Algal biomass	1.6	1.9	–0.3
11	Grass	1.6	3.1	–1.5
12	Urban wastes	1.6	1.0	0.6
13	Hydrolysates	1.3	2.8	–1.5
14	Food wastes	0.9	0.6	0.3
15	Forestry wastes	0.6	0.8	–0.2
16	Plants	0.6	0.3	0.3
17	Starch feedstocks	0.6	1.7	–1.1
18	Cellobiose	0.3	0.4	–0.1
19	Lignocellulosic wastes	0.3	1.8	–1.5
20	Sugar feedstocks	0.3	2.2	–1.9
21	Xylan	0.3	0.8	–0.5
22	Hemicellulose	0.0	0.8	–0.8
23	Syngas	0.0	0.4	–0.4
	Sample size	318	775	

N paper (%) sample I, the number of papers in the Chinese sample of 318 papers; N paper (%) sample II, the number of papers in the global sample of 775 papers.

with 2%–9% of the sample papers each. Further, the other minor research fronts are the hydrolysates, food wastes, forestry wastes, plants, starch feedstocks, cellobiose, lignocellulosic wastes, sugar feedstocks, and xylan with 0.3%–1.3% of the sample papers each.

On the other hand, bioethanol fuels are over-represented in the sample paper in relation to the global sample by 20%. Similarly, starch feedstock residues are over-represented in the sample paper in relation to the global sample by 5%. On the contrary, lignocellulosic biomass at large is under-represented in the sample paper in relation to the global sample by 14%. Additionally, a large number of feedstocks are under-represented in the sample paper in relation to the global sample by 0%–2% each.

Information about the thematic research fronts for the sample papers in the bioethanol fuels is given in Table 97.10. As this table shows, the most-prolific research front is the production of bioethanol fuels with 69% of the sample papers, followed by the utilization and evaluation of bioethanol fuels with 30% and 1% of the sample papers, respectively.

On the other hand, the most-prolific research front for the production of bioethanol fuels is the biomass pretreatment with 51% of the sample papers, followed by biomass hydrolysis, bioethanol production, and hydrolysate fermentation with 27%, 11%, and 11% of the sample papers. The other minor research front is separation and distillation of bioethanol fuels with 2% of the sample papers.

Further, the most-prolific research fronts for the utilization of bioethanol fuels are the utilization of bioethanol fuels in fuel cells and bioethanol fuel cells with 9% of the sample papers each. The other research fronts are the utilization of bioethanol fuels in the transport engines, bioethanol-based biohydrogen fuels, and bioethanol-based biochemicals with 6%, 4%, and 2% of the sample papers, respectively.

On the other hand, the field of the utilization of bioethanol fuels is over-represented in the Chinese sample in relation to the global sample by 11%. On the contrary, the fields of evaluation and production of bioethanol fuels are under-represented in the Chinese sample in relation to the global sample by 6% and 5%, respectively.

Further, on the individual terms, the fields of bioethanol sensors, bioethanol fuels in fuel cells, bioethanol-based biochemicals, and bioethanol fuel distillation are over-represented in the Chinese sample in relation to the global sample by 7%, 2%, 1%, and 1%, respectively. On the contrary,

TABLE 97.10
The Most-Prolific Thematic Research Fronts for the Chinese Experience of Bioethanol Fuels

No.	Research Fronts	N Paper (%) Sample I	N Paper (%) Sample II	Surplus (%)
1	Bioethanol fuel production	69.2	74.1	–4.9
	Biomass pretreatments	50.9	58.5	–7.6
	Biomass hydrolysis	27.4	29.8	–2.4
	Bioethanol production	10.7	21.5	–10.8
	Hydrolysate fermentation	11.3	17.9	–6.6
	Bioethanol fuel distillation	1.9	0.8	1.1
2	Bioethanol fuel evaluation	0.6	6.2	–5.6
3	Bioethanol fuel utilization	30.2	19.7	10.5
	Bioethanol fuels in fuel cells	9.1	6.8	2.3
	Bioethanol sensors	9.1	2.2	6.9
	Bioethanol fuels in engines	6.0	6.1	–0.1
	Bioethanol-based biohydrogen fuels	4.4	4.3	0.1
	Bioethanol-based biochemicals	1.6	0.4	1.2

N paper (%) sample II, the number of papers in the global sample of 775 papers.

the fields of bioethanol production, biomass pretreatments, hydrolysate fermentation, and biomass hydrolysis are under-represented in the Chinese sample in relation to the global sample by 8%, 7%, and 2%, respectively.

97.4 DISCUSSION

97.4.1 Introduction

The crude oil-based gasoline fuels have been widely used in the transportation sector since the 1920s. However, there have been great public concerns over the adverse environmental and human impact of these fuels. Hence, biomass-based bioethanol fuels have increasingly been used in blending gasoline fuels, in the fuel cells, and in the biochemicals and biohydrogen production in a biorefinery context.

However, it is necessary to pretreat the biomass to enhance the yield of the bioethanol prior to the bioethanol production through the hydrolysis and fermentation. China has been one of the most-prolific countries engaged in the bioethanol fuel research. The Chinese research in the field of the bioethanol fuels has intensified in this context in the key research fronts of the production of bioethanol fuels, utilization of bioethanol fuels, and to a lesser extent the evaluation of bioethanol fuels. For the first research front, pretreatment and hydrolysis of the feedstocks, fermentation of the feedstock-based hydrolysates, production of bioethanol fuels, and to a lesser extent distillation of bioethanol fuels are the key research areas, while for the second research front, utilization of bioethanol fuels in fuel cells and transport engines, bioethanol-based biohydrogen fuels, bioethanol sensors, and to a lesser extent bioethanol-based biochemicals are the key research areas. The research in this field has also intensified for the feedstocks of bioethanol fuels for their utilization, starch feedstock residues, cellulose, wood biomass, lignocellulosic biomass at large, and to a lesser extent industrial waste, lignin, grass, hydrolysates, biomass, sugar feedstocks, sugar feedstock residues, and algal biomass.

However, it is essential to develop efficient incentive structures for the primary stakeholders to enhance the research in this field. This is especially important to maintain energy security in the cases of supply shocks such as oil price shocks, war-related shocks as in the case of Russian invasion of Ukraine, or COVID-19 shocks.

The scientometric analysis has been used in this context to inform the primary stakeholders about the current state of the research in a selected research field. As there has been no current scientometric study in this field, this book chapter presents a scientometric study of the Chinese research in the bioethanol fuels. It examines the scientometric characteristics of both the sample and population data presenting scientometric characteristics of these both datasets in the order of documents, authors, publication years, institutions, funding bodies, source titles, countries, Scopus subject categories, Scopus keywords, and research fronts.

As a first step for the search of the relevant literature, the keywords were selected using the most-cited first 300 population papers for each research front. The selected keyword list was then optimized to obtain a representative sample of papers for the each research field. These keyword lists were then integrated to obtain the keyword list for this research field (Konur, 2023a,b).

As a second step, two sets of data were used for this study. First, a population sample of 15,913 papers was used to examine the scientometric characteristics of the population data. Second, a sample of 318 most-cited papers, corresponding to 2% of the population papers, was used to examine the scientometric characteristics of these citation classics.

The scientometric characteristics of these sample and population datasets were presented in the order of documents, authors, publication years, institutions, funding bodies, source titles, countries, Scopus subject categories, Scopus keywords, and research fronts.

Lastly, the key scientometric findings for both datasets were discussed to highlight the research landscape for bioethanol fuels. Additionally, a number of brief conclusions were drawn, and a number of relevant recommendations were made to enhance the future research landscape.

97.4.2 The Most-Prolific Documents in the Chinese Experience of Bioethanol Fuels

Articles (together with conference papers) dominate both the sample (87%) and population (97%) papers with 10% deficit (Table 97.1). Further, review papers have a surplus (11%) and the representation of the reviews in the sample papers is quite extraordinary (13%).

Scopus differs from the Web of Science database in differentiating and showing articles (85%) and conference papers (2%) published in the journals separately. However, it should be noted that these conference papers are also published in journals as articles, compared to those published only in the conference proceedings. Hence, the total number of articles and review papers in the sample dataset are 87% and 13%, respectively.

It is observed during the search process that there has been inconsistency in the classification of the documents in Scopus as well as in other databases such as Web of Science. This is especially relevant for the classification of articles as reviews or articles as the papers not involving a literature review may be erroneously classified as a review paper. There is also a case of review papers being classified as articles. Nevertheless, the total number of the reviews in the sample data set was manually found as nearly 13% compared to 13% as indexed by Scopus, keeping the number of articles and conference papers at 87% for the sample dataset. It is notable that many techno-economic and life cycle studies were often indexed as reviews by the Scopus database, while many articles were indexed as reviews.

In this context, it would be helpful to provide a classification note for the published papers in the books and journals at the first instance. It would also be helpful to use the document types listed in Table 97.1 for this purpose. Book chapters may also be classified as articles or reviews as an additional classification to differentiate review chapters from the experimental chapters as it is done by the Web of Science. It would be further helpful to additionally classify the conference papers as articles or review papers as well as it is done in the Web of Science database.

97.4.3 The Most-Prolific Authors in the Chinese Experience of Bioethanol Fuels

There have been most-prolific 28 authors with at least 1.3% of the sample papers each as given in Table 97.2. These authors have shaped the development of the research in this field.

The most-prolific authors are Runcang Sun, Gongquan Sun, Quin Xin, Peikang Shen, and to a lesser extent Feng Xu, Luhua Jiang, Hongzhang Chen, Changwei Xu, Shuqin Song, and Shijin Shuai. Further, the most influential authors are Gongquan Sun, Runcang Sun, Quin Xin, and to a lesser extent Peikang Shen, Luhua Jiang, Shijin Shuai, Shuqin Song, Changwei Xu, Feng Xu, and Hongzhang Chen.

It is important to note the inconsistencies in indexing of the author names in Scopus and other databases. It is especially an issue for the names with more than two components such as 'Blake Sam de Hyun Sun'. The probable outcomes are 'Sun, B.S.D.H.', 'de Hyun Sun, B.S.', or 'Hyun Sun, B.S.D.'. The first choice is the gold standard of the publishing sector as the last word in the name is taken as the last name. In most of the academic databases such as PUBMED and EBSCO databases, this version is used predominantly. The second choice is a strong alternative, while the last choice is an undesired outcome as two last words are taken as the last name. It is a good practice to combine the words of the last name by a hyphen: 'Hyun-Sun, B.S.D.'. It is notable that inconsistent indexing of the author names may cause substantial inefficiencies in the search process for the papers as well as allocating credit to the authors as there are sometimes different author entries for each outcome in the databases.

There are also inconsistencies in the shortening Chinese names. For example, 'Hongzhang Sun' is often shortened as 'Sun, H.Z.', 'Sun, H.-Z.', and 'Sun, H', as it is done in the Web of Science database as well. However, the gold standard in this case is 'Sun, H', where the last word is taken as the last name and the first word is taken as a single forename. In most of the academic databases such as

PUBMED and EBSCO, this first version is used predominantly. Nevertheless, it makes sense to use the first option to differentiate Chinese names efficiently: 'Sun, H.Z.'. Therefore, there have been difficulties in locating papers for the Chinese authors. In such cases, the use of the unique author codes provided for each author by the Scopus database has been helpful.

There is also a difficulty in allowing credit for the authors especially for the authors with common names such as 'Sun, X.' in conducting scientometric studies. These difficulties strongly influence the efficiency of the scientometric studies as well as allocating credit to the authors as there are the same author entries for different authors with the same name, e.g., 'Sun, X.' in the databases.

In this context, the coding of authors in Scopus database is a welcome innovation compared to the other databases such as Web of Science. In this process, Scopus allocates a unique number to each author in the database (Aman, 2018). However, there might still be substantial inefficiencies in this coding system especially for common names. For example, some of the papers for a certain author maybe allocated to another researcher with a different author code. It is possible that Scopus uses a number of software programs to differentiate the author names, and the program may not be false-proof (Kim, 2018).

In this context, it does not help that author names are not given in full in some journals and books. This makes difficult to differentiate authors with common names and makes the scientometric studies further difficult in the author domain. Therefore, the author names should be given in full in all books and journals at the first instance. There is also a cultural issue where some authors do not use their full names in their papers. Instead, they use initials for their forenames: 'Sun, H.J.', 'Sun, H.', or 'Sun, J.' or just plain 'Sun' instead of 'Sun, Hyun Jae'.

There are also inconsistencies in naming of the authors with more than two components by the authors themselves in journal papers and book chapters. For example, 'Parajo, A.P.C.' might be given as 'Parajo, A.' or 'Parajo, A.C.' or 'Parajo, A.P.' or 'Parajo, C' in the journals and books. This also makes the scientometric studies difficult in the author domain. Hence, contributing authors should use their name consistently in their publications.

The other critical issue regarding the author names is the inconsistencies in the spelling of the author names in the national spellings (e.g., Özğümüş, Gökçe) rather than in the English spellings (e.g., Ozgumus, Gokce) in Scopus database. Scopus differs from the Web of Science database and many other databases in this respect where the author names are given only in the English spellings. It is observed that national spellings of the author names do not help much in conducting scientometric studies as well in allocating credits to the authors as sometimes there are often the different author entries for the English and national spellings in the Scopus database.

The most-prolific institutions for the sample dataset are Chinese Academy of Sciences, Tsinghua University, and Beijing Forestry University. On the other hand, pretreatments and hydrolysis of the feedstocks, utilization of bioethanol fuels in fuel cells, and to a lesser extent fermentation of the feedstock-based hydrolysates, bioethanol production, utilization of bioethanol fuels for biohydrogen fuel production, and utilization of bioethanol fuels in transport engines are the key research fronts studied by these top authors.

Further, the most-prolific collaborating authors are Junyong Zhu, Bruce E. Dale, Venkatesh Balan, Panagiotis Tsiakaras, Jinlong Gong, Rolf J. Behm, Shishir P. S. Chundawat, Roland Gleisner, Xuejun Pan, Jeremy Tomkinson, Nirmal Uppugundhla, Charles E. Wyman, Weijiang Zhou, and Zhenhua Zhou.

It is also notable that there is a significant gender deficit for the sample dataset as surprisingly with a representation rate of 8%. This finding is the most thought-provoking with strong public policy implications. Hence, institutions, funding bodies, and policymakers should take efficient measures to reduce the gender deficit in this field as well as other scientific fields with strong gender deficit. In this context, it is worth to note the level of representation of the researchers from the minority groups in science on the basis of race, sexuality, age, and disability, besides the gender (Blankenship, 1993; Dirth and Branscombe, 2017; Konur, 2000, 2002a,b,c, 2006a,b, 2007a,b).

97.4.4 The Most-Prolific Research Output by Years in the Chinese Experience of Bioethanol Fuels

The research output observed between 1970 and 2022 is illustrated in Figure 97.1. This figure clearly shows that the bulk of the research papers in the population dataset were published primarily in the 2010s and the early 2020s. Similarly, the bulk of the research papers in the sample dataset were published in the 2010s and 2000s. Further, 94% of the population and sample papers were published between 2008 and 2022 and between 2005 and 2019, respectively. These findings suggest that the most-prolific sample and population papers were primarily published in the late 2000s, 2010s, and the early 2020s.

These are the thought-provoking findings as there has been a significant research boom since 2008 and 2005 for the population and sample papers, respectively. The rising trend of the research output for the population papers since 2003 is a clear contrast to the declining trend of the US research output after 2011 (Konur, 2023c).

In this context, the increasing public concerns about climate change (Change, 2007), greenhouse gas emissions (Carlson et al., 2017), and global warming (Kerr, 2007) have been certainly behind the boom in the research in this field since 2007. Furthermore, the recent supply shocks experienced due to the COVID-19 pandemics and the Ukrainian war might also be behind the research boom in this field since 2019. In this context, it is notable that 40% of the population papers were published in the last 4 years. Based on these findings, the size of the population papers is likely to more than double in the current decade, provided that the public concerns about climate change, greenhouse gas emissions, and global warming, as well as the supply shocks are translated efficiently to the research funding in this field.

97.4.5 The Most-Prolific Institutions in the Chinese Experience of Bioethanol Fuels

The most-prolific 31 Chinese institutions publishing papers on the bioethanol fuels with at least 1.3% of the sample papers each given in Table 97.3 have shaped the development of the research in this field.

The most-prolific institutions are the Chinese Academy of Sciences, and to a lesser extent Tsinghua University, South China University of Technology, Tianjin University, Sun Yat-Sen University, Beijing Forestry University, Beijing University of Chemical Technology, Shanghai Jiao Tong University, and Xiamen University.

On the other hand, the institutions with the most citation impact are the Chinese Academy of Sciences, and to a lesser extent Tsinghua University, Sun Yat-Sen University, Xiamen University, Henan Normal University, South China University of Technology, and Beijing Institute of Technology.

Further, the most-prolific overseas collaborating institutions are USDA Forest Service, University of Wisconsin-Madison, National University of Singapore, Oak Ridge National Laboratory, Michigan State University, Bangor University, University of Thessaly, NC State University, Nanyang Technological University of Illinois Urbana-Champaign, NC State University, University of New Brunswick, University of Tennessee Knoxville, SUNY, University of British Columbia, and University of Minnesota Twin Cities. It is notable that most of these overseas institutions are from the USA.

97.4.6 The Most-Prolific Funding Bodies in the Chinese Experience of Bioethanol Fuels

The most-prolific 17 Chinese funding bodies funding at least 1.3% of the sample papers each is given in Table 97.4. It is notable that only 55% and 70% of the sample and population papers were funded, respectively.

The most-prolific funding bodies are the National Natural Science Foundation of China (NNSFC), and to a lesser extent National Key Research and Development Program of China, Chinese Academy of Sciences, National High-tech Research and Development Program, China Postdoctoral Science Foundation, Fundamental Research Funds for the Central Universities, and Ministry of Education of China. It appears that the NNFSC is the key funding body of China (Hu, 2020). Further, a large number of funding bodies have operated at the national and state level to fund the bioethanol research in China.

The funding bodies with the most citation impact are the Chinese Academy of Sciences, National High-Tech Research and Development Program, Natural Science Foundation of Guangdong Province, Ministry of Science and Technology of China, Project 211, and Scientific Research Foundation for Returned Scholars of Ministry of Education. Further, the funding bodies with the least citation impact are the National Science Foundation and Natural Sciences and to a lesser extent Fundamental Research Funds for the Central Universities, National Key Research and Development Program of China, and Priority Academic Program Development of Jiangsu Higher Education Institutions.

On the other hand, the prolific overseas collaborating funding bodies are the Office of Science, U.S. Department of Energy, Great Lakes Bioenergy Research Center, U.S. Forest Service, Basic Energy Sciences, Engineering and Physical Sciences Research Council, European Commission, USDA Forest Service, National Science Foundation, Natural Sciences and Engineering Research Council of Canada, U.S. Department of Agriculture, Alexander von Humboldt-Stiftung, Army Research Office, Ministry of Science and Technology Taiwan, Oak Ridge National Laboratory, and Research Grants Council.

These findings on the funding of the research in this field suggest that the funding rate is exceptionally high, especially higher than the funding rate of 36% and 39% for the sample and population papers, respectively, for the US research (Konur, 2023c). It appears that the extensive funding by China is behind the recent surge of China since 2008 in this field considering the close funding-research output relationships (Ebadi and Schiffauerova, 2016; Hu, 2020) in the light of North's institutional framework (North, 1991). Further, it is expected that this high funding rate would continue in the light of the recent supply shocks.

97.4.7 The Most-Prolific Source Titles in the Chinese Experience of Bioethanol Fuels

The most-prolific 23 source titles publishing at least 1.3% of the sample papers each in the bioethanol fuels have shaped the development of the research in this field (Table 97.5).

The most-prolific source titles are the Bioresource Technology, and to a lesser extent Green Chemistry, Journal of Power Sources, Sensors and Actuators B Chemical, and Electrochimica Acta. On the other hand, the source titles with the most impact are the Green Chemistry, Journal of Power Sources, Bioresource Technology, and to a lesser extent Sensors and Actuators B Chemical, Energy and Environmental Science, Electrochimica Acta, Applied Catalysis B Environmental, ACS Applied Materials and Interfaces, and Advanced Materials.

It is notable that these top source titles are primarily related to the bioresources, energy, sensors, and electrochemistry. This finding suggests that Bioresource Technology and the other prolific journals in these fields have significantly shaped the development of the research in this field as they focus primarily on the production and utilization of bioethanol fuels. In this context, the influence of the top journal is quite extraordinary.

This top journal covers more the technical issues such as waste treatment, biofuels, bioprocesses, bioproducts, and physicochemical and thermochemical processes, excluding the social science- and humanities-based interdisciplinary studies with the only exceptions for the 'circular bioeconomy and energy and environmental sustainability' and 'system analysis and technoeconomics of biofuels and chemicals production'. According to its scope, it thus does not cover other social science-based interdisciplinary studies such as scientometric (Konur, 2012a,b,c,d,e,f,g,h,i), user (Huijts et al., 2012; Wustenhagen et al., 2007), policy (Greening et al., 2000; Hook and Tang, 2013), environmental (Popp et al., 2014; Tilman et al., 2009), and economic (Asafu-Adjaye, 2000; Stern, 1993) studies as in the case of the most journals in the field of energy and fuels excepting social-science oriented journals.

97.4.8 The Most-Prolific Collaborating Countries in the Chinese Experience of Bioethanol Fuels

The most-prolific 13 collaborating countries publishing at least 0.6% of the sample papers each have significantly shaped the development of the research in this field (Table 97.6).

The most-prolific countries are the USA and to a lesser extent the UK, Japan, Greece, Canada, Singapore, and Germany. Further, nine European countries listed in Table 97.6 as a whole produce 12.5% and 4.5% of the sample and population papers, respectively, with 8% surplus. On the other hand, the countries with the most citation impact are the USA and to a lesser extent the UK, Japan, and Greece.

It is notable that the USA and Europe produce 9% and 5% of the population papers, respectively, while they produce 17% and 13% of the sample papers. The USA and Europe are the key competitors of China in the bioethanol research, and they have a first mover advantage in the bioethanol research in relation to China considering the fact that 94% of the population papers were published by China between 2008 and 2022, and there were no publications in the 1970s and 1980s (Lieberman and Montgomery, 1988). On the contrary, the USA has been active in this field since 1970s (Keeney, 2009). As Beaver and Rosen (1978) state the 'collaboration is a typical research style associated with professionalization'. Thus, the collaboration between researchers improves their research output (Miramontes and Gonzalez-Brambila, 2016).

The USA, Europe, China, and to a lesser extent Canada, India, Japan, and Brazil are the major producers of research in this field (Konur, 2023a,b). It is a fact that the USA has been a major player in science (Leydesdorff and Wagner, 2009). The USA has further developed a strong research infrastructure to support its corn and grass-based bioethanol industry.

However, China has been a rising mega star in scientific research in competition with the USA and Europe (Leydesdorff and Zhou, 2005). China is also a major player in this field as a major producer of bioethanol (Fang et al., 2010).

Next, Europe has been a persistent player in the scientific research in competition with both the USA and China (Leydesdorff, 2000). Europe has also been a persistent producer of bioethanol along with the USA and Brazil (Gnansounou, 2010).

Further, Canada (Tahmooresnejad et al., 2015) and Japan (Negishi et al., 2004) are the other countries with substantial research activities in bioethanol fuels.

97.4.9 The Most-Prolific Scopus Subject Categories in the Chinese Experience of Bioethanol Fuels

The most-prolific 11 Scopus subject categories indexing at least 2.2% of the sample papers each, given in Table 97.7, have shaped the development of the Chinese research in this field.

The most-prolific Scopus subject categories in the Chinese experience of bioethanol fuels are Chemical Engineering, Energy, Environmental Science, Chemistry, and to a lesser extent Engineering, Materials Science, Biochemistry, Genetics and Molecular Biology, and Immunology and Microbiology. It is also notable that Social Sciences including Economics and Business have a minimal presence in both sample and population studies.

On the other hand, the Scopus subject categories with the most citation impact are Energy, Environmental Science, and Chemistry. Similarly, the least influential subject categories are Agricultural and Biological Sciences and Biochemistry, Genetics, and Molecular Biology.

These findings are thought-provoking suggesting that the primary subject categories are related to chemical engineering, energy, environmental science, and chemistry as the core of the research in this field concerns with production and utilization of the bioethanol fuels. The other finding is that social sciences are not well represented in both the sample and population papers as in line with the most fields in bioethanol fuels. The social, environmental, and economics studies account for the field of social sciences.

As discussed briefly in Section 97.4.7, the scope of the most journals in this field does not allow for social science- and humanities-based interdisciplinary studies. This development has been in contrast to the interdisciplinarity (Jacobs and Frickel, 2009; Nissani, 1997) of this field in the light of the pressures for increasing incentives for the primary stakeholders (North, 1991).

97.4.10 The Most-Prolific Keywords in the Chinese Experience of Bioethanol Fuels

A limited number of keywords have shaped the development of the research in this field as shown in Table 97.8. These keywords are grouped under the six headings: feedstocks, pretreatments of the biomass, fermentation of the hydrolysates, hydrolysis of the biomass and hydrolysates, utilization of bioethanol fuels, and products.

The most-prolific keywords across all of the research fronts are ethanol, cellulose, lignin, hydrolysis, biomass, biofuels, sugar, enzymatic hydrolysis, ILs, fermentation, enzyme activity, lignocellulose, glucose, pretreatment, catalysts, and solvents. Similarly, the most influential keywords are ethanol, cellulose, sugar, lignin, biofuels, biomass, ILs, lignocellulosic biomass, hydrolysis, pretreatment, solvents, fuel cells, lignocellulose, and electrocatalysts.

These findings suggest that it is necessary to determine the keyword set carefully to locate the relevant research in each of these research fronts. Additionally, the size of the samples for each keyword highlights the intensity of the research in the relevant research areas for both sample and population datasets. These findings also highlight different spelling of some strategic keywords such as pretreatment v. pre-treatment v. treatment and bioethanol v. ethanol v. bio-ethanol V. bioethanol fuels, etc. However, there is a tendency toward the use of the connected keywords without using a hyphen: bioethanol fuels or pretreatment. It is particularly notable that the use of treatment and ethanol instead of pretreatment and bioethanol in the paper titles, respectively, makes the literature search less efficient and time-consuming.

97.4.11 The Most-Prolific Research Fronts in the Chinese Experience of Bioethanol Fuels

Information about the research fronts for the sample papers in the Chinese experience of bioethanol fuels is given in Table 97.9. As this table shows, the most-prolific research front for this field is the bioethanol fuels as feedstocks for the utilization of bioethanol fuels, followed by starch feedstock residues and cellulose. The other prolific research fronts are the lignocellulosic biomass at large, wood biomass, industrial waste, lignin, sugar feedstock residues, biomass at large, algal biomass, grass, and urban wastes, while the other minor research fronts are the hydrolysates, food wastes, forestry wastes, plants, starch feedstocks, cellobiose, lignocellulosic wastes, sugar feedstocks, and xylan.

Thus, the first six research fields have substantial importance, complementing the remaining bioethanol fuel research fields. It is important to note that the research on the production of the first generation bioethanol fuels from sugar and starch feedstocks for bioethanol fuels comprises only 0.9% of the sample papers in total. These first generation bioethanol fuels are not much desirable as they undermine the food security (Makenete et al., 2008; Wu et al., 2012).

On the other hand, bioethanol fuels as feedstocks are over-represented in the sample paper in relation to the global sample. Similarly, starch feedstock residues over-represented in the sample paper in relation to the global sample. On the contrary, lignocellulosic biomass at large are under-represented in the sample paper in relation to the global sample by 14%. Additionally, a large number of feedstocks are under-represented in the sample paper in relation to the global sample.

These findings are thought-provoking in understanding the strength and weaknesses of the Chinese research for the feedstocks used in this field. It appears that China differs largely from the USA and the rest of the world in using bioethanol fuels as feedstocks for their utilization in engines, fuel cells, biochemicals, biohydrogen fuels, as well as developing bioethanol sensors (Konur, 2023c).

Table 97.10 shows that the most-prolific thematic research front is the production of bioethanol fuels, utilization of bioethanol fuels, and to a lesser extent the evaluation of bioethanol fuels. On the other hand, the most-prolific research front for the production of bioethanol fuels is the biomass pretreatment and hydrolysis and to a lesser extent bioethanol fuel production, hydrolysate fermentation, and separation and distillation of bioethanol fuels.

Further, the most-prolific research fronts for the utilization of bioethanol fuels are the utilization of bioethanol fuels in fuel cells and bioethanol sensors and to a lesser extent utilization of bioethanol fuels in the transport engines, bioethanol-based biohydrogen fuels, and bioethanol-based biochemical.

On the other hand, the field of the utilization of bioethanol fuels is over-represented in the Chinese sample in relation to the global sample. On the contrary, the fields of evaluation and production of bioethanol fuels are under-represented in the Chinese sample in relation to the global sample.

Further, on the individual terms, the fields of bioethanol sensors, bioethanol fuels in fuel cells, bioethanol-based biochemicals, and bioethanol fuel distillation are over-represented in the Chinese sample in relation to the global sample. On the contrary, the fields of bioethanol production, biomass pretreatments, hydrolysate fermentation, and biomass hydrolysis are under-represented in the Chinese sample in relation to the global sample.

These findings are thought-provoking in understanding the strength and weaknesses of the Chinese research for the production, evaluation, and utilization of bioethanol in this field. It appears that China differs largely from the USA and the rest of the world in utilizing bioethanol fuels in engines, fuel cells, biochemicals, biohydrogen fuels, as well as developing bioethanol sensors (Konur, 2023c). China is also under-weighted in the production and evaluation of bioethanol fuels in relation to the USA and the rest of the world.

It is clear that all of these research fronts have public importance and merit substantial funding and other incentives. Further, it is notable that the utilization and evaluation of bioethanol fuels have also become a core unit of the bioethanol research, besides the production of bioethanol fuels to make it more competitive with the crude oil-based gasoline and diesel fuels, especially for the USA, Europe and China.

It is notable that the pretreatment and hydrolysis of the feedstocks emerge as primary research fronts for the field of the production of bioethanol fuels. These processes are required to improve the ethanol yield. However, the research fronts of the fermentation of the feedstock-based hydrolysates and the bioethanol production from the feedstock-based hydrolysates are also important. However, the research in the field of separation and distillation of bioethanol fuels is negligible.

Further, the field of the evaluation of bioethanol fuels is also a neglected area. This suggests that the primary stakeholders have been primarily interested in these key processes of the bioethanol production. It is also notable that evaluation of the bioethanol fuels such as techno-economics, life cycle, economics, social, land use, labor, and environment-related studies emerges as a case study for the bioethanol fuels.

In the end, these most-cited papers in this field hint that the production and utilization of bioethanol fuels could be optimized using the structure, processing, and property relationships of feedstocks and bioethanol fuels in the fronts of the feedstock pretreatment and hydrolysis, hydrolysate fermentation, and the utilization of bioethanol fuels and their derivatives, and coproducts (Formela et al., 2016; Konur, 2018a, 2020b, 2021a,b,c,d; Konur and Matthews, 1989).

97.5 CONCLUSION AND FUTURE RESEARCH

The research on the Chinese experience of bioethanol fuels has been mapped through a scientometric study of both sample (318 papers) and population (15,913 papers) datasets.

The critical issue in this study has been to obtain a representative sample of the research as in any other scientometric study. Therefore, the keyword set has been carefully devised and optimized after a number of runs in the Scopus database. It is a representative sample of the wider population studies. This keyword set was provided in the appendix of the related studies, and the relevant keywords are presented in Table 97.8. However, it should be noted that it has been very difficult to compile a representative keyword set since this research field has been connected closely with many other fields. Therefore, it has been necessary to compile a keyword list to exclude papers concerned with the other research fields.

The other issue has been the selection of a multidisciplinary database to carry out the scientometric study of the research in this field. For this purpose, Scopus database has been selected. The journal coverage of this database has been notably wider than that of Web of Science and other multisubject databases.

The key scientometric properties of the research in this field have been determined and discussed in this book chapter. It is evident that a limited number of documents, authors, institutions, publication years, institutions, funding bodies, source titles, countries, Scopus subject categories, Scopus keywords, and research fronts have shaped the development of the research in this field.

There is ample scope to increase the efficiency of the scientometric studies in this field in the author and document domains by developing consistent policies and practices in both domains across all the academic databases. In this respect, it seems that authors, journals, and academic databases have a lot to do. Furthermore, the significant gender deficit as in most scientific fields emerges as a public policy issue. The potential deficits on the basis of age, race, disability, and sexuality need also to be explored in this field as in other scientific fields.

The research in this field has boomed since 2008 and 2005 for the population and sample papers, respectively, possibly promoted by the public concerns on global warming, greenhouse gas emissions, and climate change as China is the largest polluter of the environment. Furthermore, the recent COVID-19 pandemics and Russian invasion of Ukraine have resulted in a global supply shocks shifting the recent focus of the stakeholders from the crude oil-based fuels to biomass-based fuels such as bioethanol fuels. It is expected that there would be further incentives for the key stakeholders to carry out the research for the bioethanol fuels to increase the ethanol yield and their utilization and to make it more competitive with the crude oil-based gasoline and petrodiesel fuels. This might be truer for the crude oil-deficient countries like China to maintain the energy and food security at the face of the global supply shocks.

The relatively high funding rates of 55% and 70% for the sample and population papers, respectively (comparable funding rates for the US sample are 36% and 39%), suggest that funding in this field significantly enhanced the research in this field primarily since 2008, possibly more than doubling in the current decade. However, it is evident that there is ample room for more funding and other incentives to enhance the research in this field further.

It is recommended that the term ‘bioethanol fuels’ are used instead of ethanol or bio-ethanol in the future. On the other hand, the Scopus keywords are grouped under the six headings: biomass, pretreatments, fermentation, hydrolysis and hydrolysates, bioethanol utilization, and products. These prolific keywords highlight the major fields of the research in this field for both sample and population papers.

Table 97.9 shows that the most-prolific biomass used for this field are the bioethanol fuels for their utilization, cellulose, starch feedstock residues, lignocellulosic biomass at large, wood biomass, and to a lesser extent industrial wastes, lignin, grass biomass, hydrolysates, biomass at large, sugar feedstocks, sugar feedstock residues, algal biomass, lignocellulosic wastes at large, starch feedstocks, urban wastes, forestry wastes, hemicellulose, xylan, food wastes, cellobiose, and plants.

It is important to note that the research on the production of the first generation bioethanol fuels from sugar and starch feedstocks for bioethanol fuels comprises only a small part of the sample papers in total. These first generation bioethanol fuels are not much desirable as they undermine the food security.

Further, Table 97.10 shows that the most-prolific thematic research fronts are the production of bioethanol fuels, utilization of bioethanol fuels, and to a lesser extent evaluation of bioethanol fuels. On the other hand, the most-prolific research fronts for the production of bioethanol fuels are the biomass pretreatment, biomass hydrolysis and to a lesser extent bioethanol production, hydrolysate fermentation, and separation and distillation of bioethanol fuels. Further, the most-prolific research fronts for the utilization of bioethanol fuels are the utilization of bioethanol fuels in fuel cells, bioethanol sensors, utilization of bioethanol fuels and gasoline and diesel engines, and to a lesser extent bioethanol-based biohydrogen fuels and bioethanol fuel-based biochemicals.

It appears that China has a unique strength in the research front of the utilization of bioethanol fuels compared to the USA and the rest of world at large, while it has a relative weakness in the production and evaluation of bioethanol fuels.

In this context, it is notable that there is ample room for the improvement of the research on social and humanitarian aspects of the research on the Chinese experience of bioethanol fuels. As discussed briefly in Section 97.4.7, the scope of most journals in this field excludes social science- and humanities-based interdisciplinary studies. This development has been in contrast to the interdisciplinarity of this field in the light of the pressures for increasing incentives for the primary stakeholders. Thus, for the healthy development of this research field, social science- and humanities-based interdisciplinary disciplines have a lot to contribute to the bioethanol fuel research.

Thus, the scientometric analysis has a great potential to gain valuable insights into the evolution of the Chinese research in this field as in other scientific fields especially in the aftermath of the significant global supply shocks such as COVID-19 pandemics and the Russian invasion of Ukraine, considering the fact that 40% of the population papers were published in the last 4 years.

It is recommended that further scientometric studies are carried out for the primary research fronts. It is further recommended that reviews of the most-cited papers are carried out for each primary research front to complement these scientometric studies. Next, the scientometric studies of the hot papers in these primary fields are carried out.

ACKNOWLEDGMENTS

The contribution of the highly cited researchers in the field of the Chinese experience of bioethanol fuels has been gratefully acknowledged.

REFERENCES

Alvira, P., E. Tomas-Pejo, M. Ballesteros and M. J. Negro. 2010. Pretreatment technologies for an efficient bioethanol production process based on enzymatic hydrolysis: A review. *Bioresource Technology* 101:4851–4861.

Aman, V. 2018. Does the Scopus author ID suffice to track scientific international mobility? A case study based on Leibniz laureates. *Scientometrics* 117:705–720.

Angelici, C., B. M. Weckhuysen and P. C. A. Bruijnincx. 2013. Chemocatalytic conversion of ethanol into butadiene and other bulk chemicals. *ChemSusChem* 6:1595–1614.

Antolini, E. 2007. Catalysts for direct ethanol fuel cells. *Journal of Power Sources* 170:1–12.

Antolini, E. 2009. Palladium in fuel cell catalysis. *Energy and Environmental Science* 2:915–931.

Asafu-Adjaye, J. 2000. The relationship between energy consumption, energy prices and economic growth: Time series evidence from Asian developing countries. *Energy Economics* 22:615–625.

Bai, F. W., W. A. Anderson and M. Moo-Young. 2008. Ethanol fermentation technologies from sugar and starch feedstocks. *Biotechnology Advances* 26:89–105.

Beaudry, C. and V. Lariviere. 2016. Which gender gap? Factors affecting researchers' scientific impact in science and medicine. *Research Policy* 45:1790–1817.

Beaver, D. and R. Rosen. 1978. Studies in scientific collaboration: Part I. The professional origins of scientific co-authorship. *Scientometrics*, 1:65–84.

Binod, P., R. Sindhu and R. R. Singhania, et al. 2010. Bioethanol production from rice straw: An overview. *Bioresource Technology* 101:4767–4774.

Blankenship, K. M. 1993. Bringing gender and race in: US employment discrimination policy. *Gender & Society* 7:204–226.

Bourbonnais, R. and M. G. Paice. 1990. Oxidation of non-phenolic substrates. An expanded role for laccase in lignin biodegradation. *FEBS Letters* 267:99–102.

Burnham, J. F. 2006. Scopus database: A review. *Biomedical Digital Libraries* 3:1–8.

Cardona, C. A., J. A. Quintero and I. C. Paz. 2010. Production of bioethanol from sugarcane bagasse: Status and perspectives. *Bioresource Technology* 101:4754–4766.

Carlson, K. M., J. S. Gerber and D. Mueller, et al. 2017. Greenhouse gas emissions intensity of global croplands. *Nature Climate Change* 7:63–68.

Change, C. 2007. Climate change impacts, adaptation and vulnerability. *Science of the Total Environment* 326:95–112.

Dirth, T. P. and N. R. Branscombe. 2017. Disability models affect disability policy support through awareness of structural discrimination. *Journal of Social Issues* 73:413–442.

Duff, S. J. B. and W. D. Murray. 1996. Bioconversion of forest products industry waste cellulosics to fuel ethanol: A review. *Bioresource Technology* 55:1–33.

Ebadi, A. and A. Schiffauerova. 2016. How to boost scientific production? A statistical analysis of research funding and other influencing factors. *Scientometrics* 106:1093–1116.

Fang, X., Y. Shen, J. Zhao, X. Bao and Y. Qu. 2010. Status and prospect of lignocellulosic bioethanol production in China. *Bioresource Technology* 101:4814–4819.

Fauci, A. S., H. C. Lane and R. R. Redfield. 2020. Covid-19-navigating the uncharted. *New England Journal of Medicine* 382:1268–1269.

Fernando, S., S. Adhikari, C. Chandrapal and M. Murali. 2006. Biorefineries: Current status, challenges, and future direction. *Energy & Fuels* 20:1727–1737.

Formela, K., A. Hejna, L. Piszczyk, M. R. Saeb and X. Colom. 2016. Processing and structure-property relationships of natural rubber/wheat bran biocomposites. *Cellulose* 23:3157–3175.

Galbe, M. and G. Zacchi. 2002. A review of the production of ethanol from softwood. *Applied Microbiology and Biotechnology* 59:618–628.

Garfield, E. 1955. Citation indexes for science. *Science* 122:108–111.

Gnansounou, E. 2010. Production and use of lignocellulosic bioethanol in Europe: Current situation and perspectives. *Bioresource Technology* 101:4842–4850.

Greening, L. A., D. L. Greene and C. Difiglio. 2000. Energy efficiency and consumption-the rebound effect-a survey. *Energy Policy* 28:389–401.

Hahn-Hagerdal, B., M. Galbe, M. F. Gorwa-Grauslund, G. Liden and G. Zacchi. 2006. Bio-ethanol - The fuel of tomorrow from the residues of today. *Trends in Biotechnology* 24:549–556.

Hamelinck, C. N., G. van Hooijdonk and A. P. C. Faaij. 2005. Ethanol from lignocellulosic biomass: Techno-economic performance in short-, middle- and long-term. *Biomass and Bioenergy*, 28:384–410.

Hamilton, J. D. 1983. Oil and the macroeconomy since World War II. *Journal of Political Economy* 91:228–248.

Hamilton, J. D. 2003. What is an oil shock? *Journal of Econometrics* 113:363–398.

Hansen, A. C., Q. Zhang and P. W. L. Lyne. 2005. Ethanol-diesel fuel blends - A review. *Bioresource Technology* 96:277–285.

Hendriks, A. T. W. M. and G. Zeeman. 2009. Pretreatments to enhance the digestibility of lignocellulosic biomass. *Bioresource Technology* 100:10–18.

Henstra, A. M., J. Sipma, A. Rinzema and A. J. Stams. 2007. Microbiology of synthesis gas fermentation for biofuel production. *Current Opinion in Biotechnology* 18:200–206.

Hill, J., E. Nelson, D. Tilman, S. Polasky and D. Tiffany. 2006. Environmental, economic, and energetic costs and benefits of biodiesel and ethanol biofuels. *Proceedings of the National Academy of Sciences of the United States of America* 103:11206–11210.

Hill, J., S. Polasky and E. Nelson, et al. 2009. Climate change and health costs of air emissions from biofuels and gasoline. *Proceedings of the National Academy of Sciences of the United States of America* 106:2077–2082.

Hook, M. and X. Tang. 2013. Depletion of fossil fuels and anthropogenic climate change-A review. *Energy Policy* 52:797–809.

Hsieh, W. D., R. H. Chen, T. L. Wu and T. H. Lin. 2002. Engine performance and pollutant emission of an SI engine using ethanol-gasoline blended fuels. *Atmospheric Environment* 36:403–410.

Hu, A. G. 2020. Public funding and the ascent of Chinese science: Evidence from the National Natural Science Foundation of China. *Research Policy* 49:103983.

Huang, H. J., S. Ramaswamy, U. W. Tschirner and B. V. Ramarao. 2008. A review of separation technologies in current and future biorefineries. *Separation and Purification Technology* 62:1–21.

Huijts, N. M. A., E. J. E. Molin and L. Steg. 2012. Psychological factors influencing sustainable energy technology acceptance: A review-based comprehensive framework. *Renewable and Sustainable Energy Reviews* 16:525–531.

Jacobs, J. A. and S. Frickel. 2009. Interdisciplinarity: A critical assessment. *Annual Review of Sociology* 33:43–65.

John, R. P., G. S. Anisha, K. M. Nampoothiri and A. Pandey. 2011. Micro and macroalgal biomass: A renewable source for bioethanol. *Bioresource Technology* 102:186–193.

Jones, T. C. 2012. America, oil, and war in the Middle East. *Journal of American History* 99:208–218.

Jonsson, L. J. and C. Martin. 2016. Pretreatment of lignocellulose: Formation of inhibitory by-products and strategies for minimizing their effects. *Bioresource Technology* 199:103–112.

Keeney, D. 2009. Ethanol USA. *Environmental Science & Technology* 43:8–11.

Kerr, R. A. 2007. Global warming is changing the world. *Science* 316:188–190.

Kilian, L. 2008. Exogenous oil supply shocks: How big are they and how much do they matter for the US economy? *Review of Economics and Statistics* 90:216–240.

Kilian, L. 2009. Not all oil price shocks are alike: Disentangling demand and supply shocks in the crude oil market. *American Economic Review*, 99:1053–1069.

Kim, J. 2018. Evaluating author name disambiguation for digital libraries: A case of DBLP. *Scientometrics* 116:1867–1886.

Konur, O. 2000. Creating enforceable civil rights for disabled students in higher education: An institutional theory perspective. *Disability & Society* 15:1041–1063.

Konur, O. 2002a. Access to nursing education by disabled students: Rights and duties of nursing programs. *Nurse Education Today* 22:364–374.

Konur, O. 2002b. Assessment of disabled students in higher education: Current public policy issues. *Assessment and Evaluation in Higher Education* 27:131–152.

Konur, O. 2002c. Access to employment by disabled people in the UK: Is the Disability Discrimination Act working? *International Journal of Discrimination and the Law* 5:247–279.

Konur, O. 2006a. Participation of children with dyslexia in compulsory education: Current public policy issues. *Dyslexia* 12:51–67.

Konur, O. 2006b. Teaching disabled students in higher education. *Teaching in Higher Education* 11:351–363.

Konur, O. 2007a. A judicial outcome analysis of the *Disability Discrimination Act*: A windfall for the employers? *Disability & Society* 22:187–204.

Konur, O. 2007b. Computer-assisted teaching and assessment of disabled students in higher education: The interface between academic standards and disability rights. *Journal of Computer Assisted Learning* 23:207–219.

Konur, O. 2011. The scientometric evaluation of the research on the algae and bio-energy. *Applied Energy* 88:3532–3540.

Konur, O. 2012a. The evaluation of the biogas research: A scientometric approach. *Energy Education Science and Technology Part A: Energy Science and Research* 29:1277–1292.

Konur, O. 2012b. The evaluation of the educational research: A scientometric approach. *Energy Education Science and Technology Part B: Social and Educational Studies* 4:1935–1948.

Konur, O. 2012c. The evaluation of the global energy and fuels research: A scientometric approach. *Energy Education Science and Technology Part A: Energy Science and Research* 30:613–628.

Konur, O. 2012d. The evaluation of the research on the biodiesel: A scientometric approach. *Energy Education Science and Technology Part A: Energy Science and Research* 28:1003–1014.

Konur, O. 2012e. The evaluation of the research on the bioethanol: A scientometric approach. *Energy Education Science and Technology Part A: Energy Science and Research* 28:1051–1064.

Konur, O. 2012f. The evaluation of the research on the biofuels: A scientometric approach. *Energy Education Science and Technology Part A: Energy Science and Research* 28:903–916.

Konur, O. 2012g. The evaluation of the research on the biohydrogen: A scientometric approach. *Energy Education Science and Technology Part A: Energy Science and Research* 29:323–338.

Konur, O. 2012h. The evaluation of the research on the microbial fuel cells: A scientometric approach. *Energy Education Science and Technology Part A: Energy Science and Research* 29:309–322.

Konur, O. 2012i. The scientometric evaluation of the research on the production of bioenergy from biomass. *Biomass and Bioenergy* 47:504–515.

Konur, O. 2015. Current state of research on algal bioethanol. In *Marine Bioenergy: Trends and Developments*, Ed. S. K. Kim and C. G. Lee, pp. 217–244. Boca Raton, FL: CRC Press.

Konur, O., Ed. 2018a. *Bioenergy and Biofuels*. Boca Raton, FL: CRC Press.

Konur, O. 2018b. Bioenergy and biofuels science and technology: Scientometric overview and citation classics. In *Bioenergy and Biofuels*, Ed. O. Konur, pp. 3–63. Boca Raton: CRC Press.

Konur, O. 2019. Cyanobacterial bioenergy and biofuels science and technology: A scientometric overview. In *Cyanobacteria: From Basic Science to Applications*, Ed. A. K. Mishra, D. N. Tiwari and A. N. Rai, pp. 419–442. Amsterdam: Elsevier.

Konur, O. 2020a. The scientometric analysis of the research on the bioethanol production from green macroalgae. In *Handbook of Algal Science, Technology and Medicine*, Ed. O. Konur, pp. 385–401. London: Academic Press.

Konur, O., Ed. 2020b. *Handbook of Algal Science, Technology and Medicine*. London: Academic Press.

Konur, O., Ed. 2021a. *Handbook of Biodiesel and Petrodiesel Fuels: Science, Technology, Health, and Environment*. Boca Raton, FL: CRC Press.

Konur, O., Ed. 2021b. *Handbook of Biodiesel and Petrodiesel Fuels: Science, Technology, Health, and Environment. Volume 1. Biodiesel Fuels: Science, Technology, Health, and Environment*. Boca Raton, FL: CRC Press.

Konur, O., Ed. 2021c. *Handbook of Biodiesel and Petrodiesel Fuels: Science, Technology, Health, and Environment. Volume 2. Biodiesel Fuels based on the Edible and Nonedible Feedstocks, Wastes, and Algae: Science, Technology, Health, and Environment*. Boca Raton, FL: CRC Press.

Konur, O., Ed. 2021d. *Handbook of Biodiesel and Petrodiesel Fuels: Science, Technology, Health, and Environment. Volume 3. Petrodiesel Fuels: Science, Technology, Health, and Environment*. Boca Raton, FL: CRC Press.

Konur, O. 2023a. Bioethanol fuels: Scientometric study. In *Bioethanol Fuel Production Processes. I: Biomass Pretreatments. Handbook of Bioethanol Fuels Volume 1*, Ed. O. Konur, pp. 47–76. Boca Raton, FL: CRC Press.

Konur, O. 2023b. Bioethanol fuels: Review. In *Bioethanol Fuel Production Processes. I: Biomass Pretreatments. Handbook of Bioethanol Fuels Volume 1*, Ed. O. Konur, pp. 77–98. Boca Raton, FL: CRC Press.

Konur, O. 2023c. The US experience of bioethanol fuels: Scientometric study. In *Evaluation and Utilization of Bioethanol Fuels. II.: Biohydrogen Fuels, Fuel Cells, Biochemicals, and Country Experiences. Handbook of Bioethanol Fuels Volume 6*, Ed. O. Konur, pp. 110–138. Boca Raton, FL: CRC Press.

Konur, O. and F. L. Matthews. 1989. Effect of the properties of the constituents on the fatigue performance of composites: A review. *Composites* 20:317–328.

Kruyt, B., D. P. van Vuuren, H. J. de Vries and H. Groenenberg. 2009. Indicators for energy security. *Energy Policy* 37:2166–2181.

Leydesdorff, L. 2000. Is the European Union becoming a single publication system? *Scientometrics* 47:265–280.

Leydesdorff, L. and C. Wagner. 2009. Is the United States losing ground in science? A global perspective on the world science system. *Scientometrics* 78:23–36.

Leydesdorff, L. and P. Zhou. 2005. Are the contributions of China and Korea upsetting the world system of science? *Scientometrics* 63:617–630.

Li, H., S. M. Liu, X. H. Yu, S. L. Tang and C. K. Tang. 2020. Coronavirus disease 2019 (COVID-19): Current status and future perspectives. *International Journal of Antimicrobial Agents* 55:105951.

Lieberman, M. B. and D. B. Montgomery. 1988. First-mover advantages. *Strategic Management Journal* 9:41–58.

Limayem, A. and S. C. Ricke. 2012. Lignocellulosic biomass for bioethanol production: Current perspectives, potential issues and future prospects. *Progress in Energy and Combustion Science*, 38:449–467.

Lin, Y. and S. Tanaka. 2006. Ethanol fermentation from biomass resources: Current state and prospects. *Applied Microbiology and Biotechnology* 69:627–642.

Ma, X., L. Sun and C. Song. 2002. A new approach to deep desulfurization of gasoline, diesel fuel and jet fuel by selective adsorption for ultra-clean fuels and for fuel cell applications. *Catalysis Today* 77:107–116.

Makenete, A. L., W. J. Lemmer and J. Kupka. 2008. The impact of biofuel production on food security: A briefing paper with a particular emphasis on maize-to-ethanol production. *International Food and Agribusiness Management Review* 11:101–110.

Miramontes, J. R. and C. N. Gonzalez-Brambila. 2016. The effects of external collaboration on research output in engineering. *Scientometrics* 109:661–675.

Morschbacker, A. 2009. Bio-ethanol based ethylene. *Polymer Reviews* 49:79–84.

Najafi, G., B. Ghobadian and T. Tavakoli, et al. 2009. Performance and exhaust emissions of a gasoline engine with ethanol blended gasoline fuels using artificial neural network. *Applied Energy* 86:630–639.

Negishi, M., Y. Sun and K. Shigi. 2004. Citation database for Japanese papers: A new bibliometric tool for Japanese academic society. *Scientometrics* 60:333–351.

Newman, P. W. G. and J. R. Kenworthy. 1989. Gasoline consumption and cities: A comparison of U.S. cities with a global survey. *Journal of the American Planning Association* 55:24–37.

Nissani, M. 1997. Ten cheers for interdisciplinarity: The case for interdisciplinary knowledge and research. *Social Science Journal* 34:201–216.

North, D. C. 1991. Institutions. *Journal of Economic Perspectives* 5:97–112.

Olsson, L. and B. Hahn-Hagerdal. 1996. Fermentation of lignocellulosic hydrolysates for ethanol production. *Enzyme and Microbial Technology* 18:312–331.

Palmqvist, E. and B. Hahn-Hagerdal. 2000. Fermentation of lignocellulosic hydrolysates. I: Inhibition and detoxification. *Bioresource Technology* 74:17–24.

Pimentel, D. and T. W. Patzek. 2005. Ethanol production using corn, switchgrass, and wood; biodiesel production using soybean and sunflower. *Natural Resources Research* 14:65–76.

Pinkert, A., K. N. Marsh, S. Pang and M. P. Staiger. 2009. Ionic liquids and their interaction with cellulose. *Chemical Reviews* 109:6712–6728.

Popp, J., Z. Lakner, M. Harangi-Rakos and M. Fari. 2014. The effect of bioenergy expansion: Food, energy, and environment. *Renewable and Sustainable Energy Reviews* 32:559–578.

Ravindran, R. and A. K. Jaiswal. 2016. A comprehensive review on pre-treatment strategy for lignocellulosic food industry waste: Challenges and opportunities. *Bioresource Technology* 199:92–102.

Reeves, S. 2014. To Russia with love: How moral arguments for a humanitarian intervention in Syria opened the door for an invasion of the Ukraine. *Michigan State University International Law Review* 23:199.

Sanchez, C. 2009. Lignocellulosic residues: Biodegradation and bioconversion by fungi. *Biotechnology Advances* 27:185–194.

Sanchez, O. J. and C. A. Cardona. 2008. Trends in biotechnological production of fuel ethanol from different feedstocks. *Bioresource Technology* 99:5270–5295.

Sano, T., H. Yanagishita, Y. Kiyozumi, F. Mizukami and K. Haraya. 1994. Separation of ethanol/water mixture by silicalite membrane on pervaporation. *Journal of Membrane Science* 95:221–228.

Stern, D. I. 1993. Energy and economic growth in the USA: A multivariate approach. *Energy Economics* 15:137–150.

Sukumaran, R. K., R. R. Singhania, G. M. Mathew and A. Pandey. 2009. Cellulase production using biomass feed stock and its application in lignocellulose saccharification for bio-ethanol production. *Renewable Energy* 34:421–424.

Sun, Y. and J. Cheng. 2002. Hydrolysis of lignocellulosic materials for ethanol production: A review. *Bioresource Technology* 83:1–11.

Taherzadeh, M. J. and K. Karimi. 2007. Enzyme-based hydrolysis processes for ethanol from lignocellulosic materials: A review. *Bioresources* 2:707–738.

Taherzadeh, M. J. and K. Karimi. 2008. Pretreatment of lignocellulosic wastes to improve ethanol and biogas production: A review. *International Journal of Molecular Sciences* 9:1621–1651.

Tahmooresnejad, L., C. Beaudry and A. Schiffauerova. 2015. The role of public funding in nanotechnology scientific production: Where Canada stands in comparison to the United States. *Scientometrics* 102:753–787.

Talebnia, F., D. Karakashev and I. Angelidaki. 2010. Production of bioethanol from wheat straw: An overview on pretreatment, hydrolysis and fermentation. *Bioresource Technology* 101:4744–4753.

Tilman, D., R. Socolow and J. A. Foley, et al. 2009. Beneficial biofuels-the food, energy, and environment trilemma. *Science* 325:270–271.

Vane, L. M. 2005. A review of pervaporation for product recovery from biomass fermentation processes. *Journal of Chemical Technology and Biotechnology* 80:603–629.

Winzer, C. 2012. Conceptualizing energy security. *Energy Policy* 46:36–48.

Wu, F., D. Zhang and J. Zhang. 2012. Will the development of bioenergy in China create a food security problem? Modeling with fuel ethanol as an example. *Renewable Energy* 47:127–134.

Wustenhagen, R., M. Wolsink and M. J. Burer. 2007. Social acceptance of renewable energy innovation: An introduction to the concept. *Energy Policy* 35:2683–2691.

Yang, B. and C. E. Wyman. 2008. Pretreatment: The key to unlocking low-cost cellulosic ethanol. *Biofuels, Bioproducts and Biorefining* 2:26–40.

Zhang, Y. H. P. and L. R. Lynd. 2004. Toward an aggregated understanding of enzymatic hydrolysis of cellulose: Noncomplexed cellulase systems. *Biotechnology and Bioengineering* 88:797–824.

Zhu, J. Y. and X. J. Pan. 2010. Woody biomass pretreatment for cellulosic ethanol production: Technology and energy consumption evaluation. *Bioresource Technology* 101:4992–5002.

98 The Chinese Experience of Bioethanol Fuels
Review

Ozcan Konur
Ankara Yildirim Beyazit University

98.1 INTRODUCTION

Crude oil-based gasoline fuels (Ma et al., 2002; Newman and Kenworthy, 1989) have been widely used in the transportation sector since the 1920s. However, there have been great public concerns over the adverse environmental and human impact of these fuels (Hill et al., 2006, 2009). Hence, biomass-based bioethanol fuels (Hill et al., 2006; Konur, 2012, 2015, 2019, 2020) have increasingly been used in blending gasoline fuels (Hsieh et al., 2002; Najafi et al., 2009), in fuel cells (Antolini, 2007, 2009), and in the biochemical (Angelici et al., 2013; Morschbacker, 2009) and biohydrogen fuel production (Haryanto et al., 2005; Murdoch et al., 2011) in a biorefinery context (Fernando et al., 2006; Huang et al., 2008).

However, it is necessary to pretreat biomass (Alvira et al., 2010; Taherzadeh and Karimi, 2008) to enhance the yield of the bioethanol (Hahn-Hagerdal et al., 2006; Sanchez and Cardona, 2008) prior to the bioethanol fuel production from the feedstocks through the hydrolysis (Sun and Cheng, 2002; Taherzadeh and Karimi, 2007) and fermentation (Lin and Tanaka, 2006; Olsson and Hahn-Hagerdal, 1996) of biomass.

China has been one of the most prolific countries engaged in bioethanol fuel research (Konur, 2023a,b). The Chinese research in the field of bioethanol fuels has intensified in this context in the key research fronts of the production of bioethanol fuels, utilization of bioethanol fuels (Antolini, 2007; Hansen et al., 2005), and to a lesser extent evaluation (Hamelinck et al., 2005; Pimentel and Patzek, 2005) of bioethanol fuels. For the first research front, pretreatment (Alvira et al., 2010; Hendriks and Zeeman, 2009) and hydrolysis (Alvira et al., 2010; Sun and Cheng, 2002) of the feedstocks, fermentation of the feedstock-based hydrolysates (Jonsson and Martin, 2016; Lin and Tanaka, 2006), production of bioethanol fuels (Limayem and Ricke, 2012, Lin and Tanaka, 2006), and to a lesser extent separation and distillation of bioethanol fuels (Sano et al., 1994; Vane, 2005) are the key research areas while for the second research front, utilization of bioethanol fuels in fuel cells (Antolini, 2007, 2009) and transport engines (Hsieh et al., 2002; Najafi et al., 2009), bioethanol-based biohydrogen fuels (Haryanto et al., 2005; Murdoch et al., 2011), bioethanol sensors (Liu et al., 2005; Wan et al., 2004), and bioethanol-based biochemicals (Angelici et al., 2013; Morschbacker, 2009) are the key research areas.

The research in this field has also intensified for the feedstocks of lignocellulosic biomass at large (Hendriks and Zeeman, 2009; Sun and Cheng, 2002), bioethanol fuels for their utilization (Antolini, 2007; Hansen et al., 2005), cellulose (Pinkert et al., 2009; Zhang and Lynd, 2004), starch feedstock residues (Binod et al., 2010; Talebnia et al., 2010), wood biomass (Galbe and Zacchi, 2002; Zhu and Pan, 2010), and to a lesser extent industrial waste (Cardona et al., 2010), lignin (Bourbonnais and Paice, 1990), grass biomass (Pimentel and Patzek, 2005), hydrolysates (Palmqvist and Hahn-Hagerdal, 2000), biomass at large (Lin and Tanaka, 2006), sugar feedstocks (Bai et al., 2008), sugar feedstock residues (Cardona et al., 2010), algal biomass (John et al., 2011), lignocellulosic wastes

DOI: 10.1201/9781003226574-131

(Sanchez, 2009), starch feedstocks (Bai et al., 2008), urban wastes (Ravindran and Jaiswal, 2016), forestry wastes (Duff and Murray, 1996), food wastes (Ravindran and Jaiswal, 2016), biosyngas (Henstra et al., 2007), and plants (Sukumaran et al., 2009).

However, it is essential to develop efficient incentive structures (North, 1991) for the primary stakeholders to enhance the research in this field (Konur, 2000, 2002a,b,c, 2006a,b, 2007a,b). Although there has been a number of review papers on bioethanol fuels (Hendriks and Zeeman, 2009; Mosier et al., 2005; Sun and Cheng, 2002), there has been no review of the most cited 25 papers by Chinese researchers in this field.

Thus, this book chapter presents a review of the most-cited 25 Chinese articles in the field of bioethanol fuels. Then, it discusses the key findings of these highly influential papers and comments on future research priorities in this field.

98.2 MATERIALS AND METHODS

The search for this study was carried out using the Scopus database (Burnham, 2006) in November 2022.

As a first step for the search of the relevant literature, the keywords were selected using the most-cited first 300 population papers for each research front. The selected keyword list was then optimized to obtain a representative sample of papers for each research field. These keyword lists were then integrated to obtain the keyword list for this research field (Konur (2023a,b). The keyword set for China is 'AFFILCOUNTRY (chin* OR "hong kong" OR macau) OR TITLE (chin*) OR AUTHKEY (chin*) OR SRCTITLE (chin*)' in addition to the keyword list for bioethanol fuels (Konur (2023a,b).

As a second step, a sample dataset was used for this study. The first 25 articles with at least 306 citations each were selected for the review study. Key findings from each paper were taken from the abstracts of these papers and were briefly discussed. Additionally, a number of brief conclusions were drawn and a number of relevant recommendations were made to enhance the future research landscape.

98.3 RESULTS

The brief information about the 25 most-cited papers with at least 306 citations each on the bioethanol fuels produced by China is given below. The primary research fronts are the production of bioethanol fuels and utilization of bioethanol fuels with 7 and 18 highly cited papers (HCPs), respectively. The secondary research fronts for the production of bioethanol fuels are the pretreatments and hydrolysis of biomass with three and four HCPs, respectively. On the other hand, the secondary research fronts for the utilization of bioethanol fuels are the utilization of bioethanol fuels in fuel cells and transport engines, development of the bioethanol sensors, and the utilization of bioethanol fuels for the production of biohydrogen fuels with 10, 4, 3, and 1 HCPs, respectively.

98.3.1 Bioethanol Fuel Production

The secondary research fronts for the production of bioethanol fuels are the pretreatments and hydrolysis of biomass with three and four HCPs, respectively. However, there are no HCPs for the research fronts of the fermentation of the hydrolysates, production of the bioethanol fuels, and separation and distillation of the bioethanol fuels, produced by the Chinese researchers.

98.3.1.1 Feedstock Pretreatments for Bioethanol Fuel Production

The brief information about the three most-cited papers on the pretreatments of feedstocks with at least 329 citations each is given in Table 98.1.

TABLE 98.1
The Pretreatment of Feedstocks for the Bioethanol Production: Chinese Research

No.	Papers	Biomass	Prts.	Parameters	Keywords	Lead Authors	Affil.	Cits
1	Cai and Zhang (2005)	Cellulose	IL	Cellulose dissolution, alkaline solvents, cellulose dissolution behavior, solubility, hydrogen bonding	Cellulose, dissolution	Zhang, Lina* 55917992100	Wuhan Univ. China	761
2	Sun et al. (2005)	Wheat straw	Steam, H_2O2	Cellulose yield, delignification, cellulose crystallinity	Cellulose, straw, steam	Sun, Runcang 55661525600	Dalian Polytech. Univ. China	551
3	Xu et al. (2010)	Cellulose	IL	Cellulose dissolution, anionic structure, lithium salts addition, cellulose solvent, IL synthesis, cellulose solubility	Cellulose, IL, solvent	Wang, Jianji 55904673200	Henan Normal Univ. China	329

*, Female; Cits., Number of citations received for each paper; Na, non available; Prt, Biomass pretreatments.

Cai and Zhang (2005) dissolved cellulose in LiOH/urea and NaOH/urea aqueous solutions in a paper with 761 citations. They evaluated the dissolution behavior and solubility of cellulose. They found that cellulose having a viscosity-average molecular weight of 11.4×10^4 and 37.2×10^4 could be dissolved, respectively, in 7% NaOH/12% urea and 4.2% LiOH/12% urea aqueous solutions pre-cooled to −10°C within 2 min, whereas all of them could not be dissolved in KOH/urea aqueous solution. Further, the dissolution power of the solvent systems was in the order of LiOH/urea >NaOH/urea ≫KOH/urea aqueous solution. LiOH/urea and NaOH/urea aqueous solutions as non-derivatizing solvents broke the intra- and inter-molecular hydrogen bonding of cellulose and prevented the approach toward each other of the cellulose molecules, leading to the good dispersion of cellulose to form an actual solution.

Sun et al. (2005) isolated cellulose from wheat straw using a two-stage process based on steam explosion pretreatment followed by alkaline hydrogen peroxide posttreatment in a paper with 551 citations. They steamed straw at 200°C, 15 bar for 10 and 33 min, and 220°C, 22 bar for 3, 5 and 8 min with a solid-to-liquid ratio of 2:1 (w/w) and 220°C, 22 bar for 5 min with a solid-to-liquid ratio of 10:1, respectively. They then washed this steamed straw with hot water to yield a solution rich in hemicelluloses-derived mono- and oligo-saccharides and with 61.3%, 60.2%, 66.2%, 63.1%, 60.3% and 61.3% of the straw residue, respectively. They next delignified this washed fiber and bleached by 2% H_2O_2 at 50°C for 5 h under pH 11.5, which yielded 34.9%, 32.6%, 40.0%, 36.9%, 30.9% and 36.1% (% dry wheat straw) of the cellulose preparation, respectively. They obtained the optimum cellulose yield (40.0%) when the steam explosion pretreatment was performed at 220°C, 22 bar for 3 min with a solid-to-liquid ratio of 2:1, in which the cellulose fraction obtained had a viscosity average degree of polymerization of 587 and contained 14.6% hemicelluloses and 1.2% klason lignin. The steam explosion pretreatment led to a significant loss in hemicelluloses and alkaline peroxide posttreatment resulted in substantial dissolution of lignin and an increase in cellulose crystallinity.

Xu et al. (2010) evaluated the effects of anionic structure and lithium salt addition on the dissolution of cellulose in 1-N-butyl-3-methylimidazolium $[C_4mim]$-based IL solvent systems in a paper with 329 citations. They synthesized a series of ILs coupling the $[C_4mim]^+$ with the Bronsted basic anions $[CH_3COO]^-$, $[HSCH_2COO]^-$, $[HCOO]^-$, $[(C_6H_5]COO]^-$, $[H_2NCH_2COO]^-$, $[HOCH_2COO]^-$, $[CH_3CHOHCOO]^-$ and $[N(CN)_2]^-$. They found that the solubility of cellulose increased almost linearly with increasing hydrogen bond accepting ability of anions in the ILs. At the same time, they developed $[C_4mim][CH_3COO]$/lithium salt (LiCl, LiBr, LiAc, $LiNO_3$, or $LiClO_4$) solvent systems by adding 1.0 wt% of lithium salt into $[C_4mim]$[CH3COO]. They showed that the addition of lithium salts significantly increased the solubility of the cellulose and reasoned that the enhanced solubility of cellulose originated from the disruption of the intermolecular hydrogen bond, O(6)H···O(3) owing to the interaction of Li^+ with the hydroxyl oxygen O(3) of cellulose. Finally, they determined the thermal stability of the regenerated cellulose.

98.3.1.2 Hydrolysis of the Feedstocks for Bioethanol Fuel Production

There are four HCPs for the hydrolysis of feedstocks with at least 343 citations each (Table 98.2).

Zhu et al. (2009) performed sulfite pretreatment (SPORL) for enzymatic saccharification of spruce and red pine in a paper with 451 citations. The SPORL process consisted of sulfite pretreatment of wood chips under acidic conditions followed by mechanical size reduction using disk refining. After the SPORL pretreatment of spruce chips with 8%–10% bisulfite and 1.8%–3.7% sulfuric acid on oven-dry (OD) wood at 180°C for 30 min, they obtained more than 90% cellulose conversion of the substrate with an enzyme loading of about 14.6 FPU cellulase plus 22.5 CBU β-glucosidase per gram of OD substrate after 48 h hydrolysis. Glucose yield from enzymatic hydrolysis of the substrate per 100 g of untreated OD spruce wood (glucan content 43%) was about 37 g (excluding the dissolved glucose during pretreatment). Hemicellulose removal was critical as lignin sulfonation for cellulose conversion in the SPORL process. Pretreatment altered the wood chips, which reduced electric energy consumption for size reduction to about 19 Wh/kg OD untreated wood or about 19 g glucose/Wh electricity. Furthermore, the SPORL produced low amounts of fermentation inhibitors, hydroxymethylfurfural (HMF) and furfural, of about 5 and 1 mg/g of untreated OD wood, respectively. They obtained similar results when the SPORL was applied to red pine. They concluded that the SPORL process had very few technological barriers and risks for commercialization.

Li et al. (2008) showed that acid in IL was an efficient system for the hydrolysis of lignocellulosic biomass with improved total reducing sugars (TRS) yield under mild conditions in a paper

TABLE 98.2
The Hydrolysis of Feedstocks for the Bioethanol Production: Chinese Research

No.	Papers	Biomass	Prts.	Parameters	Keywords	Lead Authors	Affil.	Cits
1	Zhu et al. (2009)	Spruce, pine	Sulfite, acids, milling, enzymes	SPORL pretreatment, cellulose conversion, enzymatic hydrolysis, glucose yield, hemicellulose removal	Spruce, pine, pretreatment, saccharification	Zhu, Junyong 7405692678	USDA Forest Serv. USA	451
2	Li et al. (2008)	Lignocellulosic biomass, corn stalk, rice straw, pine, bagasse	Acids, IL	Acid in IL, TRS yield, acids, ILs, acid hydrolysis	Lignocellulose, acid, hydrolysis, IL	Zhao, Zongbao K. 56972812400	Chinese Acad. Sci. China	425
3	Zhao et al. (2009)	Cellulose	IL, enzymes	Cellulose regeneration, enzymatic hydrolysis, crystallinity, accessibility, cellulase/substrate ratio, stability	Cellulose, hydrolysis, ILs	Zhao, Hua 7404778309	Univ. N. Colorado USA	403
4	Li and Zhao (2007)	Cellulose	IL, acids	Cellulose hydrolysis, acids, IL	Cellulose, hydrolysis, IL, acid	Zhao, Zongbao K. 56972812400	Chinese Acad. Sci. China	343

Cits., Number of citations received for each paper; Na, non available; Prt, Biomass pretreatments..

with 425 citations. They found that TRS yields were up to 66%, 74%, 81%, and 68% for hydrolysis of corn stalk, rice straw, pine wood, and bagasse, respectively, in [C_4mim]Cl in the presence of 7 wt% hydrogen chloride [HCl] at 100°C under atmospheric pressure within 60 min. Further, different combinations between ILs, such as [C_6mim]Cl, [C_4mim]Br, [Amim]Cl, [C_4mim]HSO_4, and [Sbmim]HSO_4, and acids, including sulfuric acid, nitric acid, phosphoric acid, as well as maleic acid, afforded similar results albeit longer reaction time was generally required comparing with the combination of [C_4mim]Cl and HCl. Further, modification of lignin occurred during sulfuric acid-catalyzed hydrolysis and the hydrolysis likely followed a consecutive first-order reaction sequence, where $k1$ and $k2$, the rate constants for TRS formation and TRS degradation, were 0.068 and 0.007 min^{-1}, respectively.

Zhao et al. (2009) regenerated cellulose from ILs for accelerated enzymatic hydrolysis in a paper with 403 citations. They explored a number of chloride- and acetate-based ILs for cellulose regeneration. They observed that all regenerated celluloses were less crystalline (58–75% lower) and more accessible to cellulase (>2 times) than untreated substrates. Thus, they found that regenerated Avicel® cellulose, filter paper, and cotton were hydrolyzed 2–10 times faster than the respective untreated celluloses. They obtained complete hydrolysis of Avicel® cellulose in 6 h given the *Trichoderma reesei* cellulase/substrate ratio (w/w) of 3:20 at 50°C. In addition, they observed that cellulase was more thermally stable (up to 60°C) in the presence of regenerated cellulose. Since the presence of various ILs during the hydrolysis induced different degrees of cellulase inactivation they recommended a thorough removal of IL residues after cellulose regeneration.

Li and Zhao (2007) developed a method for cellulose hydrolysis catalyzed by mineral acids in the ionic liquid 1-butyl-3-methylimidazolium chloride ([C_4mim]Cl) that facilitated the hydrolysis of cellulose with dramatically accelerated reaction rates at 100 °C under atmospheric pressure and without pretreatment.

98.3.2 Bioethanol Fuel Utilization

The secondary research fronts for the utilization of bioethanol fuels are the utilization of bioethanol fuels in fuel cells and transport engines, development of the bioethanol sensors, and the utilization of bioethanol fuels for the production of biohydrogen fuels with 10, 4, 3, and 1 HCPs, respectively. However, there are no HCPs produced by Chinese researchers on the production of the biochemicals such as bioethylene from bioethanol fuels.

98.3.2.1 Bioethanol Fuel Utilization in Fuel Cells

There are 10 HCPs for the utilization of bioethanol fuels in fuel cells with at least 308 citations each (Table 98.3).

Liang et al. (2009) studied the mechanism of the ethanol oxidation reaction (EOR) on palladium (Pd) electrodes in alkaline media using the cyclic voltammetry method in a paper with 713 citations. They found that the dissociative adsorption of ethanol proceeded rather quickly and the rate-determining step was the removal of the adsorbed ethoxi by the adsorbed hydroxyl on the Pd electrode. The Tafel slope was 130 mV/dec at lower potentials suggesting that the adsorption of OH^- ions followed the Temkin-type isotherm on the Pd electrode. In comparison, the Tafel slope increased gradually to 250 mV/dec at higher potentials. Thus at higher potentials, the kinetics was not only affected by the adsorption of the OH^- ions but also by the formation of the inactive oxide layer on the Pd electrode.

Zhou et al. (2003) prepared platinum-(Pt) based anode/carbon catalyst systems for direct ethanol fuel cells (DEFCs) in a paper with 600 citations. They found that all these catalysts consisted of uniform nanosized particles (NPs) with sharp distribution and the Pt lattice parameter decreased with the addition of ruthenium (Ru) or Pd and increased with the addition of tin (Sn) or tungsten (W). Cyclic voltammetry (CV) measurements and single DEFC tests jointly showed that the presence of Sn, Ru, and W enhanced the activity of Pt towards ethanol electrooxidation in the following order: Pt_1Sn_1/C

TABLE 98.3
The Utilization of Bioethanol Fuels in the Fuel Cells: Chinese Research

No.	Papers	Catalysts	Res. Fronts	Parameters	Keywords	Lead Authors	Affil.	Cits
1	Liang et al. (2009)	Pd	Fuel cells	Mechanism of the ethanol oxidation reaction, Pd electrode, dissociative adsorption of ethanol, Tafel slope	Ethanol, oxidation	Zhao, Tianshou 13004121800	Southern Univ. Sci. Technol. China	713
2	Zhou et al. (2003)	PtSn/C, PtRu/C, PtW/C, PtPd/C, Pt/C	Fuel cells	Catalyst systems, lattice parameter, ethanol electrooxidation, DEFC performance PT-additive interactions, bifunctional mechanism	Ethanol, fuel cells	Qin Xin 28167866000	Chinese Acad. Sci. China	600
3	Xu et al. (2007a)	Pt/C, Pd/C	Fuel cells	Ethanol electooxidation, electrocatalysts, catalyst supports, electrocatalytic properties	Ethanol, electrooxidation	Liu, Yingliang 57192569158	Shenzhen Univ. China	481
4	Xu et al. (2007b)	Pd NWA, Pd film, PtRu/C	Fuel cells	Ethanol oxidation, Pd metal nanowire arrays, DAFCs, electrocatalytic activity and stability	Ethanol, oxidation, electrocatalysts, fuel cells	Jiang, San Ping 7404452780	Curtin Univ. Australia	466
5	Tian et al. (2010)	Pd NCs, Pd black	Fuel cells	Ethanol electrooxidation, catalysts, catalytic activity, high-index-faceted metal nanocatalysts	Ethanol electrooxidation	Sun, Shi-Gang 7404510197	Univ. Sci. Technol. China	413
6	Shen et al. (2010)	PdNi/C, Pd/C	Fuel cells	PdNi catalyst synthesis, ethanol oxidation reaction, DEFCs, catalytic activity and stability, power density	Ethanol, oxidation, catalysts, fuel cells	Zhao, Tianshou 13004121800	Southern Univ. Sci. Technol. China	398
7	Xu et al. (2007c)	Pt/C, Pd/C, PtRu/C	Fuel cells	Ethanol electrooxidation, Pd-based catalysts, catalytic activity and stability, CeO2 and NiO additives	Ethanol electrooxidation, catalysts	Shen, Peikang 7201767641	Guangxi Univ. China	367
8	Wang and Liu (2008)	Pt	Fuel cells	Ethanol oxidation selectivity, Pt catalysts, reaction network, structure-sensitivity and selectivity, dehydrogenation	Ethanol oxidation	Liu, Zhi-Pan	Fudan Univ. China	363
9	Wang et al. (2013)	Pd/PANI/Pd SNTAs, Pd NTAs, Pd/c	Fuel cells	Ethanol electrooxidation, electrocatalysts, electrocatalytic activity and durability, Pd-PANI composites	Ethanol electrooxidation, catalysts	Li, Gao-Ren 7407053622	Sun Yat-Sen Univ. China	327
10	Chen et al. (2017)	Pd-Ni-P, Pd/C	Fuel cells	Ethanol electrooxidation, Pd-based nanocatalysts, Ni, P additives, electrocatalytic activity and stability	Ethanol electrooxidation, nanocatalysts	Wang, Leyu 7409176968	Beijing Univ. Chem. Technol. China	308

Cits., Number of citations received for each paper; Na, non available; Prt, Biomass pretreatments.

>Pt_1Ru_1/C >Pt_1W_1/C >Pt_1Pd_1/C >Pt/C. Moreover, Pt_1Ru_1/C further modified by W and molybdenum (Mo) showed improved ethanol electrooxidation activity, but its DEFC performance was inferior to that measured for Pt_1Sn_1/C. They then found that the single DEFC having Pt_1Sn_1/C or Pt_3Sn_2/C or Pt_2Sn_1/C as anode catalyst showed better performances than those with Pt_3Sn_1/C or Pt_4Sn_1/C. They also found that the latter two cells exhibited higher performances than the single cell using Pt_1Ru_1/C, which was exclusively used in proton-exchange membrane fuel cells (PEMFC) as an anode catalyst for both methanol electrooxidation and CO tolerance. They attributed this distinct difference in DEFC performance between these to the so-called bifunctional mechanism and to the electronic interaction between Pt and additives. An amount of $-OH_{ads}$, an amount of surface Pt active sites and the conductivity effect of PtSn/C catalysts would determine the activity of PtSn/C with different Pt/Sn ratios. At lower temperature values or at low current density regions where the electrooxidation of ethanol was considered not so fast and its chemisorption was not the rate-determining step, the Pt_3Sn_2/C was more suitable for the DEFC. At 75°C, the single DEFC with Pt_3Sn_2/C as anode catalyst showed comparable performance to that with Pt_2Sn_1/C, but at a higher temperature of 90°C, the latter presented much better performance. They concluded that Pt_2Sn_1/C, supplying sufficient $-OH_{ads}$ and having adequate active Pt sites and acceptable ohmic effect, could be the appropriate anode catalyst for DEFC.

Xu et al. (2007a) used Pt and Pd electrocatalysts supported on carbon microspheres (CMS) for methanol and ethanol electrooxidation in alkaline media in a paper with 481 citations. They found that these electrocatalysts supported on carbon microspheres gave better performance than that supported on carbon black. Pd showed excellently higher activity and better steady-state electrolysis than Pt for ethanol electrooxidation in alkaline media. They observed a synergistic effect by the interaction between Pd and carbon microspheres. The Pd supported on carbon microspheres possessed excellent electrocatalytic properties.

Xu et al. (2007b) fabricated highly ordered Pd metal nanowire arrays (NWA) electrodes by the anodized aluminum oxide (AAO) template-electrodeposition method for ethanol oxidation in direct alcohol fuel cells (DAFC) in a paper with 466 citations. They observed that these nanowires were highly ordered, with uniform diameter and length. They also observed that the NWs were uniform, well isolated, parallel to one another, and standing vertically to the electrode substrate surface and Pd NWAs exhibited a face-centered cubic (FCC) lattice structure. The electrocatalytic activity and stability of the Pd NWAs for ethanol electrooxidation was not only significantly higher than that of conventional Pd film electrodes but also higher than that of well-established commercial PtRu/C electrocatalysts. The Pd NWAs with high electrochemically active surface area showed great potential as electrocatalysts for ethanol electrooxidation in alkaline media in DAFCs as they had high electrocatalytic activity and stability.

Tian et al. (2010) produced tetrahexahedral Pd nanocrystals (THH Pd NCs) with {730} high-index facets on a glassy carbon substrate in a dilute $PdCl_2$ solution by a programmed electrodeposition method in a paper with 413 citations. They found that these THH Pd NCs exhibited 4–6 times higher catalytic activity than commercial Pd black catalyst toward ethanol electrooxidation in alkaline solutions due to their high density of surface atomic steps. This straightforward method provides a promising route to the facile preparation of high-index-faceted metal nanocatalysts with high catalytic activity.

Shen et al. (2010) synthesized carbon-supported PdNi (PdNi/C) catalysts by the simultaneous reduction method using $NaBH_4$ as a reductant for the oxidation of ethanol in alkaline DEFCs in a paper with 398 citations. They observed the formation of the FCC crystalline Pd and $Ni(OH)_2$ on the carbon powder for the PdNi/C catalysts. Further, the metal particles were well-dispersed on the carbon powder with a uniform distribution of Ni around Pd. They showed the chemical states of Ni, including metallic Ni, NiO, $Ni(OH)_2$ and NiOOH. Thus, they observed that the Pd_2Ni_3/C catalyst exhibited higher activity and stability for the ethanol oxidation reaction (EOR) in an alkaline medium than the Pd/C catalyst. They finally found that the application of Pd_2Ni_3/C as the anode catalyst of an alkaline DEFC with an anion-exchange membrane could yield a maximum power density of 90 mW/cm^2 at 60°C.

Xu et al. (2007c) studied the Pd-based catalysts, Pt/C and Pd/C catalysts, promoted with CeO_2 and NiO, as a replacement for Pt-based catalysts for ethanol electrooxidation in alkaline media in a paper with 367 citations. They found that the Pd/C electrocatalyst had a higher catalytic activity and better stability for ethanol oxidation than that of Pt/C electrocatalyst. They further studied the effect of the addition of CeO_2 and NiO additives to the Pt/C and Pd/C electrocatalysts on ethanol oxidation in alkaline media. They then found that the electrocatalysts with a weight ratio of noble metal (Pt, Pd) to CeO_2 of 2:1 and a noble metal to NiO ratio of 6:1 showed the highest catalytic activity for ethanol oxidation. Further, the oxide-promoted Pt/C and Pd/C electrocatalysts showed higher catalytic activity than the commercial PtRu/C electrocatalyst for ethanol oxidation in alkaline media.

Wang and Liu (2008) explored the mechanism and structure-sensitivity of ethanol oxidation on different Pt surfaces, including close-packed Pt{111}, stepped Pt{211}, and open Pt{100}, thoroughly with an efficient reaction path-searching method in a paper with 363 citations. They located the transition state and saddle points for most surface reactions simply and efficiently by optimization of local minima. They showed that the selectivity of ethanol oxidation on Pt depended markedly on the surface structure, which could be attributed to the structure sensitivity of two key reaction steps: the initial dehydrogenation of ethanol and the oxidation of acetyl (CH_3CO). Further, on open surface sites, ethanol preferred C–C bond cleavage via strongly adsorbed intermediates (CH_2CO or CHCO), which led to complete oxidation to CO_2. However, only partial oxidization to CH_3CHO and CH_3COOH occurred on Pt{111}. They further found that the open surface Pt{100} was the best facet to fully oxidize ethanol at low coverages, which shed light on the origin of the remarkable catalytic performance of Pt tetrahexahedral nanocrystals found recently. Finally, they identified two fundamental quantities that dictated the selectivity of ethanol oxidation: the ability of surface metal atoms to bond with unsaturated C-containing fragments and the relative stability of hydroxyl at surface atop sites with respect to other sites.

Wang et al. (2013) designed hybrid Pd/polyaniline (PANI)/Pd sandwich-structured nanotube array (SNTA) catalysts to exploit shape effects and synergistic effects of Pd-PANI composites for ethanol electrooxidation for DEFCs in a paper with 327 citations. They found that these synthesized SNTAs exhibited significantly improved electrocatalytic activity and durability compared with Pd NTAs and commercial Pd/C catalysts. The unique SNTAs provided fast transport and short diffusion paths for electroactive species and a high utilization rate of catalysts. Besides the merits of nanotube arrays, they attributed the improved electrocatalytic activity and durability, especially to the special Pd/PANI/Pd sandwich-like nanostructures, which resulted in electron delocalization between Pd d orbitals and PANI π-conjugated ligands and in electron transfer from Pd to PANI.

Chen et al. (2017) improved ethanol electrooxidation performance by shortening Pd- oxophilic Ni active site distance in Pd-Ni-P nanocatalyst in DEFCs in a paper with 308 citations. They used ultrasmall (~5 nm) Pd–Ni–P ternary nanoparticles for ethanol electrooxidation. They found that the electrocatalytic activity was improved up to 4.95 A per mgPd, which was 6.88 times higher than commercial Pd/C (0.72 A per mgPd), by shortening the distance between Pd and Ni active sites, achieved through shape transformation from Pd/Ni–P heterodimers into Pd–Ni–P nanoparticles and tuning the Ni/Pd atomic ratio to 1:1. The improved activity and stability stemmed from the promoted production of free OH radicals (on Ni active sites) which facilitated the oxidative removal of carbonaceous poison and combination with CH_3CO radicals on adjacent Pd active sites.

98.3.2.2 Other Bioethanol Fuel Utilization

There are four, three, and one HCPs for the utilization of bioethanol fuels in transport engines, development of bioethanol sensors, and utilization of bioethanol fuels for biohydrogen fuel production, respectively (Table 98.4). However, there is no CHP for the production of biochemicals from bioethanol fuels.

TABLE 98.4
The Other Utilization of Bioethanol Fuels: Chinese Research

No.	Papers	Cats. Mats.	Res. Fronts	Parameters	Keywords	Lead Authors	Affil.	Cits
1	Wan et al. (2004)	ZnO nanowires	Sensors	Ethanol sensor fabrication and evaluation, ZnO sensors, ethanol sensitivity	Ethanol sensing, sensors	Wang, Tai-Hong 35241217600	Southern Univ. Sci. Technol. China	1908
2	Liu et al. (2005)	V_2O_5 nanobelts	Sensors	Ethanol sensors, single crystalline divanadium pentoxide nanobelts, ethanol selectivity	Ethanol, sensor	Li, Yadong 57192004602	Tsinghua Univ. China	526
3	Wang et al. (2012)	ZnO nanorods	Sensors	ZnO nanorod fabrication, gas sensors, ethanol detection	Ethanol sensor, ZnO	Wang, Shurong 57202355405	Nankai Univ. China	406
4	Zhang et al. (2007)	Ir/CeO_2, Co/ CeO_2, Ni/ CeO_2	Biohydrogen fuels	Ethanol steam reforming, biohydrogen production, ethanol dehydrogenation and decomposition, water gas shift, catalysts, catalytic performance	Ethanol, hydrogen, reforming, catalysts	Shen, Wenjie 7403601371	Chinese Acad. Sci. China	376
5	Li et al. (2005)	Na	Engines	Bioethanol-petrodiesel blends, DI diesel engine, performance, emissions, bioethanol contents, E0-E20	Ethanol, diesel, blends, emissions, engines	Li, De-Gang 55982700800	Jilin Univ. China	361
6	He et al. (2003)	Na	Engines	Bioethanol-petrodiesel blends, diesel engine, fuel properties, emissions, ethanol content, E10, E30, smoke, NOx, CO_2, UHC, acetaldehyde	Ethanol, blended, diesel, engine, emissions	He, Bang-Quan 7402047862	Tianjin Univ. China	331
7	Xing-Cai et al. (2004)	Na	Engines	Bioethanol-.petrodiesel blends, CN enhancer, emissions, BSFC, HRR, NOx, smoke, ignition delay, combustion	Ethanol, diesel, blend, cetane, engine, emissions	Xingcai, Lu 55571103600	Shanghai Jiaotong Univ. china	327
8	Zhu et al. (2011)	Na	Engines	Bioethanol-biodiesel blends, combustion, performance, emission, bioethanol content, E5, NO_x, PM, UHC, CO	Combustion, emissions, diesel, engine, ethanol, biodiesel, blends	Cheung, Chun Shun 57191305782	Hong Kong Polytech. Univ. China	306

Cats., Catalysts; Cits., Number of citations received for each paper; Mats., Materials; Na, non available; Prt, Biomass pretreatments.

98.3.2.2.1 Utilization of Bioethanol in Transport Engines

There are four highly cited papers for the field of utilization of bioethanol in transport engines produced by China with at least 306 citations each.

Li et al. (2005) evaluated the effects of bioethanol–petrodiesel blended fuels with different bioethanol contents (E0-E20) on the performance and emissions of diesel engines to find the optimum percentage of ethanol that gave simultaneously better performance and lower emissions in a paper with 361 citations. They performed the experiments on a water-cooled single-cylinder direct injection (DI) diesel engine using 0% (neat petrodiesel fuel, E0), 5% (E5), 10% (E10), 15% (E15), and

20% (E20) ethanol–petrodiesel blended fuels. They found that the brake-specific fuel consumption (BSFC) and brake thermal efficiency (BTE) increased with an increase of ethanol contents in the blended fuel at overall operating conditions while smoke emissions decreased with ethanol–petrodiesel blended fuel, especially with E10 and E15. Further, CO and NO_x emissions were reduced for ethanol–petrodiesel blends, but total hydrocarbons (THC) increased significantly when compared to neat petrodiesel fuel.

He et al. (2003) evaluated the effect of bioethanol-blended petrodiesel fuels (E10, E30) on the fuel properties and emissions from a diesel engine in a paper with 331 citations. They found that the addition of ethanol to petrodiesel fuel simultaneously decreased cetane number (CN), higher heating value (HHV), aromatics fractions, and kinematic viscosity of ethanol-blended petrodiesel fuels and changed distillation temperatures. Further, an additive used to keep the blends homogenous and stable, and an ignition improver, which could enhance the CN of the blends, had favorable effects on the physicochemical properties related to ignition and combustion of the blends with 10% (E10) and 30% ethanol (E30) by volume. On the other hand, at high loads, the blends reduced smoke significantly with a small penalty on CO, acetaldehyde, and unburned ethanol emissions (UE) compared to petrodiesel fuel while NO_x and CO_2 emissions of the blends were decreased somewhat. However, at low loads, the blends had slight effects on smoke reduction due to the overall leaner mixture. They finally found that with the aid of additive and ignition improver, CO, UE, and acetaldehyde emissions of the blends could be decreased moderately, even though total hydrocarbon emissions (THC) were less than those of petrodiesel fuel.

Xing-Cai et al. (2004) evaluated the effect of CN improver on heat release rate (HRR) and emissions of a high-speed diesel engine fueled with ethanol–petrodiesel blend fuels in a paper with 327 citations. They added different percentages of CN enhancer (0%, 0.2%, and 0.4%) to these blends and performed the engine tests on a four-cylinder high-speed DI diesel engine. They found that the BSFC increased, the diesel equivalent BSFC decreased, the thermal efficiency improved remarkably, and NO_x and smoke emissions decreased simultaneously when diesel engine fueled with ethanol–petrodiesel blend fuels while NO_x and smoke emissions further reduced when CN improver was added to blends. They further found that the ignition delay was prolonged, and the total combustion duration shortened for ethanol–petrodiesel blend fuels compared to petrodiesel fuel while the combustion characteristics of ethanol–petrodiesel blend fuel at large load might be resumed to diesel fuel by CN improver, but a large difference existed at lower load yet.

Zhu et al. (2011) evaluated the combustion, performance, and emission characteristics of a DI diesel engine fueled with ethanol-biodiesel blends in a paper with 306 citations. They tested Euro V diesel fuel, biodiesel, and ethanol–biodiesel blends (BE) in a four-cylinder DI diesel engine under five engine loads at the maximum torque engine speed of 1,800 rpm. When compared with biodiesel, they found that the combustion characteristics of ethanol–biodiesel blends changed while the engine performance improved slightly with 5% ethanol in biodiesel (E5). Further, in comparison with Euro V diesel fuel, the biodiesel and BE blends had higher BTE. On the whole, compared with Euro V diesel fuel, the BE blends could lead to a reduction of both NO_x and particulate matter (PM) emissions of the diesel engine. The effectiveness of NO_x and PM reductions increased with increasing ethanol in the blends. With a high percentage of ethanol in the BE blends, the THC and CO emissions could increase while the use of BE5 could reduce the HC and CO emissions as well.

98.3.2.2.2 Bioethanol Sensors

There are three highly cited papers for the field of development and utilization of bioethanol sensors produced by China with at least 406 citations each.

Wan et al. (2004) synthesized ZnO nanowire ethanol gas sensors with microelectromechanical system technology and evaluated their ethanol sensing characteristics in a paper with 1,908 citations. They found that this sensor exhibited high sensitivity and fast response to ethanol gas at a work temperature of 300°C.

Liu et al. (2005) developed highly selective and stable bioethanol sensors based on single crystalline divanadium pentoxide (V_2O_5) nanobelts in a paper with 526 citations. They obtained these nanobelts by a simple mild hydrothermal method with a high yield. The gas sensors fabricated using these nanobelts showed great potential for the detection of ethanol molecules at a relatively low temperature. The experiments with variations in relative humidity and tests with other gases indicated no problems of interference with ethanol.

Wang et al. (2012) fabricated ZnO nanorods by a simple low-temperature hydrothermal process in high yield (about 85%), starting with aqueous solution in the presence of CTAB for ethanol detection in a paper with 406 citations. They then used these ZnO nanorods to construct a gas sensor for ethanol detection at different operating temperatures. They found that this ZnO nanorod gas sensor exhibited a high, reversible, and fast response to ethanol, indicating its potential application as a gas sensor to detect ethanol.

98.3.2.2.3 Bioethanol Fuel-based Biohydrogen Fuels

There is only one highly cited paper for the field of the utilization of bioethanol fuels for the production of biohydrogen fuels for fuel cells, produced by China with 376 citations.

Zhang et al. (2007) produced biohydrogen fuel from steam reforming of ethanol and glycerol over ceria (CeO_2)-supported iridium (Ir), cobalt (Co), and nickel (Ni) catalysts in a paper with 376 citations. For ethanol steam reforming, they found that ethanol dehydrogenation to acetaldehyde and ethanol decomposition to methane and CO were the primary reactions at low temperatures, depending on the active metals. At higher temperatures where all the ethanol and the intermediate compounds, like acetaldehyde and acetone, were completely converted into hydrogen, carbon oxides and methane, steam reforming of methane and water gas shift became the major reactions. They further found that the Ir/CeO_2 catalyst was significantly more active and selective toward hydrogen production and interpreted its superior catalytic performance in terms of the intimated contact between Ir particles and ceria based on the ceria-mediated redox process.

98.4 DISCUSSION

98.4.1 Introduction

Crude oil-based gasoline fuels have been widely used in the transportation sector since the 1920s. However, there have been great public concerns over the adverse environmental and human impact of these fuels. Hence, biomass-based bioethanol fuels have increasingly been used in blending gasoline and petrodiesel fuels, in fuel cells, and in biochemical and biohydrogen fuel production in a biorefinery context.

However, it is necessary to pretreat biomass to enhance the yield of the bioethanol prior to the bioethanol fuel production from the feedstocks through the hydrolysis and fermentation of biomass and hydrolysates respectively.

China has been one of the most prolific countries engaged in bioethanol fuel research. The Chinese research in the field of bioethanol fuels has intensified in this context in the key research fronts of the production of bioethanol fuels, utilization of bioethanol fuels and to a lesser extent the evaluation of bioethanol fuels. For the first research front, pretreatment and hydrolysis of the feedstocks, fermentation of the feedstock-based hydrolysates, production of bioethanol fuels, and to a lesser extent distillation of bioethanol fuels are the key research areas while for the second research front, utilization of bioethanol fuels in fuel cells and transport engines, bioethanol-based biohydrogen fuels, bioethanol sensors, and to a lesser extent bioethanol-based biochemicals are the key research areas. The research in this field has also intensified for the feedstocks of bioethanol fuels for their utilization, starch feedstock residues, cellulose, wood biomass, lignocellulosic biomass at large, and to a lesser extent industrial waste, lignin, grass, hydrolysates, biomass at large, sugar feedstocks, sugar feedstock residues, algal biomass, lignocellulosic wastes, starch feedstocks, urban wastes, forestry wastes, food wastes, and plants.

However, it is essential to develop efficient incentive structures for the primary stakeholders to enhance the research in this field. Although there has been a number of review papers for this field, there has been no review of the most-cited 25 Chinese articles in this field. Thus, this book chapter presents a review of the most-cited 25 articles on the Chinese experience of bioethanol fuels. Then, it discusses the key findings of these highly influential papers and comments on future research priorities in this field.

As a first step for the search of the relevant literature, the keywords were selected using the most-cited first 300 population papers for each research front. The selected keyword list was then optimized to obtain a representative sample of papers for each research field. These keyword lists were then integrated to obtain the keyword list for this research field (Konur, 2023a,b).

As a second step, a sample dataset was used for this study. The first 25 articles with at least 306 citations each were selected for the review study. Key findings from each paper were taken from the abstracts of these papers and were discussed. Additionally, a number of brief conclusions were drawn and a number of relevant recommendations were made to enhance the future research landscape.

Information about the research fronts for the sample papers in bioethanol fuels is given in Table 98.5. As this table shows, the most prolific research front for this field is bioethanol fuels as feedstocks for their utilization with 72% of the reviewed papers. The other feedstocks used in these

TABLE 98.5
The Most Prolific Research Fronts for the Chinese Experience of Bioethanol Fuels

No.	Research Fronts	N Paper (%) Review	N Paper (%) Sample	Surplus (%)
1	Bioethanol fuels	72	19.6	52.4
2	Cellulose	16	15.1	0.9
3	Starch feedstock residues	8	9.7	−1.7
4	Wood biomass	8	9.0	−1.0
5	Lignocellulosic biomass at large	4	23.0	−19.0
6	Industrial waste	4	6.5	−2.5
7	Sugar feedstock residues	4	2.2	1.8
8	Other feedstocks	0	27.4	−27.4
	Lignin	0	6.2	−6.2
	Grass	0	3.1	−3.1
	Hydrolysates	0	2.8	−2.8
	Biomass	0	2.6	−2.6
	Sugar feedstocks	0	2.2	−2.2
	Algal biomass	0	1.9	−1.9
	Lignocellulosic wastes	0	1.8	−1.8
	Starch feedstocks	0	1.7	−1.7
	Urban wastes	0	1.0	−1.0
	Forestry wastes	0	0.8	−0.8
	Hemicellulose	0	0.8	−0.8
	Xylan	0	0.8	−0.8
	Food wastes	0	0.6	−0.6
	Cellobiose	0	0.4	−0.4
	Syngas	0	0.4	−0.4
	Plants	0	0.3	−0.3
	Sample size	25	775	

N Paper (%) review, The number of papers in the sample of 25 reviewed papers for the Chinese; N paper (%) sample, The number of papers in the world sample of 775 papers.

reviewed papers are cellulose, starch feedstock residues, wood biomass, lignocellulosic biomass at large, industrial waste, and sugar feedstock residues with 4%–16% of the sample papers.

On the other hand, it is notable that there are no HCPs for the other feedstocks listed in Table 98.5. Since only seven HCPs have focused on the production of bioethanol fuels, the share of the feedstocks used in the production of bioethanol fuels from biomass has been necessarily low. However, as 18 HCPs primarily have focused on the utilization of bioethanol fuels, the share of bioethanol fuels in the sample papers has been high.

Further, it appears that bioethanol fuels as feedstocks are over-represented in the Chinese sample in relation to the world sample at large by 52.4%. This finding highlights the strength of China in the research front of the utilization of bioethanol fuels in relation to the USA and the rest of the world.

Similarly, a large number of feedstocks used in the production of bioethanol fuels from biomass are under-represented in the Chinese sample in relation to the USA and the rest of the world. For example, lignocellulosic biomass at large, lignin, grass, hydrolysates, biomass at large, and industrial waste are under-represented in the Chinese sample by 3%–19% each.

Information about the thematic research fronts for the sample papers in the Chinese experience of bioethanol fuels is given in Table 98.6. As this table shows, the most prolific research fronts for this field are the utilization and production of bioethanol fuels with 72% and 28% of the HCPs, respectively. It is notable that there is no HCP on the evaluation of bioethanol fuels such as techno-economics and environmental impact of bioethanol fuels.

Further, the most prolific research fronts for the first group are the hydrolysis and pretreatments of the biomass with 16% and 12% of the HCPs, respectively. It is notable that there are no HCPs for the other research fronts related to bioethanol fuel production.

Similarly, the most prolific research fronts for the second group are the utilization of bioethanol fuels in fuel cells and transport engines, the development of bioethanol sensors, and bioethanol-based biohydrogen fuels with 49%, 16%, 12%, and 4% of the sample papers, respectively. Further, there is no HCP for biochemical production from bioethanol fuels.

China has strength in the field of the utilization of bioethanol fuels as this field has been over-represented in relation to the world sample by 52.3%. On the contrary, China has a relative research weakness in the fields of production and evaluation of bioethanol fuels in relation to the world sample by 46.1% and 6.2%, respectively.

TABLE 98.6
The Most Prolific Thematic Research Fronts for the Chinese Experience of Bioethanol Fuels

No.	Research Fronts	N Paper (%) Review	N Paper (%) Sample	Surplus (%)
1	Bioethanol fuel production	28	74.1	–46.1
	Biomass hydrolysis	16	29.8	–13.8
	Biomass pretreatments	12	58.5	–46.5
	Bioethanol production	0	21.5	–21.5
	Hydrolysate fermentation	0	17.9	–17.9
	Bioethanol fuel distillation	0	0.8	–0.8
2	Bioethanol fuel evaluation	0	6.2	–6.2
3	Bioethanol fuel utilization	72	19.7	52.3
	Bioethanol fuels in fuel cells	40	6.8	33.2
	Bioethanol fuels in engines	16	6.1	9.9
	Bioethanol sensors	12	2.2	9.8
	Bioethanol-based biohydrogen fuels	4	4.3	–0.3
	Bioethanol-based biochemicals	0	0.4	–0.4

N Paper (%) review, The number of papers in the sample of 25 reviewed papers for the Chinese sample. N paper (%) sample, The number of papers in the world sample of 775 papers.

Further, on individual terms, China has strength in the secondary fields of the utilization of bioethanol fuels in fuel cells and engines, and bioethanol sensors in relation to the world sample with 33.2%, 9.9%, and 9.8%, respectively. On the contrary, China has a relative weakness in the secondary fields of biomass pretreatments, bioethanol production, hydrolysate fermentation, and biomass hydrolysis in relation to the world sample with 46.5%, 21.5%, 17.9%, and 13.8%, respectively.

98.4.2 Bioethanol Fuel Production

The secondary research fronts for the production of bioethanol fuels are the pretreatments and hydrolysis of biomass with three and four HCPs, respectively. However, there are no HCPs for the research fronts of the fermentation of the hydrolysates (Lin and Tanaka, 2006; Olsson and Hahn-Hagerdal, 1996), production of bioethanol fuels (Hahn-Hagerdal et al., 2006; Sanchez and Cardona, 2008), and separation and distillation of bioethanol fuels (Sano et al., 1994; Vane, 2005), produced by the Chinese researchers.

98.4.2.1 Feedstock Pretreatments for Bioethanol Fuel Production

The brief information about the three most-cited papers on the pretreatments of feedstocks with at least 329 citations each is given in Table 98.1.

The pretreatment of biomass is the first step in the production of bioethanol fuels from biomass (Alvira et al., 2010; Taherzadeh and Karimi, 2008). A number of pretreatments are used to fractionate the biomass to its constituents such as cellulose, hemicellulose, and lignin. This process makes cellulose accessible for the acid or enzymatic hydrolysis process to convert cellulose to the sugars such as glucose or xylose.

One of these pretreatments is the alkaline pretreatment and Cai and Zhang (2005) dissolved cellulose in LiOH/urea and NaOH/urea aqueous solutions to evaluate the dissolution behavior and solubility of cellulose.

It is often beneficial to combine alkaline pretreatments of the biomass with the steam explosion pretreatment and Sun et al. (2005) isolated cellulose from wheat straw using a two-stage process based on steam explosion pretreatment followed by alkaline hydrogen peroxide posttreatment.

One of the most used pretreatments in bioethanol production from biomass is the use of ILs and Xu et al. (2010) evaluated the effects of anionic structure and lithium salt addition on the dissolution of cellulose in [C_4mim]-based IL solvent systems.

Although these papers focus on a small number of pretreatments, pretreatments are grouped in general as chemical, biological, hydrothermal, and mechanical pretreatments (Konur, 2023c). In this context, alkaline and IL pretreatments are the chemical pretreatments while steam explosion pretreatment is the hydrothermal pretreatment.

98.4.2.2 Hydrolysis of the Feedstocks for Bioethanol Fuel Production

There are four HCPs for the hydrolysis of feedstock with at least 343 citations each (Table 98.2).

The second step in the production of bioethanol fuels from biomass is the hydrolysis of the biomass constituents, liberated in the first step of the pretreatments (Sun and Cheng, 2002; Taherzadeh and Karimi, 2007). The most used hydrolytic process is the enzymatic hydrolysis of biomass in addition to the acid hydrolysis of biomass.

For example, Zhu et al. (2009) performed sulfite pretreatment (SPORL) for enzymatic saccharification of spruce and red pine while Li et al. (2008) showed that acid in IL was an efficient system for hydrolysis of lignocellulosic materials with improved total reducing sugar yield under mild conditions. Further Zhao et al. (2009) regenerated cellulose from ILs for accelerated enzymatic hydrolysis while Li and Zhao (2007) developed a method for cellulose hydrolysis catalyzed by mineral acids in the IL that facilitated the hydrolysis of cellulose.

98.4.3 Bioethanol Fuel Utilization

The secondary research fronts for the utilization of bioethanol fuels are the utilization of bioethanol fuels in fuel cells and transport engines, development of the bioethanol sensors, and the utilization of bioethanol fuels for the production of biohydrogen fuels with 10, 4, 3, and 1 HCPs, respectively. However, there are no HCPs produced by Chinese researchers on the production of biochemicals such as bioethylene from bioethanol fuels.

In general, there are five secondary research fronts for the field of the utilization of bioethanol fuels: Utilization of bioethanol fuels in fuel cells (Antolini, 2007, 2009) and transport engines (Hsieh et al., 2002; Najafi et al., 2009), bioethanol-based biohydrogen fuels (Haryanto et al., 2005; Murdoch et al., 2011), bioethanol sensors (Liu et al., 2005; Wan et al., 2004), and bioethanol-based biochemicals (Angelici et al., 2013; Morschbacker, 2009).

98.4.3.1 Bioethanol Fuel Utilization in Fuel Cells

There are 10 HCPs for the direct utilization of bioethanol fuels in the fuel cells with at least 308 citations each (Table 98.3).

In general, bioethanol fuels are increasingly used in the DEFCs for the production of bioelectricity to power electric or hybrid cars. The Chinese researchers have also focused on this research front using a wide variety of catalyst systems.

For example, Liang et al. (2009) studied the mechanism of the ethanol oxidation reaction on the Pd electrode in alkaline media while Zhou et al. (2003) prepared Pt-based anode/carbon catalyst systems for DEFC.

On the other hand, Xu et al. (2007a) used both Pt and Pd electrocatalysts supported on carbon microspheres for methanol and ethanol electrooxidation in alkaline media while Xu et al. (2007b) fabricated highly ordered Pd metal nanowire arrays electrodes for ethanol oxidation in DAFCs.

Tian et al. (2010) produced tetrahexahedral Pd nanocrystals with {730} high-index facets on a glassy carbon substrate in a dilute $PdCl_2$ solution while Shen et al. (2010) synthesized carbon-supported PdNi (PdNi/C) catalysts for the oxidation of ethanol in alkaline DEFCs while Xu et al. (2007c) studied the Pd-based catalysts, Pt/C and Pd/C catalysts, promoted with CeO_2 and NiO, as a replacement for Pt-based catalysts for ethanol electrooxidation in alkaline media.

Wang and Liu (2008) explored the mechanism and structure-sensitivity of ethanol oxidation on different Pt surfaces while Wang et al. (2013) designed hybrid Pd/polyaniline (PANI)/Pd sandwich-structured nanotube array catalysts to exploit shape effects and synergistic effects of Pd-PANI composites for ethanol electrooxidation for DEFCs while Chen et al. (2017) improved ethanol electrooxidation performance by shortening Pd- oxophilic Ni active site distance in Pd-Ni-P nanocatalyst in DEFCs.

Thus, it appears that the Chinese researchers have focused on the development and utilization of both Pt- and Pd-based electrocatalyst systems for the electrooxidation of bioethanol fuels in the DEFCs to improve and electrocatalytic activity and stability of this electrocatalyst for the efficient operation of the DEFCs.

These Chinese researchers have also explored the application of nanotechnology in electrocatalyst development successfully (Chen et al., 2017; Wang et al., 2013, Xu et al., 2007b). However, it appears that there is ample room for the further application of one- and two-dimensional nanomaterials in the development of electrocatalysts for the electrooxidation of bioethanol fuels in the DEFCs.

98.4.3.2 Other Bioethanol Fuel Utilization

There are four, three, and one HCPs for the utilization of bioethanol fuels in transport engines, development of bioethanol sensors, and utilization of bioethanol fuels for biohydrogen fuel production, respectively (Table 98.4).

98.4.3.2.1 Utilization of Bioethanol in Transport Engines

There are four highly cited papers for the field of utilization of bioethanol in transport engines produced by Chinese researchers, with at least 306 citations each.

In general, gasoline or petrodiesel fuels have been primarily used to power gasoline or diesel engines, respectively. However, due to the adverse impact of these petrofuels, there has been a large number of public initiatives to replace these petrofuels with biodiesel or bioethanol fuels. In this context, bioethanol fuels have been blended with gasoline or petrodiesel fuels at certain content such as 20% of the fuels, E20. The research in this field has focused on the combustion, performance, and emissions of these blends.

For example, Li et al. (2005) evaluated the effects of bioethanol–petrodiesel blended fuels with different bioethanol contents (E0-E20) on the performance and emissions of diesel engines to find the optimum percentage of ethanol that gave simultaneously better performance and lower emissions while He et al. (2003) evaluated the effect of bioethanol-blended petrodiesel fuels (E10, E30) on the fuel properties and emissions from a diesel engine.

Similarly, Xing-Cai et al. (2004) evaluated the effect of CN improver on HRR and emissions of a high-speed diesel engine fueled with ethanol–petrodiesel blend fuels while Zhu et al. (2011) evaluated the combustion, performance, and emission characteristics of a DI diesel engine fueled with ethanol-biodiesel blends.

98.4.3.2.2 Bioethanol Sensors

There are three HCPs for the field of development and utilization of bioethanol sensors produced by China with at least 406 citations each.

With the increasing use of bioethanol fuels in cars, fuel cells, and biochemicals in addition to the drinking of ethanol-based alcoholic drinks by humans, there has been a need to develop bioethanol sensors with high bioethanol sensitivity.

Chinese researchers in this area have focused on the development of materials for sensors with high bioethanol sensitivity. For example, Wan et al. (2004) synthesized ZnO nanowire ethanol gas sensors with microelectromechanical system technology and evaluated their ethanol sensing characteristics while Liu et al. (2005) developed highly selective and stable bioethanol sensors based on single crystalline divanadium pentoxide (V_2O_5) nanobelts. Further, Wang et al. (2012) fabricated ZnO nanorods in high yield (about 85%) for ethanol detection.

98.4.3.2.3 Bioethanol Fuel-based Biohydrogen Fuels

There is only one highly cited paper for the field of the utilization of bioethanol fuels for the production of biohydrogen fuels for fuel cells, produced by China with 376 citations.

Although bioethanol fuels could be used directly as feedstocks in the DEFCs, the alternative has been the use of biohydrogen fuels produced through the reforming of bioethanol fuels. Chinese research in this field has primarily focused on the development of materials for the reforming of bioethanol fuels for the efficient operation of fuel cells. For example, Zhang et al. (2007) produced biohydrogen fuel from steam reforming of ethanol and glycerol over CeO_2-supported Ir, Co, and Ni catalysts.

98.5 CONCLUSION AND FUTURE RESEARCH

The brief information about the key research fronts covered by the 25 most-cited papers with at least 306 citations each is given under two primary headings: The production and utilization of bioethanol fuels.

The usual characteristics of these HCPs for the production of bioethanol fuels are that the pretreatments and hydrolysis of the feedstocks, fermentation of the resulting hydrolysates, and separation and distillation of bioethanol are the primary processes for the bioethanol fuel production from various feedstocks to improve the ethanol yield. However, there have been only seven HCPs for the first two research fronts in this context.

The key findings on these research fronts should be read in light of the increasing public concerns about climate change, GHG emissions, and global warming as these concerns have been certainly behind the boom in the research on bioethanol fuels as an alternative to crude oil-based gasoline and petrodiesel fuels in the last decades.

The recent supply shocks caused by the COVID-19 pandemic and the Russian invasion of Ukraine also highlight the importance of the production and utilization of bioethanol fuels as an alternative to crude oil-based gasoline and diesel fuels.

Since only seven HCPs have focused on the production of bioethanol fuels, the share of the feedstocks used in the production of bioethanol fuels from biomass has been necessarily low. However, as 18 HCPs have primarily focused on the utilization of bioethanol fuels, the share of bioethanol fuels as feedstocks in the sample papers has been high.

Further, it appears that bioethanol fuels as feedstocks are over-represented in the Chinese sample in relation to the world sample at large by 52.4%. This finding highlights the strength of China in the research front of the utilization of bioethanol fuels in relation to the USA and the rest of the world.

Similarly, a large number of feedstocks used in the production of bioethanol fuels from biomass are under-represented in the Chinese sample in relation to the USA and the rest of the world. For example, lignocellulosic biomass at large, lignin, grass biomass, hydrolysates, biomass at large, and industrial waste are under-represented in the Chinese sample by 3%–19% each.

As Table 98.6 shows, the most prolific thematic research fronts for this field are the utilization and production of bioethanol fuels with 72% and 28% of the HCPs, respectively. It is notable that there is no HCP on the evaluation of bioethanol fuels such as technoeconomics and environmental impact of bioethanol fuels.

Further, the most prolific research fronts for the first group are the hydrolysis and pretreatments of the biomass with 16% and 12% of the HCPs, respectively. It is notable that there are no HCPs for the other research fronts related to bioethanol fuel production.

Similarly, the most prolific research fronts for the second group are the utilization of bioethanol fuels in fuel cells and transport engines, the development of bioethanol sensors, and bioethanol-based biohydrogen fuels with 49%, 16%, 12%, and 4% of the sample papers, respectively. Further, there is no HCP for biochemical production from bioethanol fuels.

China has strength in the field of the utilization of bioethanol fuels as this field has been over-represented in relation to the world sample by 52.3%. On the contrary, China has a relative research weakness in the fields of production and evaluation of bioethanol fuels in relation to the world sample by 46.1% and 6.2%, respectively.

Further, on individual terms, China has strength also in the secondary fields of the utilization of bioethanol fuels in fuel cells and engines, and bioethanol sensors in relation to the world sample with 33.2%, 9.9%, and 9.8%, respectively. On the contrary, China has a relative weakness in the secondary fields of biomass pretreatments, bioethanol production, hydrolysate fermentation, and biomass hydrolysis in relation to the world sample with 46.5%, 21.5%, 17.9%, and 13.8%, respectively.

These studies emphasize the importance of proper incentive structures for the efficient production and utilization of bioethanol fuels in light of North's institutional framework (North, 1991). In this context, China has developed strong incentive structures for research on the efficient production and utilization of bioethanol fuels. In the light of the recent supply shocks caused primarily by the COVID-19 pandemic and the Russian invasion of Ukraine, it is expected that the incentive structures such as public funding would be enhanced to increase the share of bioethanol fuels in the Chinese fuel portfolio as a strong alternative to crude oil-based gasoline and petrodiesel fuels.

In this context, it is expected that the most prolific researchers, institutions, funding bodies, and journals in this field would have a first-mover advantage to benefit from such potential incentives. It is expected the research would focus more on the algal, wood, grass, and lignocellulosic biomass-based bioethanol fuels such as the agricultural residues- and waste biomass-based bioethanol fuels at the expense of the first generation starch and sugar-based bioethanol fuels due to the large societal concerns about the food security in the future. Similarly, it is expected that the research would focus more on the pretreatments and hydrolysis of the feedstocks for the production of bioethanol fuels and utilization and evaluation of bioethanol fuels in the future.

It is recommended that further review studies are performed for the primary research fronts of bioethanol fuels.

ACKNOWLEDGMENTS

The contribution of the highly cited researchers in the field of the bioethanol fuels has been gratefully acknowledged.

REFERENCES

Alvira, P., E. Tomas-Pejo, M. Ballesteros and M. J. Negro. 2010. Pretreatment technologies for an efficient bioethanol production process based on enzymatic hydrolysis: A review. *Bioresource Technology* 101:4851–4861.

Angelici, C., B. M. Weckhuysen and P. C. A. Bruijnincx. 2013. Chemocatalytic conversion of ethanol into butadiene and other bulk chemicals. *ChemSusChem* 6:1595–1614.

Antolini, E. 2007. Catalysts for direct ethanol fuel cells. *Journal of Power Sources* 170:1–12.

Antolini, E. 2009. Palladium in fuel cell catalysis. *Energy and Environmental Science* 2:915–931.

Bai, F. W., W. A. Anderson and M. Moo-Young. 2008. Ethanol fermentation technologies from sugar and starch feedstocks. *Biotechnology Advances* 26:89–105.

Binod, P., R. Sindhu and R. R. Singhania, et al. 2010. Bioethanol production from rice straw: An overview. *Bioresource Technology* 101:4767–4774.

Bourbonnais, R. and M. G. Paice. 1990. Oxidation of non-phenolic substrates. An expanded role for laccase in lignin biodegradation. *FEBS Letters* 267:99–102.

Burnham, J. F. 2006. Scopus database: A review. *Biomedical Digital Libraries* 3:1–8.

Cai, J. and L. Zhang. 2005. Rapid dissolution of cellulose in LiOH/urea and NaOH/urea aqueous solutions. *Macromolecular Bioscience* 5:539–548.

Cardona, C.A., J. A. Quintero and I. C. Paz. 2010. Production of bioethanol from sugarcane bagasse: Status and perspectives. *Bioresource Technology* 101:4754–4766.

Chen, L., L. Lu and H. Zhu, et al. 2017. Improved ethanol electrooxidation performance by shortening Pd-Ni active site distance in Pd-Ni-P nanocatalysts. *Nature Communications* 8:14136.

Duff, S. J. B. and W. D. Murray. 1996. Bioconversion of forest products industry waste cellulosics to fuel ethanol: A review. *Bioresource Technology* 55:1–33.

Fernando, S., S. Adhikari, C. Chandrapal and M. Murali. 2006. Biorefineries: Current status, challenges, and future direction. *Energy & Fuels* 20:1727–1737.

Galbe, M. and G. Zacchi. 2002. A review of the production of ethanol from softwood. *Applied Microbiology and Biotechnology* 59:618–628.

Hahn-Hagerdal, B., M. Galbe, M. F. Gorwa-Grauslund, G. Liden and G. Zacchi. 2006. Bio-ethanol - The fuel of tomorrow from the residues of today. *Trends in Biotechnology* 24:549–556.

Hamelinck, C. N., G. van Hooijdonk and A. P. C. Faaij. 2005. Ethanol from lignocellulosic biomass: Techno-economic performance in short-, middle- and long-term. *Biomass and Bioenergy*, 28:384–410.

Hansen, A. C., Q. Zhang and P. W. L. Lyne. 2005. Ethanol-diesel fuel blends - A review. *Bioresource Technology* 96:277–285.

Haryanto, A., S. Fernando, N. Murali and S. Adhikari. 2005. Current status of hydrogen production techniques by steam reforming of ethanol: A review. *Energy and Fuels* 19:2098–2106.

He, B. Q., S. J. Shuai, J. X. Wang and H. He. 2003. The effect of ethanol blended diesel fuels on emissions from a diesel engine. *Atmospheric Environment* 37:4965–4971.

Hendriks, A. T. W. M and G. Zeeman. 2009. Pretreatments to enhance the digestibility of lignocellulosic biomass. *Bioresource Technology* 100:10–18.

Henstra, A. M., J. Sipma, A. Rinzema and A. J. Stams. 2007. Microbiology of synthesis gas fermentation for biofuel production. *Current Opinion in Biotechnology* 18:200–206.

Hill, J., E. Nelson, D. Tilman, S. Polasky and D. Tiffany. 2006. Environmental, economic, and energetic costs and benefits of biodiesel and ethanol biofuels. *Proceedings of the National Academy of Sciences of the United States of America* 103:11206–11210.

Hill, J., S. Polasky and E. Nelson, et al. 2009. Climate change and health costs of air emissions from biofuels and gasoline. *Proceedings of the National Academy of Sciences of the United States of America* 106:2077–2082.

Hsieh, W. D., R. H. Chen, T. L. Wu and T. H. Lin. 2002. Engine performance and pollutant emission of an SI engine using ethanol-gasoline blended fuels. *Atmospheric Environment* 36:403–410.

Huang, H. J., S. Ramaswamy, U. W. Tschirner and B. V. Ramarao. 2008. A review of separation technologies in current and future biorefineries. *Separation and Purification Technology* 62:1–21.

John, R. P., G. S. Anisha, K. M. Nampoothiri and A. Pandey. 2011. Micro and macroalgal biomass: A renewable source for bioethanol. *Bioresource Technology* 102:186–193.

Jonsson, L. J. and C. Martin. 2016. Pretreatment of lignocellulose: Formation of inhibitory by-products and strategies for minimizing their effects. *Bioresource Technology* 199:103–112.

Konur, O. 2000. Creating enforceable civil rights for disabled students in higher education: An institutional theory perspective. *Disability & Society* 15:1041–1063.

Konur, O. 2002a. Access to nursing education by disabled students: Rights and duties of nursing programs. *Nurse Education Today* 22:364–374.

Konur, O. 2002b. Assessment of disabled students in higher education: Current public policy issues. *Assessment and Evaluation in Higher Education* 27:131–152.

Konur, O. 2002c. Access to employment by disabled people in the UK: Is the Disability Discrimination Act working? *International Journal of Discrimination and the Law* 5:247–279.

Konur, O. 2006a. Participation of children with dyslexia in compulsory education: Current public policy issues. *Dyslexia* 12:51–67.

Konur, O. 2006b. Teaching disabled students in higher education. *Teaching in Higher Education* 11:351–363.

Konur, O. 2007a. A judicial outcome analysis of the *Disability Discrimination Act*: A windfall for the employers? *Disability & Society* 22:187–204.

Konur, O. 2007b. Computer-assisted teaching and assessment of disabled students in higher education: The interface between academic standards and disability rights. *Journal of Computer Assisted Learning* 23:207–219.

Konur, O. 2012. The evaluation of the research on the bioethanol: A scientometric approach. *Energy Education Science and Technology Part A: Energy Science and Research* 28:1051–1064.

Konur, O. 2015. Current state of research on algal bioethanol. In *Marine Bioenergy: Trends and Developments*, Ed. S. K. Kim and C. G. Lee, pp. 217–244. Boca Raton, FL: CRC Press.

Konur, O. 2019. Cyanobacterial bioenergy and biofuels science and technology: A scientometric overview. In *Cyanobacteria: From Basic Science to Applications*, Ed. A. K. Mishra, D. N. Tiwari and A. N. Rai, pp. 419–442. Amsterdam: Elsevier.

Konur, O. 2020. The scientometric analysis of the research on the bioethanol production from green macroalgae. In *Handbook of Algal Science, Technology and Medicine*, Ed. O. Konur, pp. 385–401. London: Academic Press.

Konur, O. 2023a. Bioethanol fuels: Scientometric study. In *Bioethanol Fuel Production Processes. I: Biomass Pretreatments. Handbook of Bioethanol Fuels Volume 1*, Ed. O. Konur, pp. 47–76. Boca Raton, FL: CRC Press.

Konur, O. 2023b. Bioethanol fuels: Review. In *Bioethanol Fuel Production Processes. I: Biomass Pretreatments. Handbook of Bioethanol Fuels Volume 1*, Ed. O. Konur, pp. 77–98. Boca Raton, FL: CRC Press.

Konur, O. 2023c. Biomass pretreatments: Review. In *Bioethanol Fuel Production Processes. I: Biomass Pretreatments. Handbook of Bioethanol Fuels Volume 1*, Ed. O. Konur, pp. 125–141. Boca Raton, FL: CRC Press.

Li, C., Q. Wang and Z. K. Zhao. 2008. Acid in ionic liquid: An efficient system for hydrolysis of lignocellulose. *Green Chemistry* 10:177–182.

Li, C. and Z. K. Zhao. 2007. Efficient acid-catalyzed hydrolysis of cellulose in ionic liquid. *Advanced Synthesis and Catalysis* 349:1847–1850.

Li, D. G., H. Zhen, L. Xingcai, Z. Wu-Gao and Y. Jian-Guang. 2005. Physico-chemical properties of ethanol-diesel blend fuel and its effect on performance and emissions of diesel engines. *Renewable Energy* 30:967–976.

Liang, Z. X., T. S. Zhao, J. B. Xu and L. D. Zhu. 2009. Mechanism study of the ethanol oxidation reaction on palladium in alkaline media. *Electrochimica Acta* 54:2203–2208.

Limayem, A. and S. C. Ricke. 2012. Lignocellulosic biomass for bioethanol production: Current perspectives, potential issues and future prospects. *Progress in Energy and Combustion Science*, 38:449–467.

Lin, Y. and S. Tanaka. 2006. Ethanol fermentation from biomass resources: Current state and prospects. *Applied Microbiology and Biotechnology* 69:627–642.

Liu, J., X. Wang, Q. Peng and Y. Li. 2005. Vanadium pentoxide nanobelts: Highly selective and stable ethanol sensor materials. *Advanced Materials* 17:764–767.

Ma, X., L. Sun and C. Song. 2002. A new approach to deep desulfurization of gasoline, diesel fuel and jet fuel by selective adsorption for ultra-clean fuels and for fuel cell applications. *Catalysis Today* 77:107–116.

Morschbacker, A. 2009. Bio-ethanol based ethylene. *Polymer Reviews* 49:79–84.
Mosier, N., C. Wyman and B. Dale, et al. 2005. Features of promising technologies for pretreatment of lignocellulosic biomass. *Bioresource Technology* 96:673–686.
Murdoch, M., G. I. N. Waterhouse and M. A. Nadeem, et al. 2011. The effect of gold loading and particle size on photocatalytic hydrogen production from ethanol over Au/TiO_2 nanoparticles. *Nature Chemistry* 3:489–492.
Najafi, G., B. Ghobadian and T. Tavakoli, et al. 2009. Performance and exhaust emissions of a gasoline engine with ethanol blended gasoline fuels using artificial neural network. *Applied Energy* 86:630–639.
Newman, P. W. G. and J. R. Kenworthy. 1989. Gasoline consumption and cities: A comparison of U.S. cities with a global survey. *Journal of the American Planning Association* 55:24–37.
North, D. C. 1991. Institutions. *Journal of Economic Perspectives* 5:97–112.
Olsson, L. and B. Hahn-Hagerdal. 1996. Fermentation of lignocellulosic hydrolysates for ethanol production. *Enzyme and Microbial Technology* 18:312–331.
Palmqvist, E. and B. Hahn-Hagerdal. 2000. Fermentation of lignocellulosic hydrolysates. I: Inhibition and detoxification. *Bioresource Technology* 74:17–24.
Pimentel, D. and T. W. Patzek. 2005. Ethanol production using corn, switchgrass, and wood; Biodiesel production using soybean and sunflower. *Natural Resources Research* 14:65–76.
Pinkert, A., K. N. Marsh, S. Pang and M. P. Staiger. 2009. Ionic liquids and their interaction with cellulose. *Chemical Reviews* 109:6712–6728.
Ravindran, R. and A. K. Jaiswal. 2016. A comprehensive review on pre-treatment strategy for lignocellulosic food industry waste: Challenges and opportunities. *Bioresource Technology* 199:92–102.
Sanchez, C. 2009. Lignocellulosic residues: Biodegradation and bioconversion by fungi. *Biotechnology Advances* 27:185–194.
Sanchez, O. J. and C. A. Cardona. 2008. Trends in biotechnological production of fuel ethanol from different feedstocks. *Bioresource Technology* 99:5270–5295.
Sano, T., H. Yanagishita, Y. Kiyozumi, F. Mizukami and K. Haraya. 1994. Separation of ethanol/water mixture by silicalite membrane on pervaporation. *Journal of Membrane Science* 95:221–228.
Shen, S. Y., T. S. Zhao, J. B. Xu and Y. S. Li. 2010. Synthesis of PdNi catalysts for the oxidation of ethanol in alkaline direct ethanol fuel cells. *Journal of Power Sources* 195:1001–1006.
Sukumaran, R. K., R. R. Singhania, G. M. Mathew and A. Pandey. 2009. Cellulase production using biomass feed stock and its application in lignocellulose saccharification for bio-ethanol production. *Renewable Energy* 34:421–424.
Sun, X. F., F. Xu, R. C. Sun, P. Fowler and M. S. Baird. 2005. Characteristics of degraded cellulose obtained from steam-exploded wheat straw. *Carbohydrate Research* 340:97–106.
Sun, Y. and J. Cheng. 2002. Hydrolysis of lignocellulosic materials for ethanol production: A review. *Bioresource Technology* 83:1–11.
Taherzadeh, M. J. and K. Karimi. 2007. Enzyme-based hydrolysis processes for ethanol from lignocellulosic materials: A review. *Bioresources* 2:707–738.
Taherzadeh, M. J. and K. Karimi. 2008. Pretreatment of lignocellulosic wastes to improve ethanol and biogas production: A review. *International Journal of Molecular Sciences* 9:1621–1651.
Talebnia, F., D. Karakashev and I. Angelidaki. 2010. Production of bioethanol from wheat straw: An overview on pretreatment, hydrolysis and fermentation. *Bioresource Technology* 101:4744–4753.
Tian, N., Z. Y. Zhou, N. F. Yu, L. Y. Wang and S. G. Sun. 2010. Direct electrodeposition of tetrahexahedral Pd nanocrystals with high-index facets and high catalytic activity for ethanol electrooxidation. *Journal of the American Chemical Society* 132:7580–7581.
Vane, L. M. 2005. A review of pervaporation for product recovery from biomass fermentation processes. *Journal of Chemical Technology and Biotechnology* 80:603–629.
Wan, Q., Q. H. Li and Y. J. Chen, et al. 2004. Fabrication and ethanol sensing characteristics of ZnO nanowire gas sensors. *Applied Physics Letters* 84:3654–3656.
Wang, A. L., H. Xu and J. X. Feng, et al. 2013. Design of Pd/PANI/Pd sandwich-structured nanotube array catalysts with special shape effects and synergistic effects for ethanol electrooxidation. *Journal of the American Chemical Society* 135:10703–10709.
Wang, H. F. and Z. P. Liu. 2008. Comprehensive mechanism and structure-sensitivity of ethanol oxidation on platinum: New transition-state searching method for resolving the complex reaction network. *Journal of the American Chemical Society* 130:10996–11004.
Wang, L., Y. Kang and X. Liu, et al. 2012. ZnO nanorod gas sensor for ethanol detection. *Sensors and Actuators, B: Chemical* 162:237–243.

Xing-Cai, L., Y. Jian-Guang, Z. Wu-Gao and H. Zhen. 2004. Effect of cetane number improver on heat release rate and emissions of high speed diesel engine fueled with ethanol-diesel blend fuel. *Fuel* 83:2013–2020.

Xu, A., J. Wang and H. Wang. 2010. Effects of anionic structure and lithium salts addition on the dissolution of cellulose in 1-butyl-3-methylimidazolium-based ionic liquid solvent systems. *Green Chemistry* 12:268–275.

Xu, C., L. Cheng, P. Shen and Y. Liu. 2007a. Methanol and ethanol electrooxidation on Pt and Pd supported on carbon microspheres in alkaline media. *Electrochemistry Communications* 9:997–1001.

Xu, C., P. K. Shen and Y. Liu. 2007c. Ethanol electrooxidation on Pt/C and Pd/C catalysts promoted with oxide. *Journal of Power Sources* 164:527–531.

Xu, C., H. Wang, P. K. Shen and S. P. Jiang. 2007b. Highly ordered Pd nanowire arrays as effective electrocatalysts for ethanol oxidation in direct alcohol fuel cells. *Advanced Materials* 19:4256–4259.

Zhang, B., X. Tang, Y. Li, Y. Xu and W. Shen. 2007. Hydrogen production from steam reforming of ethanol and glycerol over ceria-supported metal catalysts. *International Journal of Hydrogen Energy* 32:2367–2373.

Zhang, Y. H. P. and L. R. Lynd. 2004. Toward an aggregated understanding of enzymatic hydrolysis of cellulose: Noncomplexed cellulase systems. *Biotechnology and Bioengineering* 88:797–824.

Zhao, H., C. L. Jones and G. A. Baker, et al. 2009. Regenerating cellulose from ionic liquids for an accelerated enzymatic hydrolysis. *Journal of Biotechnology* 139:47–54.

Zhou, W., Z. Zhou and S. Song, et al. 2003. Pt based anode catalysts for direct ethanol fuel cells. *Applied Catalysis B: Environmental* 46:273–285.

Zhu, J. Y. and X. J. Pan. 2010. Woody biomass pretreatment for cellulosic ethanol production: Technology and energy consumption evaluation. *Bioresource Technology* 101:4992–5002.

Zhu, J. Y., X. J. Pan, G. S. Wang and R. Gleisner. 2009. Sulfite pretreatment (SPORL) for robust enzymatic saccharification of spruce and red pine. *Bioresource Technology* 100:2411–2418.

Zhu, L., C. S. Cheung, W. G. Zhang and Z. Huang. 2011. Combustion, performance and emission characteristics of a di diesel engine fueled with ethanol-biodiesel blends. *Fuel* 90:1743–1750.

99 The US Experience of Bioethanol Fuels
Scientometric Study

Ozcan Konur
(Formerly) Ankara Yildirim Beyazit University

99.1 INTRODUCTION

The crude oil-based gasoline fuels (Ma et al., 2002; Newman and Kenworthy, 1989) have been widely used in the transportation sector since the 1920s. However, there have been great public concerns over the adverse environmental and human impact of these fuels (Hill et al., 2006, 2009). Hence, biomass-based bioethanol fuels (Hill et al., 2006; Konur, 2012e, 2015, 2020a) have increasingly been used in blending gasoline fuels (Hsieh et al., 2002; Najafi et al., 2009), in the fuel cells (Antolini, 2007, 2009), and in the biochemical production (Angelici et al., 2013; Morschbacker, 2009) in a biorefinery context (Fernando et al., 2006; Huang et al., 2008).

Bioethanol fuels also play a critical role in maintaining the energy security (Kruyt et al., 2009; Winzer, 2012) in the supply shocks (Kilian, 2008, 2009) related to oil price shocks (Hamilton, 1983, 2003), COVID-19 pandemics (Fauci et al., 2020; Li et al., 2020), or wars (Hamilton, 1983; Jones, 2012) in the aftermath of the Russian invasion of Ukraine (Reeves, 2014).

However, it is necessary to pretreat the biomass (Mosier et al., 2005; Taherzadeh and Karimi, 2008; Yang and Wyman, 2008) to enhance the yield of the bioethanol (Hahn-Hagerdal et al., 2006; Sanchez and Cardona, 2008) prior to the bioethanol production through the hydrolysis (Sun and Cheng, 2002; Taherzadeh and Karimi, 2007) and fermentation (Lin and Tanaka, 2006; Olsson and Hahn-Hagerdal, 1996) of the biomass and the resulting hydrolysates, respectively.

The USA has been one of the most prolific countries engaged in the bioethanol fuel research (Konur, 2023a,b). The US research in the field of the bioethanol fuels has intensified in this context in the key research fronts of the production of bioethanol fuels, and to a lesser extent, utilization (Antolini, 2007; Hansen et al., 2005) and evaluation (Hamelinck et al., 2005; Pimentel and Patzek, 2005) of bioethanol fuels. For the first research front, pretreatment (Alvira et al., 2010; Hendriks and Zeeman, 2009) and hydrolysis (Alvira et al., 2010; Sun and Cheng, 2002) of the feedstocks, fermentation of the feedstock-based hydrolysates (Jonsson and Martin, 2016; Lin and Tanaka, 2006), production of bioethanol fuels (Limayem and Ricke, 2012, Lin and Tanaka, 2006), and to a lesser extent distillation of bioethanol fuels (Sano et al., 1994; Vane, 2005) are the key research areas while for the second research front, direct utilization of bioethanol fuels in fuel cells (Antolini, 2007, 2009) and transport engines (Hsieh et al., 2002; Najafi et al., 2009), bioethanol-based biohydrogen fuels (Haryanto et al., 2005; Ni et al., 2007), bioethanol sensors (Liu et al., 2005; Wan et al., 2004), and bioethanol-based biochemicals (Angelici et al., 2013; Morschbacker, 2009) are the key research areas.

The research in this field has also intensified for the feedstocks of lignocellulosic biomass at large (Hendriks and Zeeman, 2009; Sun and Cheng, 2002), bioethanol fuels for their utilization (Antolini, 2007; Haryanto et al., 2005), cellulose (Pinkert et al., 2009; Zhang and Lynd, 2004), starch feedstock residues (Binod et al., 2010; Talebnia et al., 2010), wood biomass (Galbe and Zacchi, 2002; Zhu and Pan, 2010), and to a lesser extent industrial waste (Cardona et al., 2010; Prasad et al., 2007), lignin

DOI: 10.1201/9781003226574-132

(Bourbonnais and Paice, 1990; Kirk and Farrell, 1987), grass (Keshwani and Cheng, 2009; Pimentel and Patzek, 2005), hydrolysates (Palmqvist and Hahn-Hagerdal, 2000a,b), biomass at large (Lin and Tanaka, 2006; Zabed et al., 2017), sugar feedstocks (Bai et al., 2008; Canilha et al., 2012), sugar feedstock residues (Cardona et al., 2010; Laser et al., 2002), algal biomass (Ho et al., 2013; John et al., 2011), lignocellulosic wastes (Sanchez, 2009; Saini et al., 2015), starch feedstocks (Bai et al., 2008; Bothast and Schlicher, 2005), urban wastes (Prasad et al., 2007; Ravindran and Jaiswal, 2016), forestry wastes (Duff and Murray, 1996), food wastes (Guimaraes et al., 2010; Ravindran and Jaiswal, 2016), biosyngas (Henstra et al., 2007; Munasinghe and Khanal, 2010), and plants (Nigam, 2002; Sukumaran et al., 2009).

However, it is essential to develop efficient incentive structures (North, 1991) for the primary stakeholders to enhance the research in this field (Konur, 2000, 2002a,b,c, 2006a,b, 2007a,b). The scientometric analysis has been used in this context to inform the primary stakeholders about the current state of the research in a selected research field (Garfield, 1955; Konur, 2011, 2012a,b,c,d,e,f,g,h,i, 2015, 2018b, 2019, 2020a).

As there have been no published current scientometric studies in this field, this book chapter presents a scientometric study of the research in the bioethanol fuels. It examines the scientometric characteristics of both the sample and population data presenting scientometric characteristics of these both datasets in the order of documents, authors, publication years, institutions, funding bodies, source titles, countries, Scopus subject categories, Scopus keywords, and research fronts.

99.2 MATERIALS AND METHODS

The search for this study was carried out using Scopus database (Burnham, 2006) in November 2022.

As a first step for the search of the relevant literature, the keywords were selected using the most-cited first 300 population papers for each research front. The selected keyword list was then optimized to obtain a representative sample of papers for the each research field. These keyword lists were then integrated to obtain the keyword list for the research field of bioethanol fuels research field (Konur, 2023a,b). Additionally, a second keyword set was developed for the USA: 'AFFILCOUNTRY (united states) OR TITLE ("united states" OR usa OR us) OR AUTHKEY ("united states" OR usa OR us) OR SRCTITLE ("united states" OR usa OR us)'. These two keyword sets were integrated with the use of 'AND' keyword. Thus, 12,992 references were located for this field.

As a second step, two sets of data were used for this study. First, a population sample of 12,992 papers was used to examine the scientometric characteristics of the population data. Secondly, a sample of 260 most-cited papers, corresponding to 2% of the population papers, was used to examine the scientometric characteristics of these citation classics.

The scientometric characteristics of these both sample and population datasets were presented in the order of documents, authors, publication years, institutions, funding bodies, source titles, countries, Scopus subject categories, Scopus keywords, and research fronts.

Lastly, the key scientometric findings for both datasets were discussed to highlight the research landscape for the bioethanol fuels. Additionally, a number of brief conclusions were drawn and a number of relevant recommendations were made to enhance the future research landscape.

99.3 RESULTS

99.3.1 The Most-Prolific Documents in the US Experience of Bioethanol Fuels

The information on the types of documents for both datasets is given in Table 99.1. The articles and conference papers, published in journals, dominate both the sample (80%) and population (93%) papers with 13% deficit. Further, review papers and short surveys have a 17% surplus as

TABLE 99.1
Documents in the US Experience of Bioethanol Fuels

Documents	Sample Dataset (%)	Population Dataset (%)	Surplus (%)
Article	78.5	88.6	−10.1
Review	18.8	3.2	15.6
Short survey	1.5	0.2	1.3
Conference paper	1.2	4.6	−3.4
Book chapter	0.0	2.1	−2.1
Letter	0.0	0.7	−0.7
Note	0.0	0.4	−0.4
Book	0.0	0.1	−0.1
Editorial	0.0	0.1	−0.1
Sample size	260	12,992	

Population dataset: the number of papers (%) in the set of the 12,992 population papers; sample dataset: the number of papers (%) in the set of 260 highly cited papers.

they are overrepresented in the sample papers as they constitute 20% and 3% of the sample and population papers, respectively. Additionally, 2% of the sample papers were published as book chapters.

It is further notable that 97%, 2%, and 1% of the population papers were published in journals, book series, and books, respectively. Similarly, 98% and 2% of the sample papers were published in the journals and book series, respectively.

99.3.2 The Most-Prolific Authors in the US Experience of Bioethanol Fuels

The information about the most-prolific 31 authors with at least 1.5% of sample papers each is given in Table 99.2. The most-prolific authors are Charles E. Wyman and Bruce E. Dale with 6.9% and 5.4% of the sample papers, respectively, followed by Lee R. Lynd, Blake A. Simmons, Arthur J. Ragauskas, Michael E. Himmel, and Michael R. Ladisch with 3.1%–3.9% of the sample papers each.

On the other hand, the most influential authors are Charles E. Wyman and Bruce E. Dale with 5.8% and 4.3% surplus, respectively, followed by Lee R. Lynd, Michael R. Ladisch, Blake A. Simmons, Michael E. Himmel, Robin D. Rogers, and Igor V. Grigoriev with 2.2%–3.1% surplus each.

The most-prolific institution for the sample dataset is the Joint Genome Institute with four authors, followed by National Renewable Energy Laboratory (NREL) with three authors. The other prolific institutions are Purdue University, USDA Agricultural Service, and USDA Forest Service with two authors each.

The most-prolific research front for these top authors is the pretreatments of the feedstocks with 31 authors followed by the hydrolysis of the feedstocks with 21 authors. The other prolific research fronts are the fermentation of the feedstock-based hydrolysates and the bioethanol production with ten and nine authors, respectively. The other minor research front is evaluation of bioethanol fuels with one author.

On the other hand, there is significant gender deficit (Beaudry and Lariviere, 2016) for the sample dataset as surprisingly only two of these top researchers are female with a representation rate of 6%.

Additionally, there are other USA-based authors with the relatively low citation impact and with 0.2%–0.9% of the population papers each: Venkatesh Balan, Yong-Su Jin, Hasan Jameel, Vijay

TABLE 99.2
Most-Prolific Authors in the US Experience of Bioethanol Fuels

No.	Author Name	Author Code	Sample Papers (%)	Population Papers (%)	Surplus	Institution	HI	N	Res. Front
1	Wyman, Charles E.	7004396809	6.9	1.1	5.8	Univ. Calf. Riverside	80	287	P, H, F, R, E
2	Dale, Bruce E.	7201511969	5.4	1.1	4.3	Michigan State Univ.	92	430	P, H, F, R
3	Lynd, Lee R.	35586183800	3.8	0.7	3.1	Dartmouth Coll.	75	287	P, H, F, R
4	Simmons, Blake A.	7102183263	3.5	1.0	2.5	Lawrence Berkeley Natl. Lab.	76	446	P, H
5	Ragauskas, Arthur J.	7006265204	3.1	1.4	1.7	Univ. Tennessee	95	766	P, H, F, R
6	Himmel, Michael E.	7007125552	3.1	0.6	2.5	Natl. Renew. Ener. Lab.	74	423	P, H
7	Ladisch, Michael R.	7005670397	3.1	0.5	2.6	Purdue Univ.	60	292	P, H, F, R
8	Holtzapple, Mark T.	7004167004	2.7	0.6	2.1	Texas A&M Univ.	47	199	P
9	Ingram, Lonnie O.	7102962097	2.3	0.8	1.5	Univ. Florida	75	279	P, F, R
10	Singh, Seema*	35264950300	2.3	0.6	1.7	Sandia Natl. Lab.	59	186	P, H
11	Lee, Yoon Y.	8948274900	2.3	0.5	1.8	Auburn Univ.	45	102	P, H
12	Cullen, Dan	7202109135	2.3	0.2	2.1	USDA Forest Serv.	46	120	P, H
13	Gold, Michael H.	7402444296	2.3	0.2	2.1	Oregon Hlth. Sci. Univ.	57	148	P, H
14	Grigoriev, Igor V.	25027225800	2.3	0.1	2.2	Jnt. Genome Inst.	116	354	P, H
15	Rogers, Robin D.	35474829200	2.3	0.0	2.3	Univ. Alabama	118	892	P
16	Zhu, Junyong	7405692678	1.9	0.6	1.3	USDA Forest Serv.	64	311	P, H
17	Cotta, Michael A.	7006656876	1.9	0.5	1.4	USDA Agr. Res. Serv.	53	186	P, H, F, R
18	Beckham, Gregg T.	16240926200	1.9	0.3	1.6	Natl. Renew. Ener. Lab.	73	271	P
19	Yang, Bin	7404473046	1.9	0.3	1.6	Washington State Univ.	41	98	P, H
20	Elander, Richard T.	6603931116	1.9	0.2	1.7	Natl. Renew. Ener. Lab.	31	61	P, H
21	Dien, Bruce, S	6603685796	1.5	0.7	0.8	USDA Agr. Serv.	63	188	P, H,F, R
22	Jeffries, Thomas W.	7005806269	1.5	0.5	1.0	Xylome Inc.	59	156	P, F
23	Mosier, Nathan S.	6602426392	1.5	0.3	1.2	Purdue Univ.	43	115	P, H
24	Pan, Xuejun	57203296000	1.5	0.3	1.2	Univ. Wisconsin	45	118	P, H
25	Baker, Scott E.	35232609200	1.5	0.2	1.3	Pacific NW Natl Lab.	51	149	P
26	Blanch, Harvey W.	7006259341	1.5	0.2	1.3	Univ. Calif. Riverside	80	336	P, H, F, R
27	Lindquist, Erika*	34571433600	1.5	0.1	1.4	Jnt. Genome Inst.	71	116	P
28	Martinez, Diego	7202958664	1.5	0.1	1.4	Mass. Inst. Technol.	29	35	P
29	Salamov, Asaf	53981779600	1.5	0.1	1.4	Jnt. Genome Inst.	79	136	P
30	Schmutz, Jeremy	56549573500	1.5	0.1	1.4	Jnt. Genome Inst.	92	274	P
31	Vogel, Kenneth P.	7102498441	1.5	0.1	1.4	Univ. Nebraska Lincoln	65	148	P, H

*: female; author code: the unique code given by Scopus to the authors; C: utilization of bioethanol fuels in fuel cells; E: evaluation of bioethanol fuels; F, fermentation of the feedstock-based hydrolysates; H: hydrolysis of the feedstock; HI: H index; I: utilization of bioethanol fuels in the transport engines; N: number of papers published by each author; P: pretreatment of the feedstock; population papers: the number of papers authored in the population dataset; R: bioethanol fuel production; sample papers: the number of papers authored in the sample dataset.

Singh, Yunqiao Pu, Chang Geun Yoo, Xianzhi Meng, Shijie Liu, Rajeev Kumar, John Ralph, Rodney J. Bothast, Jian Shi, Badal C. Saha, George T. Tsao, Stephen R. Decker, Sunkyu Park, Melvin P. Tucker, Shang-Tian Yang, Hou-Min Chang, Sishir P. S. Chundawat, David B. Hodge, Daniel J. Schell, Mark R. Wilkins, Shulin Chen, Karel G. Grohmann, Nancy N. Nichols, Keelnatham T. Shanmugam, Bryon S. Donohoe, Tae Hyun Kim, and Birgitte K. Ahring.

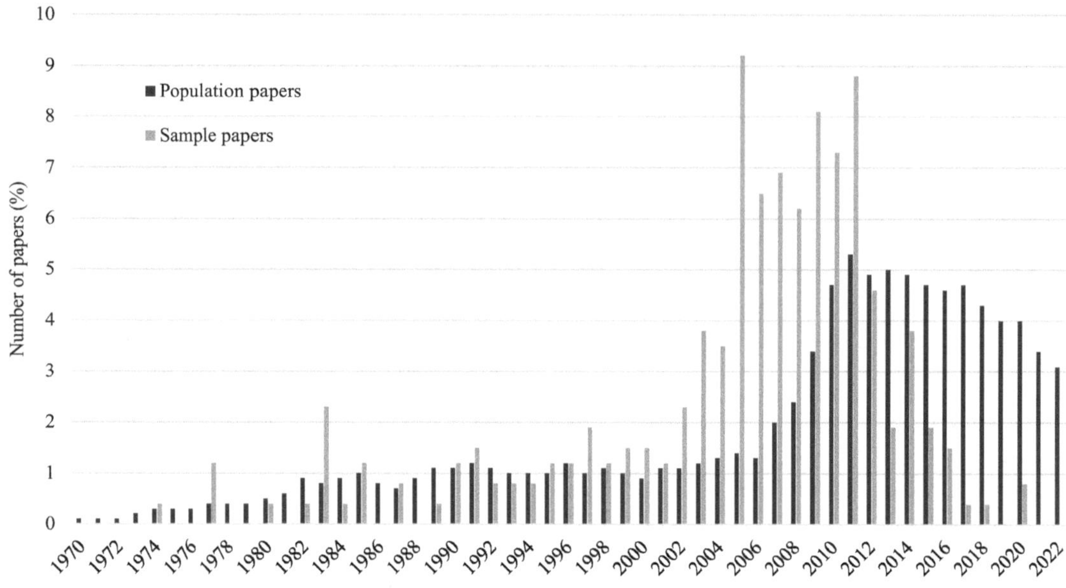

FIGURE 99.1 The research output by years regarding the US experience of bioethanol fuels.

99.3.3 The Most-Prolific Research Output by Years in the US Experience of Bioethanol Fuels

Information about papers published between 1970 and 2022 is given in Figure 99.1. This figure clearly shows that the bulk of the research papers in the population dataset were published primarily in the 2010s, the early 2020s, and 2000s with 47%, 11%, and 16% of the population dataset, respectively. Similarly, the publication rates for the 1990s, 1980s, and 1970s were 11%, 8%, and 3%, respectively. Further, the rate for the pre-1970s was 4%. It is notable that there has been a rising trend for the population papers between 2006 and 2011 and thereafter it declined linearly.

Similarly, the bulk of the research papers in the sample dataset were published in the 2000s, 2010s, and 1990s with 49%, 24%, and 12% of the sample dataset, respectively. Similarly, the publication rates for the 1980s and 1970s were 6% and 2% of the sample papers, respectively. Further, the rate for the pre-1970s was 0%.

The most-prolific publication year for the population dataset was 2011 with 5.3% of the dataset, while 65% of the population papers were published between 2007 and 2022. Similarly, 73% of the sample papers were published between 2002 and 2014 while the most-prolific publication years were 2005 and 2011 with 89.2% and 8.8% of the sample papers, respectively.

99.3.4 The Most-Prolific Institutions in the US Experience of Bioethanol Fuels

Information about the most-prolific 26 US institutions publishing papers on the bioethanol fuels with at least 2.3% of the sample papers each is given in Table 99.3.

The most-prolific institutions are the NREL with 11.5% of the sample papers, followed by the Dartmouth College, Michigan State University, and USDA Forest Service with 8.1%, 6.9%, and 6.9% of the sample papers, respectively. The other prolific institutions are University of Wisconsin Madison, Purdue University, Oak Ridge National Laboratory, NC State University, and University of California Berkeley with 4.6%–5.8% of the sample papers each.

On the other hand, the institutions with the most impact are the NREL and Dartmouth College with 7.8% and 6.9% surplus, respectively. The other influential institutions are Michigan State

TABLE 99.3
The Most-Prolific Institutions in the US Experience of Bioethanol Fuels

No.	Institutions	Sample Papers (%)	Population Papers (%)	Surplus (%)
1	Natl. Renew. Ener. Lab.	11.5	3.7	7.8
2	Dartmouth Coll.	8.1	1.2	6.9
3	Michigan State Univ.	6.9	3.0	3.9
4	USDA Forest Serv.	6.9	3.0	3.9
5	Univ. Wisconsin Madison	5.8	3.6	2.2
6	Purdue Univ.	5.8	2.7	3.1
7	Oak Ridge Natl. Lab.	5.0	3.5	1.5
8	NC State Univ.	4.6	3.2	1.4
9	Univ. Calif. Berkeley	4.6	2.1	2.5
10	USDA Agr. Res. Serv.	4.2	4.6	−0.4
11	Georgia Inst. Technol.	3.5	1.4	2.1
12	Jnt. Bioener. Inst.	3.5	1.3	2.2
13	Sandia Natl. Lab.	3.5	1.2	2.3
14	Univ. Ill. U. C.	3.1	3.4	−0.3
15	Pacific NW Natl. Lab.	3.1	0.9	2.2
16	Univ. Minnesota	2.7	2.1	0.6
17	Lawrence Berkeley Natl. Lab.	2.7	1.7	1.0
18	Univ. Calif. Riverside	2.7	1.4	1.3
19	Joint Genome Inst.	2.7	0.5	2.2
20	Novozymes Biotech Inc.	2.7	0.4	2.3
21	Univ. Florida	2.3	2.2	0.1
22	Texas A&M Univ.	2.3	1.8	0.5
23	Auburn Univ.	2.3	1.4	0.9
24	Massachusetts Inst. Technol.	2.3	0.9	1.4
25	Univ. Alabama	2.3	0.2	2.1

University, USDA Forest Service, Purdue University, and University of California Berkeley with 2.5%–3.9% surplus each.

Additionally, there are other US institutions with the relatively low citation impact and with 0.8%–2.2% of the population papers each: University of Tennessee Knoxville, Iowa State University, University of California Davis, University of Georgia, Cornell University, DOE Bioenergy Research Centers, Ohio State University, Kansas State University, Virginia Polytechnic Institute and State University, State University of New York, Pennsylvania State University, Washington State University Pullman, Oregon State University, University of Nebraska–Lincoln, University of Kentucky, University of Maine, Colorado State University, Oklahoma State University, University of Michigan Ann Arbor, and Argonne National Laboratory.

On the other hand, the most-prolific collaborating overseas institutions are University of Sao Paulo, Chinese Academy of Sciences, Nanjing Forestry University, University of British Columbia, South China University of Technology, and State University of Campinas.

99.3.5 The Most-Prolific Funding Bodies in the US Experience of Bioethanol Fuels

Information about the most-prolific 12 funding bodies funding at least 1.2% of the sample papers each is given in Table 99.4. Further, only 36% and 39% of the sample and population papers each were funded, respectively.

TABLE 99.4
The Most-Prolific Funding Bodies in the US Experience of Bioethanol Fuels

No.	Funding Bodies	Sample Paper No. (%)	Population Paper No. (%)	Surplus (%)
1	U.S. Department of Energy	10.4	8.4	2.0
2	National Science Foundation	6.5	7.2	−0.7
3	Office of Science	3.8	3.9	−0.1
4	U.S. Department of Agriculture	2.7	2.8	−0.1
5	National Institute of General Medical Sciences	2.3	1.6	0.7
6	NREL	2.3	1.0	1.3
7	Basic Energy Sciences	1.5	1.0	0.5
8	National Institute of Standards and Technology	1.5	0.1	1.4
9	Biological and Environmental Research	1.2	2.2	−1.0
10	Energy Biosciences Institute	1.2	0.6	0.6
11	Lawrence Livermore National Laboratory	1.2	0.6	0.6
12	University of California	1.2	0.1	1.1

The most-prolific funding body is the U.S. Department of Energy with 10.4% of the sample papers, followed by the National Science Foundation with 6.5% of the sample papers. The other prolific funding bodies are the Office of Science, U.S. Department of Agriculture, National Institute of General Medical Sciences, and NREL with 2.3%–3.8% of the sample papers each.

The funding body with the most citation impact is the U.S. Department of Energy with 2% surplus, followed by the National Institute of Standards and Technology, NREL, and University of California with 1.1%–1.4% surplus each. Further, the funding bodies with the least citation impact are the Biological and Environmental Research, National Science Foundation, Office of Science, and U.S. Department of Agriculture with 0.1%–1% deficit each.

Additionally, a number of US funding bodies with relatively low citation impact have also substantially contributed to the research in this field: National Institute of Food and Agriculture, Office of Energy Efficiency and Renewable Energy, Oak Ridge National Laboratory, Bioenergy Technologies Office, Great Lakes Bioenergy Research Center, Center for Bioenergy Innovation, U.S. Forest Service, Directorate for Engineering, U.S. Environmental Protection Agency, Agricultural Research Service, Alabama Agricultural Experiment Station, National Center for Research Resources, Ford Motor Company, National Aeronautics and Space Administration, Northwest Advanced Renewables Alliance, U.S. Department of Transportation, Office of Naval Research, Directorate for Biological Sciences, and Laboratory Directed Research and Development.

On the other hand, a number of overseas funding bodies jointly funded the research in this field together with the US funding bodies: National Natural Science Foundation of China, China Scholarship Council, National Council for Scientific and Technological Development, National Key Research and Development Program of China, Higher Education Personnel Improvement Coordination, National Research Foundation of Korea, Research Support Foundation of the State of Sao Paulo, Fundamental Research Funds for the Central Universities, Natural Sciences and Engineering Research Council of Canada, Priority Academic Program Development of Jiangsu Higher Education Institutions, Natural Science Foundation of Jiangsu Province, Japan Society for the Promotion of Science, Seventh Framework Program, China Postdoctoral Science Foundation, Chemical Sciences, Geosciences, and Biosciences Division, Chinese Academy of Sciences, and Ministry of Education, Science and Technology.

99.3.6 The Most-Prolific Source Titles in the US Experience of Bioethanol Fuels

Information about the most-prolific 14 source titles publishing at least 1.5% of the sample papers each in the bioethanol fuels is given in Table 99.5.

The most-prolific source title is the Bioresource Technology with 15.8% of the sample papers, followed by Biotechnology and Bioengineering with 10.4% of the sample papers. The other prolific titles are the Proceedings of the National Academy of Sciences of the United States of America, Applied and Environmental Microbiology, and Science with 3.8%–5% of the sample papers each.

On the other hand, the source title with the most citation impact is the Bioresource Technology with 10.3% surplus, followed by Biotechnology and Bioengineering with 6.5% surplus. The other influential titles are the Proceedings of the National Academy of Sciences of the United States of America, Applied and Environmental Microbiology, and Science with 3.8%–5% surplus each.

In addition, a number of source titles with relatively low citation impact have also contributed to this field substantially: Applied Biochemistry and Biotechnology, SAE Technical Papers, Journal of the American Chemical Society, ACS Sustainable Chemistry and Engineering, Journal of Agricultural and Food Chemistry, Industrial and Engineering Chemistry Research, Industrial Crops and Products, Fuel, Biotechnology Letters, Bioresources, Bioenergy Research, Energy and Fuels, Holzforschung, Journal of Biological Chemistry, Plos One, Journal of Industrial Microbiology and Biotechnology, Environmental Science and Technology, Journal of Bacteriology, Enzyme and Microbial Technology, Industrial and Engineering Chemistry, Cellulose, Metabolic Engineering, Scientific Reports, Process Biochemistry, Biochemistry, ACS Symposium Series, Applied Energy, Journal of Membrane Science, RSC Advances, and Biofuels, Bioproducts, and Biorefining.

99.3.7 The Most-Prolific Collaborating Countries in the US Experience of Bioethanol Fuels

Information about the most-prolific 12 collaborating countries publishing at least 1.5% of sample papers each in the bioethanol fuels is given in Table 99.6.

TABLE 99.5
The Most-Prolific Source Titles in the US Experience of Bioethanol Fuels

No.	Source Titles	Sample Papers (%)	Population Papers (%)	Surplus (%)
1	Bioresource Technology	15.8	5.5	10.3
2	Biotechnology and Bioengineering	10.4	3.9	6.5
3	Proceedings of the National Academy of Sciences of the United States of America	5.0	0.7	4.3
4	Applied and Environmental Microbiology	4.6	2.3	2.3
5	Science	3.8	0.2	3.6
6	Biomass and Bioenergy	2.7	1.7	1.0
7	Green Chemistry	2.7	0.9	1.8
8	Current Opinion in Biotechnology	2.7	0.2	2.5
9	Biotechnology Progress	2.3	1.1	1.2
10	Nature Biotechnology	1.9	0.0	1.9
11	Biotechnology for Biofuels	1.5	2.1	–0.6
12	Applied Biochemistry and Biotechnology Part A	1.5	1.2	0.3
13	Applied Microbiology and Biotechnology	1.5	1.1	0.4
14	Journal of Biotechnology	1.5	0.4	1.1

TABLE 99.6
The Most-Prolific Collaborating Countries in the US Experience of Bioethanol Fuels

No.	Countries	Sample Papers (%)	Population Papers (%)	Surplus (%)
1	China	3.8	10.5	–6.7
2	Canada	3.1	2.7	0.4
3	S. Korea	2.7	2.6	0.1
4	UK	2.7	1.2	1.5
5	France	2.7	1.0	1.7
6	Spain	2.7	0.9	1.8
7	Germany	2.3	1.3	1.0
8	Finland	1.9	0.7	1.2
9	Chile	1.9	0.3	1.6
10	Denmark	1.5	0.8	0.7
11	Sweden	1.5	0.8	0.7
12	S. Africa	1.5	0.2	1.3

The most collaborating prolific country is China with 3.8% of the sample papers, followed by Canada with 3.1% of the sample papers. The other prolific countries are S. Korea, the UK, France, and Spain with 2.7% of the sample papers each. Seven European countries listed in Table 99.6 as a whole produce 15% and 7% of the sample and population papers, respectively, with 8% surplus.

On the other hand, the country with the most citation impact is Spain with 1.8% surplus, followed by France, Chile, and the UK with 1.5%–1.7% surplus each. Similarly, the country with the least citation impact is China with 6.7% deficit.

Additionally, there are other collaborating countries with relatively low citation impact: Brazil, India, Japan, Australia, Mexico, Taiwan, Italy, Netherlands, Turkey, Israel, Thailand, Belgium, Egypt, Iran, Saudi Arabia, Switzerland, and Russia.

99.3.8 The Most-Prolific Scopus Subject Categories in the US Experience of Bioethanol Fuels

Information about the most-prolific ten Scopus subject categories indexing at least 6.2% of the sample papers each is given in Table 99.7.

The most-prolific Scopus subject category in the US experience of bioethanol fuels is the Chemical Engineering with 49% of the sample papers, followed by Biochemistry, Genetics, and Molecular Biology with 43% of the sample papers. The other prolific subject categories are Environmental Science, Immunology and Microbiology, and Energy with 28%–34% of the sample papers each. It is notable that Social Sciences including Economics and Business account for only 1% and 4% of the sample and population studies, respectively, mostly published outside the energy and fuel journals.

On the other hand, the Scopus subject category with the most citation impact is the Chemical Engineering with 13% surplus. The other influential subject areas are Multidisciplinary, Immunology and Microbiology, Environmental Science, Energy, Materials Science, and Biochemistry, Genetics and Molecular Biology with 5%–7% surplus each. Similarly, the least influential subject category is the Agricultural and Biological Sciences with 10% deficit. The other least influential subjects are Chemistry and Engineering with 7% and 3% deficits, respectively.

TABLE 99.7
The Most-Prolific Scopus Subject Categories in the US Experience of Bioethanol Fuels

No.	Scopus Subject Categories	Sample Papers (%)	Population Papers (%)	Surplus (%)
1	Chemical Engineering	49.2	35.8	13.4
2	Biochemistry, Genetics, and Molecular Biology	42.7	37.6	5.1
3	Environmental Science	34.2	28.4	5.8
4	Immunology and Microbiology	29.2	22.9	6.3
5	Energy	27.7	22.2	5.5
6	Chemistry	14.2	20.7	–6.5
7	Agricultural and Biological Sciences	12.3	22.1	–9.8
8	Multidisciplinary	10.0	2.6	7.4
9	Engineering	8.5	11.5	–3.0
10	Materials Science	6.2	0.9	5.3

99.3.9 The Most-Prolific Keywords in the US Experience of Bioethanol Fuels

Information about the Scopus keywords used with at least 6.9% or 3.8% of the sample or population papers, respectively, is given in Table 99.8. For this purpose, keywords related to the keyword set given in the appendix of the related papers are selected from a list of the most-prolific keyword set provided by Scopus database.

These keywords are grouped under the five headings: feedstocks, pretreatments, fermentation, hydrolysis and hydrolysates, and products. Further, the most-prolific keyword related to the biomass and biomass constituents is cellulose with 48% of the sample papers, followed by lignin, biomass, lignocellulose, zea, hemicellulose, and corn with 11%–43% of the sample papers each.

Further, the most-prolific keyword related to the pretreatments is cellulases with 21% of the sample papers, followed by enzymes, pretreatments, temperature, and pretreatment with 9%–19% of the sample papers each. Further, the most-prolific keyword related to the fermentation is fermentation with 29% of the sample papers, followed by fungi, bacteria, and saccharomyces with 7%–17% of the sample papers each.

Further, the most-prolific keyword related to the hydrolysis and hydrolysates is hydrolysis with 37% of the sample papers, followed by sugar, enzyme activity, enzymatic hydrolysis, glucose, and saccharification with 13%–25% of the sample papers each. Finally, the most-prolific keyword related to the products is ethanol with 39% of the sample papers, followed by biofuels, oxidation, and biofuel production with 8 to 30% of the sample papers each.

On the other hand, the most-prolific keywords across all of the research fronts are cellulose, lignin, biomass, ethanol, hydrolysis, biofuels, fermentation, sugar, lignocellulose, cellulases, enzymes, enzyme activity, zea, fungi, enzymatic hydrolysis, and hemicellulose with 16%–48% of the sample papers each.

Similarly, the most influential keywords are cellulose, lignin, biomass, hydrolysis, lignocellulose, cellulases, biofuels, sugar, hemicellulose, enzymes, zea, pretreatment, and corn with 9%–29% surplus each.

99.3.10 The Most-Prolific Research Fronts in the US Experience of Bioethanol Fuels

Information about the research fronts for feedstocks using the sample papers in the US experience of bioethanol fuels is given in Table 99.9. As this table shows, the most-prolific research front for this field is the lignocellulosic biomass at large with 25% of the sample papers, followed by cellulose, starch feedstock residues, and wood biomass with 16%, 14%, and 12% of the sample papers, respectively. The other prolific research fronts are the bioethanol fuels, lignin,

TABLE 99.8
The Most-Prolific Keywords in the US Experience of Bioethanol Fuels

No.	Keywords	Sample Papers (%)	Population Papers (%)	Surplus (%)
1	Feedstocks			
	Cellulose	48.1	19.4	28.7
	Lignin	43.1	17.1	26.0
	Biomass	42.7	17.8	24.9
	Lignocellulose	23.8	6.8	17.0
	Zea	17.7	8.8	8.9
	Hemicellulose	15.8	3.8	12.0
	Corn	11.2	2.6	8.6
	Corn stover	9.2	3.0	6.2
	Lignocellulosic biomass	8.1	5.0	3.1
	Wood	7.7	4.8	2.9
	Xylan	6.9	2.4	4.5
	Maize	5.0	4.8	0.2
2	Pretreatments			
	Cellulases	21.2	6.7	14.5
	Enzymes	19.2	10.0	9.2
	Pretreatment	14.2	5.6	8.6
	Temperature	8.8	5.2	3.6
	Pre-treatment	8.5	5.0	3.5
	Ammonia	7.3	2.7	4.6
	Sulfuric acid	6.9	2.4	4.5
	Genetics	6.5	5.1	1.4
	pH	4.2	4.7	–0.5
3	Fermentation			
	Fermentation	29.2	22.8	6.4
	Fungi	17.3	9.2	8.1
	Bacteria	13.1	10.9	2.2
	Saccharomyces	7.3	6.2	1.1
	Escherichia	6.9	4.9	2.0
	Acetic acid	5.4	3.9	1.5
	Yeast	3.8	6.3	–2.5
4	Hydrolysis and hydrolysates			
	Hydrolysis	36.9	16.9	20.0
	Sugar	25.4	11.6	13.8
	Enzyme activity	17.7	9.9	7.8
	Enzymatic hydrolysis	16.9	8.7	8.2
	Glucose	13.5	11.7	1.8
	Saccharification	12.7	6.2	6.5
	Xylose	9.2	6.0	3.2
5	Products			
	Ethanol	39.2	33.1	6.1
	Biofuels	29.6	15.8	13.8
	Oxidation	8.1	3.2	4.9
	Biofuel production	8.1	2.2	5.9
	Bioethanol	3.8	4.9	–1.1
	Ethanol production	3.8	3.8	0.0

TABLE 99.9
The Most-Prolific Research Fronts for the US Experience of Bioethanol Fuels

No.	Research Fronts	N Paper (%) Sample I	N Paper (%) Sample II	Surplus (%)
1	Lignocellulosic biomass at large	25.4	23.0	2.4
2	Cellulose	15.8	15.1	0.7
3	Starch feedstock residues	14.2	9.7	4.5
4	Wood biomass	11.9	9.0	2.9
5	Bioethanol fuels	8.8	19.6	−10.8
6	Lignin	8.5	6.2	2.3
7	Other feedstocks	31.9	29.8	2.1
	Grass	5.8	3.1	2.7
	Industrial waste	5.4	6.5	−1.1
	Sugars	5.0	5.0	0.0
	Biomass at large	4.6	2.6	2.0
	Starch feedstocks	3.1	1.7	1.4
	Hydrolysates	1.9	2.8	−0.9
	Lignocellulosic wastes	1.9	1.8	0.1
	Sugar feedstock residues	1.5	2.2	−0.7
	Sugar feedstocks	1.5	2.2	−0.7
	Algal biomass	1.2	1.9	−0.7
8.	Minor feedstocks	2.8	5.1	2.3
	Forestry wastes	0.8	0.8	0.0
	Xylan	0.8	0.8	0.0
	Cellobiose	0.4	0.4	0.0
	Syngas	0.4	0.4	0.0
	Urban wastes	0.4	1.0	−0.6
	Food wastes	0.0	0.6	−0.6
	Hemicellulose	0.0	0.8	−0.8
	Plants	0.0	0.3	−0.3
	Sample size	260	775	

N paper (%) sample I: the number of papers in the sample of 260 papers; N paper (%) sample II: the number of papers in the sample of 775 papers for the whole world.

grass, industrial wastes, sugars, and biomass at large with 5%–9% of the sample papers each. The other minor research fronts are the starch feedstocks, hydrolysates, lignocellulosic wastes, sugar feedstock residues, sugar feedstocks, algal biomass, and forestry wastes with 1%–3% of the sample papers.

On the other hand, starch feedstock residues, wood biomass, grass, lignocellulosic biomass at large, lignin, biomass at large are overrepresented in the sample of 260 papers in relative to a sample of 765 papers for the whole world with 2%–5% surplus each. Similarly, bioethanol fuels are underrepresented in the sample of 260 papers in relative to a sample of 765 papers for the whole world by 10%. The other underrepresented biomass are industrial wastes, hydrolysates, hemicellulose, sugar feedstock residues, sugar feedstocks, algal biomass, urban wastes, and food wastes with 1% deficit each.

TABLE 99.10
The Most-Prolific Thematic Research Fronts for the US Experience of the Bioethanol Fuels

No.	Research Fronts	N Paper (%) Sample I	N Paper (%) Sample II	Surplus (%)
1	Bioethanol fuel production	80.8	74.1	6.7
	Biomass pretreatments	70.4	58.5	11.9
	Biomass hydrolysis	38.1	29.8	8.3
	Hydrolysate fermentation	18.1	17.9	0.2
	Bioethanol production	17.7	21.5	–3.8
	Bioethanol fuel distillation	0.8	0.8	0.0
2	Bioethanol fuel evaluation	10.0	6.2	3.8
3	Bioethanol fuel utilization	9.6	19.7	–10.1
	Bioethanol fuels in fuel cells	4.2	6.8	–2.6
	Bioethanol fuels in engines	3.5	6.1	–2.6
	Bioethanol-based biohydrogen fuels	1.9	4.3	–2.4
	Bioethanol sensors	0.0	2.2	–2.2
	Bioethanol-based biochemicals	0.0	0.4	–0.4
	Sample size	260	775	

N paper (%) sample I: the number of papers in the sample of 260 papers; N paper (%) sample II: the number of papers in the sample of 775 papers for the whole world.

Information about the thematic research fronts for the sample papers in the bioethanol fuels is given in Table 99.10. As this table shows, the most-prolific research front is the production of bioethanol fuels with 81% of the sample papers, followed by the utilization and evaluation of bioethanol fuels with 10% of the sample papers each.

On the other hand, the most-prolific research front for the production of bioethanol fuels is the biomass pretreatment with 70% of the sample papers, followed by biomass hydrolysis, bioethanol production, and hydrolysate fermentation with 38%, 18%, and 18% of the sample papers, respectively. The other minor research front is separation and distillation of bioethanol fuels with 1% of the sample papers.

Further, the most-prolific research fronts for the utilization of bioethanol fuels are the utilization of bioethanol fuels in fuel cells and transport engines, followed by bioethanol-based biohydrogen fuels with 2% of the sample papers. It appears that there are no papers for the research fronts of bioethanol sensors and bioethanol-based biochemicals.

99.4 DISCUSSION

99.4.1 Introduction

The crude oil-based gasoline fuels have been widely used in the transportation sector since the 1920s. However, there have been great public concerns over the adverse environmental and human impact of these fuels. Hence, biomass-based bioethanol fuels have increasingly been used in blending gasoline fuels, in the fuel cells, and in the biochemical production in a biorefinery context.

However, it is necessary to pretreat the biomass to enhance the yield of the bioethanol prior to the bioethanol production through the hydrolysis and fermentation processes. The USA has been one of the most-prolific countries engaged in the bioethanol fuel research. The US research in the field of the bioethanol fuels has intensified in this context in the key research fronts of the production of bioethanol fuels, and to a lesser extent, utilization and evaluation of bioethanol fuels. For the first research front, pretreatment and hydrolysis of the feedstocks, fermentation of the feedstock-based hydrolysates, production of bioethanol fuels, and to a lesser extent distillation of bioethanol fuels are the key research areas while for the second research front, utilization of bioethanol fuels in fuel cells and transport engines, bioethanol-based biohydrogen fuels, bioethanol sensors, and bioethanol-based biochemicals are the key research areas. The research in this field has also intensified for the feedstocks of lignocellulosic biomass at large, bioethanol fuels for their utilization, cellulose, starch feedstock residues, wood biomass, and to a lesser extent industrial waste, lignin, grass, hydrolysates, biomass, sugar feedstocks, sugar feedstock residues, algal biomass, lignocellulosic wastes, starch feedstocks, urban wastes, forestry wastes, food wastes, syngas, and plants.

However, it is essential to develop efficient incentive structures for the primary stakeholders to enhance the research in this field. This is especially important to maintain energy security in the cases of supply shocks such as oil price shocks, war-related shocks as in the case of Russian invasion of Ukraine, or COVID-19 shocks.

The scientometric analysis has been used in this context to inform the primary stakeholders about the current state of the research in a selected research field. As there has been no current scientometric study in this field, this book chapter presents a scientometric study of the research in the bioethanol fuels. It examines the scientometric characteristics of both the sample and population data presenting scientometric characteristics of these both datasets in the order of documents, authors, publication years, institutions, funding bodies, source titles, countries, Scopus subject categories, Scopus keywords, and research fronts.

As a first step for the search of the relevant literature, the keywords were selected using the most-cited first 300 population papers for each research front. The selected keyword list was then optimized to obtain a representative sample of papers for each research field. These keyword lists were then integrated to obtain the keyword list for the research field of bioethanol fuels research field (Konur, 2023a,b). Additionally, a second keyword set was developed for the USA. These two keyword sets were then integrated with the use of 'AND' keyword. Thus, 12,992 references were located for this field, corresponding to nearly 17% of the global bioethanol research output.

As a second step, two sets of data were used for this study. First, a population sample of 12,992 papers was used to examine the scientometric characteristics of the population data. Secondly, a sample of 260 most-cited papers, corresponding to 2% of the population papers, was used to examine the scientometric characteristics of these citation classics.

The scientometric characteristics of these sample and population datasets were presented in the order of documents, authors, publication years, institutions, funding bodies, source titles, countries, Scopus subject categories, Scopus keywords, and research fronts.

Lastly, the key scientometric findings for both datasets were discussed to highlight the research landscape for bioethanol fuels. Additionally, a number of brief conclusions were drawn and a number of relevant recommendations were made to enhance the future research landscape.

99.4.2 The Most-Prolific Documents in the US Experience of Bioethanol Fuels

Articles (together with conference papers) dominate both the sample (80%) and population (93%) papers with 13% deficit (Table 99.1). Further, review papers have a surplus (17%), and the representation of the reviews in the sample papers is quite extraordinary (20%).

Scopus differs from the Web of Science database in differentiating and showing articles (79%) and conference papers (1%) published in the journals separately. However, it should be noted that these conference papers are also published in journals as articles, compared to those published only in the conference proceedings. Hence, a total number of articles and review papers in the sample dataset are 80% and 20%, respectively.

It is observed during the search process that there has been inconsistency in the classification of the documents in Scopus as well as in other databases such as Web of Science. This is especially relevant for the classification of papers as reviews or articles as the papers not involving a literature review may be erroneously classified as a review paper. There is also a case of review papers being classified as articles. For example, the total number of the reviews in the sample dataset was manually found as nearly 25% compared to 20% as indexed by Scopus, decreasing the number of articles and conference papers to 75% for the sample dataset.

In this context, it would be helpful to provide a classification note for the published papers in the books and journals at the first instance. It would also be helpful to use the document types listed in Table 99.1 for this purpose. Book chapters may also be classified as articles or reviews as an additional classification to differentiate review chapters from the experimental chapters as it is done by the Web of Science. It would be further helpful to additionally classify the conference papers as articles or review papers as well as it is done in the Web of Science database.

99.4.3 The Most-Prolific Authors in the US Experience of Bioethanol Fuels

There have been most-prolific 31 authors with at least 1.5% of the sample papers each as given in Table 99.2. These authors have shaped the development of the research in this field.

The most-prolific authors are Charles E. Wyman, Bruce E. Dale, and to a lesser extent Lee R. Lynd, Blake A. Simmons, Arthur J. Ragauskas, Michael E. Himmel, and Michael R. Ladisch. Further, the most influential authors are Charles E. Wyman, Bruce E. Dale, and to a lesser extent Lee R. Lynd, Michael R. Ladisch, Blake A. Simmons, Michael E. Himmel, Robin D. Rogers, and Igor V. Grigoriev.

It is important to note the inconsistencies in indexing of the author names in Scopus and other databases. It is especially an issue for the names with more than two components such as 'Blake Sam de Hyun Simmons'. The probable outcomes are 'Simmons, B.S.D.H.', 'de Hyun Simmons, B.S.', or 'Hyun Simmons, B.S.D.'. The first choice is the gold standard of the publishing sector as the last word in the name is taken as the last name. In most of the academic databases such as PUBMED and EBSCO, this version is used predominantly. The second choice is a strong alternative while the last choice is an undesired outcome as two last words are taken as the last name. It is good practice to combine the words of the last name by a hyphen: 'Hyun Simmons, B.S.D.'. It is notable that inconsistent indexing of the author names may cause substantial inefficiencies in the search process for the papers as well as allocating credit to the authors as there are different author entries for each outcome in the databases.

There are also inconsistencies in the shortening Chinese names. For example, 'Hongzhang Chen' is often shortened as 'Chen, H.Z.', 'Chen, H.-Z.', and 'Chen H.' as it is done in the Web of Science database as well. However, the gold standard in this case is 'Chen, H' where the last word is taken as the last name and the first word is taken as a single forename. In most of the academic databases such as PUBMED and EBSCO, this first version is used predominantly. Nevertheless, it makes sense to sue the first option to differentiate Chinese names efficiently: 'Chen, H.Z.'. Therefore, there have been difficulties in locating papers for the Chinese authors. In such cases, the use of the unique author codes provided for each author by the Scopus database has been helpful.

There is also a difficulty in allowing credit for the authors especially for the authors with common names such as 'Wang, X.' in conducting scientometric studies. These difficulties strongly influence the efficiency of the scientometric studies as well as allocating credit to the authors as there are the same author entries for different authors with the same name, for example, 'Wang, X.' in the databases.

In this context, the coding of authors in Scopus database is a welcome innovation compared to the other databases such as Web of Science. In this process, Scopus allocates a unique number to each author in the database (Aman, 2018). However, there might still be substantial inefficiencies in this coding system especially for common names. For example, some of the papers for a certain author may be allocated to another researcher with a different author code. It is possible that Scopus uses a number of software programs to differentiate the author names and the program may not be false-proof (Kim, 2018).

In this context, it does not help that author names are not given in full in some journals and books. This makes difficult to differentiate authors with common names and makes the scientometric studies further difficult in the author domain. Therefore, the author names should be given in full in all books and journals at the first instance. There is also a cultural issue where some authors do not use their full names in their papers. Instead they use initials for their forenames: 'Parajo, H.J.', 'Parajo, H.', or 'Parajo, J.' instead of 'Parajo, Hyun Jae'.

There are also inconsistencies in naming of the authors with more than two components by the authors themselves in journal papers and book chapters. For example, 'Parajo, A.P.C.' might be given as 'Parajo, A.' or 'Parajo, A.C.' or 'Parajo, A.P.' or 'Parajo, C' in the journals and books. This also makes the scientometric studies difficult in the author domain. Hence, contributing authors should use their name consistently in their publications.

The other critical issue regarding the author names is the inconsistencies in the spelling of the author names in the national spellings (e.g., Özgümüş, Söğüt) rather than in the English spellings (e.g., Ozgumus, Sogut) in Scopus database. Scopus differs from the Web of Science database and many other databases in this respect where the author names are given only in the English spellings. It is observed that national spellings of the author names do not help much in conducting scientometric studies as well in allocating credits to the authors as sometimes there are often the different author entries for the English and National spellings in the Scopus database.

The most-prolific institutions for the sample dataset are the Joint Genome Institute, NREL, and to a lesser extent Purdue University, USDA Agricultural Service, and USDA Forest Service. On the other hand, pretreatments and hydrolysis of the feedstocks, fermentation of feedstock-based hydrolysates, the bioethanol production, and to a lesser extent evaluation of bioethanol fuels are the key research fronts studied by these top authors.

However, it is notable that there are no authors focusing on the utilization of bioethanol fuels such as biohydrogen fuel production from bioethanol fuels, direct utilization of bioethanol fuels in the gasoline engines or fuel cells, and the development of bioethanol sensors.

It is also notable that there is significant gender deficit for the sample dataset as surprisingly with a representation rate of 6%. This finding is the most thought-provoking with strong public policy implications. Hence, institutions, funding bodies, and policy-makers should take efficient measures to reduce the gender deficit in this field as well as other scientific fields with strong gender deficit. In this context, it is worth to note the level of representation of the researchers from the minority groups in science on the basis of race, sexuality, age, and disability, besides the gender (Blankenship, 1993; Dirth and Branscombe, 2017; Konur, 2000, 2002a,b,c, 2006a,b, 2007a,b).

99.4.4 The Most-Prolific Research Output by Years in the US Experience of Bioethanol Fuels

The research output observed between 1970 and 2022 is illustrated in Figure 99.1. This figure clearly shows that the bulk of the research papers in the population dataset were published primarily in the 2010s, and to a lesser extent in the early 2020s and 2000s. Similarly, the bulk of the research papers in the sample dataset were published in the 2000s, and to a lesser extent 2010s and 1990s.

These findings suggest that the most-prolific sample and population papers were primarily published in the 2000s and 2010s, respectively. Further, a significant portion of the sample and population papers were published in the early 2010s and 2020s, respectively.

These are the thought-provoking findings as there has been significant research boom between 2007 and 2022 and 2002 and 2014 for the population and sample papers, respectively. In this context, the increasing public concerns about climate change (Change, 2007), greenhouse gas emissions (Carlson et al., 2017), and global warming (Kerr, 2007) have been certainly behind the boom in the research in this field since 2007. Furthermore, the recent supply shocks experienced due to the COVID-19 pandemics and the Ukrainian war might also be behind the research boom in this field since 2019. However, it is notable that a number of publications for the population papers fell after 2011 linearly. This finding is much thought-provoking to ponder upon the determinants of this extraordinary fall in the research output.

Based on these findings, the size of the population papers likely to more than double in the current decade provided that the public concerns about climate change, greenhouse gas emissions, and global warming, as well as the supply shocks, are translated efficiently to the research funding in this field and depending on whether the decline in the number of population papers would continue.

99.4.5 The Most-Prolific Institutions in the US Experience of Bioethanol Fuels

The most-prolific 26 institutions publishing papers on the US experience of bioethanol fuels with at least 2.3% of the sample papers each given in Table 99.3 have shaped the development of the research in this field.

The most-prolific US institutions are the NREL, Dartmouth College, Michigan State University, USDA Forest Service, and to a lesser extent University of Wisconsin Madison, Purdue University, Oak Ridge National Laboratory, NC State University, and University of California Berkeley.

On the other hand, the US institutions with the most impact are the NREL, Dartmouth College, and to a lesser extent Michigan State University, USDA Forest Service, Purdue University, and University of California Berkeley.

Further, the most-prolific collaborating overseas institutions are University of Sao Paulo, Chinese Academy of Sciences, Nanjing Forestry University, University of British Columbia, South China University of Technology, and State University of Campinas.

99.4.6 The Most-Prolific Funding Bodies in the US Experience of Bioethanol Fuels

The most-prolific 12 funding bodies funding at least 1.2% of the sample papers each are given in Table 99.4. It is notable that only 36% and 39% of the sample and population papers were funded, respectively.

The most-prolific funding bodies are the U.S. Department of Energy, National Science Foundation, and to a lesser extent the Office of Science, U.S. Department of Agriculture, National Institute of General Medical Sciences, and NREL. It appears that U.S. Department of Energy and National Science Foundation are the key funding bodies of the USA in this field.

The funding bodies with the most citation impact are the U.S. Department of Energy, and to a lesser extent National Institute of Standards and Technology, NREL, and University of California. Further, the funding bodies with the least citation impact are the Biological and Environmental Research, National Science Foundation, Office of Science, and U.S. Department of Agriculture.

On the other hand, a large number of US funding bodies with relatively low citation impact have also funded the research in this area. In addition, a number of overseas funding bodies jointly funded the research in this field together with the US funding bodies. It appears that the Chinese funding bodies have heavily joint-funded the US research in this field.

These findings on the funding of the research in this field suggest that the level of the funding for the population papers, mostly since 2007, is modest and it has been largely instrumental in enhancing the research in this field (Ebadi and Schiffauerova, 2016) in light of North's institutional framework (North, 1991). However, the funding rate for the sample papers is relatively low. Further, it is expected that this funding rate would improve in light of the recent supply shocks.

99.4.7 The Most-Prolific Source Titles in the US Experience of Bioethanol Fuels

The most-prolific 14 source titles publishing at least 1.5% of the sample papers each in the bioethanol fuels have shaped the development of the research in this field (Table 99.5).

The most-prolific source titles are the Bioresource Technology, Biotechnology and Bioengineering, and to a lesser extent National Academy of Sciences of the United States of America, Applied and Environmental Microbiology, and Science. On the other hand, the source titles with the most impact are the Bioresource Technology, Biotechnology and Bioengineering, and to a lesser extent the Proceedings of the National Academy of Sciences of the United States of America, Applied and Environmental Microbiology, and Science.

It is notable that these top source titles are primarily related to the Bioresources, Biotechnology, Microbiology, and Multidisciplinary Sciences. This finding suggests that Bioresource Technology, Biotechnology and Bioengineering, and the other prolific journals in these fields have significantly shaped the development of the research in this field as they focus primarily on the production of bioethanol fuels with a high yield. In this context, the influence of two top journals is quite extraordinary.

This top journal covers more the technical issues such as waste treatment, biofuels, bioprocesses, bioproducts, and physicochemical and thermochemical processes, excluding the social science- and humanities-based interdisciplinary studies with the only exceptions for the 'circular bioeconomy and energy & environmental sustainability' and 'system analysis & technoeconomics of biofuels and chemicals production'. It appears that it thus excludes by its editorial policies other social science- and humanities-based interdisciplinary studies such as scientometric (Konur, 2012a,b,c,d,e,f,g,h,i), user (Huijts et al., 2012; Wustenhagen et al., 2007), policy (Greening et al., 2000; Hook and Tang, 2013), environmental (Popp et al., 2014; Tilman et al., 2009), and economic (Asafu-Adjaye, 2000; Stern, 1993) studies as in the case of the most journals in the field of energy and fuels excepting social-science-oriented journals.

99.4.8 The Most-Prolific Collaborating Countries in the US Experience of Bioethanol Fuels

The most-prolific 12 countries publishing at least 1.5% of the sample papers each have significantly shaped the development of the research in this field (Table 99.6).

The most-prolific countries are China, Canada, and to a lesser extent S. Korea, the UK, France, and Spain. On the other hand, the countries with the most citation impact are Spain, and to a lesser extent France, Chile, and the UK. Similarly, the country with the least citation impact is China. Further, seven European countries listed in Table 99.6 as a whole produce 15% and 7% of the sample and population papers, respectively, with 8% surplus.

It is notable that all of these collaborating countries are the major producers of the research in the field of bioethanol fuels at large. As Beaver and Rosen (1978) state, the 'collaboration is a typical research style associated with professionalization'. Thus, the collaboration between researchers improves their research output (Miramontes and Gonzalez-Brambila, 2016).

China has been a rising mega star in scientific research in competition with the USA and Europe (Leydesdorff and Zhou, 2005). China is also a major player in this field as a major producer of bioethanol (Fang et al., 2010).

Next, European countries have been a persistent player in the scientific research in competition with both the USA and China (Leydesdorff, 2000). European Union has also been a persistent producer of bioethanol along with the USA and Brazil (Gnansounou, 2010).

Further, Canada (Tahmooresnejad et al., 2015) and Japan (Negishi et al., 2004) are the other countries with substantial research activities in bioethanol fuels.

99.4.9 The Most-Prolific Scopus Subject Categories in the US Experience of Bioethanol Fuels

The most-prolific ten Scopus subject categories indexing at least 6.2% of the sample papers each, given in Table 99.7, have shaped the development of the research in this field.

The most-prolific Scopus subject categories in the bioethanol fuels are Chemical Engineering, Biochemistry, Genetics, and to a lesser extent Environmental Science, Immunology and Microbiology, and Energy. It is also notable that Social Sciences including Economics and Business have a minimal presence in both sample and population studies.

On the other hand, the Scopus subject categories with the most citation impact are Chemical Engineering, and to a lesser extent Multidisciplinary, Immunology and Microbiology, Environmental Science, Energy, Materials Science, and Biochemistry, Genetics and Molecular Biology. Similarly, the least influential subject categories are Agricultural and Biological Sciences, and to a lesser extent Chemistry and Engineering.

These findings are thought-provoking suggesting that the primary subject categories are related to Chemical Engineering, Biochemistry, and to a lesser extent Microbiology, Environmental Science, and Energy as the core of the US research in this field concerns with the production and utilization of bioethanol fuels. The other finding is that Social Sciences are not well represented in both the sample and population papers as in line with the most fields in bioethanol fuels. The social, environmental, and economics studies account for the field of Social Sciences.

As discussed briefly in Section 99.4.7, most journals in this field do not exclude social science- and humanities-based interdisciplinary studies by their editorial policies. This development has been in contrast to the interdisciplinarity (Jacobs and Frickel, 2009; Nissani, 1997) of this field in light of the pressures for increasing incentives for the primary stakeholders (North, 1991). Thus, for the healthy development of this research field, journals in this field should have inclusive editorial policies toward social science- and humanities-based interdisciplinary studies.

99.4.10 The Most-Prolific Keywords in the US Experience of Bioethanol Fuels

A limited number of keywords have shaped the development of the research in this field as shown in Table 99.8. These keywords are grouped under the five headings: feedstocks, pretreatments, fermentation, hydrolysis and hydrolysates, and products.

The most-prolific keywords across all of the research fronts are cellulose, lignin, biomass, ethanol, hydrolysis, biofuels, fermentation, sugar, lignocellulose, cellulases, enzymes, enzyme activity, zea, fungi, enzymatic hydrolysis, and hemicellulose. Similarly, the most influential keywords across all of the research fronts are cellulose, lignin, biomass, hydrolysis, lignocellulose, cellulases, biofuels, sugar, hemicellulose, enzymes, zea, pretreatment, and corn.

These findings suggest that it is necessary to determine the keyword set carefully to locate the relevant research in each of these research fronts. Additionally, the size of the samples for each keyword highlights the intensity of the research in the relevant research areas for both sample and population datasets. These findings also highlight different spelling of some strategic keywords such as pretreatment v. pretreatment v. treatment and bioethanol v. ethanol v. bioethanol v. bioethanol fuels, etc. However, there is tendency toward the use of the connected keywords without using a hyphen: bioethanol fuels or pretreatment. It is particularly notable that the use of treatment and ethanol instead of pretreatment and bioethanol in the paper titles, respectively, makes the literature search less efficient and time-consuming.

The other interesting finding is that the US research focuses more on the production of bioethanol fuels rather than the utilization of bioethanol fuels such as feedstocks for fuels cells, biohydrogen fuels, biochemicals, gasoline and diesel engines as well as the development and utilization of bioethanol sensors as the related keywords are missing in the last section of Table 99.8.

99.4.11 The Most-Prolific Research Fronts in the US Experience of Bioethanol Fuels

Information about the research fronts for the sample papers in the US experience of bioethanol fuels is given in Table 99.9. As this table shows, the most-prolific research front for this field is the lignocellulosic biomass at large, cellulose, starch feedstock residues such as corn stover, wood biomass, and to a lesser extent bioethanol fuels, lignin, grass, industrial wastes, sugars as a product of hydrolysis of the biomass, biomass at large, starch feedstocks such as corn grains, hydrolysates, lignocellulosic wastes, sugar feedstock residues such as sugarcane bagasse, sugar feedstocks such as sugarcane biomass, algal biomass, and forestry wastes.

On the other hand, starch feedstock residues, wood biomass, grass, lignocellulosic biomass at large, lignin, and biomass at large are overrepresented in the US sample of 260 papers in relation to a global sample of 765 papers for the whole world. Similarly, bioethanol fuels and to a lesser extent industrial wastes, hydrolysates, hemicellulose, sugar feedstock residues, sugar feedstocks, algal biomass, urban wastes, and food wastes are underrepresented in the US sample in relation to a global sample for the whole world.

Thus, the first six research fields have substantial importance, complementing the remaining bioethanol fuel research fields. It is important to note that the research on the production of the First Generation Bioethanol Fuels from sugar and starch feedstocks for bioethanol fuels comprises only 4.6% of the sample papers in total. These First Generation Bioethanol Fuels are not much desirable as they undermine the food security (Makenete et al., 2008; Wu et al., 2012).

As the USA is the major producer and consumer of the bioethanol fuels from starch feedstock residues such as corn stover, this is an important research front for the USA. Similarly, as the USA has vast forests, the production of bioethanol fuels from wood biomass is also an important research front. In addition, the research on the bioethanol fuels from lignocellulosic biomass at large and cellulose, a key constituent of the lignocellulosic biomass, is extremely important. Grass and industrial wastes especially from the ethanol and sugar industry are also important feedstocks for the bioethanol production in the USA.

However, it appears that the utilization of bioethanol fuels as a feedstock for the biohydrogen and biochemical production, gasoline and diesel engines and fuel cells directly, and for the development and utilization of bioethanol sensors is not much pronounced as only 9% of sample papers are related to bioethanol fuels as feedstocks compared to 20% of the global sample for the whole world.

Table 99.10 shows that the most-prolific research fronts are the production of bioethanol fuels, and to a lesser extent utilization and evaluation of bioethanol fuels. On the other hand, the most-prolific research fronts for the production of bioethanol fuels are the biomass pretreatment, biomass hydrolysis, and to a lesser extent bioethanol production, hydrolysate fermentation, and separation and distillation of bioethanol fuels.

Further, the most-prolific research fronts for the utilization of bioethanol fuels are the utilization of bioethanol fuels in fuel cells and the transport engines, and to a lesser extent bioethanol-based biohydrogen fuels, while there are no papers for the fields of bioethanol sensors and bioethanol fuel-based biochemicals.

It appears that the research fronts of production and evaluation of bioethanol fuels are overrepresented in the US sample in relation to the global sample for the whole world by 7% and 4%, respectively. On the contrary, research front of utilization of bioethanol fuels is underrepresented in the US sample in relation to the global sample for the whole world by 10%.

Further, the research fronts of the biomass pretreatments and hydrolysis are overrepresented in the US sample in relation to the global sample by 12% and 8%, respectively, while the front of bioethanol production is underrepresented by 4%. Similarly, the research fronts of utilization of bioethanol fuels in fuel cells and engines, bioethanol-based biohydrogen fuels, bioethanol sensors, and bioethanol-based biochemical are underrepresented in the US sample in relation to the global sample by 0.4%–2.6% each.

These findings are thought-provoking in seeking ways to evaluate the strength and weaknesses of the USA in terms of both feedstock- and theme-based research fronts. It appears that the USA is strong in the research fronts of production and evaluation of bioethanol fuels but is weak in the research front of the utilization of bioethanol fuels. Further, it is highly strong in the research fronts of the pretreatments and hydrolysis of the biomass.

Further, in relation to the feedstocks, the USA is strong in the research fronts of starch feedstock residues such as corn stover, wood biomass, grass, lignocellulosic biomass at large, lignin, biomass at large, starch feedstocks such as corn grains, and cellulose in relation to the global sample. On the contrary, USA is weak in the research fronts of bioethanol fuels, industrial wastes, hydrolysates, hemicellulose, sugar feedstock residues, sugar feedstocks, algal biomass, urban wastes, and food wastes in relation to the global sample. Its weakness in bioethanol fuels as feedstocks is substantial. Thus, it appears that the USA is also weak in the research front of lignocellulosic wastes such as industrial wastes with the exception of starch feedstock residues by 3% in relation to the global sample. Further, the research in the third generation algal bioethanol production forms only 1% of the US research on the bioethanol fuels and is weak in relation to the global sample by 1%.

Considering the fact that the funding rate for the sample and population papers is only 36% and 39%, respectively, and landscape of the research funders is highly fragmented with many funding bodies operating in this field, it appears that the USA might lose its top research producer status with its first-mover advantage (Lieberman and Montgomery, 1988) in this field to its key competitors such as Europe, China, India, Brazil, and Japan in the future.

The weakness of the USA in the field of utilization of bioethanol fuels is highly problematic as there is a transition from the gasoline and petrodiesel fueled cars to fuel cell-based electric or hybrid cars at a global scale (Wilberforce et al., 2017) in light of the public concerns regarding the climate change, greenhouse gas emissions, and global warming, intensified by the recent supply shock caused, for example, by the COVID-19 pandemics and Ukrainian war. In this context, the heavy research funding by the USA's competitors is notable.

In the end, these most-cited papers in this field hint that the production and utilization of bioethanol fuels could be optimized using the structure, processing, and property relationships of these bioethanol fuels in the fronts of the feedstock pretreatment and hydrolysis, hydrolysate fermentation, and the utilization of bioethanol fuels (Formela et al., 2016; Konur, 2018a, 2020b, 2021a,b,c,d; Konur and Matthews, 1989).

99.5 CONCLUSION AND FUTURE RESEARCH

The US research on the bioethanol fuels has been mapped through a scientometric study of both sample (260 papers) and population (12,992 papers) datasets.

The critical issue in this study has been to obtain a representative sample of the research as in any other scientometric study. Therefore, the keyword set has been carefully devised and optimized after a number of runs in the Scopus database. It is a representative sample of the wider population studies. This keyword set was provided in the appendix of the related studies, and the relevant keywords are also presented in Table 99.8. However, it should be noted that it has been very difficult to compile a representative keyword set since this research field has been connected closely with many other fields. Therefore, it has been necessary to compile a keyword list to exclude papers concerned with the other research fields.

The other issue has been the selection of a multidisciplinary database to carry out the scientometric study of the research in this field. For this purpose, Scopus database has been selected. The journal coverage of this database has been notably wider than that of Web of Science and other multi-subject databases.

The key scientometric properties of the US research in this field have been determined and discussed in this book chapter. It is evident that a limited number of documents, authors, institutions, publication years, institutions, funding bodies, source titles, countries, Scopus subject categories, Scopus keywords, and research fronts have shaped the development of the research in this field.

There is ample scope to increase the efficiency of the scientometric studies in this field in the author and document domains by developing consistent policies and practices in both domains across all the academic databases. In this respect, it seems that authors, journals, and academic databases have a lot to do. Furthermore, the significant gender deficit as in most scientific fields emerges as a public policy issue. The potential deficits on the basis of age, race, disability, and sexuality also need to be explored in this field as in other scientific fields.

The research in this field has boomed since 2007 and 2002 for the population and sample papers, respectively, possibly promoted by the public concerns on global warming, greenhouse gas emissions, and climate change. Furthermore, the recent COVID-19 pandemics and Russian invasion of Ukraine have resulted in a global supply shocks shifting the recent focus of the stakeholders from the crude oil-based fuels and other fossil-based fuels to biomass-based fuels such as bioethanol fuels. However, there has been a declining trend for the number of population papers starting from 2011.

It is expected that there would be further incentives for the key stakeholders to carry out the research for the bioethanol fuels to increase the ethanol yield and their utilization and to make it more competitive with the crude oil-based gasoline and diesel fuels and the other feedstocks for the fuel cells.

The relatively modest funding rate of 36% and 39% for the sample and population papers, respectively, suggests that funding significantly enhanced the research in this field primarily since 2007, possibly more than doubling in the current decade. However, it is evident that there is ample room for more funding and other incentives to enhance the research in this field further considering the persistent decline in the research output since 2011.

The public and private research institutions such as NREL, USDA Forest Service, Oak Ridge National Laboratory, USDA Agricultural Research Service, Joint Bioenergy Institute, Sandia National Laboratory, Pacific Northwest National Laboratory, Lawrence Berkeley National Laboratory, and Joint Genome Institute have immensely contributed to the US research on the bioethanol fuels. Further, the key universities contributing to the US research in this field have been Dartmouth College, Michigan State University, University of Wisconsin Madison, Purdue University, NC State University, University of California Berkeley, Georgia Institute of Technology, University of Minnesota, and University of California Riverside.

It emerges that ethanol is more popular than bioethanol as a keyword with strong implications for the search strategy. In other words, the search strategy using only bioethanol keyword would not be much helpful. It is recommended that the term 'bioethanol fuels' is used instead of ethanol or bioethanol in the future for the integration of the research in this field. On the other hand, the Scopus keywords are grouped under the five headings: feedstocks, pretreatments, fermentation, hydrolysis and hydrolysates, and products. These prolific keywords highlight the major fields of the research in this field for both sample and population papers.

Table 99.9 shows that the most prolific research front for this field is the lignocellulosic biomass at large, cellulose, starch feedstock residues, wood biomass, and to a lesser extent bioethanol fuels, lignin, grass, industrial wastes, sugars, biomass at large, starch feedstocks, hydrolysates, lignocellulosic wastes, sugar feedstock residues, sugar feedstocks, algal biomass, and forestry wastes.

On the other hand, starch feedstock residues, wood biomass, grass, lignocellulosic biomass at large, lignin, and biomass at large are overrepresented in the US sample in relation to the global sample. Similarly, bioethanol fuels and to a lesser extent industrial wastes, hydrolysates, hemicellulose, sugar feedstock residues, sugar feedstocks, algal biomass, urban wastes, and food wastes are underrepresented in the US sample in relation to the global sample.

Thus, the first six research fields have substantial importance, complementing the remaining bioethanol fuel research fields. It is important to note that the research on the production of the First Generation Bioethanol Fuels from sugar and starch feedstocks for bioethanol fuels comprises only 4.6% of the sample papers in total. These First Generation Bioethanol Fuels are not much desirable as they undermine the food security.

However, it appears that the utilization of bioethanol fuels as a feedstock for the biohydrogen and biochemical production, gasoline and diesel engines, and fuel cells directly, and for the development and utilization of bioethanol sensors, is not much pronounced with 11% deficit in relation to the global sample.

Table 99.10 shows that the most-prolific research fronts are the production of bioethanol fuels, and to a lesser extent utilization and evaluation of bioethanol fuels. On the other hand, the most-prolific research fronts for the production of bioethanol fuels are the biomass pretreatment, biomass hydrolysis, and to a lesser extent bioethanol production, hydrolysate fermentation, and separation and distillation of bioethanol fuels.

Further, the most-prolific research fronts for the utilization of bioethanol fuels are the utilization of bioethanol fuels in fuel cells, the transport engines, and to a lesser extent bioethanol-based biohydrogen fuels while there are no papers for the fields of bioethanol sensors and bioethanol fuel-based biochemicals.

It appears that the research fronts of production and evaluation of bioethanol fuels are overrepresented in the US sample in relation to the global sample for the whole world. On the contrary, research front of utilization of bioethanol fuels is underrepresented in the US sample in relation to the global sample for the whole world.

These findings are thought-provoking in seeking ways to evaluate the strength and weaknesses of the USA in terms of both feedstock and theme-based research fronts. It appears that the USA is strong in the research fronts of production and evaluation of bioethanol fuels but is weak in the research front of the utilization of bioethanol fuels. Further, it is highly strong in the research fronts of the pretreatments and hydrolysis of the biomass.

Further, in relation to the feedstocks, the USA is strong in the research fronts of starch feedstock residues such as corn stover, wood biomass, grass, lignocellulosic biomass at large, lignin, biomass at large, starch feedstocks such as corn grains, and cellulose in relation to the global sample. On the contrary, USA is weak in the research fronts of bioethanol fuels, industrial wastes, hydrolysates, hemicellulose, sugar feedstock residues, sugar feedstocks, algal biomass, urban wastes, and food wastes in relation to the global sample. Its weakness in bioethanol fuels as feedstocks is substantial. Thus, it appears that the USA is also weak in the research front of lignocellulosic wastes such as industrial wastes with the exception of starch feedstock residues in relation to the global sample. Further, the US research in the third generation algal bioethanol production is also weak in relation to the global sample.

Considering the fact that the funding rate for the sample and population papers is only 36% and 39%, respectively, and landscape of the research funders is highly fragmented with many funding bodies operating in this field, and it appears that the USA might lose its top research producer status with its first-mover advantage in this field to its key competitors such as Europe, China, India, Brazil, and Japan in the future.

The weakness of the USA in the field of utilization of bioethanol fuels is highly problematic as there is a transition from the gasoline and petrodiesel fueled cars to fuel cell-based electric or hybrid cars at a global scale in light of the public concerns regarding the climate change, greenhouse gas emissions, and global warming, intensified by the recent supply shock cases, for example, by the COVID-19 pandemics and Ukrainian war. In this context, the heavy research funding by the USA's competitors is notable.

Further, the field of the evaluation of bioethanol fuels is also relatively a neglected area. It is also notable that evaluation of the bioethanol fuels such as techno-economics, life cycle, economics, social, land use, labor, and environment-related studies emerges as a case study for the bioethanol fuels.

In the end, these most-cited papers in this field hint that the production and utilization of bioethanol fuels could be optimized using the structure, processing, and property relationships of the bioethanol fuels and their feedstocks.

In this context, it is notable that there is ample room for the improvement of the research on social and humanitarian aspects of the research on the bioethanol fuels.

Thus, the scientometric analysis has a great potential to gain valuable insights into the evolution of the research in this field as in other scientific fields especially in the aftermath of the significant global supply shocks such as COVID-19 pandemics and the Russian invasion of Ukraine.

It is recommended that further scientometric studies are carried out for the primary research fronts. It is further recommended that reviews of the most-cited papers are carried out for each primary research front to complement these scientometric studies. Next, the scientometric studies of the hot papers in these primary fields are carried out.

ACKNOWLEDGMENTS

The contribution of the highly cited researchers in the field of the bioethanol fuels has been gratefully acknowledged.

REFERENCES

Alvira, P., E. Tomas-Pejo, M. Ballesteros and M. J. Negro. 2010. Pretreatment technologies for an efficient bioethanol production process based on enzymatic hydrolysis: A review. *Bioresource Technology* 101:4851–4861.

Aman, V. 2018. Does the Scopus author ID suffice to track scientific international mobility? A case study based on Leibniz laureates. *Scientometrics* 117:705–720.

Angelici, C., B. M. Weckhuysen and P. C. A. Bruijnincx. 2013. Chemocatalytic conversion of ethanol into butadiene and other bulk chemicals. *ChemSusChem* 6:1595–1614.

Antolini, E. 2007. Catalysts for direct ethanol fuel cells. *Journal of Power Sources* 170:1–12.

Antolini, E. 2009. Palladium in fuel cell catalysis. *Energy and Environmental Science* 2:915–931.

Asafu-Adjaye, J. 2000. The relationship between energy consumption, energy prices and economic growth: Time series evidence from Asian developing countries. *Energy Economics* 22:615–625.

Bai, F. W., W. A. Anderson and M. Moo-Young. 2008. Ethanol fermentation technologies from sugar and starch feedstocks. *Biotechnology Advances* 26:89–105.

Beaudry, C. and V. Lariviere. 2016. Which gender gap? Factors affecting researchers' scientific impact in science and medicine. *Research Policy* 45:1790–1817.

Beaver, D. and R. Rosen, R. 1978. Studies in scientific collaboration: Part I. The professional origins of scientific co-authorship. *Scientometrics*, 1:65–84.

Binod, P., R. Sindhu and R. R. Singhania, et al. 2010. Bioethanol production from rice straw: An overview. *Bioresource Technology* 101:4767–4774.

Blankenship, K. M. 1993. Bringing gender and race in: US employment discrimination policy. *Gender & Society* 7:204–226.

Bothast, R. J. and M. A. Schlicher. 2005. Biotechnological processes for conversion of corn into ethanol. *Applied Microbiology and Biotechnology* 67:19–25.

Bourbonnais, R. and M. G. Paice. 1990. Oxidation of non-phenolic substrates. An expanded role for laccase in lignin biodegradation. *FEBS Letters* 267:99–102.

Burnham, J. F. 2006. Scopus database: A review. *Biomedical Digital Libraries* 3:1–8.

Canilha, L., A. K. Chandel and T. S. dos Santos Milessi, et al. 2012. Bioconversion of sugarcane biomass into ethanol: An overview about composition, pretreatment methods, detoxification of hydrolysates, enzymatic saccharification, and ethanol fermentation. *Journal of Biomedicine and Biotechnology* 2012:989572.

Cardona, C.A., J. A. Quintero and I. C. Paz. 2010. Production of bioethanol from sugarcane bagasse: Status and perspectives. *Bioresource Technology* 101:4754–4766.

Carlson, K. M., J. S. Gerber and D. Mueller, et al. 2017. Greenhouse gas emissions intensity of global croplands. *Nature Climate Change* 7:63–68.

Change, C. 2007. Climate change impacts, adaptation and vulnerability. *Science of the Total Environment* 326:95–112.

Dirth, T. P. and N. R. Branscombe. 2017. Disability models affect disability policy support through awareness of structural discrimination. *Journal of Social Issues* 73:413–442.

Duff, S. J. B. and W. D. Murray. 1996. Bioconversion of forest products industry waste cellulosics to fuel ethanol: A review. *Bioresource Technology* 55:1–33.

Ebadi, A. and A. Schiffauerova. 2016. How to boost scientific production? A statistical analysis of research funding and other influencing factors. *Scientometrics* 106:1093–1116.

Fang, X., Y. Shen, J. Zhao, X. Bao and Y. Qu. 2010. Status and prospect of lignocellulosic bioethanol production in China. *Bioresource Technology* 101:4814–4819.

Fauci, A. S., H. C. Lane and R. R. Redfield. 2020. Covid-19-navigating the uncharted. *New England Journal of Medicine* 382:1268–1269.

Fernando, S., S. Adhikari, C. Chandrapal and M. Murali. 2006. Biorefineries: Current status, challenges, and future direction. *Energy & Fuels* 20:1727–1737.

Formela, K., A. Hejna, L. Piszczyk, M. R. Saeb and X. Colom. 2016. Processing and structure-property relationships of natural rubber/wheat bran biocomposites. *Cellulose* 23:3157–3175.

Galbe, M. and G. Zacchi. 2002. A review of the production of ethanol from softwood. *Applied Microbiology and Biotechnology* 59:618–628.

Garfield, E. 1955. Citation indexes for science. *Science* 122:108–111.

Gnansounou, E. 2010. Production and use of lignocellulosic bioethanol in Europe: Current situation and perspectives. *Bioresource Technology* 101:4842–4850.

Greening, L. A., D. L. Greene and C. Difiglio. 2000. Energy efficiency and consumption-the rebound effect-a survey. *Energy Policy* 28:389–401.

Guimaraes, P. M. R., J. A. Teixeira and L. Domingues. 2010. Fermentation of lactose to bio-ethanol by yeasts as part of integrated solutions for the valorisation of cheese whey. *Biotechnology Advances* 28:375–384.

Hahn-Hagerdal, B., M. Galbe, M. F. Gorwa-Grauslund, G. Liden and G. Zacchi. 2006. Bio-ethanol - The fuel of tomorrow from the residues of today. *Trends in Biotechnology* 24:549–556.

Hamelinck, C. N., G. van Hooijdonk and A. P. C. Faaij. 2005. Ethanol from lignocellulosic biomass: Techno-economic performance in short-, middle- and long-term. *Biomass and Bioenergy*, 28:384–410.

Hamilton, J. D. 1983. Oil and the macroeconomy since World War II. *Journal of Political Economy* 91:228–248.

Hamilton, J. D. 2003. What is an oil shock? *Journal of Econometrics* 113:363–398.

Hansen, A. C., Q. Zhang and P. W. L. Lyne. 2005. Ethanol-diesel fuel blends - A review. *Bioresource Technology* 96:277–285.

Haryanto, A., S. Fernando, N. Murali and S. Adhikari. 2005. Current status of hydrogen production techniques by steam reforming of ethanol: A review. *Energy and Fuels* 19:2098–2106.

Hendriks, A. T. W. M. and G. Zeeman. 2009. Pretreatments to enhance the digestibility of lignocellulosic biomass. *Bioresource Technology* 100:10–18.

Henstra, A. M., J. Sipma, A. Rinzema and A. J. Stams. 2007. Microbiology of synthesis gas fermentation for biofuel production. *Current Opinion in Biotechnology* 18:200–206.

Hill, J., E. Nelson, D. Tilman, S. Polasky and D. Tiffany. 2006. Environmental, economic, and energetic costs and benefits of biodiesel and ethanol biofuels. *Proceedings of the National Academy of Sciences of the United States of America* 103:11206–11210.

Hill, J., S. Polasky and E. Nelson, et al. 2009. Climate change and health costs of air emissions from biofuels and gasoline. *Proceedings of the National Academy of Sciences of the United States of America* 106:2077–2082.

Ho, S. H., S. W. Huang and C. Y. Chen, et al. 2013. Bioethanol production using carbohydrate-rich microalgae biomass as feedstock. *Bioresource Technology* 135:191–198.

Hook, M. and X. Tang. 2013. Depletion of fossil fuels and anthropogenic climate change-A review. *Energy Policy* 52:797–809.

Hsieh, W. D., R. H. Chen, T. L. Wu and T. H. Lin. 2002. Engine performance and pollutant emission of an SI engine using ethanol-gasoline blended fuels. *Atmospheric Environment* 36:403–410.

Huang, H. J., S. Ramaswamy, U. W. Tschirner and B. V. Ramarao. 2008. A review of separation technologies in current and future biorefineries. *Separation and Purification Technology* 62:1–21.

Huijts, N. M. A., E. J. E. Molin and L. Steg. 2012. Psychological factors influencing sustainable energy technology acceptance: A review-based comprehensive framework. *Renewable and Sustainable Energy Reviews* 16:525–531.

Jacobs, J. A. and S. Frickel. 2009. Interdisciplinarity: A critical assessment. *Annual Review of Sociology* 33:43–65.

John, R. P., G. S. Anisha, K. M. Nampoothiri and A. Pandey. 2011. Micro and macroalgal biomass: A renewable source for bioethanol. *Bioresource Technology* 102:186–193.

Jones, T. C. 2012. America, oil, and war in the Middle East. *Journal of American History* 99:208–218.

Jonsson, L. J. and C. Martin. 2016. Pretreatment of lignocellulose: Formation of inhibitory by-products and strategies for minimizing their effects. *Bioresource Technology* 199:103–112.

Kerr, R. A. 2007. Global warming is changing the world. *Science* 316:188–190.

Keshwani, D. R. and J. J. Cheng. 2009. Switchgrass for bioethanol and other value-added applications: A review. *Bioresource Technology* 100:1515–1523.

Kilian, L. 2008. Exogenous oil supply shocks: How big are they and how much do they matter for the US economy? *Review of Economics and Statistics* 90:216–240.

Kilian, L. 2009. Not all oil price shocks are alike: Disentangling demand and supply shocks in the crude oil market. *American Economic Review*, 99:1053–1069.

Kim, J. 2018. Evaluating author name disambiguation for digital libraries: A case of DBLP. *Scientometrics* 116:1867–1886.

Kirk, T. K. and R. L. Farrell. 1987. Enzymatic "combustion": The microbial degradation of lignin. *Annual Review of Microbiology* 41:465–505.

Konur, O. 2000. Creating enforceable civil rights for disabled students in higher education: An institutional theory perspective. *Disability & Society* 15:1041–1063.

Konur, O. 2002a. Access to nursing education by disabled students: Rights and duties of nursing programs. *Nurse Education Today* 22:364–374.

Konur, O. 2002b. Assessment of disabled students in higher education: Current public policy issues. *Assessment and Evaluation in Higher Education* 27:131–52.

Konur, O. 2002c. Access to employment by disabled people in the UK: Is the Disability Discrimination Act working? *International Journal of Discrimination and the Law* 5:247–279.

Konur, O. 2006a. Participation of children with dyslexia in compulsory education: Current public policy issues. *Dyslexia* 12:51–67.

Konur, O. 2006b. Teaching disabled students in higher education. *Teaching in Higher Education* 11:351–363.

Konur, O. 2007a. A judicial outcome analysis of the *Disability Discrimination Act*: A windfall for the employers? *Disability & Society* 22:187–204.

Konur, O. 2007b. Computer-assisted teaching and assessment of disabled students in higher education: The interface between academic standards and disability rights. *Journal of Computer Assisted Learning* 23:207–219.

Konur, O. 2011. The scientometric evaluation of the research on the algae and bio-energy. *Applied Energy* 88:3532–3540.

Konur, O. 2012a. The evaluation of the biogas research: A scientometric approach. *Energy Education Science and Technology Part A: Energy Science and Research* 29:1277–1292.

Konur, O. 2012b. The evaluation of the educational research: A scientometric approach. *Energy Education Science and Technology Part B: Social and Educational Studies* 4:1935–1948.

Konur, O. 2012c. The evaluation of the global energy and fuels research: A scientometric approach. *Energy Education Science and Technology Part A: Energy Science and Research* 30:613–628.

Konur, O. 2012d. The evaluation of the research on the biodiesel: A scientometric approach. *Energy Education Science and Technology Part A: Energy Science and Research* 28:1003–1014.

Konur, O. 2012e. The evaluation of the research on the bioethanol: A scientometric approach. *Energy Education Science and Technology Part A: Energy Science and Research* 28:1051–1064.

Konur, O. 2012f. The evaluation of the research on the biofuels: A scientometric approach. *Energy Education Science and Technology Part A: Energy Science and Research* 28:903–916.

Konur, O. 2012g. The evaluation of the research on the biohydrogen: A scientometric approach. *Energy Education Science and Technology Part A: Energy Science and Research* 29:323–338.

Konur, O. 2012h. The evaluation of the research on the microbial fuel cells: A scientometric approach. *Energy Education Science and Technology Part A: Energy Science and Research* 29:309–322.

Konur, O. 2012i. The scientometric evaluation of the research on the production of bioenergy from biomass. *Biomass and Bioenergy* 47:504–515.

Konur, O. 2015. Current state of research on algal bioethanol. In *Marine Bioenergy: Trends and Developments*, Ed. S. K. Kim and C. G. Lee, pp. 217–244. Boca Raton, FL: CRC Press.

Konur, O., Ed. 2018a. *Bioenergy and Biofuels*. Boca Raton, FL: CRC Press.

Konur, O. 2018b. Bioenergy and biofuels science and technology: Scientometric overview and citation classics. In *Bioenergy and Biofuels*, Ed. O. Konur, pp. 3–63. Boca Raton: CRC Press.

Konur, O. 2019. Cyanobacterial bioenergy and biofuels science and technology: A scientometric overview. In *Cyanobacteria: From Basic Science to Applications*, Ed. A. K. Mishra, D. N. Tiwari and A. N. Rai, pp. 419–442. Amsterdam: Elsevier.

Konur, O. 2020a. The scientometric analysis of the research on the bioethanol production from green macroalgae. In *Handbook of Algal Science, Technology and Medicine,* Ed. O. Konur, pp. 385–401. London: Academic Press.

Konur, O., Ed. 2020b. *Handbook of Algal Science, Technology and Medicine.* London: Academic Press.

Konur, O., Ed. 2021a. *Handbook of Biodiesel and Petrodiesel Fuels: Science, Technology, Health, and Environment*. Boca Raton, FL: CRC Press.

Konur, O., Ed. 2021b. *Handbook of Biodiesel and Petrodiesel Fuels: Science, Technology, Health, and Environment. Volume 1. Biodiesel Fuels: Science, Technology, Health, and Environment.* Boca Raton, FL: CRC Press.

Konur, O., Ed. 2021c. *Handbook of Biodiesel and Petrodiesel Fuels: Science, Technology, Health, and Environment. Volume 2. Biodiesel Fuels based on the Edible and Nonedible Feedstocks, Wastes, and Algae: Science, Technology, Health, and Environment*. Boca Raton, FL: CRC Press.

Konur, O., Ed. 2021d. *Handbook of Biodiesel and Petrodiesel Fuels: Science, Technology, Health, and Environment. Volume 3. Petrodiesel Fuels: Science, Technology, Health, and Environment*. Boca Raton, FL: CRC Press.

Konur, O. 2023a. Bioethanol fuels: Scientometric study. In *Bioethanol Fuel Production Processes. I: Biomass Pretreatments. Handbook of Bioethanol Fuels Volume 1*, Ed. O. Konur, pp. 47–76. Boca Raton, FL: CRC Press.

Konur, O. 2023b. Bioethanol fuels: Review. In *Bioethanol Fuel Production Processes. I: Biomass Pretreatments. Handbook of Bioethanol Fuels Volume 1*, Ed. O. Konur, pp. 77–98. Boca Raton, FL: CRC Press.

Konur, O. and F. L. Matthews. 1989. Effect of the properties of the constituents on the fatigue performance of composites: A review. *Composites* 20:317–328.

Kruyt, B., D. P. van Vuuren, H. J. de Vries and H. Groenenberg. 2009. Indicators for energy security. *Energy Policy* 37:2166–2181.

Laser, M., D. Schulman and S. G. Allen, et al. 2002. A comparison of liquid hot water and steam pretreatments of sugar cane bagasse for bioconversion to ethanol. *Bioresource Technology* 81:33–44.

Leydesdorff, L. 2000. Is the European Union becoming a single publication system? *Scientometrics* 47:265–280.

Leydesdorff, L. and P. Zhou. 2005. Are the contributions of China and Korea upsetting the world system of science? *Scientometrics* 63:617–630.

Li, H., S. M. Liu, X. H. Yu, S. L. Tang and C. K. Tang. 2020. Coronavirus disease 2019 (COVID-19): Current status and future perspectives. *International Journal of Antimicrobial Agents* 55:105951.

Lieberman, M. B. and D. B. Montgomery. 1988. First-mover advantages. *Strategic Management Journal* 9:41–58.

Limayem, A. and S. C. Ricke. 2012. Lignocellulosic biomass for bioethanol production: Current perspectives, potential issues and future prospects. *Progress in Energy and Combustion Science*, 38:449–467.

Lin, Y. and S. Tanaka. 2006. Ethanol fermentation from biomass resources: Current state and prospects. *Applied Microbiology and Biotechnology* 69:627–642.

Liu, J., X. Wang, Q. Peng and Y. Li. 2005. Vanadium pentoxide nanobelts: Highly selective and stable ethanol sensor materials. *Advanced Materials* 17:764–767.

Ma, X., L. Sun and C. Song. 2002. A new approach to deep desulfurization of gasoline, diesel fuel and jet fuel by selective adsorption for ultra-clean fuels and for fuel cell applications. *Catalysis Today* 77:107–116.

Makenete, A. L., W. J. Lemmer and J. Kupka, J. 2008. The impact of biofuel production on food security: A briefing paper with a particular emphasis on maize-to-ethanol production. *International Food and Agribusiness Management Review* 11:101–110.

Miramontes, J. R. and C. N. Gonzalez-Brambila. 2016. The effects of external collaboration on research output in engineering. *Scientometrics* 109:661–675.

Morschbacker, A. 2009. Bio-ethanol based ethylene. *Polymer Reviews* 49:79–84.

Mosier, N., C. Wyman and B. Dale, et al. 2005. Features of promising technologies for pretreatment of lignocellulosic biomass. *Bioresource Technology* 96:673–686.

Munasinghe, P. C. and S. K. Khanal. 2010. Biomass-derived syngas fermentation into biofuels: Opportunities and challenges. *Bioresource Technology* 101:5013–5022.

Najafi, G., B. Ghobadian and T. Tavakoli, et al. 2009. Performance and exhaust emissions of a gasoline engine with ethanol blended gasoline fuels using artificial neural network. *Applied Energy* 86:630–639.

Negishi, M., Y. Sun and K. Shigi. 2004. Citation database for Japanese papers: A new bibliometric tool for Japanese academic society. *Scientometrics* 60:333–351.

Newman, P. W. G. and J. R. Kenworthy. 1989. Gasoline consumption and cities: A comparison of U.S. cities with a global survey. *Journal of the American Planning Association* 55:24–37.

Ni, M., D. Y. C. Leung and M. K. H. Leung. 2007. A review on reforming bio-ethanol for hydrogen production. *International Journal of Hydrogen Energy*, 32:3238–3247.

Nigam, J. N. 2002. Bioconversion of water-hyacinth (*Eichhornia crassipes*) hemicellulose acid hydrolysate to motor fuel ethanol by xylose-fermenting yeast. *Journal of Biotechnology* 97:107–116.

Nissani, M. 1997. Ten cheers for interdisciplinarity: The case for interdisciplinary knowledge and research. *Social Science Journal* 34:201–216.

North, D. C. 1991. Institutions. *Journal of Economic Perspectives* 5:97–112.

Olsson, L. and B. Hahn-Hagerdal. 1996. Fermentation of lignocellulosic hydrolysates for ethanol production. *Enzyme and Microbial Technology* 18:312–331.

Palmqvist, E. and B. Hahn-Hagerdal. 2000a. Fermentation of lignocellulosic hydrolysates. I: Inhibition and detoxification. *Bioresource Technology*, 74:17–24.

Palmqvist, E. and B. Hahn-Hagerdal. 2000b. Fermentation of lignocellulosic hydrolysates. II: Inhibitors and mechanisms of inhibition. *Bioresource Technology* 74:25–33.

Pimentel, D. and T. W. Patzek. 2005. Ethanol production using corn, switchgrass, and wood; biodiesel production using soybean and sunflower. *Natural Resources Research* 14:65–76.

Pinkert, A., K. N. Marsh, S. Pang and M. P. Staiger. 2009. Ionic liquids and their interaction with cellulose. *Chemical Reviews* 109:6712–6728.

Popp, J., Z. Lakner, M. Harangi-Rakos and M. Fari. 2014. The effect of bioenergy expansion: Food, energy, and environment. *Renewable and Sustainable Energy Reviews* 32:559–578.

Prasad, S., A. Singh and H. C. Joshi. 2007. Ethanol as an alternative fuel from agricultural, industrial and urban residues. *Resources, Conservation and Recycling* 50:1–39.

Ravindran, R. and A. K. Jaiswal. 2016. A comprehensive review on pre-treatment strategy for lignocellulosic food industry waste: Challenges and opportunities. *Bioresource Technology* 199:92–102.

Reeves, S. 2014. To Russia with love: How moral arguments for a humanitarian intervention in Syria opened the door for an invasion of the Ukraine. *Michigan State University International Law Review* 23:199.

Saini, J. K., R. Saini and L. Tewari. 2015. Lignocellulosic agriculture wastes as biomass feedstocks for second-generation bioethanol production: Concepts and recent developments. *3 Biotech* 5:337–353.

Sanchez, C. 2009. Lignocellulosic residues: Biodegradation and bioconversion by fungi. *Biotechnology Advances* 27:185–194.

Sanchez, O. J. and C. A. Cardona. 2008. Trends in biotechnological production of fuel ethanol from different feedstocks. *Bioresource Technology* 99:5270–5295.

Sano, T., H. Yanagishita, Y. Kiyozumi, F. Mizukami and K. Haraya. 1994. Separation of ethanol/water mixture by silicalite membrane on pervaporation. *Journal of Membrane Science* 95:221–228.

Stern, D. I. 1993. Energy and economic growth in the USA: A multivariate approach. *Energy Economics* 15:137–150.

Sukumaran, R. K., R. R. Singhania, G. M. Mathew and A. Pandey. 2009. Cellulase production using biomass feed stock and its application in lignocellulose saccharification for bio-ethanol production. *Renewable Energy* 34:421–424.

Sun, Y. and J. Cheng. 2002. Hydrolysis of lignocellulosic materials for ethanol production: A review. *Bioresource Technology* 83:1–11.

Taherzadeh, M. J. and K. Karimi. 2007. Enzyme-based hydrolysis processes for ethanol from lignocellulosic materials: A review. *Bioresources* 2:707–738.

Taherzadeh, M. J. and K. Karimi. 2008. Pretreatment of lignocellulosic wastes to improve ethanol and biogas production: A review. *International Journal of Molecular Sciences* 9:1621–1651.

Tahmooresnejad, L., C. Beaudry and A. Schiffauerova. 2015. The role of public funding in nanotechnology scientific production: Where Canada stands in comparison to the United States. *Scientometrics* 102:753–787.

Talebnia, F., D. Karakashev and I. Angelidaki. 2010. Production of bioethanol from wheat straw: An overview on pretreatment, hydrolysis and fermentation. *Bioresource Technology* 101:4744–4753.

Tilman, D., R. Socolow and J. A. Foley, et al. 2009. Beneficial biofuels-the food, energy, and environment trilemma. *Science* 325:270–271.

Vane, L. M. 2005. A review of pervaporation for product recovery from biomass fermentation processes. *Journal of Chemical Technology and Biotechnology* 80:603–629.

Wan, Q., Q. H., Li and Y. J. Chen, et al. 2004. Fabrication and ethanol sensing characteristics of ZnO nanowire gas sensors. *Applied Physics Letters*, 84:3654–3656.

Wilberforce, T., Z. El-Hassan and F. N. Khatib, et al. 2017. Developments of electric cars and fuel cell hydrogen electric cars. *International Journal of Hydrogen Energy* 42:25695–25734.

Winzer, C. 2012. Conceptualizing energy security. *Energy Policy* 46:36–48.

Wu, F., D. Zhang and J. Zhang. 2012. Will the development of bioenergy in China create a food security problem? Modeling with fuel ethanol as an example. *Renewable Energy* 47:127–134.

Wustenhagen, R., M. Wolsink and M. J. Burer. 2007. Social acceptance of renewable energy innovation: An introduction to the concept. *Energy Policy* 35:2683–2691.

Yang, B. and C. E. Wyman. 2008. Pretreatment: The key to unlocking low-cost cellulosic ethanol. *Biofuels, Bioproducts and Biorefining* 2:26–40.

Zabed, H., J. N. Sahu, A. Suely, A. N. Boyce and Q. Faruq. 2017. Bioethanol production from renewable sources: Current perspectives and technological progress. *Renewable and Sustainable Energy Reviews* 71:475–501.

Zhang, Y. H. P. and L. R. Lynd. 2004. Toward an aggregated understanding of enzymatic hydrolysis of cellulose: Noncomplexed cellulase systems. *Biotechnology and Bioengineering* 88:797–824.

Zhu, J. Y. and X. J. Pan. 2010. Woody biomass pretreatment for cellulosic ethanol production: Technology and energy consumption evaluation. *Bioresource Technology* 101:4992–5002.

100 The US Experience of Bioethanol Fuels
Review

Ozcan Konur
(Formerly) Ankara Yildirim Beyazit University

100.1 INTRODUCTION

Crude oil-based gasoline fuels (Ma et al., 2002; Newman and Kenworthy, 1989) have been widely used in the transportation sector since the 1920s. However, there have been great public concerns over the adverse environmental and human impact of these fuels (Hill et al., 2006, 2009). Hence, biomass-based bioethanol fuels (Hill et al., 2006; Konur, 2012, 2015, 2020) have increasingly been used in blending gasoline fuels (Hsieh et al., 2002; Najafi et al., 2009), in fuel cells (Antolini, 2007, 2009), and in biochemical production (Angelici et al., 2013; Morschbacker, 2009) in a biorefinery context (Fernando et al., 2006; Huang et al., 2008).

However, it is necessary to pretreat the biomass (Alvira et al., 2010; Taherzadeh and Karimi, 2008) to enhance the yield of the bioethanol (Hahn-Hagerdal et al., 2006; Sanchez and Cardona, 2008) prior to bioethanol fuel production from the feedstocks through the hydrolysis (Sun and Cheng, 2002; Taherzadeh and Karimi, 2007) and fermentation (Lin and Tanaka, 2006; Olsson and Hahn-Hagerdal, 1996) of the biomass and hydrolysates, respectively.

The USA has been one of the most prolific countries engaged in bioethanol fuel research (Konur, 2023a,b). US research in the field of bioethanol fuels has intensified in this context in key research fronts of the production of bioethanol fuels, and to a lesser extent to the utilization (Antolini, 2007; Hansen et al., 2005) and evaluation (Hamelinck et al., 2005; Pimentel and Patzek, 2005) of bioethanol fuels. For the first research front, pretreatment (Alvira et al., 2010; Hendriks and Zeeman, 2009) and hydrolysis (Alvira et al., 2010; Sun and Cheng, 2002) of the feedstocks, fermentation of the feedstock-based hydrolysates (Jonsson and Martin, 2016; Lin and Tanaka, 2006), production of bioethanol fuels (Limayem and Ricke, 2012, Lin and Tanaka, 2006), and to a lesser extent distillation of bioethanol fuels (Sano et al., 1994; Vane, 2005) are the key research areas while for the second research front, the direct utilization of bioethanol fuels in fuel cells (Antolini, 2007, 2009) and transport engines (Hsieh et al., 2002; Najafi et al., 2009), bioethanol-based biohydrogen fuels (Haryanto et al., 2005; Ni et al., 2007), bioethanol sensors (Liu et al., 2005; Wan et al., 2004), and bioethanol-based biochemicals (Angelici et al., 2013; Morschbacker, 2009) are the key research areas.

Research in this field has also intensified for the feedstocks of lignocellulosic biomass at large (Hendriks and Zeeman, 2009; Sun and Cheng, 2002), bioethanol fuels for their utilization (Antolini, 2007; Haryanto et al., 2005), cellulose (Pinkert et al., 2009; Zhang and Lynd, 2004), starch feedstock residues (Binod et al., 2010; Talebnia et al., 2010), wood biomass (Galbe and Zacchi, 2002; Zhu and Pan, 2010), and to a lesser extent industrial waste (Cardona et al., 2010), lignin (Bourbonnais and Paice, 1990), grass (Pimentel and Patzek, 2005), hydrolysates (Palmqvist et al., 2000), biomass at large (Lin and Tanaka, 2006), sugar feedstocks (Bai et al., 2008), sugar feedstock residues (Cardona et al., 2010), algal biomass (John et al., 2011), lignocellulosic wastes (Sanchez, 2009), starch feedstocks (Bai et al., 2008), urban wastes (Ravindran and Jaiswal, 2016), forestry wastes (Duff and

DOI: 10.1201/9781003226574-133

Murray, 1996), food wastes (Ravindran and Jaiswal, 2016), biosyngas (Henstra et al., 2007), and plants (Nigam, 2002).

However, it is essential to develop efficient incentive structures (North, 1991) for the primary stakeholders to enhance the research in this field (Konur, 2000, 2002a,b,c, 2006a,b, 2007a,b). Although there have been a number of review papers on bioethanol fuels (Hendriks and Zeeman, 2009; Mosier et al., 2005; Sun and Cheng, 2002), there has been no review of the 25 most-cited papers in this field.

Thus, this chapter presents a review of the 25 most-cited articles in the field of bioethanol fuels. Then, it discusses the key findings of these highly influential papers and comments on future research priorities in this field.

100.2 MATERIALS AND METHODS

Search for this study was carried out using the Scopus database (Burnham, 2006) in October 2022.

As the first step for the search of the relevant literature, keywords were selected using the 300 most-cited population papers for each research front. The selected keyword list was then optimized to obtain a representative sample of papers in the research field. These keyword lists were then integrated to obtain a keyword list for the research field of bioethanol fuels research field (Konur, 2023a,b). Additionally, a second keyword set was developed for the USA: 'AFFILCOUNTRY ("united states") OR TITLE ("united states" OR usa OR us) OR AUTHKEY ("united states" OR usa OR us) OR SRCTITLE ("united states" OR usa OR us)'. These two keyword sets were integrated with the use of 'AND' keyword. Thus, 12,992 references were located for this field.

As the second step, two sets of data were used in this study. First, a population sample of 12,992 papers was used to examine the scientometric characteristics of the population data. Second, a sample of the 25 most-cited papers was used to consider the key findings from these influential papers.

100.3 RESULTS

Brief information about the 25 most-cited papers with at least 659 citations each on bioethanol fuels is given below. The primary research fronts are the pretreatments and hydrolysis of the feedstocks with nine and seven HCPs, respectively, while the other research fronts are the production, evaluation, and utilization of bioethanol fuels with one, five, and three HCPs, respectively.

100.3.1 Feedstock Pretreatments

Brief information about the nine most-cited papers on the pretreatments of feedstocks with at least 659 citations each is given below (Table 100.1).

Bourbonnais and Paice (1990) oxidized non-phenolic substrates for lignin biodegradation in a paper with 1,189 citations. They used substrates such as Remazol Blue and 2,2'-azinobis(3-ethylbenzthiazoline-6-sulphonate) (ABTS) and found that laccases could also oxidize non-phenolic lignin model compounds. They oxidized Veratryl alcohol (I) and 1-(3,4-dimethoxyphenyl)-2-(2-methoxyphenoxy)-propane-1, 3-diol (III) with laccase and a mediator to give α-carbonyl derivatives. The β-1 lignin model dimer, 1-(3,4-dimethoxyphenyl)-2-phenoxy-ethane-1, 2-diol(II) was cleaved by laccase in the presence of ABTS to give veratraldehyde and benzaldehyde. In conclusion, laccase was capable of oxidizing both phenolic and non-phenolic moieties of lignin but the latter was dependent on the co-presence of a primary laccase substrate.

Floudas et al. (2012) mapped the detailed evolution of wood-degrading enzymes in a paper with 1,104 citations. They noted that wood was highly resistant to decay, largely due to the presence of lignin and the only organisms capable of substantial lignin decay were white rot fungi in the agaricomycetes, which also contained non-lignin-degrading brown rot and ectomycorrhizal species. They analyzed 31 fungal genomes and found that lignin-degrading peroxidases expanded in the

TABLE 100.1
The Pretreatment of Feedstocks

No.	Papers	Biomass	Prts.	Parameters	Keywords	Lead Authors	Affil.	Cits
1	Bourbonnais and Pace (1990)	Lignin	Enzymes	Lignin biodegradation, laccases, non-phenolic substrates, phenolic substrates	Lignin, biodegradation	Bourbonnais, Robert 7003682574	McGill Univ. Canada	1189
2	Floudas et al. (2012)	Wood, lignin	Enzymes	Wood degradation, wood degrading enzyme evolution, genome, peroxidases	Lignin, decomposition	Grigoriev, Igor V. 25027225800	Jnt. Genome Inst. USA	1104
3	Tien and Kirk (1983)	Wood, lignin	Enzymes	Lignin-degrading enzymes, fungi	Lignin, degrading, enzyme	Tien, Ming 7006146774	Pennsylvania State Univ. USA	978
4	Martinez et al. (2008)	Lignocellulosic biomass	Enzymes	Biomass-degrading fungi, genome, enzymes	Biomass, fungus, degrading	Martinez, Diego 7202958664	Massachusetts Inst. Technol. USA	899
5	Sun et al. (2009)	Wood Pine, oak	IL	IL types, dissolution rates, wood types, delignification, cellulose recovery	Wood, delignification, IL	Rogers, Robin D. 35474829200	Univ. Alabama USA	839
6	Fort et al. (2007)	Wood	IL	Wood dissolution profiles, [C_4mim]Cl	Wood, dissolve, ILs	Moyna, Guillermo 6701636629	Univ. Republic Uruguay	725
7	Martinez et al. (2004)	Wood	Enzymes	Wood degrading fungi, genome sequencing	Lignocellulose, fungus, degrading	Cullen, Dan 7202109135	USDA Forest Serv. USA	688
8	Quinlan et al. (2011)	Cellulose	Enzymes	Oxidative cellulose degrading, cellulase-enhancing factors, copper metalloenzyme, GH6b1	Cellulose, biomass, degradation	Johansen, Katja Salomon* 36473579400	Univ. Copenhagen Denmark	679
9	Remsing et al. (2006)	Cellulose	IL	Cellulose dissolution mechanisms, NMR study, [C_4mim]Cl	Cellulose, ILs	Moyna, Guillermo 6701636629 Rogers, Robin D. 35474829200	Univ. Republic Uruguay Univ. Alabama USA	659

*, Female; Cits., Number of citations received for each paper; Prt, Biomass pretreatments.

lineage leading to the ancestor of the agaricomycetes, which was reconstructed as a white rot species, and then contracted in parallel lineages leading to brown rot and mycorrhizal species. Thus, the origin of lignin degradation might have coincided with the sharp decrease in the rate of organic carbon burial around the end of the Carboniferous period.

Tien and Kirk (1983) determined the lignin-degrading enzyme from the *Phanerochaete chrysasporium* Burds in a paper with 978 citations. They found that the extracellular fluid of ligninolytic cultures of this basidiomycete contained an enzyme that degraded lignin substructure model compounds as well as spruce and birch lignins. It had a molecular size of 42,000 daltons and required hydrogen peroxide for activity.

Martinez et al. (2008) performed the genome sequencing and analysis of *Trichoderma reesei* in a paper with 899 citations. They assembled 89 scaffolds to generate 34 Mbp of a nearly contiguous

T. reesei genome sequence comprising 9,129 predicted gene models. They found that its genome encoded fewer cellulases and hemicellulases than any other sequenced fungus able to hydrolyze plant cell wall polysaccharides. Further, many *T. reesei* genes encoding carbohydrate-active enzymes were distributed nonrandomly in clusters that lay between regions of synteny with other sordariomycetes. Hence, numerous genes encoding biosynthetic pathways for secondary metabolites might promote the survival of *T. reesei* in its competitive soil habitat.

Sun et al. (2009) dissolved and partially delignified yellow pine and red oak in 1-ethyl-3-methylimidazolium acetate ([C_2mim]OAc) after mild grinding in a paper with 839 citations. They showed that [C_2mim]OAc was a better solvent for wood than 1-butyl-3-methylimidazolium chloride ([C_4mim]Cl) and that the type of wood, initial wood load, and particle size, etc. affected dissolution and dissolution rates. For example, red oak dissolved better and faster than yellow pine. They obtained carbohydrate-free lignin and cellulose-rich materials by using proper reconstitution solvents (e.g., acetone/water 1:1 v/v) and they achieved approximately 26.1% and 34.9% reductions of lignin content in the reconstituted cellulose-rich materials from pine and oak, respectively, in one dissolution/reconstitution cycle. Further, for pine, they recovered 59% of the holocellulose from the original wood in the cellulose-rich reconstituted material while they recovered 31% and 38% of the original lignin, respectively, as carbohydrate-free lignin and as carbohydrate-bonded lignin from the cellulose-rich materials.

Fort et al. (2007) dissolved wood in ionic liquids (ILs) in a paper with 725 citations. They presented a simple and novel alternative approach for the processing of lignocellulosic materials that rely on their solubility in solvent systems based on 1-n-butyl-3-methylimidazolium chloride ([C_4mim]Cl). They then presented dissolution profiles for woods of different hardness, focusing on the direct analysis of cellulosic material and lignin content in the resulting liquors by means of conventional 1^{3C}NMR techniques. They also showed that cellulose could be readily reconstituted from IL-based wood liquors in fair yields with the addition of a variety of precipitating solvents. The polysaccharide obtained in this manner was virtually free of lignin and hemicellulose and had characteristics that were comparable to those of pure cellulose samples subjected to similar processing conditions.

Martinez et al. (2004) sequenced the 30-million base-pair genome of *Phanerochaete chrysosporium* strain RP78 using a whole genome shotgun approach in a paper with 688 citations. They found that this genome had an impressive array of genes encoding secreted oxidases, peroxidases, and hydrolytic enzymes that cooperate in wood decay. This genome provided a high-quality draft sequence of a basidiomycete, a major fungal phylum that included important plant and animal pathogens.

Quinlan et al. (2011) studied the oxidative degradation of cellulose with a copper metalloenzyme that exploited biomass components in a paper with 679 citations. They showed that glycoside hydrolase (CAZy) GH61 enzymes were a unique family of copper-dependent oxidases. They then demonstrated that copper was needed for GH61 maximal activity and the formation of cellodextrin and oxidized cellodextrin products by GH61 was enhanced in the presence of small molecule redox-active cofactors such as ascorbate and gallate. They found that the active site of GH61 contained a type II copper and, uniquely, a methylated histidine in the copper's coordination sphere, thus providing an innovative paradigm in bioinorganic enzymatic catalysis.

Remsing et al. (2006) studied the mechanism of cellulose dissolution in 1-n-butyl-3-methylimidazolium chloride ([C_4mim]Cl) with a ^{13}C and $^{35/37}$Cl NMR relaxation study on model systems in a paper with 659 citations. They found that the solvation of cellulose with [C_4mim]Cl involved hydrogen-bonding between the carbohydrate hydroxyl protons and the IL chloride ions in a 1:1 stoichiometry.

100.3.2 Hydrolysis of the Feedstocks

There are seven HCPs for the hydrolysis of feedstock with at least 674 citations each (Table 100.2).

TABLE 100.2
The Hydrolysis of Feedstocks

No.	Papers	Biomass	Prts.	Parameters	Keywords	Lead Authors	Affil.	Cits
1	Park et al. (2010)	Cellulose	Na	Cellulose digestibility, crystallinity index, measurement, XRD, NMR, amorphous and crystalline celluloses	Cellulose, cellulase	Johnson, David K. 24550868900	NREL USA	2050
2	Chen and Dixon (2007)	Alfalfa	Acids, enzymes	Biomass engineering, pretreatment, hydrolysis, lignin content, sugar yield	Lignin, sugar, fermentable	Chen, Fang 57188570847	Univ. N. Texas USA	1037
3	Li et al. (2010)	Switchgrass	Acids, IL, enzymes	Pretreatment types, acid hydrolysis, delignification, saccharification, sugar yields, crystallinity, surface area, lignin content, enzymatic hydrolysis	Switchgrass, pretreatment, saccharification, recalcitrance, IL	Singh, Seema* 35264950300	Sandia Natl. Lab. USA	861
4	Kilpelainen et al. (2007)	Wood Spruce, pine	IL, enzymes	IL pretreatment, enzymatic hydrolysis, IL types, wood types	Wood, IL, dissolution	Kilpelainen, Ilkka 7006830888	Univ. Helsinki Finland	834
5	Lee et al. (2009)	Wood	IL, enzymes	Enzymatic hydrolysis, IL pretreatment, delignification, cellulose crystallinity index, enzyme recycling	Wood, IL, hydrolysis	Dordick, Jonathan S. 7102545507	Rensselaer Polytech Inst. USA	818
6	Kumar et al. (2009)	Corn stover, poplar	Enzymes	Biomass hydrolysis and pretreatments, biomass crystallinity, cellulose degree of polymerization, cellulase adsorption capacity	Corn stover, poplar, pretreatment	Charles E. Wyman 7004396809	Univ. Calif. Riverside USA	728
7	Silverstein et al. (2007)	Cotton stalks	H_2SO_4, NaOH, and H_2O_2, enzymes	Hydrolysis, pretreatment types, solid loading, temperature, residence time, glucose conversion	Stalks, pretreatment, saccharification	Sharma-Shivappa, Ratna R.* 16231216900	NC State Univ. USA	674

*, Female; Cits., Number of citations received for each paper; Na, non available; Prt, Biomass pretreatments.

Park et al. (2010) compared four different techniques incorporating X-ray diffraction (XRD) and solid-state ^{13}C nuclear magnetic resonance (NMR) for the measurement of the crystallinity index (CI) using eight different cellulose preparations in a paper with 2,050 citations. They found that the simplest method which involved measurement of just two heights in an X-ray diffractogram, produced significantly higher crystallinity values than the other methods. Alternative XRD and NMR methods which considered the contributions from amorphous and crystalline cellulose to the entire XRD and NMR spectra provided a more accurate measure of the crystallinity of cellulose. Although celluloses having a high amorphous content were usually more easily digested by enzymes, it was unclear whether CI actually provided a clear indication of the digestibility of a cellulose sample. Cellulose accessibility should be affected by crystallinity but was also likely to be affected by several other parameters, such as lignin/hemicellulose contents and distribution, porosity, and particle size. Given the methodological dependency of cellulose CI values and the complex nature of cellulase interactions with amorphous and crystalline celluloses, they caution against trying to correlate relatively small changes in CI with changes in cellulose digestibility.

Chen and Dixon (2007) showed that lignin modification improved fermentable sugar yields for bioethanol production in a paper with 1,037 citations. They found that in stems of transgenic alfalfa lines independently downregulated in each of six lignin biosynthetic enzymes, recalcitrance to both acid pretreatments and enzymatic digestion was directly proportional to lignin content. Further, some transgenics yield nearly twice as much sugar from cell walls as wild-type plants. Thus, lignin modification could bypass the need for acid pretreatment and thereby facilitate bioprocess consolidation.

Li et al. (2010) compared the efficiency of dilute acid hydrolysis and dissolution in an IL with switchgrass in a paper with 861 citations. When subject to IL pretreatment they observed that switchgrass exhibited reduced cellulose crystallinity, increased surface area, and decreased lignin content compared to dilute acid pretreatment. Further, IL pretreatment enabled a significant enhancement in the rate of enzymatic hydrolysis of the cellulose component of switchgrass, with a rate increase of 16.7-fold, and a glucan yield of 96.0% obtained in 24h. In conclusion, IL pretreatment offered unique advantages compared to the dilute acid pretreatment process for switchgrass.

Kilpelainen et al. (2007) dissolved wood in ILs in a paper with 834 citations. They found that 1-butyl-3-methylimidazolium chloride ([Bmim]Cl) and 1-allyl-3-methylimidazolium chloride ([Amim]Cl) had good solvating power for Norway spruce sawdust and Norway spruce and Southern pine thermo-mechanical pulp fibers as these ILs provided solutions which permitted the complete acetylation of the wood. Alternatively, they obtained transparent amber solutions of wood when the dissolution of the same lignocellulosic samples was attempted in 1-benzyl-3-methylimidazolium chloride ([BzMim]Cl). They then digested the cellulose of the regenerated wood into glucose with a cellulase enzymatic hydrolysis treatment. Furthermore, completely acetylated wood was readily soluble in chloroform.

Lee et al. (2009) used 1-ethyl-3-methylimidazolium acetate ([Emim]CH_3COO) to extract lignin from wood flour in a paper with 818 citations. They observed that the cellulose in the pretreated wood flour became far less crystalline without undergoing solubilization. When 40% of the lignin was removed, the cellulose crystallinity index dropped below 45, resulting in >90% of the cellulose in wood flour being hydrolyzed with *Trichoderma viride* cellulase. They then reused this IL, thereby resulting in a highly concentrated solution of chemically unmodified lignin.

Kumar et al. (2009) pretreated corn stover and poplar solids using leading pretreatment technologies in a paper with 728 citations. They used ammonia fiber expansion (AFEX), ammonia recycled percolation (ARP), controlled pH, dilute acid, flowthrough, lime, and SO_2 pretreatments. They observed that lime pretreatment removed the most acetyl groups from both corn stover and poplar, while AFEX removed the least while low pH pretreatments depolymerized cellulose and enhanced biomass crystallinity much more than higher pH approaches. Further, lime-pretreated corn stover solids and flowthrough pretreated poplar solids had the highest cellulase adsorption capacity, while dilute acid pretreated corn stover solids and controlled pH pretreated poplar solids had the least. Then, enzymatically extracted AFEX lignin preparations for both corn stover and poplar had the lowest cellulase adsorption capacity while SO_2 pretreated solids had the highest surface O/C ratio for poplar, but for corn stover, the highest value was observed for dilute acid pretreatment with a Parr reactor. In conclusion, although dependent on pretreatment and substrate, along with changes in cross-linking and chemical changes, pretreatments might also decrystallize cellulose and change the ratio of crystalline cellulose polymorphs ($I\alpha/I\beta$).

Silverstein et al. (2007) compared chemical pretreatment methods for improving the saccharification of cotton stalks for ethanol production in a paper with 673 citations. They used sulfuric acid (H_2SO_4), sodium hydroxide (NaOH), hydrogen peroxide (H_2O_2), and ozone (O_3) pretreatments. They pretreated these ground stalks at a solid loading of 10% (w/v) with H_2SO_4, NaOH, and H_2O_2 at concentrations of 0.5%, 1%, and 2% (w/v). Treatment temperatures were 90°C and 121°C at 15 psi for residence times of 30, 60, and 90min. They performed ozone pretreatment at 4°C with constant sparging of stalks in water. Solids from H_2SO_4, NaOH, and H_2O_2 pretreatments (at 2%, 60min, and 121°C/15 psi) showed significant lignin degradation and/or high sugar availability. They hydrolyzed these solids with Celluclast 1.5L and Novozym 188 at 50°C. They found that sulfuric acid pretreatment resulted in the highest xylan reduction (95.23% for 2% acid, 90 min, and 121°C/15 psi) but in the lowest cellulose to glucose conversion during hydrolysis (23.85%). NaOH pretreatment resulted in the highest level of delignification (65.63% for 2% NaOH, 90 min, and 121°C/15 psi) and cellulose conversion (60.8%). H_2O_2 pretreatment resulted in significantly lower delignification (maximum of 29.51% for 2%, 30 min, and 121°C/15 psi) and cellulose conversion (49.8%) than NaOH pretreatment, but had a higher cellulose conversion than sulfuric acid pretreatment. Ozone did not cause any significant changes in lignin, xylan, or glucan contents over time.

100.3.3 Bioethanol Production, Evaluation, and Utilization

There are one, five, and three HCPs for the production, evaluation, and utilization of bioethanol fuels with at least 666 citations each (Table 100.3). For the last research front, there are one and two HCPs for biohydrogen fuel production from bioethanol fuels and the utilization of bioethanol fuels in fuel cells, respectively. However, there are no HCPs for the utilization of bioethanol fuels in the production of biochemicals, the development of bioethanol sensors, and for their use in gasoline or diesel engines.

100.3.3.1 Bioethanol Production

Saha et al. (2005) produced bioethanol from wheat straw through acid pretreatment, enzymatic hydrolysis of the biomass, and fermentation and detoxification of the resulting hydrolysate in separate hydrolysis and fermentation (SHF) and simultaneous saccharification and fermentation (SSF) processes in a paper with 666 citations. They evaluated dilute acid pretreatment at varied temperatures and enzymatic saccharification for the conversion of wheat straw cellulose and hemicellulose to monomeric sugars. Wheat straw had 48.57% cellulose and 27.70% hemicellulose on a dry solid (DS) basis. They found that the maximum yield of monomeric sugars from wheat straw (7.83%, w/v, DS) with dilute H_2SO_4 (0.75%, v/v) pretreatment and enzymatic saccharification (45°C, pH 5.0, 72 h) using cellulase, β-glucosidase, xylanase, and esterase was 565 mg/g. Under this condition, there were no measurable quantities of furfural and hydroxymethylfurfural. The yield of ethanol (per liter) from acid-pretreated enzyme saccharified wheat straw (78.3 g) hydrolysate by a recombinant *Escherichia coli* strain FBR5 was 19 g with a yield of 0.24 g/g DS. Detoxification of the acid and enzyme-treated wheat straw hydrolysate by overliming reduced the fermentation time from 118 to 39 h in the case of SHF (35°C, pH 6.5), and increased the ethanol yield from 13 to 17 g/L and decreased the fermentation time from 136 to 112 h in the case of SSF (35°C, pH 6.0).

100.3.3.2 Bioethanol Fuel Evaluation

Hill et al. (2006) developed four criteria to evaluate first generation corn grain-based bioethanol and soybean-based biodiesel fuels through life cycle accounting in a paper with 2,167 citations. They found that bioethanol yielded 25% more energy than the energy invested in its production, whereas biodiesel yielded 93% more. Compared with bioethanol, biodiesel released just 1.0%, 8.3%, and 13% of agricultural nitrogen, phosphorus, and pesticide pollutants, respectively, per net energy gain. Relative to the fossil fuels they displaced, greenhouse gas (GHG) emissions were reduced by 12% by the production and combustion of bioethanol and 41% by biodiesel. Biodiesel also released less air pollutants per net energy gain than ethanol. These advantages of biodiesel over ethanol arise from lower agricultural inputs and more efficient conversion of feedstocks to fuel. However, neither biofuel could replace too much petroleum without impacting food supplies as even dedicating all the U.S. corn and soybean production to biofuels would meet only 12% of gasoline demand and 6% of diesel demand. Until the recent increases in petroleum prices, high production costs made biofuels unprofitable without subsidies. Biodiesel provided sufficient environmental advantages to merit a subsidy. Thus, transportation biofuels such as synfuel hydrocarbons, or lignocellulosic ethanol, if produced from low-input biomass grown on agriculturally marginal land or from waste biomass, could provide much greater supplies and environmental benefits than food-based biofuels. This

TABLE 100.3
The Production, Evaluation, and Utilization of Bioethanol Fuels

No.	Papers	Biomass	Res. Fronts	Prts.	Yeasts	Parameters	Keywords	Lead Authors	Affil.	Cits
1	Hill et al. (2006)	Corn grains	Evaluation	Na	Na	Life cycle assessment, net energy gain, economic competitiveness, GHG emissions, food security, large-scale production	Ethanol, economic, environmental, energetic	Hill, Jason 14053737500	Univ. Minnesota	2167
2	Kim and Dale (2004)	Starch feedstock residues Sugar feedstock residues	Evaluation	Na	Na	Lignocellulosic wastes, sugar feedstock residues starch feedstock residues, bioethanol production, Asia, USA, Europe, wheat straw, rice straw, corn stover	Bioethanol, wasted, crop residues	Dale, Bruce E. 7201511969	Michigan State Univ. USA	1455
3	Pimentel and Patzek (2005)	Corn, switchgrass, wood	Evaluation	Na	Na	Bioethanol fuels, net energy gain, biodiesel fuels, biomass types	Ethanol, corn, switchgrass, wood	Pimentel, David 7005471319	Cornell Univ. USA	1023
4	Deluga et al. (2004)	Bioethanol	Biohydrogen	Na	Na	Biohydrogen production, ethanol autothermal reforming, hydrogen selectivity, ethanol conversion rate	Ethanol, hydrogen, reforming	Schmidt, Lanny D. 13103591600	Univ. Minnesota USA	918
5	Schmer et al. (2008)	Switchgrass	Evaluation	Na	Na	Net energy efficiency, economic feasibility, net energy yield, economic costs, biomass yield, GHG emissions	Ethanol, switchgrass, cellulosic, net energy	Vogel, Kenneth P. 7102498441	Univ. Nebraska Lincoln USA	831
6	Klein-Marcuschamer et al. (2012)	Corn stover	Evaluation	Na	Na	Ethanol production cost, enzyme costs, feedstock prices, fermentation times, techno-economics	Lignocellulosic, costs, enzymes	Blanch, Harvey W. 7006259341	Univ. Calif. R USA	709
7	(1999)	Bioethanol fuels	Fuel cells	Na	Na	Chemical kinetic model, high- temperature ethanol oxidation, ethanol decomposition kinetics	Ethanol, oxidation	Marinov, Nick M 6701797865	Sierra Eng. USA	701
8	Kowal et al. (2009)	Bioethanol fuels	Fuel cells	Na	Na	Ethanol electrooxidation, ternary $PtRhSnO_2/C$ electrocatalysts, electrocatalytic activity	Ethanol, electrocatalysts	Kowal, Andrzej 36728886000	Polish Acad. Sci. Poland	667
9	Saha et al.	Wheat straw	Production	H_2SO_4, enzymes	E. coli	Ethanol production, acid pretreatment, enzymatic hydrolysis, fermentation, detoxification, ethanol yield, fermentation time, SHF, SSF	Straw, pretreatment, ethanol, saccharification	Saha, Badal, C. 7202946302	USDA Agr. Res. Serv. USA	666

Cits., Number of citations received for each paper; Na, Non available; Prt, Biomass pretreatments.

study compared corn grain-based bioethanol fuels with crude oil-based diesel fuels and showed that they had a net energy gain, had environmental benefits, they were economically competitive, but not producible in large quantities without reducing food supplies. For this reason, they recommended lignocellulosic bioethanol fuels from wood, grass, wastes, and agricultural residues.

Kim and Dale (2004) estimated the global annual potential bioethanol production from second generation starch feedstock residues of corn, barley, oat, rice, wheat, and the sugar feedstock residues of sorghum and sugarcane in a paper with 1,455 citations. They found that there was about 1.57 petagram (10^{15} g) of dry lignocellulosic wastes in the world that could potentially produce 491 gigaliters (GL) (10^9 L)/year of bioethanol, about 16 times higher than the world ethanol production as of 2004. Thus, this potential bioethanol production could replace 353 GL of gasoline (32% of the global gasoline consumption) when bioethanol was used in E85 fuel for a midsize passenger vehicle. Furthermore, lignin-rich fermentation residue could potentially generate both 458 terawatt-hours (10^9 KWh) of bioelectricity (about 3.6% of world electricity production) and 2.6 exajoule (10^{18} J) of steam. Asia was the largest potential producer of bioethanol from lignocellulosic wastes of rice straw, wheat straw, and corn stover, and could produce up to 291 GL/year of bioethanol. The next highest potential region was Europe (69.2 GL of bioethanol), in which most bioethanol came from wheat straw. Corn stover was the main feedstock in North America from which about 38.4 GL/year of bioethanol could potentially be produced. Globally rice straw could produce 205 GL of bioethanol. The next highest potential feedstock was wheat straw, which could produce 104 GL of bioethanol.

Pimentel and David (2005) compared the net energy gain of bioethanol fuels from corn, switchgrass, and wood and biodiesel fuels from soybean and sunflower in a paper with 1,023 citations. They found that energy outputs from ethanol produced from this biomass were each less than the respective fossil energy inputs. The same was true for producing biodiesel, however, the energy cost for producing soybean biodiesel was only slightly negative compared with ethanol production. Ethanol production using corn grain, switchgrass, and wood required 29%, 50%, and 57% more fossil energy than the ethanol fuel produced, respectively. Similarly, biodiesel production using soybean and sunflower required 27% and 118% more fossil energy than the biodiesel fuel produced, respectively. The energy yield from soy oil per hectare was far lower than the ethanol yield from corn.

Schmer et al. (2008) determined the net energy yield and GHG emissions of lignocellulosic bioethanol fuels from switchgrass in a paper with 831 citations. They managed switchgrass in field trials of 3–9 ha on marginal cropland on 10 farms across a wide precipitation and temperature gradient in the mid-continental U.S. to determine net energy and economic costs based on known farm inputs and harvested yields. They found that the annual biomass yields of these fields averaged 5.2–11.1 Mg/ha with a resulting average estimated net energy yield (NEY) of 60 GJ/ha/y. Switchgrass produced 540% more renewable than nonrenewable energy consumed. Switchgrass monocultures managed for high yield produced 93% more biomass yield and an equivalent estimated NEY than previous estimates from human-made prairies that received low agricultural inputs. Estimated average GHG emissions from bioethanol fuels were 94% lower than estimated GHG emissions from gasoline. They asserted that improved genetics and agronomics might further enhance the energy sustainability and biofuel yield of switchgrass.

Klein-Marcuschamer et al. (2012) constructed a technoeconomic model for the production of fungal cellulases in a paper with 709 citations. They found that the cost of producing enzymes was much higher than that commonly assumed in the literature. For example, the cost contribution of enzymes to ethanol produced by the conversion of corn stover was \$0.68/gal if the sugars in the biomass could be converted at maximum theoretical yields, and \$1.47/gal if the yields were based on saccharification and fermentation yields that had been previously reported in the scientific literature. They performed a sensitivity analysis to study the effect of feedstock prices and fermentation times on the cost contribution of enzymes to ethanol price. They concluded that a significant effort was still required to lower the contribution of enzymes to biofuel production costs.

100.3.3.3 Bioethanol Fuel Utilization

There are one and two HCPs for biohydrogen fuel production from bioethanol fuels and the utilization of bioethanol fuels in fuel cells, respectively. However, there are no HCPs for the utilization of bioethanol fuels for the production of biochemicals, the development of bioethanol sensors, and for their use in gasoline or diesel engines.

100.3.3.3.1 Bioethanol Fuel-based Biohydrogen Fuels

Deluga et al. (2004) produced biohydrogen fuels by the autothermal reforming of bioethanol fuels in a paper with 918 citations. They converted ethanol and ethanol-water mixtures directly into hydrogen with nearly 100% selectivity and more than 95% conversion by catalytic partial oxidation, with a residence time on rhodium-ceria catalysts of less than 10 ms. They performed the rapid vaporization and mixing with air with an automotive fuel injector at temperatures sufficiently low and times sufficiently fast that homogeneous reactions producing carbon, acetaldehyde, ethylene, and total combustion products could be minimized.

100.3.3.3.2 Bioethanol Fuel Cells

Marinov (1999) developed a chemical kinetic model for high-temperature ethanol oxidation in a paper with 701 citations. He used laminar flame speed data, ignition delay data behind a reflected shock wave, and ethanol oxidation product profiles from a jet-stirred and turbulent flow reactor in this computational study. He found a good agreement in the modeling of the data sets obtained from the five different experimental systems. High-temperature ethanol oxidation exhibited strong sensitivity to the fall-off kinetics of ethanol decomposition, branching ratio selection for decomposition reaction, and reactions involving the hydroperoxyl (HO_2) radical. He then presented an empirical branching ratio estimation procedure that determined the temperature-dependent branching ratios of the three distinct sites of hydrogen abstraction from ethanol.

Kowal et al. (2009) developed electrocatalysts for oxidizing ethanol to CO_2 in a paper with 667 citations. They synthesized a ternary $PtRhSnO_2/C$ electrocatalyst by depositing platinum (Pt) and rhodium (Rh) atoms on carbon-supported tin dioxide (SnO_2) nanoparticles that was capable of oxidizing ethanol with high efficiency and held great promise for resolving the impediments to developing practical direct ethanol fuel cells (DEFCs). They found that this electrocatalyst effectively split the C–C bond in ethanol at room temperature in acid solutions, facilitating its oxidation at low potentials to CO_2, which had not been achieved with existing catalysts. They reasoned that this electrocatalyst's electrocatalytic activity was due to the specific property of each of its constituents, induced by their interactions. These findings helped explain the high activity of Pt–Ru for methanol oxidation and the lack of it for ethanol oxidation and pointed the way to accomplishing C–C bond splitting in other catalytic processes.

100.4 DISCUSSION

100.4.1 Introduction

Crude oil-based gasoline fuels have been widely used in the transportation sector since the 1920s. However, there have been great public concerns over the adverse environmental and human impact of these fuels. Hence, biomass-based bioethanol fuels have increasingly been used in blending gasoline and petrodiesel fuels, in fuel cells, and in biochemical production in a biorefinery context.

However, it is necessary to pretreat the biomass to enhance the yield of the bioethanol prior to bioethanol fuel production from the feedstocks through the hydrolysis and fermentation of the biomass and hydrolysates, respectively.

The USA has been one of the most prolific countries engaged in bioethanol fuel research. US research in the field of bioethanol fuels has intensified in this context in the key research fronts of the production of bioethanol fuels, and to a lesser extent, on the utilization and evaluation of bioethanol fuels. For the first research front, pretreatment and hydrolysis of the feedstocks, fermentation of the feedstock-based hydrolysates, production of bioethanol fuels, and to a lesser extent distillation of bioethanol fuels are the key research areas while for the second research front, the utilization of bioethanol fuels in fuel cells and transport engines, bioethanol-based biohydrogen fuels, bioethanol sensors, and bioethanol-based biochemicals are the key research areas. The research in this field has also intensified for the feedstocks of lignocellulosic biomass at large, bioethanol fuels for their utilization, cellulose, starch feedstock residues, wood biomass, and to a lesser extent industrial waste, lignin, grass, hydrolysates, biomass, sugar feedstocks, sugar feedstock residues, algal biomass, lignocellulosic wastes, starch feedstocks, urban wastes, forestry wastes, food wastes, biosyngas, and plants.

However, it is essential to develop efficient incentive structures for the primary stakeholders to enhance the research in this field. Although there have been a number of review papers for this field, there has been no review of the 25 most-cited articles in this field. Thus, this chapter presents a review of the 25 most-cited articles on bioethanol fuel. Then, it discusses the key findings of these highly influential papers and comments on future research priorities in this field.

As the first step for the search of the relevant literature, keywords were selected using the 300 most-cited population papers for each research front. The selected keyword list was then optimized to obtain a representative sample of papers for each research field. These keyword lists were then integrated to obtain the keyword list in the research field of bioethanol fuels research field (Konur, 2023a-b). Additionally, a second keyword set was developed for the USA. These two keyword sets were integrated with the use of 'AND' keyword. Thus, 12,992 references were located for this field.

Information about the research fronts for the sample papers in the US experience of bioethanol fuels is given in Table 100.4. As this table shows the most prolific research front for this field is wood biomass such as poplar and pine with 36% of the HCPs, followed by starch feedstock residues

TABLE 100.4
The Most Prolific Research Fronts for the US Bioethanol Fuels

No.	Research Fronts	N Paper (%) Review	N Paper (%) Sample	Surplus (%)
1	Wood biomass	36	9.0	27.0
2	Starch feedstock residues	20	9.7	10.3
3	Grass	16	3.1	12.9
4	Bioethanol fuels	12	19.6	−7.6
5	Cellulose	12	15.1	−3.1
6	Lignin	12	6.2	5.8
7	Starch feedstocks	8	1.7	6.3
8	Lignocellulosic biomass at large	4	23.0	−19.0
9	Sugar feedstock residues	4	2.2	1.8
10	Missing feedstocks	0	22.9	−22.9
	Industrial waste	0	6.5	−6.5
	Hydrolysates	0	2.8	−2.8
	Biomass	0	2.6	−2.6
	Sugar feedstocks	0	2.2	−2.2
	Algal biomass	0	1.9	−1.9
	Lignocellulosic wastes	0	1.8	−1.8
	Urban wastes	0	1.0	−1.0
	Forestry wastes	0	0.8	−0.8
	Hemicellulose	0	0.8	−0.8

(Continued)

TABLE 100.4 (*Continued*)
The Most Prolific Research Fronts for the US Bioethanol Fuels

No.	Research Fronts	N Paper (%) Review	N Paper (%) Sample	Surplus (%)
	Xylan	0	0.8	−0.8
	Food wastes	0	0.6	−0.6
	Cellobiose	0	0.4	−0.4
	Syngas	0	0.4	−0.4
	Plants	0	0.3	−0.3
	Sample size	25	775	

N Paper (%) review, The number of papers in the sample of 25 reviewed papers; N paper (%) sample, The number of papers in the global sample of 775 papers.

and grass with 20% and 16% of these HCPs, respectively. Other prolific feedstocks are bioethanol fuels, cellulose, lignin, starch feedstocks, lignocellulosic biomass at large, and sugar feedstock residues with 4%–12% of the HCPs each.

On the one hand, it is notable that there are no HCPs for industrial waste, hydrolysates, biomass, sugar feedstocks, algal biomass, lignocellulosic wastes, urban wastes, forestry wastes, hemicellulose, xylan, food wastes, cellobiose, syngas, and plants. Thus, the first nine feedstocks dominate the HCPs in this field where the only major waste biomass are starch feedstock residues such as corn stover or wheat straw and sugar feedstock residues such as sugarcane bagasse. On the other hand, cellulose and lignin are the key constituents of the lignocellulosic biomass. Further, bioethanol fuels are the feedstocks for the production of biohydrogen fuels and biochemical and for utilization in transport engines and fuel cells.

It is also notable that the research on first generation food-grade sugar and starch bioethanol fuels is not substantial for all of the samples studied, reflecting the adverse effect of undermining food security (Bentivoglio et al., 2016; Elobeid et al., 2016).

Further, the most influential research fronts are wood biomass with a 27% surplus, followed by grass, starch feedstock residues, starch feedstocks, and lignin with a 13%, 10%, 6%, and 6% surplus, respectively. Further, lignocellulosic biomass at large is the least influential feedstock with a 19% deficit, followed by bioethanol fuels, industrial waste, cellulose, hydrolysates, and biomass with a 3%–8% deficit each. It is also notable that a significant part of the papers for lignocellulosic biomass-based bioethanol fuels are reviews and short surveys.

Information about the thematic research fronts for the sample papers in bioethanol fuels is given in Table 100.5. As this table shows the most prolific research front for this field is the production of bioethanol fuels with 68% of the HCPs. The other prolific research fronts are the evaluation and utilization of bioethanol evaluation with 20% and 12% of the HCPs, respectively.

Further, the most prolific research fronts for the first group are the pretreatments and hydrolysis of feedstocks with 64% and 32% of the HCPs, respectively. The other research fronts are bioethanol production and hydrolysate fermentation with 4% and 8% of the HCPs, respectively. For the third group, the only prolific research fronts are bioethanol-based biohydrogen fuels and the utilization of bioethanol fuels in fuel cells with 4% and 8% of the HCPs, respectively while there are no HCPs for the utilization of bioethanol fuels in transport engines, production of biochemical from bioethanol fuels, and bioethanol sensors.

On the other hand, biofuel fuel evaluation is the most influential research field with a 14% surplus, followed by biomass pretreatments and hydrolysis with a 6% and 2% surplus, respectively. On the other hand, the least influential research front is bioethanol fuel production with an 18% deficit, followed by hydrolysate fermentation, bioethanol fuel utilization, bioethanol fuel production, and bioethanol fuels in engines with 10%, 8%, 6%, and 6% deficit, respectively.

TABLE 100.5
The Most Prolific Thematic Research Fronts for the US Experience of Bioethanol Fuels

No.	Research Fronts	N Paper (%) Review	N Paper (%) Sample	Surplus (%)
1	Bioethanol fuel production	68	74.1	–6.1
	Biomass pretreatments	64	58.5	5.5
	Biomass hydrolysis	32	29.8	2.2
	Bioethanol production	4	21.5	–17.5
	Hydrolysate fermentation	8	17.9	–9.9
	Bioethanol fuel distillation	0	0.8	–0.8
2	Bioethanol fuel evaluation	20	6.2	13.8
3	Bioethanol fuel utilization	12	19.7	–7.7
	Bioethanol-based biohydrogen fuels	4	4.3	–0.3
	Bioethanol sensors	0	2.2	–2.2
	Bioethanol fuels in fuel cells	8	6.8	1.2
	Bioethanol fuels in engines	0	6.1	–6.1
	Bioethanol-based biochemicals	0	0.4	–0.4

N Paper (%) review, The number of papers in the sample of 25 reviewed papers; N paper (%) sample, The number of papers in the global sample of 775 papers.

100.4.2 Feedstock Pretreatments

Brief information about the nine most-cited papers on the pretreatments of feedstocks with at least 659 citations each is given below (Table 100.1). It is notable that as Table 100.5 shows, 68% of these HCPs are related to the pretreatments of feedstocks.

These findings show that both pretreatments and hydrolysis of feedstocks are the fundamental processes for bioethanol production from feedstocks. These narrated studies highlight the importance of the pretreatment and hydrolysis processes for the production of bioethanol fuels from feedstocks with a high ethanol yield. These pretreatments, primarily enzymatic and chemical pretreatments, fractionate the feedstock and enhance the enzymatic digestibility of the biomass.

Bourbonnais and Paice (1990) oxidized non-phenolic substrates for lignin biodegradation in a paper and found that laccases could also oxidize non-phenolic lignin model compounds. Further, Floudas et al. (2012) mapped the detailed evolution of wood-degrading enzymes and noted that wood was highly resistant to decay, largely due to the presence of lignin and the only organisms capable of substantial lignin decay were white rot fungi in agaricomycetes.

Tien and Kirk (1983) determined the lignin-degrading enzyme from *P. chrysasporium* Burds and found that the extracellular fluid of ligninolytic cultures of this basidiomycete contained an enzyme that degraded lignin substructure model compounds as well as spruce and birch lignins. Further, Martinez et al. (2008) performed a genome sequencing and analysis of the *Trichoderma reesei* and found that its genome encoded fewer cellulases and hemicellulases than any other sequenced fungus able to hydrolyze plant cell wall polysaccharides.

Sun et al. (2009) dissolved and partially delignified yellow pine and red oak in [C_2mim]OAc after mild grinding and showed that this IL was a better solvent for wood than 1 [C_4mim]Cl. Further, Fort et al. (2007) dissolved wood in ILs and showed that cellulose could be readily reconstituted from the IL-based wood liquors in fair yields by the addition of a variety of precipitating solvents.

Martinez et al. (2004) sequenced the 30-million base-pair genome of *Phanerochaete chrysosporium* strain RP78 using the whole genome shotgun approach and found that this genome had an impressive array of genes encoding secreted oxidases, peroxidases, and hydrolytic enzymes that cooperate in wood decay. Further, Quinlan et al. (2011) studied the oxidative degradation of cellulose with a copper metalloenzyme that exploited biomass components and showed that glycoside hydrolase (CAZy) GH61 enzymes were a unique family of copper-dependent oxidases.

Finally, Remsing et al. (2006) studied the mechanism of cellulose dissolution in the $[C_4mim]Cl$ with a ^{13}C and $^{35/37}Cl$ NMR relaxation study on model systems in a paper with 659 citations.

100.4.3 Hydrolysis of the Feedstocks

There are seven HCPs for the hydrolysis of feedstocks with at least 6n74 citations each (Table 100.2). On the other hand, it is notable that as Table 100.5 shows, 32% of these HCPs are related to the hydrolysis of feedstocks.

These findings show that both pretreatments and hydrolysis of the feedstocks are the fundamental processes for bioethanol production from the feedstocks. These narrated studies highlight the importance of the pretreatment and hydrolysis processes for the production of bioethanol fuels from feedstocks with a high ethanol yield. These pretreatments, primarily enzymatic and chemical pretreatments, fractionate the feedstocks and enhance the enzymatic digestibility of the biomass.

Park et al. (2010) compared four different techniques incorporating XRD and solid-state ^{13}C NMR for the measurement of the crystallinity index (CI) using eight different cellulose preparations and found that the simplest method which involved measurement of just two heights in the X-ray diffractogram, produced significantly higher crystallinity values than did the other methods. Further, Chen and Dixon (2007) showed that lignin modification improved fermentable sugar yields for bioethanol production.

Li et al. (2010) compared the efficiency of dilute acid hydrolysis and dissolution in an IL with switchgrass and observed that switchgrass exhibited reduced cellulose crystallinity, increased surface area, and decreased lignin content compared to dilute acid pretreatment. Further, Kilpelainen et al. (2007) dissolved wood in ILs and found that [Bmim]Cl and [Amim]Cl had good solvating power for Norway spruce sawdust and Norway spruce and Southern pine thermomechanical pulp fibers.

Lee et al. (2009) used $[Emim]CH_3COO$ to extract lignin from wood flour and observed that the cellulose in the pretreated wood flour became far less crystalline without undergoing solubilization. Further, Kumar et al. (2009) pretreated corn stover and poplar solids using leading pretreatment technologies and found that enzymatically extracted AFEX lignin preparations had the lowest cellulase adsorption capacity for both biomass while SO_2 pretreated solids had the highest surface O/C ratio for poplar, but for corn stover, the highest value was observed with dilute acid pretreatment.

Finally, Silverstein et al. (2007) compared chemical pretreatment methods for improving saccharification of cotton stalks for ethanol production and found that sulfuric acid pretreatment resulted in the highest xylan reduction but the lowest cellulose to glucose conversion during hydrolysis.

100.4.4 Bioethanol Production, Evaluation, and Utilization

There are one, five, and three HCPs for the production, evaluation, and utilization of bioethanol fuels with at least 666 citations each (Table 100.3). For the last research front, there are one and two HCPs for biohydrogen fuel production from bioethanol fuels and the utilization of bioethanol fuels in fuel cells, respectively. However, there are no HCPs for the utilization of bioethanol fuels for the production of biochemicals, the development of bioethanol sensors, and for their use in gasoline or diesel engines. These narrated studies highlight the importance of studies on the evaluation and utilization of bioethanol fuels. They also show that the research fronts of the utilization of bioethanol fuels are not well represented in the US sample.

100.4.4.1 Bioethanol Production

Saha et al. (2005) produced bioethanol from wheat straw through acid pretreatment, enzymatic hydrolysis of the biomass, and fermentation and detoxification of the resulting hydrolysate in separate SHF and SSF processes and found that the yield of ethanol (per liter) from acid-pretreated enzyme-saccharified wheat straw (78.3 g) hydrolysate with the recombinant *Escherichia coli* strain FBR5 was 19 g with a yield of 0.24 g/g DS.

100.4.4.2 Bioethanol Fuel Evaluation

Hill et al. (2006) developed four criteria to evaluate corn grain-based bioethanol and soybean-based biodiesel fuels through life cycle accounting and found that bioethanol yielded 25% more energy than the energy invested in its production, whereas biodiesel yielded 93% more. Further, Kim and Dale (2004) estimated the global annual potential bioethanol production from starch feedstock residues and sugar feedstock residues and found that there were about 1.57 petagram (10^{15} g) of dry lignocellulosic wastes in the world that could potentially produce 491 gigaliter (GL) (10^9 L)/year of bioethanol, about 16 times higher than the world ethanol production as of 2004.

Pimentel and David (2005) compared the net energy gain of bioethanol fuels from corn, switchgrass, and wood and biodiesel fuels from soybean and sunflower and found that ethanol had lesser energy outputs compared to the respective fossil energy inputs. Finally, Schmer et al. (2008) determined the NEY and GHG emissions of lignocellulosic bioethanol fuels from switchgrass and found that the annual biomass yields of these fields averaged 5.2–11.1 Mg/ha with a resulting average estimated NEY of 60 GJ/ha/y.

Finally, Klein-Marcuschamer et al. (2012) constructed a technoeconomic model for the production of fungal cellulases and found that the cost contribution of enzymes to ethanol produced by the conversion of corn stover was $0.68/gal if the sugars in the biomass could be converted at maximum theoretical yields, and $1.47/gal if the yields were based on saccharification and fermentation yields that had been previously reported in the scientific literature.

100.4.4.3 Bioethanol Fuel Utilization

There are one and two HCPs for biohydrogen fuel production from bioethanol fuels and utilization of bioethanol fuels in fuel cells, respectively. However, there are no HCPs for the utilization of bioethanol fuels for the production of biochemicals, the development of bioethanol sensors, and for their use in gasoline or diesel engines.

100.4.4.3.1 Bioethanol Fuel-based Biohydrogen Fuels

Deluga et al. (2004) produced biohydrogen fuels by the autothermal reforming of bioethanol fuels and converted ethanol and ethanol-water mixtures directly into hydrogen with nearly 100% selectivity and more than 95% conversion by catalytic partial oxidation.

100.4.4.3.2 Bioethanol Fuel Cells

Marinov (1999) developed a chemical kinetic model for high-temperature ethanol oxidation in a paper and found that high-temperature ethanol oxidation exhibited strong sensitivity to the fall-off kinetics of ethanol decomposition, branching ratio selection for decomposition reaction, and reactions involving the hydroperoxyl (HO_2) radical. Further, Kowal et al. (2009) developed electrocatalysts for oxidizing ethanol to CO_2 and found that this electrocatalyst effectively split the C–C bond in ethanol at room temperature in acid solutions.

100.5 CONCLUSION AND FUTURE RESEARCH

Brief information about the key research fronts covered by the 25 most-cited papers with at least 659 citations each is given under three primary headings: The production, evaluation, and utilization of bioethanol fuels.

The usual characteristics of these HCPs for the production of bioethanol fuels are that the pretreatments and hydrolysis of the feedstocks and fermentation of the resulting hydrolysates are the primary processes for bioethanol fuel production from various feedstocks to improve ethanol yield.

The key findings on these research fronts should be read in light of the increasing public concerns about climate change, GHG emissions, and global warming as these concerns have been certainly behind the boom in the research on bioethanol fuels as an alternative to crude oil-based gasoline and diesel fuels in the past decades. Recent supply shocks caused by the COVID-19 pandemic and the Russian invasion of Ukraine also highlight the importance of the production and utilization of bioethanol fuels as an alternative to crude oil-based gasoline and petrodiesel fuels.

As Table 100.4 shows, the most prolific feedstocks for this field are wood biomass, starch feedstock residues, grass, bioethanol fuels, cellulose, lignin, and to a lesser extent starch feedstocks, lignocellulosic biomass at large, and sugar feedstock residues.

Thus, these feedstocks dominate the HCPs in this field where the only major waste biomass is starch feedstock residues such as corn stover or wheat straw and sugar feedstock residues such as sugarcane bagasse. On the other hand, cellulose and lignin are the key constituents of the lignocellulosic biomass. Further bioethanol fuels are the feedstocks for the production of biohydrogen fuels and biochemicals and for their utilization in transport engines and fuel cells. It is also notable that the research on first generation sugar and starch bioethanol fuels is not substantial for all of the samples studied, reflecting the adverse effect of undermining food security.

Further, the most influential research fronts are wood biomass, grass, starch feedstock residues, and to a lesser extent lignin, and starch feedstocks while lignocellulosic biomass at large, bioethanol fuels, industrial wastes, and to a lesser extent cellulose, hydrolysates, and biomass are the least influential feedstocks.

As Table 100.5 shows the most prolific thematic research fronts for this field are the production of bioethanol fuels, and to a lesser degree evaluation and utilization of bioethanol evaluation. Further, the most prolific research fronts for the first group are the pretreatments, hydrolysis of the feedstocks, and to a lesser degree bioethanol production and hydrolysate fermentation. For the third group, the only prolific research fronts are bioethanol-based biohydrogen fuels and the utilization of bioethanol fuels in fuel cells while there are no HCPs for the utilization of bioethanol fuels in transport engines, production of biochemicals from bioethanol fuels, and bioethanol sensors.

On the one hand, biofuel fuel evaluation is the most influential research field, followed by biomass pretreatments and hydrolysis. On the other hand, the least influential research front is bioethanol fuel production, followed by hydrolysate fermentation, bioethanol fuel utilization, bioethanol fuel production, and bioethanol fuels in engines.

These studies emphasize the importance of proper incentive structures for the efficient production and utilization of bioethanol fuels in light of North's institutional framework (North, 1991). In this context, the major producers and users of bioethanol fuels such as the USA and Brazil with vast forests and farmlands have developed strong incentive structures for the efficient production and utilization of bioethanol fuels. In light of the recent supply shocks caused primarily by the COVID-19 pandemic and the Russian invasion of Ukraine, it is expected that the incentive structures such as public funding would be enhanced to increase the share of bioethanol fuels in the global and the US fuel portfolio as a strong alternative to crude oil-based gasoline and diesel fuels.

In this context, it is expected that the most prolific researchers, institutions, countries, funding bodies, and journals in this field would have a first-mover advantage to benefit from such potential incentives. This is especially true for the US stakeholders as the USA has become the global leader in both the production and the utilization of second generation bioethanol fuels from feedstocks. It is expected the research would focus more on the algal, wood, and grass biomass-based bioethanol fuels as well as the agricultural residues- and waste biomass-based bioethanol fuels at the expense of first generation starch- and sugar-based bioethanol fuels due to the large societal concerns about food security in the future. Similarly, it is expected that the research would focus more on the pretreatments and hydrolysis of feedstocks for the production of bioethanol fuels and utilization and evaluation of bioethanol fuels in the future.

Further, there is ample room for the intensification of research in the field of the utilization of bioethanol fuels as the USA has a significant deficit in relation to the global sample. It is also notable that there are only three HCPs in this field.

It is recommended that further review studies are performed for the primary research fronts of bioethanol fuels.

ACKNOWLEDGMENTS

The contribution of the highly cited researchers in the field of US bioethanol fuels has been gratefully acknowledged.

REFERENCES

Alvira, P., E. Tomas-Pejo, M. Ballesteros and M. J. Negro. 2010. Pretreatment technologies for an efficient bioethanol production process based on enzymatic hydrolysis: A review. *Bioresource Technology* 101:4851–4861.

Angelici, C., B. M. Weckhuysen and P. C. A. Bruijnincx. 2013. Chemocatalytic conversion of ethanol into butadiene and other bulk chemicals. *ChemSusChem* 6:1595–1614.

Antolini, E. 2007. Catalysts for direct ethanol fuel cells. *Journal of Power Sources* 170:1–12.

Antolini, E. 2009. Palladium in fuel cell catalysis. *Energy and Environmental Science* 2:915–931.

Bai, F. W., W. A. Anderson and M. Moo-Young. 2008. Ethanol fermentation technologies from sugar and starch feedstocks. *Biotechnology Advances* 26:89–105.

Bentivoglio, D., A. Finco and M. R. P. Bacchi. 2016. Interdependencies between biofuel, fuel and food prices: The case of the Brazilian ethanol market. *Energies* 9:464.

Binod, P., R. Sindhu and R. R. Singhania, et al. 2010. Bioethanol production from rice straw: An overview. *Bioresource Technology* 101:4767–4774.

Bourbonnais, R. and M. G. Paice. 1990. Oxidation of non-phenolic substrates. An expanded role for laccase in lignin biodegradation. *FEBS Letters* 267:99–102.

Burnham, J. F. 2006. Scopus database: A review. *Biomedical Digital Libraries* 3:1–8.

Cardona, C. A., J. A. Quintero and I. C. Paz. 2010. Production of bioethanol from sugarcane bagasse: Status and perspectives. *Bioresource Technology* 101:4754–4766.

Chen, F. and R. A. Dixon. 2007. Lignin modification improves fermentable sugar yields for biofuel production. *Nature Biotechnology* 25:759–761.

Deluga, G. A., J. R. Salge, L. D. Schmidt and X. E. Verykios. 2004. Renewable hydrogen from ethanol by autothermal reforming. *Science* 303:993–997.

Duff, S. J. B. and W. D. Murray. 1996. Bioconversion of forest products industry waste cellulosics to fuel ethanol: A review. *Bioresource Technology* 55:1–33.

Elobeid, A. and C. Hart. 2007. Ethanol expansion in the food versus fuel debate: How will developing countries fare? *Journal of Agricultural & Food Industrial Organization* 5:7.

Fernando, S., S. Adhikari, C. Chandrapal and M. Murali. 2006. Biorefineries: Current status, challenges, and future direction. *Energy & Fuels* 20:1727–1737.

Floudas, D., M. Binder and R. Riley, et al. 2012. The paleozoic origin of enzymatic lignin decomposition reconstructed from 31 fungal genomes. *Science* 336:1715–1719.

Fort, D. A., R. C. Remsing and R. P. Swatloski, et al. 2007. Can ionic liquids dissolve wood? Processing and analysis of lignocellulosic materials with 1-n-butyl-3-methylimidazolium chloride. *Green Chemistry* 9:63–69.

Galbe, M. and G. Zacchi. 2002. A review of the production of ethanol from softwood. *Applied Microbiology and Biotechnology* 59:618–628.

Hahn-Hagerdal, B., M. Galbe, M. F. Gorwa-Grauslund, G. Liden and G. Zacchi. 2006. Bio-ethanol - The fuel of tomorrow from the residues of today. *Trends in Biotechnology* 24:549–556.

Hamelinck, C. N., G. van Hooijdonk and A. P. C. Faaij. 2005. Ethanol from lignocellulosic biomass: Techno-economic performance in short-, middle- and long-term. *Biomass and Bioenergy*, 28:384–410.

Hansen, A. C., Q. Zhang and P. W. L. Lyne. 2005. Ethanol-diesel fuel blends - A review. *Bioresource Technology* 96:277–285.

Haryanto, A., S. Fernando, N. Murali and S. Adhikari. 2005. Current status of hydrogen production techniques by steam reforming of ethanol: A review. *Energy and Fuels* 19:2098–2106.

Hendriks, A. T. W. M and G. Zeeman. 2009. Pretreatments to enhance the digestibility of lignocellulosic biomass. *Bioresource Technology* 100:10–18.

Henstra, A. M., J. Sipma, A. Rinzema and A. J. Stams. 2007. Microbiology of synthesis gas fermentation for biofuel production. *Current Opinion in Biotechnology* 18:200–206.

Hill, J., E. Nelson, D. Tilman, S. Polasky and D. Tiffany. 2006. Environmental, economic, and energetic costs and benefits of biodiesel and ethanol biofuels. *Proceedings of the National Academy of Sciences of the United States of America* 103:11206–11210.

Hill, J., S. Polasky and E. Nelson, et al. 2009. Climate change and health costs of air emissions from biofuels and gasoline. *Proceedings of the National Academy of Sciences of the United States of America* 106:2077–2082.

Hsieh, W. D., R. H. Chen, T. L. Wu and T. H. Lin. 2002. Engine performance and pollutant emission of an SI engine using ethanol-gasoline blended fuels. *Atmospheric Environment* 36:403–410.

Huang, H. J., S. Ramaswamy, U. W. Tschirner and B. V. Ramarao. 2008. A review of separation technologies in current and future biorefineries. *Separation and Purification Technology* 62:1–21.

John, R. P., G. S. Anisha, K. M. Nampoothiri and A. Pandey. 2011. Micro and macroalgal biomass: A renewable source for bioethanol. *Bioresource Technology* 102:186–193.

Jonsson, L. J. and C. Martin. 2016. Pretreatment of lignocellulose: Formation of inhibitory by-products and strategies for minimizing their effects. *Bioresource Technology* 199:103–112.

Kilpelainen, I., X. Xie and A. King, et al. 2007. Dissolution of wood in ionic liquids. *Journal of Agricultural and Food Chemistry* 55:9142–9148.

Kim, S. and B. E. Dale. 2004. Global potential bioethanol production from wasted crops and crop residues. *Biomass and Bioenergy* 26:361–375.

Klein-Marcuschamer, D., P. Oleskowicz-Popiel, B. A. Simmons and H. W. Blanch. 2012. The challenge of enzyme cost in the production of lignocellulosic biofuels. *Biotechnology and Bioengineering*, 109:1083–1087.

Konur, O. 2000. Creating enforceable civil rights for disabled students in higher education: An institutional theory perspective. *Disability & Society* 15:1041–1063.

Konur, O. 2002a. Access to nursing education by disabled students: Rights and duties of nursing programs. *Nurse Education Today* 22:364–374.

Konur, O. 2002b. Assessment of disabled students in higher education: Current public policy issues. *Assessment and Evaluation in Higher Education* 27:131–152.

Konur, O. 2002c. Access to employment by disabled people in the UK: Is the Disability Discrimination Act working? *International Journal of Discrimination and the Law* 5:247–279.

Konur, O. 2006a. Participation of children with dyslexia in compulsory education: Current public policy issues. *Dyslexia* 12:51–67.

Konur, O. 2006b. Teaching disabled students in higher education. *Teaching in Higher Education* 11:351–363.

Konur, O. 2007a. A judicial outcome analysis of the *Disability Discrimination Act*: A windfall for the employers? *Disability & Society* 22:187–204.

Konur, O. 2007b. Computer-assisted teaching and assessment of disabled students in higher education: The interface between academic standards and disability rights. *Journal of Computer Assisted Learning* 23:207–219.

Konur, O. 2012. The evaluation of the research on the bioethanol: A scientometric approach. *Energy Education Science and Technology Part A: Energy Science and Research* 28:1051–1064.

Konur, O. 2015. Current state of research on algal bioethanol. In *Marine Bioenergy: Trends and Developments*, Ed. S. K. Kim and C. G. Lee, pp. 217–244. Boca Raton, FL: CRC Press.

Konur, O. 2020. The scientometric analysis of the research on the bioethanol production from green macroalgae. In *Handbook of Algal Science, Technology and Medicine*, Ed. O. Konur, pp. 385–401. London: Academic Press.

Konur, O. 2023a. Bioethanol fuels: Scientometric study. In *Bioethanol Fuel Production Processes. I: Biomass Pretreatments. Handbook of Bioethanol Fuels Volume 1*, Ed. O. Konur, pp. 47–76. Boca Raton, FL: CRC Press.

Konur, O. 2023b. Bioethanol fuels: Review. In *Bioethanol Fuel Production Processes. I: Biomass Pretreatments. Handbook of Bioethanol Fuels Volume 1*, Ed. O. Konur, pp. 77–98. Boca Raton, FL: CRC Press.

Kowal, A., M. Li and M. Shao, et al. 2009. Ternary Pt/Rh/SnO_2 electrocatalysts for oxidizing ethanol to CO_2. *Nature Materials* 8:325–330.

Kumar, R., G. Mago, V. Balan and C. E. Wyman. 2009. Physical and chemical characterizations of corn stover and poplar solids resulting from leading pretreatment technologies. *Bioresource Technology*, 100:3948–3962.

Lee, S. H., T. V. Doherty, R. J. Linhardt and J. S. Dordick. 2009. Ionic liquid-mediated selective extraction of lignin from wood leading to enhanced enzymatic cellulose hydrolysis. *Biotechnology and Bioengineering* 102:1368–1376.

Li, C., B. Knierim and C. Manisseri, et al. 2010. Comparison of dilute acid and ionic liquid pretreatment of switchgrass: Biomass recalcitrance, delignification and enzymatic saccharification. *Bioresource Technology* 101:4900–4906.

Limayem, A. and S. C. Ricke. 2012. Lignocellulosic biomass for bioethanol production: Current perspectives, potential issues and future prospects. *Progress in Energy and Combustion Science* 38:449–467.

Lin, Y. and S. Tanaka. 2006. Ethanol fermentation from biomass resources: Current state and prospects. *Applied Microbiology and Biotechnology* 69:627–642.

Liu, J., X. Wang, Q. Peng and Y. Li. 2005. Vanadium pentoxide nanobelts: Highly selective and stable ethanol sensor materials. *Advanced Materials* 17:764–767.

Ma, X., L. Sun and C. Song. 2002. A new approach to deep desulfurization of gasoline, diesel fuel and jet fuel by selective adsorption for ultra-clean fuels and for fuel cell applications. *Catalysis Today* 77:107–116.

Marinov, N. M. 1999. A detailed chemical kinetic model for high temperature ethanol oxidation. *International Journal of Chemical Kinetics,* 31:183–220.

Martinez, D., R. M. Berka and B. Henrissat, et al. 2008. Genome sequencing and analysis of the biomass-degrading fungus *Trichoderma reesei* (syn. *Hypocrea jecorina*). *Nature Biotechnology* 26:553–560.

Martinez, D., L. F. Larrondo and N. Putnam, et al. 2004. Genome sequence of the lignocellulose degrading fungus *Phanerochaete chrysosporium* strain RP78. *Nature Biotechnology* 22:695–700.

Morschbacker, A. 2009. Bio-ethanol based ethylene. *Polymer Reviews* 49:79–84.

Mosier, N., C. Wyman and B. Dale, et al. 2005. Features of promising technologies for pretreatment of lignocellulosic biomass. *Bioresource Technology* 96:673–686.

Najafi, G., B. Ghobadian and T. Tavakoli, et al. 2009. Performance and exhaust emissions of a gasoline engine with ethanol blended gasoline fuels using artificial neural network. *Applied Energy* 86:630–639.

Newman, P. W. G. and J. R. Kenworthy. 1989. Gasoline consumption and cities: A comparison of U.S. cities with a global survey. *Journal of the American Planning Association* 55:24–37.

Ni, M., D. Y. C. Leung and M. K. H. Leung. 2007. A review on reforming bio-ethanol for hydrogen production. *International Journal of Hydrogen Energy*, 32:3238–3247.

Nigam, J. N. 2002. Bioconversion of water-hyacinth (*Eichhornia crassipes*) hemicellulose acid hydrolysate to motor fuel ethanol by xylose-fermenting yeast. *Journal of Biotechnology* 97:107–116.

North, D. C. 1991. Institutions. *Journal of Economic Perspectives* 5:97–112.

Olsson, L. and B. Hahn-Hagerdal. 1996. Fermentation of lignocellulosic hydrolysates for ethanol production. *Enzyme and Microbial Technology* 18:312–331.

Palmqvist, E. and B. Hahn-Hagerdal. 2000. Fermentation of lignocellulosic hydrolysates. I: Inhibition and detoxification. *Bioresource Technology* 74:17–24.

Park, S., J. O. Baker, M. E. Himmel, P. A. Parilla and D. K. Johnson. 2010. Cellulose crystallinity index: Measurement techniques and their impact on interpreting cellulase performance. *Biotechnology for Biofuels* 3:10.

Pimentel, D. and T. W. Patzek. 2005. Ethanol production using corn, switchgrass, and wood; biodiesel production using soybean and sunflower. *Natural Resources Research* 14:65–76.

Pinkert, A., K. N. Marsh, S. Pang and M. P. Staiger. 2009. Ionic liquids and their interaction with cellulose. *Chemical Reviews* 109:6712–6728.

Quinlan, R. J., M. D. Sweeney and L. L. Leggio, et al. 2011. Insights into the oxidative degradation of cellulose by a copper metalloenzyme that exploits biomass components. *Proceedings of the National Academy of Sciences of the United States of America* 108:15079–15084.

Ravindran, R., and A. K. Jaiswal. 2016. A comprehensive review on pre-treatment strategy for lignocellulosic food industry waste: Challenges and opportunities. *Bioresource Technology* 199:92–102.

Remsing, R. C., R. P. Swatloski, R. D. Rogers and G. Moyna. 2006. Mechanism of cellulose dissolution in the ionic liquid 1-n-butyl-3- methylimidazolium chloride: A ^{13}C and $^{35/37}Cl$ NMR relaxation study on model systems. *Chemical Communications* 2006:1271–1273.

Saha, B. C., L. B. Iten, M. A. Cotta and Y. V. Wu. 2005. Dilute acid pretreatment, enzymatic saccharification and fermentation of wheat straw to ethanol. *Process Biochemistry* 40:3693–3700.

Sanchez, C. 2009. Lignocellulosic residues: Biodegradation and bioconversion by fungi. *Biotechnology Advances* 27:185–194.

Sanchez, O. J. and C. A. Cardona. 2008. Trends in biotechnological production of fuel ethanol from different feedstocks. *Bioresource Technology* 99:5270–5295.

Sano, T., H. Yanagishita, Y. Kiyozumi, F. Mizukami and K. Haraya. 1994. Separation of ethanol/water mixture by silicalite membrane on pervaporation. *Journal of Membrane Science* 95:221–228.

Schmer, M. R., K. P. Vogel, R. B. Mitchell and R. K. Perrin. 2008. Net energy of cellulosic ethanol from switchgrass. *Proceedings of the National Academy of Sciences of the United States of America* 105:464–469.

Silverstein, R. A., Y. Chen, R. R. Sharma-Shivappa, M. D. Boyette and J. Osborne. 2007. A comparison of chemical pretreatment methods for improving saccharification of cotton stalks. *Bioresource Technology* 98:3000–3011.

Sun, N., M. Rahman and Y. Qin, et al. 2009. Complete dissolution and partial delignification of wood in the ionic liquid 1-ethyl-3-methylimidazolium acetate. *Green Chemistry* 11:646–655.

Sun, Y. and J. Cheng. 2002. Hydrolysis of lignocellulosic materials for ethanol production: A review. *Bioresource Technology* 83:1–11.

Taherzadeh, M. J. and K. Karimi. 2007. Enzyme-based hydrolysis processes for ethanol from lignocellulosic materials: A review. *Bioresources* 2:707–738.

Taherzadeh, M. J. and K. Karimi. 2008. Pretreatment of lignocellulosic wastes to improve ethanol and biogas production: A review. *International Journal of Molecular Sciences* 9:1621–1651.

Talebnia, F., D. Karakashev and I. Angelidaki. 2010. Production of bioethanol from wheat straw: An overview on pretreatment, hydrolysis and fermentation. *Bioresource Technology* 101:4744–4753.

Tien, M. and T. K. Kirk. 1983. Lignin-degrading enzyme from the hymenomycete *Phanerochaete chrysasporium* burds. *Science* 221:661–663.

Vane, L. M. 2005. A review of pervaporation for product recovery from biomass fermentation processes. *Journal of Chemical Technology and Biotechnology* 80:603–629.

Wan, Q., Q. H. Li and Y. J. Chen, et al. 2004. Fabrication and ethanol sensing characteristics of ZnO nanowire gas sensors. *Applied Physics Letters* 84:3654–3656.

Zhang, Y. H. P. and L. R. Lynd. 2004. Toward an aggregated understanding of enzymatic hydrolysis of cellulose: Noncomplexed cellulase systems. *Biotechnology and Bioengineering* 88:797–824.

Zhu, J. Y. and X. J. Pan. 2010. Woody biomass pretreatment for cellulosic ethanol production: Technology and energy consumption evaluation. *Bioresource Technology* 101:4992–5002.

101 The European Experience of Bioethanol Fuels
Scientometric Study

Ozcan Konur
(Formerly) Ankara Yildirim Beyazit University

101.1 INTRODUCTION

Crude oil-based gasoline fuels (Ma et al., 2002; Newman and Kenworthy, 1989) have been widely used in the transportation sector since the 1920s. However, there have been great public concerns over the adverse environmental and human impact of these fuels (Hill et al., 2006, 2009). Hence, biomass-based bioethanol fuels (Hill et al., 2006; Konur, 2012e, 2015, 2019, 2020a) have increasingly been used in blending gasoline fuels (Hsieh et al., 2002; Najafi et al., 2009), in fuel cells (Antolini, 2007, 2009), and in biochemical production (Angelici et al., 2013; Morschbacker, 2009) in a biorefinery context (Fernando et al., 2006; Huang et al., 2008).

Bioethanol fuels also play a critical role in maintaining energy security (Kruyt et al., 2009; Winzer, 2012) in supply shocks (Kilian, 2008, 2009) related to oil price shocks (Hamilton, 1983, 2003), COVID-19 pandemic (Fauci et al., 2020; Li et al., 2020), or wars (Hamilton, 1983; Jones, 2012) in the aftermath of the Russian invasion of Ukraine (Reeves, 2014).

However, it is necessary to pretreat the biomass (Taherzadeh and Karimi, 2008; Yang and Wyman, 2008) to enhance the yield of bioethanol (Hahn-Hagerdal et al., 2006; Sanchez and Cardona, 2008) prior to bioethanol production through hydrolysis (Sun and Cheng, 2002; Taherzadeh and Karimi, 2007) and fermentation (Lin and Tanaka, 2006; Olsson and Hahn-Hagerdal, 1996) of the biomass and the resultant hydrolysates, respectively.

Research in the field of bioethanol fuels has intensified in this context in the key research fronts of the production of bioethanol fuels, and to a lesser extent, utilization (Antolini, 2007; Hansen et al., 2005) and evaluation (Hamelinck et al., 2005; Pimentel and Patzek, 2005) of bioethanol fuels. For the first research front, pretreatment (Alvira et al., 2010; Hendriks and Zeeman, 2009) and hydrolysis (Alvira et al., 2010; Sun and Cheng, 2002) of feedstocks, fermentation of feedstock-based hydrolysates (Jonsson and Martin, 2016; Lin and Tanaka, 2006), and production of bioethanol fuels (Limayem and Ricke, 2012, Lin and Tanaka, 2006) and to a lesser extent distillation of bioethanol fuels (Sano et al., 1994; Vane, 2005) are the key research areas while for the second research front, utilization of bioethanol fuels in fuel cells and transport engines, bioethanol-based biohydrogen fuels, bioethanol sensors, and bioethanol-based biochemicals are the key research areas.

Research in this field has also intensified for the feedstocks of lignocellulosic biomass at large (Hendriks and Zeeman, 2009; Sun and Cheng, 2002), bioethanol fuels for their utilization (Antolini, 2007; Hansen et al., 2005), cellulose (Pinkert et al., 2009; Zhang and Lynd, 2004), starch feedstock residues (Binod et al., 2010; Talebnia et al., 2010), and wood biomass (Galbe and Zacchi, 2002; Zhu and Pan, 2010) and to a lesser extent industrial waste (Cardona et al., 2010), lignin (Bourbonnais and Paice, 1990), grass (Pimentel and Patzek, 2005), hydrolysates (Palmqvist and Hahn-Hagerdal, 2000), biomass (Lin and Tanaka, 2006), sugar feedstocks (Bai et al., 2008), sugar feedstock residues (Cardona et al., 2010), algal biomass (John et al., 2011), starch feedstocks (Bai et al., 2008), urban wastes (Ravindran and Jaiswal, 2016), forestry wastes (Duff and Murray, 1996), and food wastes (Ravindran and Jaiswal, 2016).

DOI: 10.1201/9781003226574-134

However, it is essential to develop efficient incentive structures (North, 1991) for the primary stakeholders to enhance research in this field (Konur, 2000, 2002a,b,c, 2006a,b, 2007a,b). Scientometric analysis has been used in this context to inform the primary stakeholders about the current state of research in a selected research field (Garfield, 1955; Konur, 2011, 2012a,b,c,d,e,f,g,h,i, 2015, 2018b, 2019, 2020a).

As there have been no published current scientometric studies in this field, this chapter presents a scientometric study of research in bioethanol fuels. It examines the scientometric characteristics of both the sample and population data presenting the scientometric characteristics of these both datasets in the order of documents, authors, publication years, institutions, funding bodies, source titles, countries, Scopus subject categories, Scopus keywords, and research fronts.

101.2 MATERIALS AND METHODS

Search for this study was carried out using the Scopus database (Burnham, 2006) in November 2022.

As a first step for the search of the relevant literature, keywords were selected using the most-cited first 300 population papers for each research front. The selected keyword list was then optimized to obtain a representative sample of papers for each research field. These keyword lists for bioethanol fuels were then integrated to obtain the keyword list for this research field (Konur, 2023a, b). Further, the keyword set for Europe was devised as 'AFFILCOUNTRY (europ* OR eu OR spain OR "united kingdom" OR germany OR france OR sweden OR italy OR finland OR netherlands OR turkey OR poland OR denmark OR portugal OR greece OR austria OR romania OR hungary OR norway OR slovakia OR ireland OR serbia OR ukraine OR slovenia OR switzerland OR belgium OR czech* OR bulgaria OR croatia OR estonia OR latvia OR lithuania OR belarus OR cyprus OR Georgia OR Russia* OR yugoslavia OR armenia OR iceland OR azerbaijan OR luxembourg OR moldova OR bosnia OR montenegro OR macedonia OR albania OR malta OR kazakhstan OR kosovo OR andorra OR monaco OR liechtenstein OR "san marino") OR TITLE (europ* OR eu) OR AUTHKEY (europ* OR eu) OR SRCTITLE (europ*)'.

As a second step, two sets of data were used for this study. First, a population sample of 22,777 papers was used to examine the scientometric characteristics of the population data. Secondly, a sample of 228 most-cited papers, corresponding to 1% of the population papers, was used to examine the scientometric characteristics of these citation classics.

The scientometric characteristics of both sample and population datasets were presented in the order of documents, authors, publication years, institutions, funding bodies, source titles, countries, Scopus subject categories, Scopus keywords, and research fronts.

Lastly, the key scientometric findings for both datasets were discussed to highlight the research landscape for bioethanol fuels. Additionally, a number of brief conclusions were drawn and a number of relevant recommendations were made to enhance the future research landscape.

101.3 RESULTS

101.3.1 The Most Prolific Documents in the European Experience of Bioethanol Fuels

Information on the types of documents for both datasets is given in Table 101.1. The articles and conference papers, published in journals, dominate both the sample (71%) and population (94%) papers with a 23% deficit. Further, review papers and short surveys have a 26% surplus as they are over-represented in the sample papers as they constitute 29% and 3% of sample and population papers, respectively. Additionally, 2% of sample papers were published as book chapters.

It is further notable that 97%, 2%, and 1% of population papers were published in journals, books, and book series, respectively. Similarly, 98%, 2%, and 0% of sample papers were published in journals, book series, and books, respectively.

TABLE 101.1
Documents in the European Experience of Bioethanol Fuels

Documents	Sample Dataset (%)	Population Dataset (%)	Surplus (%)
Article	68.0	90.6	–22.6
Conference paper	3.1	3.6	–0.5
Review	25.4	3.0	22.4
Book chapter	0.0	2.0	–2.0
Letter	0.0	0.3	–0.3
Short Survey	3.5	0.2	3.3
Note	0.0	0.2	–0.2
Book	0.0	0.1	–0.1
Editorial	0.0	0.1	–0.1
Sample size	228	22,777	

Population dataset, The number of papers (%) in the set of 22,777 population papers; Sample dataset, The number of papers (%) in the set of 228 highly cited papers.

101.3.2 The Most Prolific Authors in the European Experience of Bioethanol Fuels

Information about the most prolific 22 authors with at least 0.5% of sample papers each is given in Table 101.2.

The most prolific authors are Barbel Hahn-Hagerdal, Guido Zacchi, and Mats Galbe with 5.7%, 4.4%, and 3.1% of sample papers, respectively. The other prolific authors are Mohammad J. Taherzadeh, Leif J. Jonsson, Claude Lamy, Johannes P. van Dijken, Keikhosro Karimi, and Christophe Coutanceau with 2.2%–2.6% of sample papers each.

On the other hand, the most influential authors are Barbel Hahn-Hagerdal, Guido Zacchi, and Mats Galbe with 5.2%, 3.5%, and 2.6% surplus, respectively. The other influential authors are Claude Lamy, Johannes P. van Dijken, Leif J. Jonsson, Mohammad J. Taherzadeh, and Christophe Coutanceau with 2.1%–2.5% surplus each.

The most prolific institutions for the sample dataset are Lund University and the University of Poitiers with four and three authors, followed by Delft University of Technology, Imperial College, and the National Technical University of Athens with two authors each. In total, only nine countries house these top authors. On the other hand, the most prolific countries are France, Sweden, and Greece with five, four, and three authors, respectively, followed by the Netherlands and the United Kingdom with two authors each.

The most prolific research fronts for these top authors are pretreatment and hydrolysis of feedstocks, fermentation of hydrolysates, and bioethanol fuel production with 9–13 authors each. The other minor research fronts are the utilization of bioethanol fuels in fuel cells, evaluation of bioethanol fuels, and utilization of bioethanol fuels in transport engines with one to five authors each.

On the other hand, there is a significant gender deficit (Beaudry and Lariviere, 2016) for the sample dataset as surprisingly only three of these top researchers are female with a representation rate of 14%.

The prolific collaborating overseas researchers are Bernard Henrissat, Igor V. Grigoriev, Dan Cullen, Scott E. Baker, Erika Lindquist, Diego Martinez, Asaf A. Salamov, Jeremy Schmutz, Randy M. Berka, Shukin Song, Edward A. Bayer, Gregg T. Beckham, Paul V. Harris, Yuval Shooham, Qin Xin, Luis F. Larrondo, and Runcang Sun.

Additionally, the other Europe-based prolific authors with a relatively low citation impact are Juan J. Parajo, Mercedes Ballesteros, Lisbeth Olsson, Herbert Sixta, Eulogio Castro, Panagiotis Christakopoulos, Jose A. Teixeira, Angel T. Martinez, Ignacio Ballesteros, Jordi Llorca, Anne S.

TABLE 101.2
Most Prolific Authors in the European Experience of Bioethanol Fuels

No.	Author name	Author code	Sample papers %	Population papers %	Surplus	Institution	Country	HI	N	Res. Front
1	Hahn-Hagerdal, Barbel*	7005389381	5.7	0.5	5.2	Lund Univ.	Sweden	76	258	P, H, F, R
2	Zacchi, Guido	7006727748	4.4	0.5	3.9	Lund Univ.	Sweden	68	204	P, H, F, R
3	Galbe, Mats	7003788758	3.1	0.5	2.6	Lund Univ.	Sweden	51	131	P, H, F, R, E
4	Taherzadeh, Mohammad J.	6701407496	2.6	0.5	2.1	Univ. Boras	Sweden	67	426	P, H, F, R
5	Jonsson, Leif J.	7102349315	2.6	0.3	2.3	Umea Univ.	Sweden	41	148	P, H, F, R
6	Lamy, Claude	7007017658	2.6	0.1	2.5	Univ. Montpellier	France	73	232	C
7	Van Dijken, Johannes P.	7102979857	2.6	0.1	2.5	Delft Univ. Technol.	Netherlands	69	190	F, R
8	Karimi, Keikhosro	10046195700	2.2	0.3	1.9	Vrije Univ.	Belgium	58	226	P, H, F, R
9	Coutanceau, Christophe	8714035200	2.2	0.1	2.1	Univ. Poitiers	France	56	154	C
10	Coutinho, Pedro M	7006153340	0.6	0.0	0.6	Aix Marseille Univ.	France	73	145	P, H
11	Martinez, Angel T.*	55622506400	0.5	2.0	-1.5	CSIC	Spain	81	345	P, H
12	Jorgensen, Henning	7202554496	0.5	0.1	0.4	Univ. Copenhagen	Denmark	35	85	P, H
13	Liden, Gunnar	7004458708	0.5	0.1	0.4	Lund Univ.	Sweden	49	144	P, H, F, R
14	Pronk, Jack T.	7005313057	0.5	0.1	0.4	Delft Univ. Technol.	Netherlands	77	295	F, R
15	Belgsir, El Mustapha	6701740559	0.5	0.0	0.5	Univ. Poitiers	France	21	35	C
16	Brandt, Agnieszka*	35785816800	0.5	0.0	0.5	Imperial Coll.	UK	24	41	P

(Continued)

TABLE 101.2 (*Continued*)
Most Prolific Authors in the European Experience of Bioethanol Fuels

No.	Author name	Author code	Sample papers %	Population papers %	Surplus	Institution	Country	HI	N	Res. Front
17	Leger, Jean M.	13105762400 7201980020	0.5	0.0	0.5	Univ. Poitiers	France	49	123	C
18	Nilvebrant, Nils-Olof	57209815309	0.5	0.0	0.5	Borregaard	Norway	22	43	H, F
19	Rakopoulos, Constantine D.	35570765900	0.5	0.0	0.5	Natl. Tech. Univ. Athens	Greece	61	223	I
20	Rakopoulos, Dimitrios C	6603012578	0.5	0.0	0.5	Natl. Tech. Univ. Athens	Greece	42	95	I
21	Tsiakaras, Panagiotis	7003948427	0.5	0.0	0.5	Univ. Thessaly	Greece	60	258	C
22	Welton, Tom	7003503272	0.5	0.0	0.5	Imperial Coll.	UK	74	191	P

*, Female; Author code, the unique code given by Scopus to authors; C, Utilization of bioethanol fuels in fuel cells; E, evaluation of bioethanol fuels; F, Fermentation of feedstock-based hydrolysates; H, Hydrolysis of the feedstock; HI, H-index; I, Utilization of bioethanol fuels in transport engines; N, Number of papers published by each author; P, Pretreatment of the feedstock; Population papers, the number of papers authored in the population dataset; R, Bioethanol fuel production; Sample papers, the number of papers authored in the sample dataset.

Meyer, Gil Garrote, Lucilia Domingues, Maria J. Negro, Paloma Manzanares, Solange I. Mussatto, Jose M. Dominguez, Maria F. Gorwa-Grauslund, Jose M. Oliva, Aloia Romani, Antonio Converti, Bernd Nidetzky, Claus Felby, Athanasios A. Koutinas, Carlos Martin, Arkadij P. Sinitsiyn, Vera P. Santos, Jalel Labidi, Vincenzo Palma, Anne B. Thomsen, Jyri P. Mikkola, and Encarnacion Ruiz.

101.3.3 The Most Prolific Research Output by Years in the European Experience of Bioethanol Fuels

Information about papers published between 1970 and 2022 is given in Figure 101.1. This figure clearly shows that the bulk of research papers in the population dataset was published primarily in the 2010s, the early 2020s, and 2000s with 47%, 17%, and 18% of population dataset, respectively. Similarly, the publication rates for the 1990s, 1980s, and 1970s were 11%, 6%, and 1% respectively. Further, the rate for the pre-1970s was 1%. Further, there has been a rising trend for the population papers between 2006 and 2016 and it steadied after 2016, losing its momentum.

Similarly, the bulk of research papers in the sample dataset was published in the 2000s, 2010s, and 1990s with 52%, 25%, and 14% of the sample dataset, respectively. Similarly, the publication rates for the 1980s and 1970s were 3% and 0.4% of sample papers, respectively. Further, the rate for the pre-1970s was 0%.

The most prolific publication years for the population dataset were 2021, 2020, 2016, and 2019 with 6.1%, 5.4%, 5.3%, and 5.3% of the dataset, respectively, while 69% of population papers was published between 2008 and 2022. Similarly, 90% of sample papers was published between 1997 and 2016 while the most prolific publication years were 2007, 2009, 2002, and 2004 with 8.3%, 7.0%, 7.5%, and 7.1% of sample papers, respectively.

101.3.4 The Most Prolific Institutions in the European Experience of Bioethanol Fuels

Information about the most prolific 22 institutions publishing papers on bioethanol fuels with at least 1.8% of sample papers each is given in Table 101.3.

The most prolific institution is Lund University with 11% of sample papers, followed by the Technical University of Denmark, National Scientific Research Center (CNRS), and Aix Marseille University

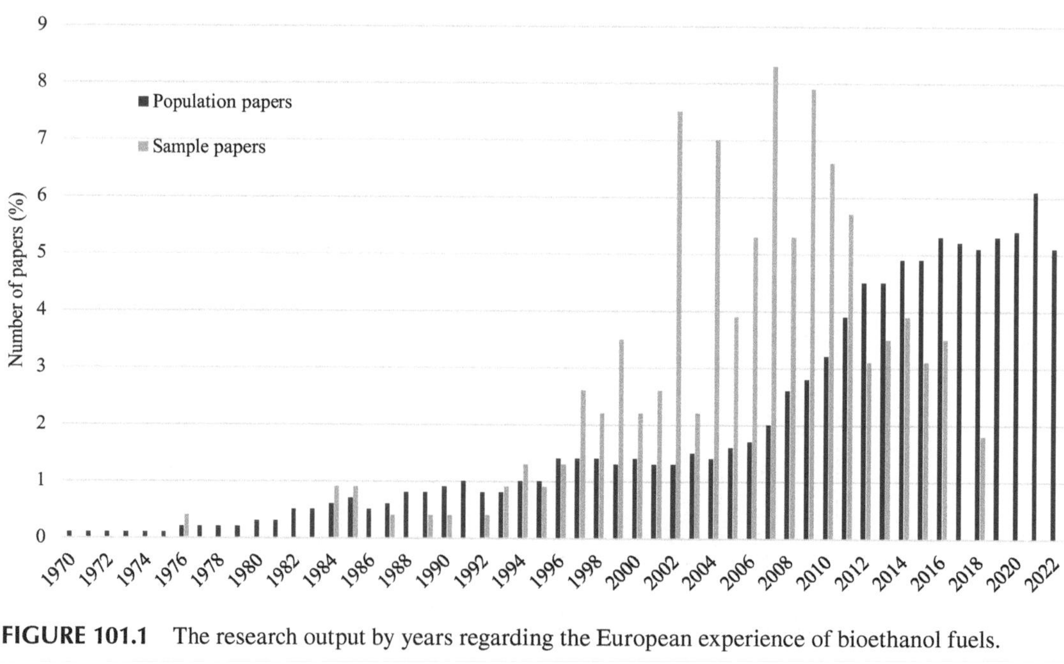

FIGURE 101.1 The research output by years regarding the European experience of bioethanol fuels.

TABLE 101.3
The Most Prolific Institutions in the European Experience of Bioethanol Fuels

No.	Institutions	Country	Sample Papers (%)	Population Papers (%)	Surplus (%)
1	Lund Univ.	Sweden	11.0	2.5	8.5
2	Tech. Univ. Denmark	Denmark	4.8	2.0	2.8
3	CNRS	France	3.9	2.4	1.5
4	Aix Marseille Univ.	France	3.5	0.7	2.8
5	CSIC	Spain	3.1	1.7	1.4
6	VTT Tech. Res. Ctr.	Finland	3.1	1.2	1.9
7	Delft Univ. Technol.	Netherlands	2.6	0.9	1.7
8	Imperial Coll.	United Kingdom	2.6	0.8	1.8
9	Univ. Poitiers	France	2.6	0.4	2.2
10	Wageningen Univ. Res.	Netherlands	2.2	1.4	0.8
11	Univ. Copenhagen	Denmark	2.2	0.9	1.3
12	Helsinki Univ.	Finland	2.2	0.6	1.6
14	Boras Univ.	Sweden	2.2	0.6	1.6
15	Univ. Patras	Greece	2.2	0.5	1.7
16	Friedrich Schiller Univ.	Germany	2.2	0.3	1.9
17	Umea Univ.	Sweden	1.8	0.6	1.2
18	CIEMAT	Spain	1.8	0.6	1.2
19	Univ. Utrecht	Netherlands	1.8	0.5	1.3
20	Inventia	Sweden	1.8	0.3	1.5
21	Bird Eng.	Netherlands	1.8	0.2	1.6
22	Univ. Thessaly	Greece	1.8	0.2	1.6

with 3.5%–4.8% of sample papers each. The other prolific institutions are the Superior Council of Scientific Investigations (CSIC), VTT Technical Research Center, Delft University of Technology, Imperial College, and the University of Poitiers with 2.6% and 3.1% of sample papers each.

Similarly, the top countries for these most prolific institutions are Netherlands and Sweden with four institutions each, followed by France with three institutions. The other prolific countries are Denmark, Finland, Greece, and Spain with two institutions each. In total, only nine European countries house these top institutions.

On the other hand, the institution with the most citation impact is Lund University with an 8.5% surplus, followed by the Technical University of Denmark and Aix Marseille University with a 2.8% surplus each. The other influential institutions are the University of Poitiers, VTT Technical Research Center, Friedrich Schiller University, Imperial College, Delft University of Technology, and the University of Patras with a 1.7%–2.2% surplus each.

The most prolific collaborating overseas institutions are the US Joint Genome Institute, USDA Forest Products Laboratory, Novozymes Biotech Inc., Isfahan University of Technology, Pacific Northwest National Laboratory, Pontifical Catholic University of Chile, National Renewable Energy Laboratory, United States Department of Agriculture, NC State University, Tel Aviv University, University of British Columbia, Chinese Academy of Sciences, Weizmann Institute of Science, Technion - Israel Institute of Technology, Clark University, University of Wisconsin-Madison, University of New Mexico, University of Sao Paulo, State University of Campinas, and HudsonAlpha Institute for Biotechnology.

Additionally, the other Europe-based prolific institutions with a relatively low citation impact are Russian Academy of Sciences, National Research Council (CNR), Chalmers University of Technology, Aalto University, University of Vigo, Swedish University of Agricultural Sciences, Royal Institute of Technology (KTH), University of Minho, National Technical University of Athens, University of Naples Federico II, KU Leuven, University of Gent, RWTH Aachen, ETH

Zurich, Abo Akademi University, University of Lorraine, University of Natural Resources and Life Sciences Vienna, Technical University of Munich, National Academy of Sciences in Ukraine, Lulea University of Technology, INRAE, Lomonosov Moscow State University, University of Santiago de Compostela, University of Belgrade, Technical University of Vienna, University of the Basque Country, University of Montpellier, University of Manchester, University of the Studies of Padova, Academy of Sciences of the Czech Republic, University of Jaen, University of Valladolid, University of the Studies of Genoa, University of Lisbon, Budapest University of Technology and Economics, University of Aveiro, University of the Studies of Milano, University College London, and University of Porto.

101.3.5 The Most Prolific Funding Bodies in the European Experience of Bioethanol Fuels

Information about the most prolific ten funding bodies funding at least 1.3% of sample papers each is given in Table 101.4. Further, only 37% and 39% of sample and population papers were funded, respectively.

The most prolific funding body is the European Commission with 6.2% of sample papers. The other prolific funding bodies are the Biotechnology and Biological Sciences Research Council and Swedish National Board for Industrial and Technical Development with 2.2% of sample papers each.

On the other hand, the most prolific countries for these top funding bodies are Sweden with five funding bodies, followed by the European Union and the United Kingdom with two funding bodies each. In total, three countries and the European Union house these top funding bodies.

The funding body with the most citation impact is the European Commission with a 4% surplus, followed by the Swedish National Board for Industrial and Technical Development with a 2.2% surplus.

On the other hand, the prolific overseas-collaborating funding bodies are National Natural Science Foundation of China, U.S. Department of Energy, National Institute of General Medical Sciences, National Science Foundation, Office of Science, Lawrence Livermore National Laboratory, University of California, Ministry of Economy and Competitiveness, National Council for Scientific and Technological Development, Higher Education Personnel Improvement Coordination, Research Support Foundation of the State of Sao Paulo, China Scholarship Council, and National Key Research and Development Program of China.

TABLE 101.4
The Most Prolific Funding Bodies in the European Experience of Bioethanol Fuels

No.	Funding Bodies	Country	Sample Paper No. (%)	Population Paper No. (%)	Surplus (%)
1	European Commission	European Union	6.6	2.6	4.0
2	Biotechnology and Biological Sciences Research Council	United Kingdom	2.2	1.0	1.2
3	Swedish National Board for Industrial and Technical Development	Sweden	2.2	0.1	2.1
4	Seventh Framework Program	European Union	1.8	1.4	0.4
5	Engineering and Physical Sciences Research Council	United Kingdom	1.8	1.1	0.7
6	Energy Agency	Sweden	1.8	1.0	0.8
7	Swedish Research Council	Sweden	1.8	0.5	1.3
8	Foundation for Science and Technology	Portugal	1.3	1.3	0.0
9	Knut and Alice Wallenberg Foundation	Sweden	1.3	0.2	1.1
10	Carl Tryggers Foundation for Scientific Research	Sweden	1.3	0.1	1.2

The other Europe-based prolific funding bodies with a relatively low citation impact are the European Regional Development Fund, Horizon 2020 Framework Program, German Research Foundation, Ministry of Science and Innovation, Russian Foundation for Basic Research, Academy of Finland, European Social Fund, Federal Ministry of Education and Research, National Research Agency, National Council of Science and Technology, Horizon 2020, Research Council of Norway, Ministry of Education, Science and Technology, Ministry of Education, Culture and Sport, Scientific and Technological Research Council of Turkey, Ministry of Education, University and Research, Council of Galicia, Swedish Research Council Formas, Tekes, Ministry of Education and Science of the Russian Federation, Russian Science Foundation, and European Research Council.

101.3.6 The Most Prolific Source Titles in the European Experience of Bioethanol Fuels

Information about the most prolific 24 source titles publishing at least 1.3% of sample papers each in bioethanol fuels is given in Table 101.5.

The most prolific source title is Bioresource Technology with 10.5% of sample papers, followed by Enzyme and Microbial Technology, Applied Microbiology and Biotechnology, and Green Chemistry with 4.4%, 3.9%, and 3.5% of sample papers, respectively. The other prolific titles are Biotechnology for Biofuels, Journal of Power Sources, Chemsuschem, and Science with 2.6% of sample papers each.

On the other hand, the source title with the most citation impact is Bioresource Technology with a 6.4% surplus, followed by Enzyme and Microbial Technology, Green Chemistry, and Science with

TABLE 101.5
The Most Prolific Source Titles in the European Experience of Bioethanol Fuels

No.	Source Titles	Sample Papers (%)	Population Papers (%)	Surplus (%)
1	Bioresource Technology	10.5	4.1	6.4
2	Enzyme and Microbial Technology	4.4	1.0	3.4
3	Applied Microbiology and Biotechnology	3.9	1.7	2.2
4	Green Chemistry	3.5	0.7	2.8
5	Biotechnology for Biofuels	2.6	1.4	1.2
6	Journal of Power Sources	2.6	0.4	2.2
7	Chemsuschem	2.6	0.3	2.3
8	Science	2.6	0.1	2.5
9	Biomass and Bioenergy	2.2	1.3	0.9
10	Fuel	2.2	1.2	1.0
11	Applied Catalysis B Environmental	1.8	0.5	1.3
12	Nature Biotechnology	1.8	0.1	1.7
13	Biotechnology and Bioengineering	1.3	1.3	0.0
14	Applied and Environmental Microbiology	1.3	1.0	0.3
15	Journal of Biotechnology	1.3	0.5	0.8
16	FEMS Yeast Research	1.3	0.3	1.0
17	Energy Conversion and Management	1.3	0.3	1.0
18	Renewable and Sustainable Energy Reviews	1.3	0.2	1.1
19	Electrochimica Acta	1.3	0.2	1.1
20	Biotechnology Advances	1.3	0.1	1.2
21	Catalysis Communications	1.3	0.1	1.2
22	Current Opinion in Biotechnology	1.3	0.1	1.2
23	Proceedings of the National Academy of Sciences of the United States of America	1.3	0.1	1.2
24	Trends in Biotechnology	1.3	0.1	1.2

3.4%, 2.8%, and 2.5% surplus, respectively. The other influential titles are Chemsuschem, Applied Microbiology and Biotechnology, and Journal of Power Sources with a 2.1%–2.2% surplus each.

The other prolific source titles with a relatively low citation impact are Industrial Crops and Products, Bioresources, Cellulose, Holzforschung, Biotechnology Letters, International Journal of Hydrogen Energy, SAE Technical Papers, Industrial and Engineering Chemistry Research, Applied Biochemistry and Biotechnology, Biomass Conversion and Biorefinery, Chemical Engineering Transactions, ACS Sustainable Chemistry and Engineering, Journal of Agricultural and Food Chemistry, Process Biochemistry, Journal of Chemical Technology and Biotechnology, Cellulose Chemistry and Technology, Energies, Chemical Engineering Journal, Carbohydrate Polymers, Journal of Cleaner Production, Wood Science and Technology, Catalysis Today, Renewable Energy, Applied Energy, International Biodeterioration and Biodegradation, and Biochemical Engineering Journal.

101.3.7 The Most Prolific Countries in the European Experience of Bioethanol Fuels

Information about the most prolific 18 countries publishing at least 1.8% of sample papers each in bioethanol fuels is given in Table 101.6.

The most prolific European country is Sweden with 20% of sample papers, followed by the United Kingdom, Spain, France, Germany, and the Netherlands with 11%–14% of sample papers each. The other prolific countries are Denmark, Greece, and Finland with 6%–9% of sample papers each.

On the other hand, the country with the most citation impact is Sweden with a 12% surplus, followed by the Netherlands, Denmark, Greece, and France 3%–6% surplus each. Similarly, the country with the least citation impact is Italy with a 5% deficit, followed by Poland, Turkey, and Belgium with 1%–2% deficit each.

Additionally, the other prolific European countries with relatively low citation impact are Russia, Czech Republic, Romania, Hungary, Bulgaria, Croatia, Estonia, Latvia, Lithuania, Slovenia, and Belarus.

TABLE 101.6
The Most Prolific Countries in the European Experience of Bioethanol Fuels

No.	Countries	Sample Papers (%)	Population Papers (%)	Surplus (%)
1	Sweden	19.7	8.2	11.5
2	United Kingdom	13.6	11.9	1.7
3	Spain	12.7	12.2	0.5
4	France	12.7	9.6	3.1
5	Germany	11.0	10.7	0.3
6	Netherlands	10.5	4.7	5.8
7	Denmark	9.2	4.0	5.2
8	Greece	6.1	2.5	3.6
9	Finland	5.7	5.2	0.5
10	Portugal	3.5	4.0	–0.5
11	Italy	3.1	8.1	–5.0
12	Turkey	3.1	4.4	–1.3
13	Norway	2.6	1.3	1.3
14	Poland	2.2	4.1	–1.9
15	Belgium	2.2	2.8	–0.6
16	Austria	2.2	2.4	–0.2
17	Switzerland	2.2	2.0	0.2
18	Ireland	1.8	1.1	0.7

On the other hand, a limited number of overseas countries collaborated with European countries in this field (Table 101.7). The most prolific overseas-collaborating country is the United States with 16% of sample papers, followed by China and Canada with 5% and 4% of sample papers, respectively. The other prolific countries are Iran, Chile, Brazil, and India with 2% of sample papers each. Further, as expected, the most influential country is the United States with a 9% surplus, followed by Canada and Chile with a 2% surplus each.

101.3.8 The Most Prolific Scopus Subject Categories in the European Experience of Bioethanol Fuels

Information about the most prolific 11 Scopus subject categories indexing at least 2.2% of sample papers each is given in Table 101.8.

TABLE 101.7
The Most Prolific Overseas-Collaborating Countries in the European Experience of Bioethanol Fuels

No.	Countries	Sample Papers (%)	Population Papers (%)	Surplus (%)
1	United States	14.5	5.6	8.9
2	China	4.8	4.3	0.5
3	Canada	3.5	1.3	2.2
4	Iran	2.2	0.9	1.3
5	Chile	2.2	0.3	1.9
6	Brazil	1.8	3.6	−1.8
7	India	1.8	1.5	0.3
8	South Korea	1.3	0.6	0.7
9	Israel	1.3	0.3	1.0
10	Japan	0.9	1.0	–0.1
11	Australia	0.9	0.9	0.0
12	New Zealand	0.9	0.2	0.7

TABLE 101.8
The Most Prolific Scopus Subject Categories in the European Experience of Bioethanol Fuels

No.	Scopus Subject Categories	Sample Papers (%)	Population Papers (%)	Surplus (%)
1	Chemical Engineering	50.0	33.8	16.2
2	Biochemistry, Genetics, and Molecular Biology	38.2	30.0	8.2
3	Energy	32.0	20.3	11.7
4	Environmental Science	31.1	24.4	6.7
5	Immunology and Microbiology	28.1	19.9	8.2
6	Chemistry	23.7	24.6	–0.9
7	Engineering	11.4	13.7	–2.3
8	Materials Science	9.2	13.3	–4.1
9	Agricultural and Biological Sciences	7.5	21.3	−13.8
10	Multidisciplinary	4.4	1.2	3.2
11	Physics and Astronomy	2.2	6.5	–4.3

The most prolific Scopus subject category is Chemical Engineering with 50% of sample papers, followed by Biochemistry, Genetics and Molecular Biology, Energy, Environmental Science, Immunology and Microbiology, and Chemistry with 24%–38% of sample papers each. It is notable that Social Science including Economics and Business accounts for only 0.5% and 2.0% of sample and population studies, respectively, mostly published outside energy and fuel journals.

On the other hand, the Scopus subject category with the most citation impact is Chemical Engineering with a 16% surplus. The other influential subject areas are Energy, Biochemistry, Genetics and Molecular Biology, Immunology and Microbiology, and Environmental Science with a 7%–12% surplus each. Similarly, the least influential subject category is Agricultural and Biological Sciences with a 14% deficit. The other least influential subjects are Physics and Astronomy, Materials Science, and Engineering with 2%–4% deficits each.

101.3.9 The Most Prolific Keywords in the European Experience of Bioethanol Fuels

Information about the Scopus keywords used with at least 6.1% or 2.6% of sample or population papers, respectively, is given in Table 101.9. For this purpose, keywords related to the keyword set given in the appendix of related papers are selected from a list of the most prolific keyword set provided by the Scopus database.

These keywords are grouped into six headings: feedstocks, pretreatments, fermentation, hydrolysis and hydrolysates, utilization of bioethanol fuels, and products. Further, the most prolific keyword related to feedstocks is cellulose with 40% of sample papers, followed by lignin, biomass, lignocellulose, hemicellulose, wood, and lignocellulosic biomass with 11%–37% of sample papers each.

TABLE 101.9
The Most Prolific Keywords in the European Experience of Bioethanol Fuels

No.	Keywords	Sample Papers (%)	Population Papers (%)	Surplus (%)
1	Feedstocks			
	Cellulose	39.5	16.5	23.0
	Lignin	36.8	13.8	23.0
	Biomass	28.1	12.3	15.8
	Lignocellulose	24.6	5.7	18.9
	Hemicellulose	13.6	3.1	10.5
	Wood	11.8	6.6	5.2
	Lignocellulosic biomass	10.5	3.6	6.9
	Straw	4.8	3.2	1.6
	Triticum	3.1	3.2	−0.1
2	Pretreatments			
	Enzymes	16.7	8.2	8.5
	Cellulases	12.3	4.1	8.2
	Pre-treatment	8.8	3.2	5.6
	Biodegradation	7.5	4.0	3.5
	Temperature	7.0	4.7	2.3
	Pretreatment	6.6	3.1	3.5
	Hypocrea	6.1	0.0	6.1
	pH	3.1	4.4	−1.3
	Genetics	3.1	3.5	−0.4
	Water	3.1	3.1	0.0

(Continued)

TABLE 101.9 (*Continued*)
The Most Prolific Keywords in the European Experience of Bioethanol Fuels

No.	Keywords	Sample Papers (%)	Population Papers (%)	Surplus (%)
3	Fermentation			
	Fungi	28.1	12.1	16.0
	Fermentation	26.8	20.4	6.4
	Saccharomyces	14.9	7.6	7.3
	Bacteria	14.5	9.2	5.3
	Yeast	12.3	7.2	5.1
	Acetic acid	4.8	3.4	1.4
	Bioreactors	3.5	5.6	–2.1
4	Hydrolysis			
	Hydrolysis	30.3	14.1	16.2
	Sugar	21.5	6.8	14.7
	Enzyme activity	17.5	8.4	9.1
	Glucose	15.4	10.0	5.4
	Enzymatic hydrolysis	11.0	6.9	4.1
	Xylose	10.5	4.0	6.5
	Saccharification	6.6	4.6	2.0
5.	Utilization			
	Oxidation	13,6	3,9	9,7
	Catalysts	12,3	5,0	7,3
	Catalysis	11,8	3,0	8,8
	Fuel cells	7,5		7,5
	Adsorption	6,6	1,7	4,9
	Carbon	3,5	3,2	0,3
	Carbon dioxide	3,1	3,6	–0,5
5	Products			
	Ethanol	48.7	30.5	18.2
	Biofuels	24.1	9.7	14.4
	Bioethanol	12.7	9.1	3.6
	Ethanol production	6.1	2.6	3.5
	Biofuel production	6.1	1.4	4.7
	Bioethanol production	3.1	3.0	0.1

Further, the most prolific keyword related to pretreatment is enzymes with 17% of sample papers, followed by cellulases, pre-treatment, biodegradation, temperature, and pretreatment with 7%–12% of sample papers each. The most prolific keywords related to fermentation are fermentation and fungi with 27% and 28% of sample papers, respectively, followed by saccharomyces, bacteria, and yeast with 12%–15% of sample papers each.

Further, the most prolific keyword related to hydrolysis and hydrolysates is hydrolysis with 30% of sample papers, followed by sugar, enzyme activity, glucose, enzymatic hydrolysis, and xylose with 11%–23% of sample papers each.

Finally, the most prolific keywords related to the utilization of bioethanol fuels are oxidation, catalysts, and catalysis with 12%–14% of sample papers each. Further, the most prolific keyword related to products is ethanol with 49% of sample papers, followed by biofuels and bioethanol with 24% and 13% of sample papers, respectively.

On the other hand, the most prolific keywords across all research fronts are ethanol, cellulose, lignin, hydrolysis, biomass, fungi, fermentation, lignocellulose, biofuels, sugar, enzyme activity, enzymes, glucose, saccharomyces, and bacteria with 15%–49% of sample papers each. Similarly, the most influential keywords are cellulose, lignin, lignocellulose, ethanol, hydrolysis, fungi, biomass, sugar, biofuels, hemicellulose, and oxidation with 10%–23% surplus each.

101.3.10 The Most Prolific Research Fronts in the European Experience of Bioethanol Fuels

Information about the research fronts for feedstocks used in sample papers in bioethanol fuels is given in Table 101.10. As this table shows, the most prolific feedstock for this field is the lignocellulosic biomass at large with 30% of sample papers, followed by bioethanol fuels, cellulose, and wood with 21%, 12%, and 11% of sample papers, respectively. The other prolific feedstocks are starch feedstock residues, lignin, sugars, industrial waste, and hydrolysates with 4%–8% of sample papers each. On the other hand, a number of minor feedstocks comprise 12% of sample papers. Further, a small number of feedstocks are not covered in this sample.

TABLE 101.10
The Most Prolific Research Fronts for the European Experience of Bioethanol Fuels

No.	Research Fronts	Sample I (%)	Sample II (%)	Surplus (%)
1	Lignocellulosic biomass at large	29.8	23.0	6.8
2	Bioethanol fuels	20.6	19.6	1.0
3	Cellulose	12.3	15.1	–2.8
4	Wood biomass	11.4	9.0	2.4
5	Starch feedstock residues	7.9	9.7	–1.8
6	Lignin	7.9	6.2	1.7
7	Sugars	7.0	4.9	2.1
8	Industrial wastes	5.3	6.5	–1.2
9	Hydrolysates	4.4	2.8	1.6
10	Biomass	2.2	2.6	–0.4
11	Minor feedstocks	12.4	16.5	4.1
	Sugar feedstock residues	1.8	2.2	–0.4
	Urban wastes	1.8	1.0	0.8
	Food wastes	1.8	0.6	1.2
	Algal biomass	1.3	1.9	–0.6
	Xylan	1.3	0.8	0.5
	Grass	0.9	3.1	–2.2
	Sugar feedstocks	0.9	2.2	–1.3
	Lignocellulosic wastes	0.9	1.8	–0.9
	Forestry wastes	0.9	0.8	0.1
	Starch feedstocks	0.4	1.7	–1.3
	Syngas	0.4	0.4	0.0
12	Missing feedstocks	0.0	1.5	–1.5
	Hemicellulose	0.0	0.8	–0.8
	Cellobiose	0.0	0.4	–0.4
	Plants	0.0	0.3	–0.3
	Sample size	228	775	

Sample I (%), The number of papers in the European sample of 228 papers; Sample II (%), The number of papers in the global sample of 775 papers.

TABLE 101.11
The Most Prolific Thematic Research Fronts for the European Experience of Bioethanol Fuels

No.	Research Fronts	Sample I (%)	Sample II (%)	Surplus (%)
1	Bioethanol fuel production	76.8	74.1	2.7
	Biomass pretreatments	57.5	58.5	–1.0
	Bioethanol fuel production	25.4	21.5	–7.0
	Biomass hydrolysis	22.8	29.8	3.9
	Hydrolysate fermentation	21.5	17.9	3.6
	Bioethanol fuel distillation	0.9	0.8	0.1
2	Bioethanol fuel evaluation	2.6	6.2	–3.6
3	Bioethanol fuel utilization	20.6	19.7	0.9
	Bioethanol fuels in fuel cells	8.3	6.8	1.5
	Bioethanol fuels in engines	5.7	6.1	–0.4
	Bioethanol-based biohydrogen fuels	5.3	4.3	1.0
	Bioethanol-based biochemicals	0.9	0.4	–1.8
	Bioethanol sensors	0.4	2.2	0.5
	Sample size	228	775	

Sample I (%), The number of papers in the European sample of 228 papers; Sample II (%), The number of papers in the global sample of 775 papers.

A number of feedstocks are over-represented in the European sample in relation to the global sample by 1%–7% of sample papers: lignocellulosic biomass at large, minor feedstocks, wood biomass, sugars, lignin, hydrolysates, food wastes, and bioethanol fuels. Similarly, a number of feedstocks are also under-represented in the European sample in relation to the global sample by 1%–3%: cellulose, grass, starch feedstock residues, missing feedstocks, sugar feedstocks, starch feedstocks, and industrial waste.

Information about the thematic research fronts for the sample papers in bioethanol fuels is given in Table 101.11. As this table shows, the most prolific research front is the production of bioethanol fuels with 77% of sample papers, followed by utilization and evaluation of bioethanol fuels with 21% and 3% of sample papers, respectively.

On the other hand, the most prolific research front for the production of bioethanol fuels is biomass pretreatment with 58% of sample papers, followed by bioethanol production, biomass hydrolysis, and hydrolysate fermentation with 22%–25% of sample papers each. The other minor research front is the separation and distillation of bioethanol fuels with 1% of sample papers.

Further, the most prolific research front for the utilization of bioethanol fuels is the utilization of bioethanol fuels in fuel cells with 8% of sample papers, followed by the utilization of bioethanol fuels in transport engines and bioethanol-based biohydrogen fuels with 6% and 5% of sample papers, respectively. The other minor research fronts are bioethanol sensors and bioethanol fuel-based biochemicals with 0.4% and 0.9% of sample papers, respectively.

On the other hand, the fields of bioethanol fuel production and utilization of bioethanol fuels are over-represented in the European sample in relation to the global sample by 3% and 1%, respectively, while the field of bioethanol fuel evaluation is under-represented in the European sample in relation to the global sample by 4%.

Similarly, in individual terms, fields of biomass hydrolysis, hydrolysate fermentation, utilization of bioethanol fuels in fuel cells, and bioethanol-based biohydrogen fuels are over-represented in the European sample in relation to the global sample by 1%–4% each. On the contrary, fields of bioethanol production, bioethanol-based biochemicals, and biomass pretreatment are under-represented in the European sample in relation to the global sample by 1%–7% each.

101.4 DISCUSSION

101.4.1 Introduction

Crude oil-based gasoline fuels have been widely used in the transportation sector since the 1920s. However, there have been great public concerns over the adverse environmental and human impact of these fuels. Hence, biomass-based bioethanol fuels have increasingly been used in blending gasoline fuels, in fuel cells, and in biochemical production in a biorefinery context.

However, it is necessary to pretreat the biomass to enhance the yield of bioethanol prior to bioethanol production through hydrolysis and fermentation processes.

Europe has been one of the most prolific countries engaged in bioethanol fuel research, producing 28% of global research output in this field. The European research in the field of bioethanol fuels has intensified in this context in the key research fronts of the production of bioethanol fuels and utilization of bioethanol fuels and to a lesser extent the evaluation of bioethanol fuels. For the first research front, pretreatment and hydrolysis of feedstocks, fermentation of feedstock-based hydrolysates, and production of bioethanol fuels and to a lesser extent distillation of bioethanol fuels are the key research areas while for the second research front, utilization of bioethanol fuels in fuel cells and transport engines, bioethanol-based biohydrogen fuels, and bioethanol sensors and to a lesser extent bioethanol-based biochemicals are the key research areas.

Research in this field has also intensified for the feedstocks of bioethanol fuels for their utilization, starch feedstock residues, cellulose, wood biomass, and lignocellulosic biomass at large and to a lesser extent industrial waste, lignin, grass, hydrolysates, biomass at large, sugar feedstocks, sugar feedstock residues, algal biomass, starch feedstocks, urban wastes, forestry wastes, and food wastes.

However, it is essential to develop efficient incentive structures for the primary stakeholders to enhance research in this field. This is especially important to maintain energy security in the cases of supply shocks such as oil price shocks, war-related chocks as in the case of the Russian invasion of Ukraine, or COVID-19 shocks.

Scientometric analysis has been used in this context to inform the primary stakeholders about the current state of research in a selected research field. As there has been no current scientometric study in this field, this chapter presents a scientometric study of research in bioethanol fuels. It examines the scientometric characteristics of both the sample and population data presenting the scientometric characteristics of these both datasets in the order of documents, authors, publication years, institutions, funding bodies, source titles, countries, Scopus subject categories, Scopus keywords, and research fronts.

As a first step for the search of the relevant literature, keywords were selected using the most-cited first 300 population papers for each research front. The selected keyword list was then optimized to obtain a representative sample of papers for each research field. These keyword lists for bioethanol fuels were then integrated to obtain the keyword list for this research field (Konur, 2023a, b). Further, a keyword set for Europe was devised.

As a second step, two sets of data were used for this study. First, a population sample of 22,777 papers was used to examine the scientometric characteristics of population data. Secondly, a sample of 228 most-cited papers, corresponding to 1% of population papers, was used to examine the scientometric characteristics of these citation classics.

The scientometric characteristics of these sample and population datasets were presented in the order of documents, authors, publication years, institutions, funding bodies, source titles, countries, Scopus subject categories, Scopus keywords, and research fronts.

Lastly, the key scientometric findings for both datasets were discussed to highlight the research landscape for bioethanol fuels. Additionally, a number of brief conclusions were drawn and a number of relevant recommendations were made to enhance the future research landscape.

101.4.2 The Most Prolific Documents in the European Experience of Bioethanol Fuels

Articles (together with conference papers) dominate both the sample (71%) and population (94%) papers with a 23% deficit (Table 101.1). Further, review papers have a surplus (26%), and the representation of reviews in sample papers is quite extraordinary (29%).

Scopus differs from the Web of Science database in differentiating and showing articles (68%) and conference papers (3%) published in journals separately. However, it should be noted that these conference papers are also published in journals as articles, compared to those published only in conference proceedings. Hence, the total number of articles and review papers in the sample dataset are 71% and 29%, respectively.

It is observed during the search process that there has been inconsistency in the classification of documents in Scopus as well as in other databases such as Web of Science. This is especially relevant for the classification of papers as reviews or articles as papers not involving a literature review may be erroneously classified as a review paper. There is also a case of review papers being classified as articles. For example, the total number of reviews in the sample dataset was manually found as nearly 35% compared to 29% as indexed by Scopus, decreasing the number of articles and conference papers to 65% for the sample dataset. It is notable that many technoeconomic and life cycle studies were often indexed as reviews by the Scopus database.

In this context, it would be helpful to provide a classification note for the published papers in books and journals in the first instance. It would also be helpful to use the document types listed in Table 101.1 for this purpose. Book chapters may also be classified as articles or reviews as an additional classification to differentiate review chapters from experimental chapters as is done by the Web of Science. It would be further helpful to additionally classify the conference papers as articles or review papers as well as it is done in the Web of Science database.

101.4.3 The Most Prolific Authors in the European Experience of Bioethanol Fuels

There have been the most prolific 22 authors with at least 0.5% of sample papers each as given in Table 101.2. These authors have shaped the development of research in this field.

The most prolific authors are Barbel Hahn-Hagerdal, Guido Zacchi, Mats Galbe, and to a lesser extent Mohammad J. Taherzadeh, Leif J. Jonsson, Claude Lamy, Johannes P. van Dijken, Keikhosro Karimi, and Christophe Coutanceau. Further, the most influential authors are Barbel Hahn-Hagerdal, Guido Zacchi, Mats Galbe, and to a lesser extent Claude Lamy, Johannes P. van Dijken, Leif J. Jonsson, Mohammad J. Taherzadeh, and Christophe Coutanceau.

It is important to note the inconsistencies in indexing the author names in Scopus and other databases. It is especially an issue for names with more than two components such as 'Blake Sam de Hyun Karimi. The probable outcomes are 'Karimi, B.S.D.H.', 'de Hyun Karimi, B.S.' or 'Hyun Karimi, B.S.D.'. The first choice is the gold standard of the publishing sector as the last word in the name is taken as the last name. In most of the academic databases such as PUBMED and EBSCO databases, this version is used predominantly. The second choice is a strong alternative while the last choice is an undesired outcome as two last words are taken as the last name. It is a good practice to combine the words of the last name with a hyphen: 'Hyun-Karimi, B.S.D.'. It is notable that inconsistent indexing of author names may cause substantial inefficiencies in the search process for the papers as well as allocating credit to the authors as there are different author entries for each outcome in the databases.

There are also inconsistencies in shortening Chinese names. For example. 'Hongzhang Chen' is often shortened as 'Chen, H.Z.', 'Chen, H.-Z.', and 'Chen H' as it is done in the Web of Science database as well. However, the gold standard in this case is 'Chen, H' where the last word is taken as the last name and the first word is taken as a single forename. In most academic databases such as

PUBMED and EBSCO, this first version is used predominantly. Nevertheless, it makes sense to use the first option to differentiate Chinese names efficiently: Chen, H.Z.'. Therefore, there have been difficulties in locating papers for Chinese authors. In such cases, the use of unique author codes provided for each author by the Scopus database has been helpful.

There is also a difficulty in allowing credit for authors especially for authors with common names such as 'Wang, X.' in conducting scientometric studies. These difficulties strongly influence the efficiency of scientometric studies as well as allocating credit to authors as there are the same author entries for different authors with the same name, e.g., 'Wang, X.' in databases.

In this context, coding of authors in the Scopus database is a welcome innovation compared to other databases such as the Web of Science. In this process, Scopus allocates a unique number to each author in the database (Aman, 2018). However, there might still be substantial inefficiencies in this coding system, especially for common names. For example, some papers for a certain author may be allocated to another researcher with a different author code. It is possible that Scopus uses a number of software programs to differentiate author names and the program may not be false-proof (Kim, 2018).

In this context, it does not help that author names are not given in full in some journals and books. This makes it difficult to differentiate authors with common names and makes the scientometric studies further difficult in the author domain. Therefore, author names should be given in full in all books and journals in the first instance. There is also a cultural issue where some authors do not use their full names in their papers. Instead, they use initials for their forenames: 'Parajo, H.J.', 'Parajo, H.', or 'Parajo, J.' instead of 'Parajo, Hyun Jae'.

There are also inconsistencies in naming authors with more than two components by the authors themselves in journal papers and book chapters. For example. 'Parajo, A.P.C.' might be given as 'Parajo, A.' or 'Parajo, A.C.' or 'Parajo, A.P.' or 'Parajo, C' in journals and books. This also makes the scientometric studies difficult in the author domain. Hence, contributing authors should use their names consistently in their publications.

The other critical issue regarding author names is the inconsistencies in the spelling of author names in national spellings (e.g. Özğümüş, Çağla) rather than in English spellings (e.g. Ozgumus, Cagla) in the Scopus database. Scopus differs from the Web of Science database and many other databases in this respect where the author names are given only in English spellings. It is observed that national spellings of author names do not help much in conducting scientometric studies as well as in allocating credits to authors as sometimes there are often different author entries for English and national spellings in the Scopus database.

The most prolific institutions for the sample dataset are Lund University and the University of Poitiers and to a lesser extent Delft University of Technology, Imperial College, and the National Technical University of Athens. On the other hand, the most prolific countries are France, Sweden, and Greece with five, four, and three authors, respectively, followed by the Netherlands and the United Kingdom with two authors each.

Further, pretreatments and hydrolysis of feedstocks, fermentation of hydrolysates, and bioethanol fuel production are the key research fronts studied by these top authors. The other minor research fronts are the utilization of bioethanol fuels in fuel cells, evaluation of bioethanol fuels, and utilization of bioethanol fuels in transport engines.

It is also notable that there is a significant gender deficit for the sample dataset surprisingly with a representation rate of 14%. This finding is the most thought provoking with strong public policy implications. Hence, institutions, funding bodies, and policymakers should take efficient measures to reduce the gender deficit in this field as well as other scientific fields with a strong gender deficit. In this context, it is worth noting the level of representation of researchers from minority groups in science on the basis of race, sexuality, age, and disability, besides gender (Blankenship, 1993; Dirth and Branscombe, 2017; Konur, 2000, 2002a,b,c, 2006a,b, 2007a,b).

There is a large number of prolific collaborating overseas researchers collaborating with European researchers, coming from the key countries engaged in bioethanol research. Similarly, there is a large number of prolific European researchers with a relatively low citation impact.

101.4.4 The Most Prolific European Research Output by Years in Bioethanol Fuels

The research output observed between 1970 and 2022 is illustrated in Figure 101.1. This figure clearly shows that the bulk of research papers in the population dataset was published primarily in the 2010s and to a lesser extent in the early 2020s, 2010s, and 1990s. Similarly, the bulk of research papers in the sample dataset was published in the 2000s and to a lesser extent in the 2010s and 1990s.

These findings suggest that the most prolific sample and population papers were primarily published in the 2010s. Further, a significant portion of sample and population papers were published in the early 2020s and 2000s, respectively. Further, there has been a rising trend for the population papers between 2006 and 2016 and it steadied after 2016, losing its momentum.

These are thought-provoking findings as there has been a significant research boom since 2008 and 1997 for the population and sample papers, respectively. In this context, the increasing public concerns about climate change (Change, 2007), greenhouse gas emissions (Carlson et al., 2017), and global warming (Kerr, 2007) have been certainly behind the boom in research in this field since 2007. Furthermore, the recent supply shocks experienced due to the COVID-19 pandemic and the Ukrainian war might also be behind the research boom in this field since 2019.

Based on these findings, the size of population papers is likely to be more than double in the current decade, provided that the public concerns about climate change, greenhouse gas emissions, global warming, as well as supply shocks are translated efficiently into research funding in this field. However, the loss of momentum in the rise of research output after 2016 is notable, similar to a significant drop in research output of the United States after 2011 (Konur, 2023d) and in clear contrast to the significant rise in the same period in research output of China (Konur, 2023c), the key competitors of Europe. It is further notable that Europe is the first mover in this field with 11% and 6% of research output published in the 1990s and 1980s, respectively, while China is the late mover with only 1% and 0% of research output in these decades, respectively. However, the United States is also the first mover with 11% and 8% of research output published in the same decades, respectively. Thus, Europe together with the United States has had a first-mover advantage in contrast to China (Lieberman and Montgomery, 1988)

Similarly, 40%, 22%, and 15% of research output of China, Europe, and the United States were published in the last 4 years between 2019 and 2022, showing the extraordinary momentum of Chinese research in this field in the last 4 years in comparison to its key competitors.

101.4.5 The Most Prolific Institutions in the European Experience of Bioethanol Fuels

The most prolific 22 institutions publishing papers on bioethanol fuels with at least 1.8% of sample papers given in Table 101.3 have shaped the development of research in this field.

The most prolific institutions are Lund University and to a lesser extent Technical University of Denmark, CNRS, Aix Marseille University, CSIC, VTT Technical Research Center, Delft University of Technology, Imperial College, and the University of Poitiers. Similarly, the top countries for these most prolific institutions are Netherlands and Sweden and to a lesser extent France, Denmark, Finland, Greece, and Spain. In total, nine European countries house these top institutions.

On the other hand, the institutions with the most citation impact are Lund University and to a lesser extent Technical University of Denmark, Aix Marseille University, University of Poitiers, VTT Technical Research Center, Friedrich Schiller University, Imperial College, Delft University of Technology, and the University of Patras.

These findings confirm the dominance of institutions from Netherlands and Sweden and to a lesser extent France, Denmark, Finland, Greece, and Spain.

Additionally, there are a large number of prolific overseas-collaborating institutions with European institutions. Similarly, there are a large number of prolific European institutions with a low citation impact contributing immensely to bioethanol fuel research in Europe.

101.4.6 The Most Prolific Funding Bodies in the European Experience of Bioethanol Fuels

The most prolific ten funding bodies funding at least 1.3% of sample papers each is given in Table 101.4. It is notable that only 37% and 39% of sample and population papers were funded, respectively.

The most prolific funding bodies are the European Commission and to a lesser extent Biotechnology and Biological Sciences Research Council and Swedish National Board for Industrial and Technical Development. On the other hand, the most prolific countries for these top funding bodies are Sweden and to a lesser extent the European Union and the United Kingdom. In total, only three countries and the European Union house these top funding bodies. Further, the funding bodies with the most citation impact are the European Commission and to a lesser extent Swedish National Board for Industrial and Technical Development.

On the other hand, there are a number of prolific overseas-collaborating funding bodies jointly funding the European papers. Further, there are a large number of European funding bodies with a relatively low citation rate.

However, the most interesting finding of this section is the fact that only 37% and 39% of sample and population papers were funded, respectively. This funding rate is close to the corresponding funding rates of the United States of 36% and 39%. However, it is relatively low in relation to China with corresponding funding rates of 55% and 70%.

A large number of studies establish that research funding improves the research output (Ebadi and Schiffauerova, 2016) providing the necessary incentives for the major stakeholders such as researchers and their institutions in line with the teachings of North's institutional framework (North, 1991).

Considering the loss of momentum in the rise of European research output after 2016 similar to a significant drop in the research output of the United States after 2011 in clear contrast to the significant rise in the same period in the research output of China, it seems that there might be problems with adequate funding of research in Europe and the United States in these time frameworks. Similarly, 40%, 22%, and 15% of research output of China, Europe, and the United States were published in the last 4 years between 2019 and 2022, showing the lack of momentum in European and US research in this field in recent years in comparison to China.

Thus, there have been calls to increase European funding for research (Pavitt, 2000). Close examination of funding bodies suggests that funding of European research is highly fragmented in the United States in contrast to China where the National Natural Science Foundation of China generously funds research in this field (Geuna, 2001; Podhora et al., 2013).

With the recent supply shocks caused by the COVID-19 pandemic and the Ukrainian war, there is ample room to increase the funding rate of European research in this field to support the strategic shift from crude oil-based gasoline and petrodiesel fuels to biomass-based bioethanol and biodiesel fuels to maintain energy security in Europe (De Jong and Sterkx, 2010).

101.4.7 The Most Prolific Source Titles in the European Experience of Bioethanol Fuels

The most prolific 24 source titles publishing at least 1.3% of sample papers each in bioethanol fuels have shaped the development of research in this field (Table 101.5).

The most prolific source titles are Bioresource Technology and to a lesser extent Enzyme and Microbial Technology, Applied Microbiology and Biotechnology, Green Chemistry, Biotechnology for Biofuels, Journal of Power Sources, Chemsuschem, and Science. On the other hand, the source titles with the most citation impact are Bioresource Technology and to a lesser extent Enzyme and Microbial Technology, Green Chemistry, Science, Chemsuschem, Applied Microbiology and Biotechnology, and Journal of Power Sources. There are also a large number of other prolific source titles with a relatively low citation impact.

It is notable that these top source titles are primarily related to bioresources, biotechnology, microbiology, and energy. This finding suggests that Bioresource Technology and other prolific journals in these fields have significantly shaped the development of research in this field as they focus primarily on the production and utilization of bioethanol fuels with a high yield.

This top journal covers most technical issues such as waste treatment, biofuels, bioprocesses, bioproducts, and physicochemical and thermochemical processes, excluding social science- and humanity-based interdisciplinary studies with only exceptions for 'circular bioeconomy and energy and environmental sustainability' and 'system analysis and techno-economics of biofuel and chemical production'. Thus, it appears that it excludes from its scope other social science- and humanity-based interdisciplinary studies such as scientometric (Konur, 2012a,b,c,d,e,f,g,h,i), user (Huijts et al., 2012; Wustenhagen et al., 2007), policy (Greening et al., 2000; Hook and Tang, 2013), environmental science (Popp et al., 2014; Tilman et al., 2009), and economic (Asafu-Adjaye, 2000; Stern, 1993) studies as in the case of most journals in the field of energy and fuels excepting social-science-oriented journals.

101.4.8 The Most Prolific Countries in the European Experience of Bioethanol Fuels

The most prolific 18 countries publishing at least 1.8% of sample papers each have significantly shaped the development of research in this field (Table 101.6).

The most prolific European countries are Sweden, the United Kingdom, Spain, France, Germany, and the Netherlands and to a lesser extent Denmark, Greece, and Finland. On the other hand, countries with the most citation impact are Sweden and to a lesser extent Netherlands, Denmark, Greece, and France. Similarly, countries with the least citation impact are Italy and to a lesser extent Poland, Turkey, and Belgium. Additionally, there are other prolific European countries with a relatively low citation impact.

It appears that although there are around 50 countries listed as European countries, only 18 European countries have produced a significant amount of influential research in this field. It is interesting that most of these prolific countries lay in the western and northern parts of Europe. Thus, these countries deviate from the other European countries in the period of 1650–1850 when these countries shifted to different and cheaper energy carriers due to the rapid growth of population and worsening climatic conditions resulting in an energy crisis and a lowering of living standards (Malanima, 2006).

On the other hand, a limited number of overseas countries collaborated with European countries in this field (Table 101.7). The most prolific collaborating overseas countries are the United States and to a lesser extent China, Canada, Iran, Chile, Brazil, and India. Further, the most influential collaborating countries are the United States and to a lesser extent Canada and Chile. It appears that the United States is the key collaborating country in this field. This country has been strong in the fields of production and evaluation of bioethanol fuels (Konur, 2023d) while China has been strong in the field of production and utilization of bioethanol fuels (Konur, 2023c). As Beaver and Rosen (1978) state, 'collaboration is a typical research style associated with professionalization'. Thus, collaboration between researchers improves their research output (Miramontes and Gonzalez-Brambila, 2016).

101.4.9 The Most Prolific Scopus Subject Categories in the Euroepan Experience of Bioethanol Fuels

The most prolific 11 Scopus subject categories indexing at least 2.2% of sample papers each, given in Table 101.8, have shaped the development of research in this field.

The most prolific Scopus subject categories in bioethanol fuels are Chemical Engineering and to a lesser extent Biochemistry, Genetics and Molecular Biology, Energy, Environmental Science, Immunology and Microbiology, and Chemistry. It is also notable that Social Sciences including Economics and Business have a minimal presence in both sample and population studies.

On the other hand, the Scopus subject categories with the most citation impact are Chemical Engineering and to a lesser extent Energy, Biochemistry, Genetics and Molecular Biology, Immunology and Microbiology, and Environmental Science. Similarly, the least influential subject categories are Agricultural and Biological Sciences and to a lesser extent Physics and Astronomy, Materials Science, and Engineering.

These findings are thought provoking suggesting that the primary subject categories are related to chemical engineering and to a lesser extent energy, biochemistry, microbiology, and environmental science as the core of research in this field concerns production, evaluation, and utilization of bioethanol fuels. The other finding is that social sciences are not well represented in both the sample and population papers as in line with most fields in bioethanol fuels. Social, environmental, and economics studies account for the field of social sciences. The restrictive scope of energy and fuel journals might be behind the inadequate coverage of social sciences and humanities in this field (Ingeborgrud et al., 2020; Royston and Foulds, 2021; Ryan et al., 2014).

As discussed briefly in Section 101.4.7, the scope of most journals in this field excludes social science- and humanity-based interdisciplinary studies from their scope. This development has been in contrast to the interdisciplinarity (Jacobs and Frickel, 2009; Nissani, 1997) of this field in light of pressures for increasing incentives for the primary stakeholders (North, 1991). Thus, for the healthy development of this research field, social science- and humanity-based interdisciplinary studies have a lot to contribute to research on bioethanol fuels (Ingeborgrud et al., 2020; Royston and Foulds, 2021; Ryan et al., 2014).

101.4.10 The Most Prolific Keywords in the European Experience of Bioethanol Fuels

A limited number of keywords have shaped the development of research in this field as shown in Table 101.9. These keywords are grouped into six headings: feedstocks, pretreatments, fermentation, hydrolysis and hydrolysates, utilization of bioethanol fuels, and products.

The most prolific keywords across all research fronts are ethanol, cellulose, lignin, hydrolysis, biomass, fungi, fermentation, lignocellulose, biofuels, sugar, enzyme activity, enzymes, glucose, saccharomyces, and bacteria. Similarly, the most influential keywords are cellulose, lignin, lignocellulose, ethanol, hydrolysis, fungi, biomass, sugar, biofuels, hemicellulose, and oxidation.

These findings suggest that it is necessary to determine the keyword set carefully to locate the relevant research in each of these research fronts. Additionally, the size of samples for each keyword highlights the intensity of research in relevant research areas for both sample and population datasets.

These findings also highlight different spelling of some strategic keywords such as pretreatment v. pre-treatment v. treatment and bioethanol v. ethanol v. bio-ethanol V. bioethanol fuels, etc. However, there is a tendency toward the use of connected keywords without using a hyphen: bioethanol fuels or pretreatment. It is particularly notable that the use of treatment and ethanol instead of pretreatment and bioethanol in the paper titles, respectively, makes the literature search less efficient and time consuming.

101.4.11 The Most Prolific Research Fronts in the European Experience of Bioethanol Fuels

Information about the research fronts for the sample papers is given in Table 101.10. As this table shows, the most prolific feedstock for this field is the lignocellulosic biomass at large, followed by bioethanol fuels, cellulose, and wood. The other prolific research fronts are starch feedstock residues, lignin, sugars, industrial waste, and hydrolysates. On the other hand, a number of feedstocks comprise 12% of sample papers. Further, a small number of feedstocks are not covered in this sample.

A number of feedstocks are over-represented in the European sample in relation to the global sample: lignocellulosic biomass at large, minor feedstocks, wood biomass, sugars, lignin, hydrolysates, food wastes, and bioethanol fuels. Similarly, a number of feedstocks are also under-represented in the European sample in relation to the global sample: cellulose, grass, starch feedstock residues, missing feedstocks, sugar feedstocks, starch feedstocks, and industrial waste.

Thus, the first nine research fields have substantial importance, complementing the remaining bioethanol fuel research fields. It is important to note that research on the production of the first generation bioethanol fuels from sugar and starch feedstocks for bioethanol fuels comprises only 1.3% of sample papers in total. These first generation bioethanol fuels are not much desirable as they undermine food security (Makenete et al., 2008; Wu et al., 2012).

On the other hand, Table 101.11 shows that the most prolific thematic research fronts are the production of bioethanol fuels and to a lesser extent the utilization and evaluation of bioethanol fuels.

On the other hand, the most prolific research fronts for the production of bioethanol fuels are biomass pretreatment and to a lesser extent bioethanol production, biomass hydrolysis, and hydrolysate fermentation. The other minor research front is the separation and distillation of bioethanol fuels.

Further, the most prolific research fronts for the utilization of bioethanol fuels are the utilization of bioethanol fuels in fuel cells, utilization of bioethanol fuels in transport engines, and bioethanol-based biohydrogen fuels. The other minor research fronts are the bioethanol sensors and bioethanol fuel-based biochemicals.

On the other hand, the fields of bioethanol fuel production and utilization of bioethanol fuels are over-represented in the European sample in relation to the global sample while the field of bioethanol fuel evaluation is under-represented in the European sample in relation to the global sample.

Similarly, in individual terms, the fields of biomass hydrolysis, hydrolysate fermentation, utilization of bioethanol fuels in fuel cells, and bioethanol-based biohydrogen fuels are over-represented in the European sample in relation to the global sample. On the contrary, the fields of bioethanol production, bioethanol-based biochemical, and biomass pretreatments are under-represented in the European sample in relation to the global sample.

These findings are thought provoking in seeking ways to increase feedstock-based bioethanol yield at the global scale as well as the utilization of bioethanol fuels. It is clear that all these research fronts have public importance and merit substantial funding and other incentives. Further, it is notable that the utilization and evaluation of bioethanol fuels have also become a core unit of bioethanol research to make it more competitive with crude oil-based gasoline and diesel fuels, especially in the United States, Europe, and China.

It is notable that pretreatment and hydrolysis of feedstocks emerge as primary research fronts for this field. These processes are required to improve ethanol yield. However, the research fronts of fermentation of feedstock-based hydrolysates and bioethanol production from feedstock-based hydrolysates are also important. Research in the field of separation and distillation of bioethanol fuels is negligible.

Further, the field of evaluation of bioethanol fuels is also a neglected area. This suggests that the primary stakeholders have been primarily interested in these key processes of bioethanol production. It is also notable that evaluation of bioethanol fuels such as technoeconomics-, life cycle-, economics-, social-, land use-, labor-, and environment-related studies emerges as a case study for bioethanol fuels.

Europe has strength in the field of bioethanol fuel production with 77% of sample papers compared to 81 and 69% of the United States and China, respectively, while it has also strength in the field of utilization of bioethanol fuels with 21% of sample papers compared to 10% and 30% of the United States and China, respectively. However, it has a relative weakness in the field of evaluation of bioethanol fuels with 3% of sample papers in relation to 10% and 1% of the United States and China (Konur 2023c, d).

In the end, these findings in this field hint that bioethanol fuels could be optimized using the structure, processing, and property relationships of these feedstocks in the fronts of feedstock pretreatment and hydrolysis, hydrolysate fermentation, and the utilization of bioethanol fuels (Formela et al., 2016; Konur, 2018a, 2020b, 2021a,b,c,d; Konur and Matthews, 1989).

101.5 CONCLUSION AND FUTURE RESEARCH

Research on the European experience of bioethanol fuels has been mapped through a scientometric study of both sample (228 papers) and population (22,777 papers) datasets. It is notable that European countries as a whole have produced 28% of research output in this field while its key competitors – the United States and China – have produced 17% and 20% of research output, respectively (Konur, 2023a,b,c,d).

The critical issue in this study has been to obtain a representative sample of research as in any other scientometric study. Therefore, the keyword set has been carefully devised and optimized after a number of runs in the Scopus database. It is a representative sample of wider population studies. This keyword set was provided in the appendix of related studies and the relevant keywords are presented in Table 101.9. However, it should be noted that it has been very difficult to compile a representative keyword set since this research field has been connected closely with many other fields. Therefore, it has been necessary to compile a keyword list to exclude papers concerned with other research fields.

The other issue has been the selection of a multidisciplinary database to carry out the scientometric study of research in this field. For this purpose, the Scopus database has been selected. Journal coverage of this database has been notably wider than that of Web of Science and other multi-subject databases.

The key scientometric properties of research in this field have been determined and discussed in this chapter. It is evident that a limited number of documents, authors, institutions, publication years, institutions, funding bodies, source titles, countries, Scopus subject categories, Scopus keywords, and research fronts have shaped the development of research in this field as in the cases of China and the United States, the key competitors of Europe.

There is ample scope to increase the efficiency of scientometric studies in this field in author and document domains by developing consistent policies and practices in both domains across all academic databases. In this respect, it seems that authors, journals, and academic databases have a lot to do. Furthermore, the significant gender deficit as in most scientific fields emerges as a public policy issue. The potential deficits on the basis of age, race, disability, and sexuality need also to be explored in this field as in other scientific fields.

Research in this field has boomed since 2005 and 1997 for the population and sample papers, respectively, possibly promoted by public concerns about global warming, greenhouse gas emissions, and climate change. Furthermore, the recent COVID-19 pandemic and the Russian invasion of Ukraine have resulted in global supply shocks shifting the recent focus of the stakeholders from crude oil-based fuels to biomass-based fuels such as bioethanol fuels.

It is thus expected that there would be further incentives for the key stakeholders to carry out research for bioethanol fuels to increase ethanol yield and their utilization and to make it more competitive with crude oil-based gasoline and petrodiesel fuels. This might be true for the Western and Northern European countries to maintain energy and food security in the face of global supply shocks.

The relatively modest funding rates of 37% and 39% for sample and population papers, respectively, suggest that funding in this field significantly enhanced research in this field primarily since 2008, possibly more than doubling in the current decade. However, it is evident that there is ample room for more funding and other incentives to enhance research in this field further. However, the most interesting finding in this matter is the fact that the European funding rate is close to the corresponding funding rates of the United States of 36% and 39%. However, it is relatively low in relation to China with corresponding funding rates of 55% and 70% (Konur, 2023c, d).

The institutions from Netherlands and Sweden and to a lesser extent Denmark, Finland, Greece, and Spain have mostly shaped research in this field. It is evident that these countries have well-developed research infrastructure in bioethanol fuels and their derivatives.

It emerges that ethanol is more popular than bioethanol as a keyword with strong implications for the search strategy. In other words, the search strategy using only the bioethanol keyword would not be much helpful. It is recommended that the term 'bioethanol fuels' is used instead of ethanol or bioethanol in the future. On the other hand, the Scopus keywords are grouped into six headings:

biomass, pretreatments, fermentation, hydrolysis and hydrolysates, utilization of bioethanol fuels, and products. These prolific keywords highlight the major fields of research in this field for both sample and population papers.

Table 101.10 shows that the most prolific biomasses used for this field are lignocellulosic biomass at large, bioethanol fuels, cellulose, and wood and to a lesser extent starch feedstock residues, lignin, sugars, industrial waste, and hydrolysates. On the other hand, a number of feedstocks comprise 12% of sample papers. Further, a small number of feedstocks are not covered in this sample.

A number of feedstocks are over-represented in the European sample in relation to the global sample: lignocellulosic biomass at large, minor feedstocks, wood biomass, sugars, lignin, hydrolysates, food wastes, and bioethanol fuels. Similarly, a number of feedstocks are also under-represented in the European sample in relation to the global sample: cellulose, grass, starch feedstock residues, missing feedstocks, sugar feedstocks, starch feedstocks, and industrial wastes.

Further, Table 101.11 shows that the most prolific research fronts are the production of bioethanol fuels and to a lesser extent the utilization and evaluation of bioethanol fuels. On the other hand, the most prolific research fronts for the production of bioethanol fuels are biomass pretreatment and to a lesser extent bioethanol production, biomass hydrolysis, and hydrolysate fermentation. The other minor research front is the separation and distillation of bioethanol fuels.

Further, the most prolific research fronts for the utilization of bioethanol fuels are the utilization of bioethanol fuels in fuel cells, utilization of bioethanol fuels in transport engines, and bioethanol-based biohydrogen fuels. The other minor research fronts are bioethanol sensors and bioethanol fuel-based biochemicals.

On the other hand, the fields of bioethanol fuel production and utilization of bioethanol fuels are over-represented in the European sample in relation to the global sample while the field of bioethanol fuel evaluation is under-represented in the European sample in relation to the global sample (Konur, 2023a, b).

Similarly, in individual terms, the fields of biomass hydrolysis, hydrolysate fermentation, utilization of bioethanol fuels in fuel cells, and bioethanol-based biohydrogen fuels are over-represented in the European sample in relation to the global sample. On the contrary, the fields of bioethanol production and bioethanol-based biochemical and biomass pretreatments are under-represented in the European sample in relation to the global sample (Konur, 2023a, b).

Europe has strength in the field of bioethanol fuel production with 77% of sample papers compared to 81% and 69% of the United States and China, respectively, while it has also strength in the field of utilization of bioethanol fuels with 21% of sample papers compared to 10% and 30% of the United States and China, respectively. However, it has a relative weakness in the field of evaluation of bioethanol fuels with 3% of sample papers in relation to 10% and 1% of the United States and China (Konur, 2023c,d).

In this context, it is notable that there is ample room for the improvement of research on social and humanitarian aspects of research on bioethanol fuels. As discussed briefly in Section 101.4.7, the scope of most journals in this field excludes social science- and humanity-based interdisciplinary studies. This development has been in clear contrast to the interdisciplinarity of this field in light of pressures for increasing incentives for the primary stakeholders. Thus, for the healthy development of this research field, social science- and humanity-based interdisciplinary studies have a lot to contribute to bioethanol fuel research.

Thus, scientometric analysis has a great potential to gain valuable insights into the evolution of research in this field as in other scientific fields, especially in the aftermath of significant global supply shocks such as the COVID-19 pandemic and the Russian invasion of Ukraine.

It is recommended that further scientometric studies are carried out for the primary research fronts. It is further recommended that reviews of the most-cited papers are carried out for each primary research front to complement these scientometric studies. Next, scientometric studies of hot papers in these primary fields are carried out. Similar studies might also be carried out by other prolific countries in the field of bioethanol fuels.

ACKNOWLEDGMENTS

The contribution of highly cited researchers in the field of the European experience of bioethanol fuels has been gratefully acknowledged.

REFERENCES

Alvira, P., E. Tomas-Pejo, M. Ballesteros and M. J. Negro. 2010. Pretreatment technologies for an efficient bioethanol production process based on enzymatic hydrolysis: A review. *Bioresource Technology* 101:4851–4861.

Aman, V. 2018. Does the Scopus author ID suffice to track scientific international mobility? A case study based on Leibniz laureates. *Scientometrics* 117:705–720.

Angelici, C., B. M. Weckhuysen and P. C. A. Bruijnincx. 2013. Chemocatalytic conversion of ethanol into butadiene and other bulk chemicals. *ChemSusChem* 6:1595–1614.

Antolini, E. 2007. Catalysts for direct ethanol fuel cells. *Journal of Power Sources* 170:1–12.

Antolini, E. 2009. Palladium in fuel cell catalysis. *Energy and Environmental Science* 2:915–931.

Asafu-Adjaye, J. 2000. The relationship between energy consumption, energy prices and economic growth: Time series evidence from Asian developing countries. *Energy Economics* 22:615–625.

Bai, F. W., W. A. Anderson and M. Moo-Young. 2008. Ethanol fermentation technologies from sugar and starch feedstocks. *Biotechnology Advances* 26:89–105.

Beaudry, C. and V. Lariviere. 2016. Which gender gap? Factors affecting researchers' scientific impact in science and medicine. *Research Policy* 45:1790–1817.

Beaver, D. and R. Rosen, R. 1978. Studies in scientific collaboration: Part I. The professional origins of scientific co-authorship. *Scientometrics*, 1:65–84.

Binod, P., R. Sindhu and R. R. Singhania, et al. 2010. Bioethanol production from rice straw: An overview. *Bioresource Technology* 101:4767–4774.

Blankenship, K. M. 1993. Bringing gender and race in: US employment discrimination policy. *Gender & Society* 7:204–226.

Bourbonnais, R. and M. G. Paice. 1990. Oxidation of non-phenolic substrates. An expanded role for laccase in lignin biodegradation. *FEBS Letters* 267:99–102.

Burnham, J. F. 2006. Scopus database: A review. *Biomedical Digital Libraries* 3:1–8.

Cardona, C. A., J. A. Quintero and I. C. Paz. 2010. Production of bioethanol from sugarcane bagasse: Status and perspectives. *Bioresource Technology* 101:4754–4766.

Carlson, K. M., J. S. Gerber and D. Mueller, et al. 2017. Greenhouse gas emissions intensity of global croplands. *Nature Climate Change* 7:63–68.

Change, C. 2007. Climate change impacts, adaptation and vulnerability. *Science of the Total Environment* 326:95–112.

De Jong, S. and S. Sterkx. 2010. The 2009 Russian-Ukrainian gas dispute: Lessons for European energy crisis management after Lisbon. *European Foreign Affairs Review* 15:511–538

Dirth, T. P. and N. R. Branscombe. 2017. Disability models affect disability policy support through awareness of structural discrimination. *Journal of Social Issues* 73:413–442.

Duff, S. J. B. and W. D. Murray. 1996. Bioconversion of forest products industry waste cellulosics to fuel ethanol: A review. *Bioresource Technology* 55:1–33.

Ebadi, A. and A. Schiffauerova. 2016. How to boost scientific production? A statistical analysis of research funding and other influencing factors. *Scientometrics* 106:1093–1116.

Fauci, A. S., H. C. Lane and R. R. Redfield. 2020. Covid-19-navigating the uncharted. *New England Journal of Medicine* 382:1268–1269.

Fernando, S., S. Adhikari, C. Chandrapal and M. Murali. 2006. Biorefineries: Current status, challenges, and future direction. *Energy & Fuels* 20:1727–1737.

Formela, K., A. Hejna, L. Piszczyk, M. R. Saeb and X. Colom. 2016. Processing and structure-property relationships of natural rubber/wheat bran biocomposites. *Cellulose* 23:3157–3175.

Galbe, M. and G. Zacchi. 2002. A review of the production of ethanol from softwood. *Applied Microbiology and Biotechnology* 59:618–628.

Garfield, E. 1955. Citation indexes for science. *Science* 122:108–111.

Geuna, A. 2001. The changing rationale for European university research funding: Are there negative unintended consequences? *Journal of Economic Issues* 35:607–632.

Greening, L. A., D. L. Greene and C. Difiglio. 2000. Energy efficiency and consumption-the rebound effect-a survey. *Energy Policy* 28:389–401.

Hahn-Hagerdal, B., M. Galbe, M. F. Gorwa-Grauslund, G. Liden and G. Zacchi. 2006. Bio-ethanol - The fuel of tomorrow from the residues of today. *Trends in Biotechnology* 24:549–556.

Hamelinck, C. N., G. van Hooijdonk and A. P. C. Faaij. 2005. Ethanol from lignocellulosic biomass: Techno-economic performance in short-, middle- and long-term. *Biomass and Bioenergy*, 28:384–410.

Hamilton, J. D. 1983. Oil and the macroeconomy since World War II. *Journal of Political Economy* 91:228–248.

Hamilton, J. D. 2003. What is an oil shock? *Journal of Econometrics* 113:363–398.

Hansen, A. C., Q. Zhang and P. W. L. Lyne. 2005. Ethanol-diesel fuel blends - A review. *Bioresource Technology* 96:277–285.

Hendriks, A. T. W. M. and G. Zeeman. 2009. Pretreatments to enhance the digestibility of lignocellulosic biomass. *Bioresource Technology* 100:10–18.

Hill, J., E. Nelson, D. Tilman, S. Polasky and D. Tiffany. 2006. Environmental, economic, and energetic costs and benefits of biodiesel and ethanol biofuels. *Proceedings of the National Academy of Sciences of the United States of America* 103:11206–11210.

Hill, J., S. Polasky and E. Nelson, et al. 2009. Climate change and health costs of air emissions from biofuels and gasoline. *Proceedings of the National Academy of Sciences of the United States of America* 106:2077–2082.

Hook, M. and X. Tang. 2013. Depletion of fossil fuels and anthropogenic climate change-A review. *Energy Policy* 52:797–809.

Hsieh, W. D., R. H. Chen, T. L. Wu and T. H. Lin. 2002. Engine performance and pollutant emission of an SI engine using ethanol-gasoline blended fuels. *Atmospheric Environment* 36:403–410.

Huang, H. J., S. Ramaswamy, U. W. Tschirner and B. V. Ramarao. 2008. A review of separation technologies in current and future biorefineries. *Separation and Purification Technology* 62:1–21.

Huijts, N. M. A., E. J. E. Molin and L. Steg. 2012. Psychological factors influencing sustainable energy technology acceptance: A review-based comprehensive framework. *Renewable and Sustainable Energy Reviews* 16:525–531.

Ingeborgrud, L., S. Heidenreich and M. Ryghaug, et al. 2020. Expanding the scope and implications of energy research: A guide to key themes and concepts from the Social Sciences and Humanities. *Energy Research & Social Science* 63:101398.

Jacobs, J. A. and S. Frickel. 2009. Interdisciplinarity: A critical assessment. *Annual Review of Sociology* 33:43–65.

John, R. P., G. S. Anisha, K. M. Nampoothiri and A. Pandey. 2011. Micro and macroalgal biomass: A renewable source for bioethanol. *Bioresource Technology* 102:186–193.

Jones, T. C. 2012. America, oil, and war in the Middle East. *Journal of American History* 99:208–218.

Jonsson, L. J. and C. Martin. 2016. Pretreatment of lignocellulose: Formation of inhibitory by-products and strategies for minimizing their effects. *Bioresource Technology* 199:103–112.

Kerr, R. A. 2007. Global warming is changing the world. *Science* 316:188–190.

Kilian, L. 2008. Exogenous oil supply shocks: How big are they and how much do they matter for the US economy? *Review of Economics and Statistics* 90:216–240.

Kilian, L. 2009. Not all oil price shocks are alike: Disentangling demand and supply shocks in the crude oil market. *American Economic Review*, 99:1053–1069.

Kim, J. 2018. Evaluating author name disambiguation for digital libraries: A case of DBLP. *Scientometrics* 116:1867–1886.

Konur, O. 2000. Creating enforceable civil rights for disabled students in higher education: An institutional theory perspective. *Disability & Society* 15:1041–1063.

Konur, O. 2002a. Access to nursing education by disabled students: Rights and duties of nursing programs. *Nurse Education Today* 22:364–374.

Konur, O. 2002b. Assessment of disabled students in higher education: Current public policy issues. *Assessment and Evaluation in Higher Education* 27:131–152.

Konur, O. 2002c. Access to employment by disabled people in the UK: Is the Disability Discrimination Act working? *International Journal of Discrimination and the Law* 5:247–279.

Konur, O. 2006a. Participation of children with dyslexia in compulsory education: Current public policy issues. *Dyslexia* 12:51–67.

Konur, O. 2006b. Teaching disabled students in higher education. *Teaching in Higher Education* 11:351–363.

Konur, O. 2007a. A judicial outcome analysis of the *Disability Discrimination Act*: A windfall for the employers? *Disability & Society* 22:187–204.

Konur, O. 2007b. Computer-assisted teaching and assessment of disabled students in higher education: The interface between academic standards and disability rights. *Journal of Computer Assisted Learning* 23:207–219.

Konur, O. 2011. The scientometric evaluation of the research on the algae and bio-energy. *Applied Energy* 88:3532–3540.

Konur, O. 2012a. The evaluation of the biogas research: A scientometric approach. *Energy Education Science and Technology Part A: Energy Science and Research* 29:1277–1292.

Konur, O. 2012b. The evaluation of the educational research: A scientometric approach. *Energy Education Science and Technology Part B: Social and Educational Studies* 4:1935–1948.

Konur, O. 2012c. The evaluation of the global energy and fuels research: A scientometric approach. *Energy Education Science and Technology Part A: Energy Science and Research* 30:613–628.

Konur, O. 2012d. The evaluation of the research on the biodiesel: A scientometric approach. *Energy Education Science and Technology Part A: Energy Science and Research* 28:1003–1014.

Konur, O. 2012e. The evaluation of the research on the bioethanol: A scientometric approach. *Energy Education Science and Technology Part A: Energy Science and Research* 28:1051–1064.

Konur, O. 2012f. The evaluation of the research on the biofuels: A scientometric approach. *Energy Education Science and Technology Part A: Energy Science and Research* 28:903–916.

Konur, O. 2012g. The evaluation of the research on the biohydrogen: A scientometric approach. *Energy Education Science and Technology Part A: Energy Science and Research* 29:323–338.

Konur, O. 2012h. The evaluation of the research on the microbial fuel cells: A scientometric approach. *Energy Education Science and Technology Part A: Energy Science and Research* 29:309–322.

Konur, O. 2012i. The scientometric evaluation of the research on the production of bioenergy from biomass. *Biomass and Bioenergy* 47:504–515.

Konur, O. 2015. Current state of research on algal bioethanol. In *Marine Bioenergy: Trends and Developments*, Ed. S. K. Kim and C. G. Lee, pp. 217–244. Boca Raton, FL: CRC Press.

Konur, O., Ed. 2018a. *Bioenergy and Biofuels*. Boca Raton, FL: CRC Press.

Konur, O. 2018b. Bioenergy and biofuels science and technology: Scientometric overview and citation classics. In *Bioenergy and Biofuels*, Ed. O. Konur, pp. 3–63. Boca Raton: CRC Press.

Konur, O. 2019. Cyanobacterial bioenergy and biofuels science and technology: A scientometric overview. In *Cyanobacteria: From Basic Science to Applications*, Ed. A. K. Mishra, D. N. Tiwari and A. N. Rai, pp. 419–442. Amsterdam: Elsevier.

Konur, O. 2020a. The scientometric analysis of the research on the bioethanol production from green macroalgae. In *Handbook of Algal Science, Technology and Medicine*, Ed. O. Konur, pp. 385–401. London: Academic Press.

Konur, O., Ed. 2020b. *Handbook of Algal Science, Technology and Medicine*. London: Academic Press.

Konur, O., Ed. 2021a. *Handbook of Biodiesel and Petrodiesel Fuels: Science, Technology, Health, and Environment*. Boca Raton, FL: CRC Press.

Konur, O., Ed. 2021b. *Handbook of Biodiesel and Petrodiesel Fuels: Science, Technology, Health, and Environment. Volume 1. Biodiesel Fuels: Science, Technology, Health, and Environment*. Boca Raton, FL: CRC Press.

Konur, O., Ed. 2021c. *Handbook of Biodiesel and Petrodiesel Fuels: Science, Technology, Health, and Environment. Volume 2. Biodiesel Fuels based on the Edible and Nonedible Feedstocks, Wastes, and Algae: Science, Technology, Health, and Environment*. Boca Raton, FL: CRC Press.

Konur, O., Ed. 2021d. *Handbook of Biodiesel and Petrodiesel Fuels: Science, Technology, Health, and Environment. Volume 3. Petrodiesel Fuels: Science, Technology, Health, and Environment*. Boca Raton, FL: CRC Press.

Konur, O. 2023a. Bioethanol fuels: Scientometric study. In *Bioethanol Fuel Production Processes. I: Biomass Pretreatments. Handbook of Bioethanol Fuels Volume 1*, Ed. O. Konur. Boca Raton, FL: CRC Press.

Konur, O. 2023b. Bioethanol fuels: Review. In *Bioethanol Fuel Production Processes. I: Biomass Pretreatments. Handbook of Bioethanol Fuels Volume 1*, Ed. O. Konur. Boca Raton, FL: CRC Press.

Konur, O. 2023c. Chinese experience of bioethanol fuels: Scientometric study. In *Evaluation and Utilization of Bioethanol Fuels. II.: Biohydrogen Fuels, Fuel Cells, Biochemicals, and Country Experiences. Handbook of Bioethanol Fuels Volume 6*, Ed. O. Konur. Boca Raton, FL: CRC Press.

Konur, O. 2023d. The US experience of bioethanol fuels: Scientometric study. In *Evaluation and Utilization of Bioethanol Fuels. II.: Biohydrogen Fuels, Fuel Cells, Biochemicals, and Country Experiences. Handbook of Bioethanol Fuels Volume 6*, Ed. O. Konur. Boca Raton, FL: CRC Press.

Konur, O. and F. L. Matthews. 1989. Effect of the properties of the constituents on the fatigue performance of composites: A review. *Composites* 20:317–328.

Kruyt, B., D. P. van Vuuren, H. J. de Vries and H. Groenenberg. 2009. Indicators for energy security. *Energy Policy* 37:2166–2181.

Li, H., S. M. Liu, X. H. Yu, S. L. Tang and C. K. Tang. 2020. Coronavirus disease 2019 (COVID-19): Current status and future perspectives. *International Journal of Antimicrobial Agents* 55:105951.

Lieberman, M. B. and D. B. Montgomery. 1988. First-mover advantages. *Strategic Management Journal* 9:41–58.

Limayem, A. and S. C. Ricke. 2012. Lignocellulosic biomass for bioethanol production: Current perspectives, potential issues and future prospects. *Progress in Energy and Combustion Science*, 38:449–467.

Lin, Y. and S. Tanaka. 2006. Ethanol fermentation from biomass resources: Current state and prospects. *Applied Microbiology and Biotechnology* 69:627–642.

Ma, X., L. Sun and C. Song. 2002. A new approach to deep desulfurization of gasoline, diesel fuel and jet fuel by selective adsorption for ultra-clean fuels and for fuel cell applications. *Catalysis Today* 77:107–116.

Makenete, A. L., W. J. Lemmer and J. Kupka. 2008. The impact of biofuel production on food security: A briefing paper with a particular emphasis on maize-to-ethanol production. *International Food and Agribusiness Management Review* 11:101–110.

Malanima, P. 2006. Energy crisis and growth 1650-1850: The European deviation in a comparative perspective. *Journal of Global History* 1:101–121.

Miramontes, J. R. and C. N. Gonzalez-Brambila. 2016. The effects of external collaboration on research output in engineering. *Scientometrics* 109:661–675.

Morschbacker, A. 2009. Bio-ethanol based ethylene. *Polymer Reviews* 49:79–84.

Najafi, G., B. Ghobadian and T. Tavakoli, et al. 2009. Performance and exhaust emissions of a gasoline engine with ethanol blended gasoline fuels using artificial neural network. *Applied Energy* 86:630–639.

Newman, P. W. G. and J. R. Kenworthy. 1989. Gasoline consumption and cities: A comparison of U.S. cities with a global survey. *Journal of the American Planning Association* 55:24–37.

Nissani, M. 1997. Ten cheers for interdisciplinarity: The case for interdisciplinary knowledge and research. *Social Science Journal* 34:201–216.

North, D. C. 1991. Institutions. *Journal of Economic Perspectives* 5:97–112.

Olsson, L. and B. Hahn-Hagerdal. 1996. Fermentation of lignocellulosic hydrolysates for ethanol production. *Enzyme and Microbial Technology* 18:312–331.

Palmqvist, E. and B. Hahn-Hagerdal. 2000. Fermentation of lignocellulosic hydrolysates. I: Inhibition and detoxification. *Bioresource Technology*, 74:17–24.

Pavitt, K. 2000. Why European Union funding of academic research should be increased: A radical proposal. *Science and Public Policy* 27:455–460.

Pimentel, D. and T. W. Patzek. 2005. Ethanol production using corn, switchgrass, and wood; biodiesel production using soybean and sunflower. *Natural Resources Research* 14:65–76.

Pinkert, A., K. N. Marsh, S. Pang and M. P. Staiger. 2009. Ionic liquids and their interaction with cellulose. *Chemical Reviews* 109:6712–6728.

Podhora, A., K. Helming and L. Adenauer, et al. 2013. The policy-relevancy of impact assessment tools: Evaluating nine years of European research funding. *Environmental Science & Policy* 31:85–95.

Popp, J., Z. Lakner, M. Harangi-Rakos and M. Fari. 2014. The effect of bioenergy expansion: Food, energy, and environment. *Renewable and Sustainable Energy Reviews* 32:559–578.

Ravindran, R. and A. K. Jaiswal. 2016. A comprehensive review on pre-treatment strategy for lignocellulosic food industry waste: Challenges and opportunities. *Bioresource Technology* 199:92–102.

Reeves, S. 2014. To Russia with love: How moral arguments for a humanitarian intervention in Syria opened the door for an invasion of the Ukraine. *Michigan State University International Law Review* 23:199.

Royston, S. and C. Foulds. 2021. The making of energy evidence: How exclusions of Social Sciences and Humanities are reproduced (and what researchers can do about it). *Energy Research & Social Science* 77:102084.

Ryan, S. E., C. Hebdon and J. Dafoe. 2014. Energy research and the contributions of the social sciences: A contemporary examination. *Energy Research & Social Science* 3:186–197.

Sanchez, O. J. and C. A. Cardona. 2008. Trends in biotechnological production of fuel ethanol from different feedstocks. *Bioresource Technology* 99:5270–5295.

Sano, T., H. Yanagishita, Y. Kiyozumi, F. Mizukami and K. Haraya. 1994. Separation of ethanol/water mixture by silicalite membrane on pervaporation. *Journal of Membrane Science* 95:221–228.

Stern, D. I. 1993. Energy and economic growth in the USA: A multivariate approach. *Energy Economics* 15:137–150.

Sun, Y. and J. Cheng. 2002. Hydrolysis of lignocellulosic materials for ethanol production: A review. *Bioresource Technology* 83:1–11.

Taherzadeh, M. J. and K. Karimi. 2007. Enzyme-based hydrolysis processes for ethanol from lignocellulosic materials: A review. *Bioresources* 2:707–738.

Taherzadeh, M. J. and K. Karimi. 2008. Pretreatment of lignocellulosic wastes to improve ethanol and biogas production: A review. *International Journal of Molecular Sciences* 9:1621–1651.

Talebnia, F., D. Karakashev and I. Angelidaki. 2010. Production of bioethanol from wheat straw: An overview on pretreatment, hydrolysis and fermentation. *Bioresource Technology* 101:4744–4753.

Tilman, D., R. Socolow and J. A. Foley, et al. 2009. Beneficial biofuels-the food, energy, and environment trilemma. *Science* 325:270–271.

Vane, L. M. 2005. A review of pervaporation for product recovery from biomass fermentation processes. *Journal of Chemical Technology and Biotechnology* 80:603–629.

Winzer, C. 2012. Conceptualizing energy security. *Energy Policy* 46:36–48.

Wu, F., D. Zhang and J. Zhang, J. 2012. Will the development of bioenergy in China create a food security problem? Modeling with fuel ethanol as an example. *Renewable Energy* 47:127–134.

Wustenhagen, R., M. Wolsink and M. J. Burer. 2007. Social acceptance of renewable energy innovation: An introduction to the concept. *Energy Policy* 35:2683–2691.

Yang, B. and C. E. Wyman. 2008. Pretreatment: The key to unlocking low-cost cellulosic ethanol. *Biofuels, Bioproducts and Biorefining* 2:26–40.

Zhang, Y. H. P. and L. R. Lynd. 2004. Toward an aggregated understanding of enzymatic hydrolysis of cellulose: Noncomplexed cellulase systems. *Biotechnology and Bioengineering* 88:797–824.

Zhu, J. Y. and X. J. Pan. 2010. Woody biomass pretreatment for cellulosic ethanol production: Technology and energy consumption evaluation. *Bioresource Technology* 101:4992–5002.

102 The European Experience of Bioethanol Fuels
Review

Ozcan Konur
(Formerly) Ankara Yildirim Beyazit University

102.1 INTRODUCTION

Crude oil-based gasoline fuels (Ma et al., 2002; Newman and Kenworthy, 1989) have been widely used in transportation sector since the 1920s. However, there have been great public concerns over adverse environmental and human impact of these fuels (Hill et al., 2006, 2009). Hence, biomass-based bioethanol fuels (Hill et al., 2006; Konur, 2012, 2015, 2019, 2020) have increasingly been used in blending gasoline fuels (Hsieh et al., 2002; Najafi et al., 2009), in fuel cells (Antolini, 2007, 2009), and in biochemical production (Angelici et al., 2013; Morschbacker, 2009) in a biorefinery context (Fernando et al., 2006; Huang et al., 2008).

However, it is necessary to pretreat biomass (Alvira et al., 2010; Taherzadeh and Karimi, 2008) to enhance yield of bioethanol (Hahn-Hagerdal et al., 2006; Sanchez and Cardona, 2008) prior to bioethanol fuel production from feedstocks through hydrolysis (Sun and Cheng, 2002; Taherzadeh and Karimi, 2007) and fermentation (Lin and Tanaka, 2006; Olsson and Hahn-Hagerdal, 1996) of biomass and hydrolysates, respectively.

Europe has been one of the most prolific producers engaged in bioethanol fuel research, producing 28% of the global research output in this field. European research in the field of bioethanol fuels has intensified in this context in the key research fronts of the production of bioethanol fuels, and, to a lesser extent, utilization (Antolini, 2007; Hansen et al., 2005) and evaluation (Hamelinck et al., 2005; Pimentel and Patzek, 2005) of bioethanol fuels.

For the first research front, pretreatment (Alvira et al., 2010; Hendriks and Zeeman, 2009) and hydrolysis (Alvira et al., 2010; Sun and Cheng, 2002) of feedstocks, fermentation of feedstock-based hydrolysates (Jonsson and Martin, 2016; Lin and Tanaka, 2006), production of bioethanol fuels (Limayem and Ricke, 2012, Lin and Tanaka, 2006), and, to a lesser extent, separation and distillation of bioethanol fuels (Sano et al., 1994; Vane, 2005) are the key research areas, while for the second research front, utilization of bioethanol fuels in fuel cells and transport engines, bioethanol-based biohydrogen fuels, bioethanol sensors, and bioethanol-based biochemicals are the key research areas.

Research in this field has also intensified for feedstocks of lignocellulosic biomass at large (Hendriks and Zeeman, 2009; Sun and Cheng, 2002), bioethanol fuels for their utilization (Antolini, 2007; Hansen et al., 2005), cellulose (Pinkert et al., 2009; Zhang and Lynd, 2004), starch feedstock residues such as corn stover (Binod et al., 2010; Talebnia et al., 2010), wood biomass such as pine (Galbe and Zacchi, 2002; Zhu and Pan, 2010), and, to a lesser extent, industrial waste such as glycerol (Cardona et al., 2010), lignin (Bourbonnais and Paice, 1990), grass such as switchgrass (Pimentel and Patzek, 2005), hydrolysates (Palmqvist and Hahn-Hagerdal, 2000), biomass at large (Lin and Tanaka, 2006), sugar feedstocks such as sugarcane (Bai et al., 2008), sugar feedstock residues such as sugarcane bagasse (Cardona et al., 2010), algal biomass such as chlorella (John et al., 2011), starch feedstocks such as corn (Bai et al., 2008), urban wastes such as municipal organic

DOI: 10.1201/9781003226574-135

waste (Ravindran and Jaiswal, 2016), forestry wastes such as tree prunings (Duff and Murray, 1996), and food wastes such as potato peels (Ravindran and Jaiswal, 2016).

However, it is essential to develop efficient incentive structures (North, 1991) for the primary stakeholders to enhance research in this field (Konur, 2000, 2002a,b,c, 2006a,b, 2007a,b). Although there has been a large number of review papers on bioethanol fuels within the European context (Hendriks and Zeeman, 2009; Mosier et al., 2005; Sun and Cheng, 2002), there has been no review of the 25 most-cited papers in this field.

Thus, this book chapter presents a review of the 25 most-cited articles in the field of bioethanol fuels. Then, it discusses the key findings of these highly influential papers and comments on future research priorities in this field.

102.2 MATERIALS AND METHODS

Search for this study was carried out using Scopus database (Burnham, 2006) in November 2022.

As the first step for the search of the relevant literature, the keywords were selected using the most-cited first 300 population papers for each feedstock. The selected keyword list was then optimized to obtain a representative sample of papers for each research field. These keyword lists for bioethanol fuels were then integrated to obtain the keyword list for this research field (Konur (2023a,b). A separate keyword set was also devised for Europe (Konur, 2023c).

As the second step, a sample dataset was used for this study. The first 25 articles with at least 546 citations each were selected for review study. Key findings from each paper were taken from the abstracts of these papers and were briefly discussed. Additionally, a number of brief conclusions were drawn and a number of relevant recommendations were made to enhance future research landscape.

102.3 RESULTS

Brief information about 25 most-cited papers with at least 546 citations each on bioethanol fuels is given below. The primary research fronts are pretreatments and hydrolysis of feedstocks and fermentation of hydrolysates with 10, 3, and 3 HCPs, respectively, for the field of bioethanol production with 16 HCPs in total. Further, there are two and seven HCPs for evaluation and utilization of bioethanol fuels with at least 546 citations, respectively.

102.3.1 Bioethanol Fuel Production: The European Experience

The primary research fronts are the pretreatments and hydrolysis of feedstocks and fermentation of hydrolysates with 10, 3, and 3 HCPs, respectively.

102.3.1.1 Feedstock Pretreatments: European Experience

Brief information about ten most-cited papers on pretreatments of feedstocks with at least 540 citations each is given below (Table 102.1).

Floudas et al. (2012) mapped the detailed evolution of wood-degrading enzymes in a paper with 1,104 citations. They noted that wood was highly resistant to decay, largely due to presence of lignin, and the only organisms capable of substantial lignin decay were white rot fungi in agaricomycetes, which also contained non-lignin-degrading brown rot and ectomycorrhizal species. They analyzed 31 fungal genomes and found that lignin-degrading peroxidases expanded in the lineage leading to ancestor of agaricomycetes, which was reconstructed as a white rot species and then contracted in parallel lineages leading to brown rot and mycorrhizal species. Thus, the origin of lignin degradation might have coincided with sharp decrease in rate of organic carbon burial around the end of carboniferous period.

TABLE 102.1
Pretreatment of Feedstocks: European Experience

No.	Papers	Biomass	Prts.	Parameters	Keywords	Lead Authors	Affil.	Cits
1	Floudas et al. (2012)	Wood, lignin	Enzymes	Wood degradation, wood degrading, enzyme evolution, genome, peroxidases	Lignin, decomposition	Grigoriev, Igor V. 25027225800	Jnt. Genome Inst. USA	1,104
2	Schwanninger et al. (2004)	Wood, cellulose	Milling	Vibratory ball milling, FTIR spectroscopy, temperature, particle size, oxygen, cellulose crystallinity, depolymerization	Wood, cellulose, milling	Schwanninger, Manfred 6602877236	Univ. Natr. Res. Appl. Life Sci. Austria	949
3	Martinez et al. (2008)	Lignocellulosic biomass	Enzymes	Biomass-degrading fungi, genome, enzymes	Biomass, fungus, degrading	Martinez, Diego 7202958664	Massachusetts Inst. Technol. USA	899
4	Pandey and Pitman (2003)	Wood Pine, sapwood, beech	Enzymes	Wood decay, fungi, lignin, carbohydrate content, fungi types, wood types, FTIR spectra	Wood, decay, fungi	Pandey, Krishna K. 56252805800	Inst. Wood Sci. Technol. India	846
5	Sun et al. (2009)	Wood Pine, oak	IL	IL types, dissolution rates, wood types, delignification, cellulose recovery	Wood, delignification, IL	Rogers, Robin D. 35474829200	Univ. Alabama USA	838
6	Martinez et al. (2004)	Wood	Fungi	Biomass-degrading fungi, genome, enzymes	Lignocellulose, degrading, fungi	Cullen, Dan 7202109135	USDA Forest Serv.	688
7	Quinlan et al. (2011)	Cellulose	Enzymes	Glycoside hydrolase GH61 enzymes, cellulose degradation, cellulase-enhancing factors	Cellulose, degradation, metalloenzyme	Johansen, Katja Salomon* 36473579400	Novozymes AS Denmark	679
8	Li et al. (2007)	Wood Aspen	Steam	Lignin depolymerization and repolymerization, delignification, phenol	Lignin, delignification, steam	Li, Jiebing 7410069979	Rise Res. Inst. Sweden Sweden	671
9	Sun et al. (2005)	Wheat straw cellulose	Steam	Pretreatments, solid-to-liquid ratio, cellulose yield, cellulose crystallinity, delignification	Straw, cellulose, steam, degraded	Sun, Runcang 55661525600	Dalian Polytech. Univ. China	557
10	Heinze et al. (2005)	Cellulose	ILs	Cellulose dissolution, ILs, non-derivatizing solvents	Cellulose, ILs	Heinze, Thomas 7006547465	Friedrich Schiller Univ. Germany	540

*, Female; Cits., Number of citations received for each paper; Prts., Biomass pretreatments.

Schwanninger et al. (2004) studied the effects of short-time vibratory ball milling on the shape of Fourier transform infrared (FTIR) spectra of wood and cellulose in a paper with 949 citations. They showed that this milling pretreatment had a strong influence on the shape of FTIR spectra of both wood and cellulose, even if samples were milled for only a short time. Further, temperature, particle size, and oxidation processes were both the causes of observed changes in the structure of FTIR spectra, but that milling pretreatment itself was the main influencing factor. They observed the most evident changes in the spectra of cellulose and wood at wave numbers 1,034, 1,059, 1,110, 1,162, 1,318, 1335, and 2,902 cm^{-1} and in the OH-stretching vibration region from 3,200 to 3,500 cm^{-1}. They concluded that these changes were mainly associated with a decrease in the degree of cellulose crystallinity and/or a decrease in the degree of polymerization of cellulose.

Martinez et al. (2008) performed genome sequencing and analysis of the *Trichoderma reesei* in a paper with 899 citations. They assembled 89 scaffolds to generate 34 Mbp of nearly contiguous *T. reesei* genome sequence comprising 9,129 predicted gene models. They found that its genome encoded fewer cellulases and hemicellulases than any other sequenced fungus able to hydrolyze plant cell wall polysaccharides. Further, many *T. reesei* genes encoding carbohydrate-active enzymes were distributed non-randomly in clusters that lied between regions of synteny with other sordariomycetes. Hence, numerous genes encoding biosynthetic pathways for secondary metabolites might promote survival of *T. reesei* in its competitive soil habitat.

Pandey and Pitman (2003) performed FTIR studies of changes in wood chemistry following decay by brown-rot and white-rot fungi in a paper with 846 citations. They used Scots pine, sapwood, and beech decayed by *Coniophora puteana*, *Coriolus versicolor*, and *Phanerochaete chrysosporium*. They exposed wood to these fungi for different durations of up to 12 weeks. They found that in wood decayed by *C. puteana*, there was a progressive increase in lignin content relative to carbohydrate evident from increases in relative height of lignin associated bands and a corresponding decrease in intensities of carbohydrate bands. At higher weight losses, spectra for wood decayed by *C. puteana* had many similarities with that of Klason lignin isolated from wood. In contrast, wood decayed by *P. chrysosporium* showed selective type decay with a reduction in peak heights associated with lignin relative to carbohydrates. Although weight losses in samples exposed to *C. versicolor* were high, simultaneous decay resulted in little change in relative intensities of lignin and carbohydrate bands, with only a slight preference for lignin.

Sun et al. (2009) dissolved and partially delignified yellow pine and red oak in 1-ethyl-3-methylimidazolium acetate ([C_2mim]OAc) after mild milling pretreatment in a paper with 838 citations. They showed that this ionic liquid (IL) was a better solvent for wood than 1-butyl-3-methylimidazolium chloride ([C_4mim]Cl), and that type of wood, initial wood load, and particle size, etc. affected dissolution and dissolution rates. For example, red oak dissolved better and faster than yellow pine. They obtained carbohydrate-free lignin and cellulose-rich materials by using proper reconstitution solvents (e.g., acetone/water 1:1 v/v), and they achieved approximately 26.1% and 34.9% reductions of lignin content in reconstituted cellulose-rich materials from pine and oak, respectively, in one dissolution/reconstitution cycle. Further, for pine, they recovered 59% of olocellulose in original wood in cellulose-rich reconstituted material, while they recovered 31% and 38% of original lignin, respectively, as carbohydrate-free lignin and as carbohydrate-bonded lignin in cellulose-rich materials.

Martinez et al. (2004) sequenced the 30-million base-pair genome of *Phanerochaete chrysosporium* strain RP78 using whole genome shotgun approach in a paper with 688 citations. They found that *P. chrysosporium* genome had an impressive array of genes encoding secreted oxidases, peroxidases, and hydrolytic enzymes that cooperated in wood decay. They noted that analysis of genome data would enhance understanding of lignocellulose degradation and provide a framework for further development of bioprocesses for biomass utilization. This genome provided a high-quality draft sequence of a basidiomycete.

Quinlan et al. (2011) performed enzymatic degradation of cellulose by a copper metalloenzyme that exploited biomass components in a paper with 679 citations. They showed that glycoside hydrolase (CAZy) GH61 enzymes were a unique family of copper (Cu)-dependent oxidases. They observed that Cu was needed for GH61 maximal activity and that formation of cellodextrin and oxidized cellodextrin products by GH61 was enhanced in the presence of small molecule redox-active cofactors such as ascorbate and gallate. Further, the active site of GH61 contained a type II Cu and, uniquely, a methylated histidine in Cu's coordination sphere, thus providing an innovative paradigm in bioinorganic enzymatic catalysis.

Li et al. (2007) explored the role of lignin depolymerization and repolymerization for delignification of aspen by steam explosion pretreatment in a paper with 671 citations. They subjected aspen and isolated lignin from aspen to steam explosion pretreatment under various conditions. They observed the competition between lignin depolymerization and repolymerization and identified conditions required for these two types of reaction. Further, they found that addition of 2-naphthol inhibited repolymerization reaction strongly, resulting in a highly improved delignification by subsequent solvent extraction and an extracted lignin of uniform structure.

Sun et al. (2005) performed both steam explosion and alkaline hydrogen peroxide pretreatments for wheat straw biomass to recover cellulose in a paper with 557 citations. They steamed straw at 200°C, 15 bar for 10 and 33 min; 220°C, 22 bar for 3, 5, and 8 min with a solid to liquid ratio of 2:1 (w/w); and 220°C, 22 bar for 5 min with a solid to liquid ratio of 10:1, respectively. They delignified washed fiber and bleached by 2% H_2O_2 at 50°C for 5 h under pH 11.5, which yielded 34.9%, 32.6%, 40.0%, 36.9%, 30.9%, and 36.1% (% dry wheat straw) of cellulose preparation, respectively. They obtained optimum cellulose yield (40.0%) when steam explosion pretreatment was performed at 220°C, 22 bar for 3 min with a solid to liquid ratio of 2:1, in which the cellulose fraction obtained had a viscosity average degree of polymerization of 587 and contained 14.6% hemicelluloses and 1.2% klason lignin. Thus, the steam explosion pretreatment led to a significant loss in hemicelluloses, and alkaline peroxide post-treatment resulted in substantial dissolution of lignin and an increase in cellulose crystallinity.

Heinze et al. (2005) used a number of ILs for dissolution of cellulose in a paper with 540 citations. These ILs were 1-N-butyl-3-methylimidazolium chloride ($[C_4mim]^+Cl^-$), 3-methyl-N-butyl-pyridinium chloride, and benzyl dimethyl (tetradecyl) ammonium chloride. They found that these ILs had the ability to dissolve cellulose with a degree of polymerization in the range from 290 to 1,200 to a very high concentration. Using $[C_4mim]^+Cl^-$, no degradation of polymer appeared. They confirmed that this IL was a so-called non-derivatizing solvent. Thus, this non-derivatizing solvent could be applied as a reaction medium for synthesis of carboxymethyl cellulose and cellulose acetate. However, without using any catalyst, cellulose derivatives with a high degree of substitution could be prepared.

102.3.1.2 Hydrolysis of Feedstocks: European Experience

There are three HCPs for hydrolysis of feedstock with at least 583 citations (Table 102.2).

Kilpelainen et al. (2007) dissolved wood in ILs in a paper with 834 citations. They found that 1-butyl-3-methylimidazolium chloride ([Bmim]Cl) and 1-allyl-3-methylimidazolium chloride ([Amim]Cl) had good solvating power for Norway spruce sawdust, Norway spruce, and Southern pine thermomechanical pulp fibers as these ILs provided solutions which permitted complete acetylation of wood. Alternatively, they obtained transparent amber solutions of wood when dissolution of same lignocellulosic samples were attempted in 1-benzyl-3-methylimidazolium chloride ([BzMim] Cl). They then digested cellulose of regenerated wood to glucose by a cellulase enzymatic hydrolysis pretreatment. Furthermore, completely acetylated wood was readily soluble in chloroform.

Eriksson et al. (2002) explored mechanism of surfactant effect in enzymatic hydrolysis of lignocellulose in a paper with 756 citations. They screened a number of surfactants for their ability to improve enzymatic hydrolysis of steam-pretreated spruce. They found that non-ionic surfactants were the most effective. Studies of adsorption of dominating cellulase of *Trichoderma reesei*, Cel7A (CBHI), during hydrolysis showed that anionic and non-ionic surfactants reduced enzyme adsorption to lignocellulose substrate. The approximate reduction of enzyme adsorption was from 90% adsorbed enzyme to 80% with surfactant addition, while surfactants had only a weak effect on cellulase temperature stability. They explained the improved conversion of lignocellulose with

TABLE 102.2
Hydrolysis of Feedstocks: European Experience

No.	Papers	Biomass	Prts.	Parameters	Keywords	Lead Authors	Affil.	Cits
1	Kilpelainen et al. (2007)	Wood Spruce, pine	IL, enzymes	IL pretreatment, enzymatic hydrolysis, IL types, wood types	Wood, IL, dissolution	Kilpelainen, Ilkka 7006830888	Univ. Helsinki Finland	834
2	Eriksson et al. (2002)	Lignocellulose Spruce	Enzymes, surfactants	Enzymatic hydrolysis, pretreatment, mechanisms, enzyme adsorption, cellulase stability, surfactant-lignin interactions	Lignocellulose, hydrolysis	Tjerneld, Folke 7006446969	Lund Univ. Sweden	756
3	Harris et al. (2010)	Lignocellulosic biomass	Enzymes	Hydrolytic activity, protein loading, hydrolytic activity, cellulolytic enzymes, metal ions, hydrolysis	Lignocellulosic biomass, hydrolysis	Harris, Paul V. 56305899400	Novozymes Inc. USA	583

Cits., Number of citations received for each paper; Prts., Biomass pretreatments.

surfactant by reduction of unproductive enzyme adsorption to the lignin part of the substrate. This was due to hydrophobic interaction of surfactant with lignin on the lignocellulose surface, which released non-specifically bound enzyme.

Harris et al. (2010) stimulated lignocellulosic biomass hydrolysis by proteins of glycoside hydrolase family 61 (GH61) and evaluated structure and function of GH61 in a paper with 583 citations. They showed that certain GH61 proteins lacked measurable hydrolytic activity by themselves but in presence of various divalent metal ions could significantly reduce the total protein loading required to hydrolyze lignocellulosic biomass. They also solved structure of one highly active GH61 protein and found that it was devoid of conserved, closely juxtaposed acidic side chains that could serve as general proton donors and nucleophiles/bases in a canonical hydrolytic reaction. They concluded that GH61 proteins were unlikely to be glycoside hydrolases. Structure-based mutagenesis showed the importance of several conserved residues for GH61 function. By incorporating gene for one GH61 protein into a commercial *Trichoderma reesei* strain producing high levels of cellulolytic enzymes, they reduced by two-fold the total protein loading (and hence the cost) required to hydrolyze lignocellulosic biomass.

102.3.1.3 Hydrolysate Fermentation: European Experience

There are three HCPs for fermentation of hydrolysates (Table 102.3).

Larsson et al. (1999) studied the effect of combined severity (CS) of dilute sulfuric acid hydrolysis of spruce on sugar yield and fermentability of hydrolysate by *Saccharomyces cerevisiae* in a paper with 891 citations. When CS of hydrolysis conditions increased, they observed that yield of fermentable sugars increased to a maximum between CS 2.0–2.7 for mannose and 3.0–3.4 for glucose above which it decreased. Further, decrease in yield of monosaccharides coincided with maximum concentrations of furfural and 5-HMF. With further increase in CS, concentrations of furfural and 5-HMF decreased while formation of formic acid and levulinic acid increased. Yield of ethanol decreased at approximately CS 3, while volumetric productivity decreased at lower CS. Ethanol yield and volumetric productivity decreased with increasing concentrations of acetic acid, formic acid, and levulinic acid. However, furfural and 5-HMF decreased volumetric productivity but did not influence the final yield of ethanol. Decrease in volumetric productivity was more pronounced when 5-HMF was added to fermentation, and this compound was depleted at a lower rate than furfural. Further, the inhibition observed in hydrolysates produced in higher CS could not be fully explained by effect of the furfural, 5-HMF, acetic acid, formic acid, and levulinic acid.

Alper et al. (2006) engineered yeast transcription machinery for improved ethanol tolerance of yeasts and bioethanol fuel production in a paper with 651 citations. They showed application of global transcription machinery engineering (gTME) to *Saccharomyces cerevisiae* for improved glucose/ethanol tolerance. They observed that mutagenesis of transcription factor Spt15p and selection led to dominant mutations that conferred increased tolerance and more efficient glucose conversion to ethanol. The desired phenotype resulted from combined effect of three separate mutations in *SPT15* gene. Thus, they concluded that gTME could provide a route to complex phenotypes that were not readily accessible by traditional methods.

Kaparaju et al. (2009) produced bioethanol, biohydrogen, and biogas fuels from wheat straw in a biorefinery concept in a paper with 569 citations. They liberated wheat straw hydrothermally to a cellulose-rich fiber fraction and a hemicellulose-rich liquid fraction. Enzymatic hydrolysis and subsequent fermentation of cellulose yielded 0.41 g ethanol/g glucose, while dark fermentation of hydrolysate produced 178.0 mL-H_2/g sugars. They then used effluents from both bioethanol and biohydrogen processes to produce biomethane gas with yields of 0.324 and 0.381 m^3/kg volatile solids, respectively. Additionally, evaluation of six different wheat straw-to-biofuel production scenario showed that either use of wheat straw for biogas production or multi-fuel production were energetically most efficient processes compared to production of mono-fuel such as bioethanol when fermenting C6 sugars alone. Thus, multiple biofuels production from wheat straw could increase efficiency of material and energy and could presumably be more economical process for biomass utilization.

TABLE 102.3
Fermentation of Hydrolysates: European Experience

No.	Papers	Biomass	Prts.	Yeasts	Parameters	Keywords	Lead Authors	Affil.	Cits
1	Larsson et al. (1999)	Softwood Spruce	Acids	*S. cerevisiae*	Acid hydrolysis severity, fermentation, sugar and ethanol yield, fermentation inhibitors, volumetric productivity	Softwood, fermentation, hydrolysis	Hahn-Hagerdal, Barbel* 7005389381	Lund Univ. Sweden	891
2	Alper et al. (20069	Glucose	Na	*S. cerevisiae*	Yeast transcription machinery, ethanol tolerance of yeasts, global transcription machinery engineering	Ethanol, yeast, engineering	Stephanopoulos, Gregory 24527470500	Massachusetts Inst. Technol. USA	651
3	Kaparaju et al. (2009)	Wheat straw	Enzymes	Yeasts	Bioethanol production, biogas, biohydrogen, ethanol yield	Straw, bioethanol, biorefinery	Angelidaki, Irini* 6603674728	Tech. Univ. Denmark Denmark	569

*, Female; Cits., Number of citations received for each paper; Na, non-available; Prts., Biomass pretreatments.

102.3.2 Bioethanol Evaluation and Utilization

There are two and seven HCPs for evaluation and utilization of bioethanol fuels with at least 546 citations, respectively (Table 102.4).

102.3.2.1 Bioethanol Fuel Evaluation: European Experience

There are two HCPs for evaluation of bioethanol fuels (Table 102.4).

Hamelinck et al. (2005) evaluated techno-economic performance of bioethanol production in short-, middle- and long-term using hydrolysis and fermentation technologies as well as developing technologies in a paper with 1,239 citations. They found that the available technology as of 2005, which was based on dilute acid hydrolysis, had about 35% efficiency (higher heating value, HHV) from biomass to ethanol. The overall efficiency, with bioelectricity coproduced from the lignin, was about 60%. However, improvements in pretreatment and advances in biotechnology, especially through process combinations, could bring the ethanol efficiency to 48% and the overall process efficiency to 68%. They estimated current investment costs at 2.1 k€/kWHHV (at 400 MWHHV input, i.e., a nominal 2,000 tonne dry/day input). However, a future technology in a five times larger plant (2 giga watts higher heating value-GWHHV) could have investments of 900 k€/kWHHV. Thus, a combined effect of higher hydrolysis-fermentation efficiency, lower specific capital investments, increase of scale, and cheaper biomass feedstock costs (from 3 to 2 €/GJHHV) could bring ethanol production costs from 22 €/giga joules higher heating value (GJHHV) in next 5 years to 13 €/GJ over the 10–15 years time scale, and down to 8.7 €/GJ in 20 or more years.

Wingren et al. (2003) performed techno-economic evaluation of bioethanol fuel production from softwood in a paper with 550 citations. They compared simultaneous saccharification and fermentation (SSF) and separate hydrolysis and fermentation (SHF) processes. They found that ethanol production costs for SSF and SHF processes were 0.57 USD/L (68 USD/barrel) and 0.63 USD/L (75 USD/barrel), respectively. The main reason for SSF being lower was that capital cost was lower and overall ethanol yield was higher. A major drawback of SSF process was problem with recirculation of yeast following SSF step. They confirmed that major economic improvements in both SSF and SHF could be achieved by increasing income from solid fuel coproduct. This was done by lowering energy consumption in process by running enzymatic hydrolysis or SSF step at a higher substrate concentration and by recycling process streams. Running SSF with use of 8% rather than 5% non-soluble solid material would result in a 19% decrease in production cost. If after distillation, 60% of the stillage stream was recycled back to SSF step, production cost would be reduced by 14%. The cumulative effect of these various improvements would result in a production cost of 0.42 USD/L (50 USD/barrel) for SSF process.

102.3.2.2 Bioethanol Fuel Utilization: European Experience

There are four and three HCPs for production of biohydrogen fuels from bioethanol fuels and direct utilization of bioethanol fuels in fuel cells with at least 546 citations each (Table 102.4).

102.3.2.2.1 Bioethanol Fuel-based Biohydrogen Fuels: European Experience

Murdoch et al. (2011) studied photocatalytic biohydrogen fuel production from ethanol fuels over Au/TiO_2 nanoparticles (NPs) in a paper with 1,013 citations. They showed that Au NPs in the size range 3–30 nm on TiO_2 were very active in hydrogen production from ethanol fuels. Further, Au NPs of similar size on anatase NPs delivered a rate two orders of magnitude higher than that recorded for Au on rutile NPs. However, Au NP size did not affect photoreaction rate over 3–12 nm range. These catalysts resulted in high biohydrogen yield.

Deluga et al. (2004) produced biohydrogen fuels by autothermal reforming of bioethanol fuels in a paper with 918 citations. They converted ethanol and ethanol-water mixtures directly into hydrogen with nearly 100% selectivity and more than 95% conversion by catalytic partial oxidation, with a residence time on rhodium-ceria catalysts of less than 10 ms. They performed rapid vaporization and mixing with air with an automotive fuel injector at temperatures sufficiently low and times sufficiently fast that homogeneous reactions producing carbon, acetaldehyde, ethylene, and total combustion products could be minimized.

TABLE 102.4
Evaluation and Utilization of Bioethanol Fuels: European Experience

No.	Papers	Cats.	Res. Fronts	Parameters	Keywords	Lead Authors	Affil.	Cits
1	Hamelinck et al. (2005)	Na	Evaluation	Bioethanol fuel production costs, techno-economic performance, available and advanced technologies, lignin, ethanol conversion efficiency	Ethanol, lignocellulosic biomass, techno-economic	Hamelinck, Carlo N 6603008025	Ecorys Netherlands	1239
2	Murdoch et al. (2011)	Au/TiO_2	Biohydrogen	Photocatalytic biohydrogen production, photocatalysts, anatase NPs, NP size	Ethanol, hydrogen	Idriss, Hicham 7006768868	Karlsruhe Inst. Technol. Germany	1013
3	Deluga et al. (2004)	Na	Biohydrogen	Biohydrogen production, ethanol autothermal reforming, hydrogen selectivity, ethanol conversion rate	Ethanol, hydrogen, reforming	Schmidt, Lanny D. 13103591600	Univ. Minnesota USA	918
4	Kowal et al. (2009)	$PtRhSnO_2/C$	Fuel cells	Ethanol oxidation, electrocatalysts, C–C bond splitting	Ethanol, electrocatalysts, oxidizing	Kowal, Adzic 36728886000	AGH Univ. Sci. Technol. Poland	667
5	Liguras et al. (20039	Rh, Ru, Pt, Pd/ Al_2O_3, MgO, TiO_2	Biohydrogen	Steam reforming, supported noble metal catalysts, hydrogen selectivity, catalytic performance, metal loading, ethanol conversion	Hydrogen, fuel cells, ethanol reforming, catalysts	Verykios, Xenophon E. 35551305100	Univ. Patras Greece	619
6	Zhou et al. (2003)	Pt/C, PtSn/C, PtRu/C, PtW/C, PtPd/C	Fuel cells	Pt-based anode catalysts, additives, ethanol electro-oxidation, DEFC performance, Pt-additive interactions	Ethanol, fuel cells, catalysts	Xin, Qin 28167866000	Chinese Acad Sci. China	601
7	Lamy et al. (2004)	PtSn/C	Fuel cells	Electrocatalytic activity, PtSn anode, additive loading, ethanol dissociative chemisorption, electrode activity, ethanol oxidation, DEFC	Ethanol fuel cells, electrocatalysts	Leger, Jean M 7201980020	Univ. Poitiers France	572
8	Wingren et al. (2003)	Na	Evaluation	Bioethanol production, techno-economic evaluation, SSF, SHF, production cost, ethanol yield, yeast recycling, substrate concentration	Softwood, ethanol, techno-economic, SSF, SHF	Zacchi, Guido 7006727748	Lund Univ. Sweden	550
9	Fatsikostas and Verykios (2004)	Ni/γ-Al_2O_3, Ni/ La_2O_3, and Ni/ La_2O_3/γ- Al_2O_3	Fuel cells	Steam reforming, ethanol fuels, biohydrogen fuels, catalyst supports, dehydrogenation, dehydration, carbon deposition, steam-to-ethanol ratio, reaction temperature	Ethanol, reforming, catalysts	Verykios, Xenophon E. 35551305100	Univ. Patras Greece	546

Cits., Number of citations received for each paper; Na, non-available; Prts., Biomass pretreatments.

Liguras et al. (2003) produced biohydrogen fuels for fuel cells by steam reforming of bioethanol fuels over supported noble metal catalysts in a paper with 619 citations. They explored catalytic performance of supported noble metal catalysts in the temperature range of 600°C–850°C with respect to nature of active metallic phase (Rhodium-Rh, Ruthenium-Ru, platinum-Pt, palladium-Pd), nature of support (alumina-Al_2O_3, magnesia-MgO, titania-TiO_2), and metal loading (0–5 wt.%). They found that for low-loaded catalysts, Rh was significantly more active and selective toward hydrogen formation compared to Ru, Pt, and Pd, which showed a similar behavior. The catalytic performance of Rh and, particularly, Ru was significantly improved with increasing metal loading, leading to higher ethanol conversions and hydrogen selectivities at given reaction temperatures. The catalytic activity and selectivity of high-loaded Ru catalysts were comparable to that of Rh. They further found that under certain reaction conditions, 5% Ru/Al_2O_3 catalyst completely converted ethanol with selectivities toward hydrogen above 95%, the only byproduct being methane. Further, this catalyst was acceptably stable and could be a good candidate for production of hydrogen by steam-reforming of ethanol for fuel cell applications.

Fatsikostas and Verykios (2004) evaluated the reaction network of steam reforming of ethanol fuels over supported nickel (Ni)-based catalysts in a paper with 546 citations. They used supports of alumina (γ-Al_2O_3), lanthana (La_2O_3), and La_2O_3/γ-Al_2O_3. They found that ethanol interacted strongly with alumina on surface of which it was dehydrated at low temperatures, and less strongly with lanthana on surface of which it was both dehydrogenated and dehydrated. They also observed cracking reactions on carriers at intermediate temperatures. In presence of Ni, catalytic activity was shifted toward lower temperatures. In addition to the above reactions, reforming, water–gas shift, and methanation contributed significantly to product distribution, while carbon deposition was also a significant route. Thus, they found that the rate of carbon deposition was a strong function of the carrier, the steam-to-ethanol ratio, and reaction temperature, while presence of lanthana on the catalyst, high steam-to-ethanol ratio, and high temperature offered enhanced resistance toward carbon deposition.

102.3.2.2.2 Direct Utilization of Bioethanol Fuels in Fuel Cells: European Experience

Kowal et al. (2009) used ternary $Pt/Rh/SnO_2$ electrocatalysts for oxidizing ethanol to CO_2 in a paper with 667 citations. They synthesized a ternary $PtRhSnO_2/C$ electrocatalyst by depositing Pt and Rh atoms on carbon-supported tin dioxide (SnO_2) nanoparticles (NPs) that were capable of oxidizing ethanol with high efficiency and held great promise for resolving the impediments to developing practical direct ethanol fuel cells. They observed that this electrocatalyst effectively split the C–C bond in ethanol at room temperature in acid solutions, facilitating its oxidation at low potentials to CO_2, which had not been achieved with existing catalysts. They reasoned that this electrocatalyst's activity was due to the specific property of each of its constituents, induced by their interactions. Their findings helped explain the lack of activity of PtRu electrocatalyst for ethanol oxidation and pointed the way to accomplishing C–C bond splitting in other catalytic processes.

Zhou et al. (2003) evaluated effect of additives on ethanol electro-oxidation activity and direct ethanol fuel cell (DEFC) performance of Pt-based anode catalysts in a paper with 601 citations. They prepared several Pt-based anode catalysts supported on carbon XC-72R. They found that all these catalysts consisted of uniform nanosized particles with sharp distribution and Pt lattice parameter decreased with addition of Ru or Pd and increased with addition of Sn or tungsten (W). They observed that presence of Sn, Ru, and W enhanced activity of Pt towards ethanol electro-oxidation in the following order: Pt_1Sn_1/C >Pt_1Ru_1/C >Pt_1W_1/C> Pt_1Pd_1/C >Pt/C. Moreover, Pt_1Ru_1/C further modified by W and molybdenum (Mo) showed improved ethanol electro-oxidation activity, but its DEFC performance was inferior to that measured for Pt_1Sn_1/C. They further found that single DEFC having Pt_1Sn_1/C or Pt_3Sn_2/C or Pt_2Sn_1/C as anode catalyst showed better performances than those with Pt_3Sn_1/C or Pt_4Sn_1/C. They also found that the latter two DEFCs exhibited higher performances than single cell using Pt_1Ru_1/C. They attributed this distinct difference in DEFC performance between these catalysts to the so-called bifunctional mechanism and to electronic interaction between Pt and additives. It is thought that an amount of $-OH_{ads}$, an amount of surface Pt active sites, and conductivity effect of PtSn/C catalysts would determine activity of PtSn/C with different Pt/Sn ratios. At lower temperature values or at low current density regions where the electro-oxidation of ethanol was not so fast and its chemisorption is not the rate-determining step, Pt_3Sn_2/C was more suitable for DEFC. At 75°C, single DEFC with Pt_3Sn_2/C as anode catalyst showed a comparable performance to that with Pt_2Sn_1/C, but at higher temperature of 90°C, the latter presented much better performance. Thus, they concluded that Pt_2Sn_1/C, supplying sufficient $-OH_{ads}$ and having adequate active Pt sites and acceptable ohmic effect, could be the appropriate anode catalyst for DEFC.

Lamy et al. (2004) developed PtSn electrocatalysts for DEFC in a paper with 572 citations. By modifying composition of Pt anode by adding Sn, they confirmed that the overall electrocatalytic activity was greatly enhanced at low potentials. The optimum composition in Sn was in the range 10–20 at.%. With this composition, they showed that poisoning by adsorbed CO coming from ethanol dissociative chemisorption was greatly reduced leading to a significant enhancement of electrode activity. However, oxidation of ethanol was not completely leading to formation of C2 products. They confirmed these observations made in half-cell experiments during electrical tests in a DEFC.

102.4 DISCUSSION

102.4.1 Introduction

Crude oil-based gasoline fuels have been widely used in transportation sector since the 1920s. However, there have been great public concerns over adverse environmental and human impact of these fuels. Hence, biomass-based bioethanol fuels have increasingly been used in blending gasoline and petrodiesel fuels, in fuel cells, and in biochemical production in a biorefinery context.

However, it is necessary to pretreat biomass to enhance yield of bioethanol prior to bioethanol fuel production from feedstocks through hydrolysis of biomass and fermentation of hydrolysates.

Europe has been one of the most prolific producers as a whole engaged in bioethanol fuel research, producing 28% of the global research output in this field. European research in the field of bioethanol fuels has intensified in this context in the key research fronts of production and utilization of bioethanol fuels, and, to a lesser extent, evaluation of bioethanol fuels.

For the first research front, pretreatment and hydrolysis of feedstocks, fermentation of hydrolysates, production of bioethanol fuels, and, to a lesser extent, separation and distillation of bioethanol fuels are the key research areas, while for the second research front, direct utilization of bioethanol fuels in fuel cells and transport engines, bioethanol-based biohydrogen fuels, bioethanol sensors, and, to a lesser extent, bioethanol-based biochemicals are the key research areas.

Research in this field has also intensified for feedstocks of bioethanol fuels for their utilization, starch feedstock residues, cellulose, wood biomass, lignocellulosic biomass at large, and, to a lesser extent, industrial waste, lignin, grass, hydrolysates, biomass at large, sugar feedstocks, sugar feedstock residues, algal biomass, starch feedstocks, urban wastes, forestry wastes, and food wastes.

However, it is essential to develop efficient incentive structures for the primary stakeholders to enhance research in this field. Although there has been a limited number of review papers for this field, there has been no review of the 25 most-cited articles in this field. Thus, this book chapter presents a review of the 25 most-cited articles on bioethanol fuels. Then, it discusses the key findings of these highly influential papers and comments on future research priorities in this field.

As the first step for search of relevant literature, the keywords were selected using the first 300 most-cited population papers for each research front. The selected keyword list was then optimized to obtain a representative sample of papers for each research field. These keyword lists bioethanol fuels were then integrated to obtain the keyword list for this research field (Konur, 2023a,b). Separately, a keyword set was devised for Europe with over 50 listed countries (Konur, 2023c).

As the second step, a sample data set was used for this study. The first 25 articles with at least 546 citations each were selected for review study. Key findings from each paper were taken from abstracts of these papers and were discussed. Additionally, a number of brief conclusions were drawn and a number of relevant recommendations were made to enhance the future research landscape.

Information about research fronts for sample papers in bioethanol fuels is given in Table 102.5. As this table shows the most prolific biomass for this field is wood biomass such as poplar and pine with 40% of the HCPs, followed by bioethanol fuels as feedstocks with 36% of the sample papers. Other prolific feedstocks are cellulose, lignocellulosic biomass at large, and starch feedstock residues with 16%, 12%, and 8% of the sample papers, respectively. Further, other minor feedstocks are lignin, sugars, and biomass at large with 4% of the sample papers each. Additionally, a relatively large number of feedstocks are not represented in this sample.

Thus, the first five feedstocks dominate HCPs in this field where only major waste biomass is starch feedstock residues such as corn stover or wheat straw. On the other hand, cellulose and lignin are the key constituents of lignocellulosic biomass. Further bioethanol fuels are feedstocks for production of biohydrogen fuels and for the utilization in the transport engines and fuel cells.

It is also notable that there are no HCPs for the first generation food-grade sugar and starch bioethanol fuels reflecting the adverse effect of undermining food security (Bentivoglio et al., 2016; Elobeid and Hart, 2007).

TABLE 102.5
The Most Prolific Research Fronts for Bioethanol Fuels: European Experience

No.	Research Fronts	N Paper (%) Review	N Paper (%) Sample	Surplus (%)
1	Wood biomass	40	9.0	31.0
2	Bioethanol fuels	36	19.6	16.4
3	Cellulose	16	15.1	0.9
4	Lignocellulosic biomass	12	23.0	−11.0
5	Starch feedstock residues	8	9.7	−1.7
6	Lignin	4	6.2	−2.2
7	Sugars	4	5.6	−1.6
8	Biomass	4	2.6	1.4
9	Missing feedstocks	0	20.8	−20.8
	Industrial waste	0	6.5	−6.5
	Hydrolysates	0	2.8	−2.8
	Sugar feedstock residues	0	2.2	−2.2
	Sugar feedstocks	0	2.2	−2.2
	Algal biomass	0	1.9	−1.9
	Lignocellulosic wastes	0	1.8	−1.8
	Starch feedstocks	0	1.7	−1.7
	Urban wastes	0	1.0	−1.0
	Forestry wastes	0	0.8	−0.8
	Hemicellulose	0	0.8	−0.8
	Xylan	0	0.8	−0.8
	Food wastes	0	0.6	−0.6
	Cellobiose	0	0.4	−0.4
	Biosyngas	0	0.4	−0.4
	Plants	0	0.3	−0.3
	Sample size	25	775	

N Paper (%) review,Number of papers in the European sample of 25 reviewed papers; N paper (%) sample, Number of papers in the world sample of 775 papers.

Further, the most influential research fronts are wood biomass with 31% surplus, followed by bioethanol fuels with 16.4% surplus in relation to world sample. Further, lignocellulosic biomass at large is the least influential feedstock with 11% deficit, followed by industrial waste, grass, hydrolysates, lignin, sugar feedstock residues, sugar feedstocks, and algal biomass with 2%–7% deficit each. It is also notable that a significant part of the papers for lignocellulosic biomass-based bioethanol fuels are reviews and short surveys.

Information about thematic research fronts for sample papers in bioethanol fuels is given in Table 102.6. As this table shows, the most prolific research front for this field is production of bioethanol fuels with 64% of the HCPs. Other prolific research fronts are utilization and evaluation of bioethanol evaluation with 28% and 8% of the HCPs, respectively.

Further, the most prolific research fronts for the first group are pretreatments of feedstocks with 60% of the HCPs. Other research fronts are biomass hydrolysis, bioethanol production, and hydrolysate fermentation with 12% of the HCPs each. However, there is no HCP for separation and distillation of bioethanol fuels.

For the third group, the only prolific research fronts are bioethanol-based biohydrogen fuels and direct utilization of bioethanol fuels in DEFCs with 16% and 12% of the HCPs, respectively, while there are no HCPs for utilization of bioethanol fuels in transport engines, and production of biochemicals from bioethanol fuels as well as bioethanol sensors.

On the other hand, bioethanol fuel production is the most influent research field with 10% surplus in relation to the global sample, followed by utilization of bioethanol fuels in DEFCs with 8% surplus.

Similarly, on individual terms, the most influential research fronts are production of bioethanol fuels for fuel cells from bioethanol fuels and utilization of bioethanol fuels in DEFCs with 12% and 5% surplus, respectively, in relation to the global sample, followed by biomass pretreatments with 2% surplus.

TABLE 102.6

The Most Prolific Thematic Research Fronts for Bioethanol Fuels: European Experience

No.	Research Fronts	N Paper (%) Review	N Paper (%) Sample	Surplus (%)
1	Bioethanol fuel production	64	74.1	−10.1
	Biomass pretreatments	60	58.5	1.5
	Biomass hydrolysis	12	29.8	−17.8
	Bioethanol production	12	21.5	−9.5
	Hydrolysate fermentation	12	17.9	−5.9
	Bioethanol fuel distillation	0	0.8	−0.8
2	Bioethanol fuel evaluation	8	6.2	1.8
3	Bioethanol fuel utilization	28	19.7	8.3
	Bioethanol-based biohydrogen fuels	16	4.3	11.7
	Bioethanol fuels in fuel cells	12	6.8	5.2
	Bioethanol fuels in engines	0	6.1	−6.1
	Bioethanol sensors	0	2.2	−2.2
	Bioethanol-based biochemicals	0	0.4	−0.4
	Sample size	25	775	

N Paper (%) review, Number of papers in the European sample of 25 reviewed papers; N paper (%) sample, Number of papers in the global sample of 775 papers.

Further, on individual terms, the least influential research fronts are biomass hydrolysis with 18% deficit, followed by bioethanol production, bioethanol fuels in engines, hydrolysate fermentation, and bioethanol sensors with 10%, 6%, 6%, and 2% deficit, separately.

102.4.2 Bioethanol Fuel Production: European Experience

The primary research fronts are pretreatments and hydrolysis of feedstocks and fermentation of hydrolysates with 10, 3, and 3 HCPs, respectively.

102.4.2.1 Feedstock Pretreatments: European Experience

The brief information about ten most-cited papers on pretreatments of feedstocks with at least 540 citations each is given below (Table 102.1).

In general, pretreatment of biomass is the first step in production of bioethanol fuels from biomass. It fractionates the biomass to its constituents such as cellulose, hemicellulose, and lignin. It thus makes these constituents ready for the second step of hydrolysis of biomass. In general, there are four types of pretreatments; chemical pretreatments such as acid or alkaline pretreatments, biological pretreatments such as enzymatic pretreatments, hydrothermal pretreatments such as steam explosion pretreatments, and mechanical pretreatments such as ultrasonic pretreatments.

Since the enzyme cost comprises a significant part of the overall cost of bioethanol production, research in the field of pretreatments focused on the development of cheap and efficient enzymes. For example, Floudas et al. (2012) mapped the detailed evolution of wood-degrading enzymes and noted that wood was highly resistant to decay, largely due to the presence of lignin and the only organisms capable of substantial lignin decay were white rot fungi in the agaricomycetes. Further, Martinez et al. (2008) performed genome sequencing and analysis of *Trichoderma reesei* and found that its genome encoded fewer cellulases and hemicellulases than any other sequenced fungus able to hydrolyze plant cell wall polysaccharides.

Similarly, Pandey and Pitman (2003) performed FTIR studies of changes in wood chemistry following decay by brown-rot and white-rot fungi and found that in wood decayed by *C. puteana* there was a progressive increase in lignin content relative to carbohydrate. Further, Quinlan et al. (2011) performed the enzymatic degradation of cellulose by a Cu metalloenzyme that exploited biomass components and showed that GH61 enzymes were a unique family of Cu-dependent oxidases. Finally, Martinez et al. (2004) sequenced 30-million base-pair genome of *Phanerochaete chrysosporium* strain RP78 using a whole genome shotgun approach and found that this genome had an impressive array of genes encoding secreted oxidases, peroxidases, and hydrolytic enzymes that co-operated in wood decay.

There is substantial focus on the use of steam explosion pretreatment in this field. For example, Li et al. (2007) explored the role of lignin depolymerization and repolymerization for delignification of aspen by steam explosion pretreatment and observed the competition between lignin depolymerization and repolymerization. Similarly, Sun et al. (2005) performed both steam explosion and alkaline hydroxide peroxide pretreatments for wheat straw biomass to recover cellulose and found that steam explosion pretreatment led to a significant loss in hemicelluloses and alkaline peroxide posttreatment resulted in substantial dissolution of lignin and an increase in cellulose crystallinity.

Mechanical pretreatment is often a must for pretreatments of biomass since it homogenizes the biomass in size. For example, Schwanninger et al. (2004) studied the effects of short-time vibratory ball milling on the shape of FTIR spectra of wood and cellulose and showed that this milling had a strong influence on the shape of FTIR spectra of wood and cellulose.

One of the most important chemical pretreatments has been the ILs with a wide range of types. For example, Sun et al. (2009) dissolved and partially delignified yellow pine and red oak in [C_2mim]OAc after mild milling and showed that this IL was a better solvent for wood than [C_4mim]

Cl. Similarly, Heinze et al. (2005) used a number of ILs for the dissolution of cellulose in a paper with 540 citations. They found that these ILs had the ability to dissolve cellulose with a degree of polymerization in the range from 290 to 1,200 to a very high concentration.

As it can be seen from these brief extracts, it is often helpful to a combine these four types of pretreatments in processing the biomass, such as using milling, steam explosion, acid, and enzymatic pretreatments altogether to fractionate a selected type of biomass in a large extent.

102.4.2.2 Hydrolysis of the Feedstocks: European Experience

There are three HCPs for hydrolysis of feedstock with at least 583 citations (Table 102.2).

Hydrolysis of biomass is generally the second step in the production of bioethanol fuels from biomass using the pretreated biomass. There are in general two types of hydrolysis: Enzymatic or acid hydrolysis. The first type of hydrolysis uses enzymes, whereas the second one uses acids to hydrolyze the pretreated biomass to produce sugars. These sugars are feedstocks for the third step of bioethanol production from biomass: fermentation of hydrolysates.

For example, Kilpelainen et al. (2007) dissolved wood in ILs and then digested the cellulose of regenerated wood to glucose by a cellulase enzymatic hydrolysis pretreatment. Further, Eriksson et al. (2002) explored the mechanism of surfactant effect in enzymatic hydrolysis of lignocellulose and found that non-ionic surfactants were the most effective. Finally, Harris et al. (2010) stimulated lignocellulosic biomass hydrolysis by proteins of GH61 and evaluated the structure and function of GH61 and showed that certain GH61 proteins lacked measurable hydrolytic activity by themselves but, in presence of various divalent metal ions, could significantly reduce the total protein loading required to hydrolyze lignocellulosic biomass.

102.4.2.3 Hydrolysate Fermentation: European Experience

There are three HCPs for production of bioethanol fuels (Table 102.3).

The third step in production of bioethanol fuels from biomass is fermentation of sugars produced in the first two steps of production process: pretreatments and hydrolysis of biomass. These sugars are fermented using a wide range of yeasts, usually engineered for an improved ethanol yield. The final step in production of bioethanol fuels from biomass is separation and distillation of bioethanol fuels from the fermentation broth. However, there is no HCP for this step.

In general, a wide range of fermentation inhibitors such as 5-HMF impede fermentation hydrolysates by yeasts decreasing ethanol yield. For example, Larsson et al. (1999) studied the effect of combined severity (SC) of dilute sulfuric acid hydrolysis of spruce on sugar yield and fermentability of hydrolysate by *S. cerevisiae* and found that the ethanol yield and volumetric productivity decreased with increasing concentrations of fermentation inhibitors such as acetic acid, formic acid, and levulinic acid.

Often the yeasts are engineered using a wide range of genomic tools to improve fermentation of hydrolysates to obtain bioethanol fuels with a high yield. For example, Alper et al. (2006) engineered yeast transcription machinery for improved ethanol tolerance of yeasts and bioethanol fuel production and observed that mutagenesis of transcription factor Spt15p and selection led to dominant mutations that conferred increased tolerance and more efficient glucose conversion to ethanol.

It is often that bioethanol fuels are produced in a biorefinery context together with a wide range of biochemicals, other biofuels, and derivatives such as biohydrogen fuels and biochemicals directly produced from bioethanol fuels. For example, Kaparaju et al. (2009) produced bioethanol, biohydrogen, and biogas fuels from wheat straw in a biorefinery concept and obtained 0.41 g ethanol/g glucose, while dark fermentation of hydrolysate produced 178.0 mL H_2/g sugars. In this way, the production cost of bioethanol fuels is largely reduced.

102.4.3 Evaluation and Utilization: European Experience

There are two and seven HCPs for the evaluation and utilization of bioethanol fuels with at least 546 citations, respectively (Table 102.4).

102.4.3.1 Bioethanol Fuel Evaluation: European Experience

There are two HCPs for evaluation of bioethanol fuels (Table 102.3).

A wide range of social science-based methods and tools are used to evaluate the energy balance and environmental and human impact of these bioethanol fuels, usually in comparison to crude oil-based gasoline or petrodiesel fuels.

One of the most widely used social science-based methods is techno-economics. For example, Hamelinck et al. (2005) evaluated the techno-economic performance of bioethanol production in short-, middle-, and long-term using hydrolysis and fermentation technologies as well as developing technologies and found that the available technology as of 2005 had about 35% efficiency (HHV) from biomass to ethanol. Similarly, Wingren et al. (2003) performed the techno-economic evaluation of bioethanol fuel production from softwood and found that ethanol production costs for SSF and SHF processes were 0.57 USD/L (68 USD/barrel) and 0.63 USD/L (75 USD/barrel), respectively and could be reduced to 0.42 USD/L (50 USD/barrel) for SSF process. Considering that crude oil prices rose to 150 USD in 2022, estimated production costs for bioethanol fuels are highly competitive in relation to gasoline and petrodiesel fuels.

The other widely used social science-based method is life cycle analysis (LCA) for evaluation of environmental, social, and human impact of production and utilization of bioethanol fuels (Von Blottnitz and Curran, 2007).

102.4.3.2 Bioethanol Fuel Utilization: European Experience

There are four and three HCPs for production of biohydrogen fuels from bioethanol fuels and utilization of bioethanol fuels in fuel cells with at least 546 citations each (Table 102.4).

In meeting the needs of consumers in line with the societal expectations for green energy and fuels, bioethanol fuels have increasingly been used in gasoline and diesel engines blending them with gasoline or petrodiesel fuels. They have also been used directly as a feedstock in DEFCs to obtain bioelectricity as a green alternative to natural gas- or coal-based power generation. Further, they have been used for the production of biohydrogen fuels for fuel cells as an alternative to direct use of bioethanol fuels in fuel cells. Finally, they have been used for the production of biochemicals such as bioethylene as a green alternative to crude oil-based chemicals. With the increasing utilization of bioethanol fuels in chemicals, fuel cells, cars, and biohydrogen fuels coupled with consumption of ethanol as an ingredient of alcoholic drinks by humans (Fadda and Rossetti, 1998), another research front has focused on the development and utilization of bioethanol sensors (Wan et al., 2004).

Although there four and three HCPs for production of biohydrogen fuels from bioethanol fuels and utilization of bioethanol fuels in fuel cells, there are no HCPs for research fronts of utilization of bioethanol fuels in transport engines and production of biochemicals from bioethanol fuels as well as bioethanol sensors.

102.4.3.2.1 Bioethanol Fuel-based Biohydrogen Fuels: European Experience

There are four HCPs for production of biohydrogen fuels from bioethanol fuels (Table 102.4).

Although biohydrogen fuels are directly produced through fermentation of biomass (Guo et al., 2010), it is also produced through, for example, reforming of bioethanol fuels for fuel cells.

One way to produce biohydrogen fuels is photocatalytic biohydrogen fuel production. For example, Murdoch et al. (2011) studied the photocatalytic biohydrogen fuel production from ethanol over Au/TiO_2 NPs and showed that Au NPs in the size range 3–30 nm on TiO_2 were very active in hydrogen production from ethanol.

Another way is autothermal reforming of bioethanol fuels. For example, Deluga et al. (2004) produced biohydrogen fuels by autothermal reforming of bioethanol fuels and converted ethanol and ethanol-water mixtures directly into hydrogen with nearly 100% selectivity and more than 95% conversion by catalytic partial oxidation.

However, steam reforming of bioethanol fuels is mostly used in biohydrogen fuel production from bioethanol fuels. For example, Liguras et al. (2003) produced biohydrogen fuels for fuel cells by steam reforming of bioethanol fuels over supported noble metal catalysts and found that for low-loaded catalysts, Rh was significantly more active and selective toward hydrogen formation compared to Ru, Pt, and Pd, which showed a similar behavior. Similarly, Fatsikostas and Verykios (2004) evaluated the reaction network of steam reforming of ethanol over supported Ni-based catalysts and found that ethanol interacted strongly with alumina on the surface of which it was dehydrated at low temperatures, and less strongly with lanthana on the surface of which it was both dehydrogenated and dehydrated.

As can been seen from these brief extracts, development of efficient catalysts with high biohydrogen selectivity is a must in biohydrogen fuel production from bioethanol fuels.

102.4.3.2.2 Direct Utilization of Bioethanol Fuels in Fuel Cells: European Experience

There are three HCPs for direct utilization of bioethanol fuels in fuel cells (Table 102.4).

In general, Pt-based electrocatalysts are used for electro-oxidation of bioethanol fuels for DEFCs. For example, Kowal et al. (2009) used ternary $Pt/Rh/SnO_2$ electrocatalysts for oxidizing ethanol to CO_2 and observed that this electrocatalyst effectively split the C–C bond in ethanol at room temperature in acid solutions, facilitating its oxidation at low potentials to CO_2. Similarly, Zhou et al. (2003) evaluated the effect of additives on ethanol electro-oxidation activity and DEFC performance of Pt-based anode catalysts and observed that presence of Sn, Ru, and W as additives enhanced the activity of Pt towards ethanol electro-oxidation. Finally, Lamy et al. (2004) developed PtSn electrocatalysts for DEFC and confirmed that overall electrocatalytic activity was greatly enhanced at low potentials by addition of Sn.

As it can been seen from these brief extracts, development and utilization of an efficient catalyst system is must for efficient electro-oxidation of bioethanol fuels in DEFCs.

102.5 CONCLUSION AND FUTURE RESEARCH

Brief information about the key research fronts covered by the 25 most-cited papers with at least 546 citations each is given under three primary headings: production, evaluation, and utilization of bioethanol fuels.

The usual characteristics of these HCPs for production of bioethanol fuels are that pretreatments and hydrolysis of feedstocks and fermentation of resulting hydrolysates are the primary processes for bioethanol fuel production from various feedstocks to improve ethanol yield. Additionally, the fourth process is separation and distillation of bioethanol fuels from fermentation broth.

The key findings on these research fronts should be read in the light of increasing public concerns about climate change, GHG emissions, and global warming as these concerns have been certainly behind the boom in the research on bioethanol fuels as an alternative to crude oil-based gasoline and petrodiesel fuels in the last two decades. The recent supply shocks caused by the COVID-19 pandemic and the Russian invasion of Ukraine also highlight the importance of production and utilization of bioethanol fuels as an alternative to crude oil-based gasoline and petrodiesel fuels in maintaining the energy security of Europe and the world at large.

As Table 102.5 shows, wood biomass, bioethanol fuels, cellulose, lignocellulosic biomass at large, and starch feedstock residues such as corn stover as the key feedstocks used for production and utilization of bioethanol fuels. Further, other minor feedstocks are lignin, sugars, and biomass. Additionally, a relatively large number of feedstocks are not represented in this sample.

Thus, the first five feedstocks dominate the HCPs in this field where only major waste biomass is starch feedstock residues such as corn stover or wheat straw. On the other hand, cellulose and lignin are the key constituents of the lignocellulosic biomass. Further bioethanol fuels are the feedstocks for production of biohydrogen fuels and for their utilization in transport engines and fuel cells. It is also notable that research on sugar and starch-based bioethanol fuels is not represented in this sample, reflecting their adverse effect on undermining food security.

Further, the most influential feedstocks are wood biomass and bioethanol fuels in relation to the global sample, while lignocellulosic biomass at large and, to a lesser extent, industrial waste such as glycerol, grass biomass, hydrolysates, lignin; sugar feedstock residues such as sugarcane bagasse, sugar feedstocks such as sugarcane and sugar beets; and algal biomass such as chlamydomonas are the least influential feedstocks. It is also notable that a significant part of the papers for lignocellulosic biomass-based bioethanol fuels are reviews and short surveys.

As Table 102.6 shows most prolific thematic research fronts for this field are production of bioethanol fuels and, to a lesser extent, utilization and evaluation of bioethanol evaluation. Further, the most prolific research fronts for the first group are pretreatments of feedstocks followed by biomass hydrolysis, bioethanol production, and hydrolysate fermentation. However, there is no HCP for separation and distillation of bioethanol fuels.

For the third group, the only prolific research fronts are bioethanol-based biohydrogen fuels and direct utilization of bioethanol fuels in DEFCs, while there are no HCPs for utilization of bioethanol fuels in transport engines and production of biochemicals from bioethanol fuels as well as bioethanol sensors.

On the other hand, bioethanol fuel production is the most influent research field in relation to the global sample, followed by utilization of bioethanol fuels in DEFCs.

Similarly, in individual terms, the most influential research fronts are production of biohydrogen fuels for fuel cells from bioethanol fuels and direct utilization of bioethanol fuels in DEFCs in relation to the global sample, followed by biomass pretreatments. Further, on individual terms, the least influential research fronts are biomass hydrolysis, followed by bioethanol production, bioethanol fuels in engines, hydrolysate fermentation, and bioethanol sensors.

These studies emphasize the importance of proper incentive structures for efficient production of bioethanol fuels in the light of North's institutional framework (North, 1991). In the light of the recent supply shocks caused primarily by COVID-19 pandemic and Russian invasion of Ukraine, it is expected that incentive structures such as public funding would be enhanced to increase the share of bioethanol fuels in the global fuel portfolio as a strong alternative to crude oil-based gasoline and petrodiesel fuels.

In this context, it is expected that the most prolific researchers, institutions, countries, funding bodies, and journals in this field would have the first-mover advantage (Lieberman and Montgomery, 1988) to benefit from such potential incentives. This is especially true for countries in the western and northern parts of Europe such as Spain, Sweden, and the UK.

It is expected the research would focus more on the algal, wood, and grass biomass-based bioethanol as well as lignocellulosic biomass-based bioethanol fuels such as agricultural residue- and waste biomass-based bioethanol fuels at the expense of starch and sugar-based bioethanol fuels due to large societal concerns about food security in future.

Similarly, it is expected that research would focus more on pretreatments and hydrolysis of the feedstocks for production of bioethanol fuels and utilization and evaluation of bioethanol fuels in future.

It is recommended that further review studies are performed for the primary research fronts of bioethanol fuels.

ACKNOWLEDGMENTS

Contribution of the highly cited researchers in the field of European experience of bioethanol fuels has been gratefully acknowledged.

REFERENCES

Alper, H., J. Moxley, E. Nevoigt, G. R. Fink and G. Stephanopoulos. 2006. Engineering yeast transcription machinery for improved ethanol tolerance and production. *Science* 314:1565–1568.

Alvira, P., E. Tomas-Pejo, M. Ballesteros and M. J. Negro. 2010. Pretreatment technologies for an efficient bioethanol production process based on enzymatic hydrolysis: A review. *Bioresource Technology* 101:4851–4861.

Angelici, C., B. M. Weckhuysen and P. C. A. Bruijnincx. 2013. Chemocatalytic conversion of ethanol into butadiene and other bulk chemicals. *ChemSusChem* 6:1595–1614.

Antolini, E. 2007. Catalysts for direct ethanol fuel cells. *Journal of Power Sources* 170:1–12.

Antolini, E. 2009. Palladium in fuel cell catalysis. *Energy and Environmental Science* 2:915–931.

Bai, F. W., W. A. Anderson and M. Moo-Young. 2008. Ethanol fermentation technologies from sugar and starch feedstocks. *Biotechnology Advances* 26:89–105.

Bentivoglio, D., A. Finco and M. R. P. Bacchi. 2016. Interdependencies between biofuel, fuel and food prices: The case of the Brazilian ethanol market. *Energies* 9:464.

Binod, P., R. Sindhu and R. R. Singhania, et al. 2010. Bioethanol production from rice straw: An overview. *Bioresource Technology* 101:4767–4774.

Bourbonnais, R. and M. G. Paice. 1990. Oxidation of non-phenolic substrates. An expanded role for laccase in lignin biodegradation. *FEBS Letters* 267:99–102.

Burnham, J. F. 2006. Scopus database: A review. *Biomedical Digital Libraries* 3:1–8.

Cardona, C. A., J. A. Quintero and I. C. Paz. 2010. Production of bioethanol from sugarcane bagasse: Status and perspectives. *Bioresource Technology* 101:4754–4766.

Deluga, G. A., J. R. Salge, L. D. Schmidt and X. E. Verykios. 2004. Renewable hydrogen from ethanol by autothermal reforming. *Science* 303:993–997.

Duff, S. J. B. and W. D. Murray. 1996. Bioconversion of forest products industry waste cellulosics to fuel ethanol: A review. *Bioresource Technology* 55:1–33.

Elobeid, A. and C. Hart. 2007. Ethanol expansion in the food versus fuel debate: How will developing countries fare? *Journal of Agricultural & Food Industrial Organization* 5:7.

Eriksson, T., J. Borjesson and F. Tjerneld. 2002. Mechanism of surfactant effect in enzymatic hydrolysis of lignocellulose. *Enzyme and Microbial Technology* 31:353–364.

Fadda, F. and Z. L. Rossetti. 1998. Chronic ethanol consumption: From neuroadaptation to neurodegeneration. *Progress in Neurobiology* 56:385–431.

Fatsikostas, A. N. and X. E. Verykios. 2004. Reaction network of steam reforming of ethanol over Ni-based catalysts. *Journal of Catalysis* 225:439–452.

Fernando, S., S. Adhikari, C. Chandrapal and M. Murali. 2006. Biorefineries: Current status, challenges, and future direction. *Energy & Fuels* 20:1727–1737.

Floudas, D., M. Binder and R. Riley, et al. 2012. The paleozoic origin of enzymatic lignin decomposition reconstructed from 31 fungal genomes. *Science* 336:1715–1719.

Galbe, M. and G. Zacchi. 2002. A review of the production of ethanol from softwood. *Applied Microbiology and Biotechnology* 59:618–628.

Guo, X. M., E. Trably, E. Latrille, H. Carrere and J. P. Steyer. 2010. Hydrogen production from agricultural waste by dark fermentation: A review. *International Journal of Hydrogen Energy* 35:10660–10673.

Hahn-Hagerdal, B., M. Galbe, M. F. Gorwa-Grauslund, G. Liden and G. Zacchi. 2006. Bio-ethanol - The fuel of tomorrow from the residues of today. *Trends in Biotechnology* 24:549–556.

Hamelinck, C. N., G. van Hooijdonk and A. P. C. Faaij. 2005. Ethanol from lignocellulosic biomass: Techno-economic performance in short-, middle- and long-term. *Biomass and Bioenergy*, 28:384–410.

Hansen, A. C., Q. Zhang and P. W. L. Lyne. 2005. Ethanol-diesel fuel blends - A review. *Bioresource Technology* 96:277–285.

Harris, P. V., D. Welner and K. C. McFarland, et al. 2010. Stimulation of lignocellulosic biomass hydrolysis by proteins of glycoside hydrolase family 61: Structure and function of a large, enigmatic family. *Biochemistry* 49:3305–3316.

Heinze, T., K. Schwikal and S. Barthel. 2005. Ionic liquids as reaction medium in cellulose functionalization. *Macromolecular Bioscience* 5:520–525.

Hendriks, A. T. W. M and G. Zeeman. 2009. Pretreatments to enhance the digestibility of lignocellulosic biomass. *Bioresource Technology* 100:10–18.

Hill, J., E. Nelson, D. Tilman, S. Polasky and D. Tiffany. 2006. Environmental, economic, and energetic costs and benefits of biodiesel and ethanol biofuels. *Proceedings of the National Academy of Sciences of the United States of America* 103:11206–11210.

Hill, J., S. Polasky and E. Nelson, et al. 2009. Climate change and health costs of air emissions from biofuels and gasoline. *Proceedings of the National Academy of Sciences of the United States of America* 106:2077–2082.

Hsieh, W. D., R. H. Chen, T. L. Wu and T. H. Lin. 2002. Engine performance and pollutant emission of an SI engine using ethanol-gasoline blended fuels. *Atmospheric Environment* 36:403–410.

Huang, H. J., S. Ramaswamy, U. W. Tschirner and B. V. Ramarao. 2008. A review of separation technologies in current and future biorefineries. *Separation and Purification Technology* 62:1–21.

John, R. P., G. S. Anisha, K. M. Nampoothiri and A. Pandey. 2011. Micro and macroalgal biomass: A renewable source for bioethanol. *Bioresource Technology* 102:186–193.

Jonsson, L. J. and C. Martin. 2016. Pretreatment of lignocellulose: Formation of inhibitory by-products and strategies for minimizing their effects. *Bioresource Technology* 199:103–112.

Kaparaju, P., M. Serrano, A. B. Thomsen, P. Kongjan and I. Angelidaki. 2009. Bioethanol, biohydrogen and biogas production from wheat straw in a biorefinery concept. *Bioresource Technology* 100:2562–2568.

Kilpelainen, I., X. Xie and A. King, et al. 2007. Dissolution of wood in ionic liquids. *Journal of Agricultural and Food Chemistry* 55:9142–9148.

Konur, O. 2000. Creating enforceable civil rights for disabled students in higher education: An institutional theory perspective. *Disability & Society* 15:1041–1063.

Konur, O. 2002a. Access to nursing education by disabled students: Rights and duties of nursing programs. *Nurse Education Today* 22:364–374.

Konur, O. 2002b. Assessment of disabled students in higher education: Current public policy issues. *Assessment and Evaluation in Higher Education* 27:131–152.

Konur, O. 2002c. Access to employment by disabled people in the UK: Is the Disability Discrimination Act working? *International Journal of Discrimination and the Law* 5:247–279.

Konur, O. 2006a. Participation of children with dyslexia in compulsory education: Current public policy issues. *Dyslexia* 12:51–67.

Konur, O. 2006b. Teaching disabled students in higher education. *Teaching in Higher Education* 11:351–363.

Konur, O. 2007a. A judicial outcome analysis of the *Disability Discrimination Act*: A windfall for the employers? *Disability & Society* 22:187–204.

Konur, O. 2007b. Computer-assisted teaching and assessment of disabled students in higher education: The interface between academic standards and disability rights. *Journal of Computer Assisted Learning* 23:207–219.

Konur, O. 2012. The evaluation of the research on the bioethanol: A scientometric approach. *Energy Education Science and Technology Part A: Energy Science and Research* 28:1051–1064.

Konur, O. 2015. Current state of research on algal bioethanol. In *Marine Bioenergy: Trends and Developments*, Ed. S. K. Kim and C. G. Lee, pp. 217–244. Boca Raton, FL: CRC Press.

Konur, O. 2019. Cyanobacterial bioenergy and biofuels science and technology: A scientometric overview. In *Cyanobacteria: From Basic Science to Applications*, Ed. A. K. Mishra, D. N. Tiwari and A. N. Rai, pp. 419–442. Amsterdam: Elsevier

Konur, O. 2020. The scientometric analysis of the research on the bioethanol production from green macroalgae. In *Handbook of Algal Science, Technology and Medicine*, Ed. O. Konur, pp. 385–401. London: Academic Press.

Konur, O. 2023a. Bioethanol fuels: Scientometric study. In *Bioethanol Fuel Production Processes. I: Biomass Pretreatments. Handbook of Bioethanol Fuels Volume 1*, Ed. O. Konur, pp. 47–76. Boca Raton, FL: CRC Press.

Konur, O. 2023b. Bioethanol fuels: Review. In *Bioethanol Fuel Production Processes. I: Biomass Pretreatments. Handbook of Bioethanol Fuels Volume 1*, Ed. O. Konur, pp. 77–98. Boca Raton, FL: CRC Press.

Konur, O. 2023c. European experience of bioethanol fuels: Scientometric study. In *Evaluation and Utilization of Bioethanol Fuels. II.: Biohydrogen Fuels, Fuel Cells, Biochemicals, and Country Experiences. Handbook of Bioethanol Fuels Volume 6*, Ed. O. Konur, pp. 159–188. Boca Raton, FL: CRC Press.

Kowal, A., M. Li and M. Shao, et al. 2009. Ternary Pt/Rh/SnO_2 electrocatalysts for oxidizing ethanol to CO_2. *Nature Materials* 8:325–330.

Lamy, C., S. Rousseau, E. M. Belgsir, C. Coutanceau and J. M. Leger. 2004. Recent progress in the direct ethanol fuel cell: Development of new platinum-tin electrocatalysts. *Electrochimica Acta* 49:3901–3908.

Larsson, S., E. Palmqvist and B. Hahn-Hagerdal, et al. 1999. The generation of fermentation inhibitors during dilute acid hydrolysis of softwood. *Enzyme and Microbial Technology* 24:151–159.

Li, J., G. Henriksson and G. Gellerstedt. 2007. Lignin depolymerization/repolymerization and its critical role for delignification of aspen wood by steam explosion. *Bioresource Technology* 98:3061–3068.

Lieberman, M. B. and D. B. Montgomery. 1988. First-mover advantages. *Strategic Management Journal* 9:41–58.

Liguras, D. K., D. I. Kondarides and V. E. Verykios. 2003. Production of hydrogen for fuel cells by steam reforming of ethanol over supported noble metal catalysts. *Applied Catalysis B: Environmental* 43:345–354.

Limayem, A. and S. C. Ricke. 2012. Lignocellulosic biomass for bioethanol production: Current perspectives, potential issues and future prospects. *Progress in Energy and Combustion Science* 38:449–467.

Lin, Y. and S. Tanaka. 2006. Ethanol fermentation from biomass resources: Current state and prospects. *Applied Microbiology and Biotechnology* 69:627–642.

Ma, X., L. Sun and C. Song. 2002. A new approach to deep desulfurization of gasoline, diesel fuel and jet fuel by selective adsorption for ultra-clean fuels and for fuel cell applications. *Catalysis Today* 77:107–116.

Martinez, D., L. F. Larrondo and N. Putnam, et al. 2004. Genome sequence of the lignocellulose degrading fungus *Phanerochaete chrysosporium* strain RP78. *Nature Biotechnology* 22:695–700.

Martinez, D., R. M. Berka and B. Henrissat, et al. 2008. Genome sequencing and analysis of the biomass-degrading fungus *Trichoderma reesei* (syn. *Hypocrea jecorina*). *Nature Biotechnology* 26:553–560.

Morschbacker, A. 2009. Bio-ethanol based ethylene. *Polymer Reviews* 49:79–84.

Mosier, N., C. Wyman and B. Dale, et al. 2005. Features of promising technologies for pretreatment of lignocellulosic biomass. *Bioresource Technology* 96:673–686.

Murdoch, M., G. I. N. Waterhouse and M. A. Nadeem, et al. 2011. The effect of gold loading and particle size on photocatalytic hydrogen production from ethanol over Au/TiO_2 nanoparticles. *Nature Chemistry* 3:489–492.

Najafi, G., B. Ghobadian and T. Tavakoli, et al. 2009. Performance and exhaust emissions of a gasoline engine with ethanol blended gasoline fuels using artificial neural network. *Applied Energy* 86:630–639.

Newman, P. W. G. and J. R. Kenworthy. 1989. Gasoline consumption and cities: A comparison of U.S. cities with a global survey. *Journal of the American Planning Association* 55:24–37.

North, D. C. 1991. Institutions. *Journal of Economic Perspectives* 5:97–112.

Olsson, L. and B. Hahn-Hagerdal. 1996. Fermentation of lignocellulosic hydrolysates for ethanol production. *Enzyme and Microbial Technology* 18:312–331.

Palmqvist, E. and B. Hahn-Hagerdal. 2000. Fermentation of lignocellulosic hydrolysates. I: Inhibition and detoxification. *Bioresource Technology* 74:17–24.

Pandey, K. K. and A. J. Pitman. 2003. FTIR studies of the changes in wood chemistry following decay by brown-rot and white-rot fungi. *International Biodeterioration and Biodegradation* 52:151–160.

Pimentel, D. and T. W. Patzek. 2005. Ethanol production using corn, switchgrass, and wood; biodiesel production using soybean and sunflower. *Natural Resources Research* 14:65–76.

Pinkert, A., K. N. Marsh, S. Pang and M. P. Staiger. 2009. Ionic liquids and their interaction with cellulose. *Chemical Reviews* 109:6712–6728.

Quinlan, R. J., M. D. Sweeney and L. Lo Leggio, et al. 2011. Insights into the oxidative degradation of cellulose by a copper metalloenzyme that exploits biomass components. *Proceedings of the National Academy of Sciences of the United States of America* 108:15079–15084.

Ravindran, R. and A. K. Jaiswal. 2016. A comprehensive review on pre-treatment strategy for lignocellulosic food industry waste: Challenges and opportunities. *Bioresource Technology* 199:92–102.

Sanchez, O. J. and C. A. Cardona. 2008. Trends in biotechnological production of fuel ethanol from different feedstocks. *Bioresource Technology* 99:5270–5295.

Sano, T., H. Yanagishita, Y. Kiyozumi, F. Mizukami and K. Haraya. 1994. Separation of ethanol/water mixture by silicalite membrane on pervaporation. *Journal of Membrane Science* 95:221–228.

Schwanninger, M., J. C. Rodrigues, H. Pereira and B. Hinterstoisser. 2004. Effects of short-time vibratory ball milling on the shape of FT-IR spectra of wood and cellulose. *Vibrational Spectroscopy* 36:23–40.

Sun, N., M. Rahman and Y. Qin, et al. 2009. Complete dissolution and partial delignification of wood in the ionic liquid 1-ethyl-3-methylimidazolium acetate. *Green Chemistry* 11:646–655.

Sun, X. F., F. Xu, R. C. Sun, P. Fowler and M. S. Baird. 2005. Characteristics of degraded cellulose obtained from steam-exploded wheat straw. *Carbohydrate Research* 340:97–106.

Sun, Y. and J. Cheng. 2002. Hydrolysis of lignocellulosic materials for ethanol production: A review. *Bioresource Technology* 83:1–11.

Taherzadeh, M. J. and K. Karimi. 2007. Enzyme-based hydrolysis processes for ethanol from lignocellulosic materials: A review. *Bioresources* 2:707–738.

Taherzadeh, M. J. and K. Karimi. 2008. Pretreatment of lignocellulosic wastes to improve ethanol and biogas production: A review. *International Journal of Molecular Sciences* 9:1621–1651.

Talebnia, F., D. Karakashev and I. Angelidaki. 2010. Production of bioethanol from wheat straw: An overview on pretreatment, hydrolysis and fermentation. *Bioresource Technology* 101:4744–4753.

Vane, L. M. 2005. A review of pervaporation for product recovery from biomass fermentation processes. *Journal of Chemical Technology and Biotechnology* 80:603–629.

Von Blottnitz, H. and M. A. Curran. 2007. A review of assessments conducted on bio-ethanol as a transportation fuel from a net energy, greenhouse gas, and environmental life cycle perspective. *Journal of Cleaner Production* 15:607–619.

Wan, Q., Q. H. Li and Y. J. Chen, et al. 2004. Fabrication and ethanol sensing characteristics of ZnO nanowire gas sensors. *Applied Physics Letters* 84:3654–3656.

Wingren, A., M. Galbe and G. Zacchi. 2003. Techno-economic evaluation of producing ethanol from softwood: Comparison of SSF and SHF and identification of bottlenecks. *Biotechnology Progress* 19:1109–1117.

Zhang, Y. H. P. and L. R. Lynd. 2004. Toward an aggregated understanding of enzymatic hydrolysis of cellulose: Noncomplexed cellulase systems. *Biotechnology and Bioengineering* 88:797–824.

Zhou, W., Z. Zhou and S. Song, et al. 2003. Pt based anode catalysts for direct ethanol fuel cells. *Applied Catalysis B: Environmental* 46:273–285.

Zhu, J. Y. and X. J. Pan. 2010. Woody biomass pretreatment for cellulosic ethanol production: Technology and energy consumption evaluation. *Bioresource Technology* 101:4992–5002.

Part 30

Bioethanol Fuel-based Biohydrogen Fuels

103 Bioethanol Fuel-based Biohydrogen Fuels *Scientometric Study*

Ozcan Konur
(Formerly) Ankara Yildirim Beyazit University

103.1 INTRODUCTION

Crude oil-based gasoline fuels (Ma et al., 2002; Newman and Kenworthy, 1989) have been widely used in the transportation sector since the 1920s. However, there have been great public concerns over the adverse environmental and human impact of these fuels (Hill et al., 2006, 2009). Hence, biomass-based bioethanol fuels (Hill et al., 2006; Konur, 2012e, 2015, 2019a, 2020a) have increasingly been used in blending gasoline fuels (Hsieh et al., 2002; Najafi et al., 2009) and in fuel cells (Antolini, 2007; Kamarudin et al., 2013).

In the meantime, research in nanomaterials and nanotechnology has intensified in recent years to become a major research field in scientific research with over one and a half million published papers (Geim, 2009; Geim and Novoselov, 2007). In this context, a large number of nanomaterials have been developed nearly for every research field. These materials offer an innovative way to increase the efficiency in the production and utilization of bioethanol fuels as in other scientific fields (Konur 2016a,b,c,d,e,f, 2017a,b,c,d,e, 2019b, 2021a,b).

Research in the field of fuel cells has also intensified in recent years with over 50,000 papers in three primary research streams of high-temperature solid oxide fuel cells (SOFCs) (Adler, 2004), low-temperature polymer electrode membrane fuel cells (PEMFCs) (Borup et al., 2007), and microbial fuel cells (MFCs) (Logan, 2009). Research in direct ethanol fuel cells (DEFCs) has been a stream of the research on the low-temperature PEMFCs. The primary focus of the research on fuel cells has been the optimization of the operating conditions and the development of catalysts to maximize fuel cell performance.

Furthermore, research in the use of both biohydrogen fuels and conventional hydrogen fuels in fuel cells has intensified in recent years in the primary research fronts of water-based photocatalytic hydrogen production (Chen et al., 2010), water-based electrocatalytic hydrogen production (Greeley et al., 2006), reforming of bioethanol fuels and other alcoholic fuels for the production of biohydrogen fuels (Haryanto et al., 2005; Ni et al., 2007), biohydrogen fuel production (Das and Veziroglu, 2001), and hydrogen storage (Orimo et al., 2007). It is also notable that the research in the fields of hydrogen production and storage has focused on nanomaterial development for hydrogen production, storage, and use.

Thus, research in the field of production of biohydrogen fuels from bioethanol fuels has progressed in the research fronts of biohydrogen production by the steam reforming of bioethanol (Haryanto et al., 2005; Ni et al., 2007) and to a lesser extent in photocatalytic biohydrogen production (Bamwenda et al., 1995; Gombac et al., 2010), biohydrogen production by oxidative steam reforming of bioethanol (Kugai et al., 2006; Velu et al., 2005), biohydrogen production by the partial oxidation of bioethanol (de Lima et al., 2008; Rabenstein and Hacker, 2008), autothermal reforming of bioethanol (Deluga et al., 2004; Ni et al., 2007), and biohydrogen production from bioethanol in general (Akande et al., 2005; Mattos et al., 2012).

DOI: 10.1201/9781003226574-137

The focus of these 100 most-cited papers is the development of materials for the most efficient processes for the reforming of bioethanol fuels for the production of biohydrogen fuels for fuel cells.

However, it is essential to develop efficient incentive structures (North, 1991) for the primary stakeholders to enhance the research in this field (Konur, 2000, 2002a,b,c, 2006a,b, 2007a,b).

Scientometric analysis has been used in this context to inform the primary stakeholders about the current state of the research in a selected research field (Garfield, 1955; Konur, 2011, 2012a,b,c,d,e,f,g,h,i, 2015, 2016a,b,c,d,e,f, 2017a,b,c,d,e, 2018b, 2019a,b, 2020a).

There has not been any scientometric study on bioethanol fuel-based biohydrogen fuels. Thus, this chapter presents a scientometric study of the research in bioethanol fuel-based biohydrogen fuels. It examines the scientometric characteristics of both the sample and population data presenting scientometric characteristics of both datasets in the order of documents, authors, publication years, institutions, funding bodies, source titles, countries, Scopus subject categories, keywords, and research fronts.

103.2 MATERIALS AND METHODS

The search for this study was carried out using the Scopus database (Burnham, 2006) in September 2021.

As the first step for the search of the relevant literature, keywords were selected using the 200 most-cited papers. The selected keyword list was optimized to obtain a representative sample of papers for this research field. This keyword list was provided in the appendix for future replication studies. Additionally, the information about the most-used keywords was given in Section 103.3.9 to highlight the key research fronts in Section 103.3.10.

As the second step, two sets of data were used for this study. First, a population sample of over 1,600 papers was used to examine the scientometric characteristics of the population data. Second, a sample of 100 most-cited papers was used to examine the scientometric characteristics of these citation classics with over 125 citations each.

Scientometric characteristics of both these sample and population datasets were presented in the order of documents, authors, publication years, institutions, funding bodies, source titles, countries, Scopus subject categories, keywords, and research fronts.

Lastly, key scientometric findings for both datasets were discussed to highlight the research landscape for bioethanol-based biohydrogen fuels. Additionally, a number of brief conclusions were drawn and a number of relevant recommendations were made to enhance the future research landscape.

103.3 RESULTS

103.3.1 The Most Prolific Documents in Bioethanol-based Biohydrogen Fuels

Information on the types of documents for both datasets is given in Table 103.1. The articles dominate both the sample and population datasets. Review papers and conference papers have a slight surplus and deficit, respectively.

It is also interesting to note that all of the papers in the sample dataset were published in journals, while only 97.1% of the papers were published in journals for the population dataset. Furthermore, 1.3% and 1.6% of the population papers were published in book series and books, respectively.

103.3.2 The Most Prolific Authors on Bioethanol-based Biohydrogen Fuels

Information about the 34 most prolific authors with at least three sample papers and five population papers each is given in Table 103.2.

TABLE 103.1
Documents in Bioethanol-based Biohydrogen Fuels

Documents	Sample Dataset (%)	Population Dataset (%)	Surplus (%)
Article	91	88.9	2.1
Review	6	1.6	4.4
Conference paper	2	7.1	–5.1
Short survey	2	0.2	1.8
Book chapter	0	1.8	–1.8
Note	0	0.3	–0.3
Book	0	0.0	0
Editorial	0	0.0	0
Letter	0	0.0	0
Sample size	100	1626	

TABLE 103.2
The Most Prolific Authors on Bioethanol-based Biohydrogen Fuels

No.	Author Name	Author Code	Sample Papers	Population Papers	Institution	Country	Res. Front
1	Cavallaro, Stefano	7006497260	9	18	Univ. Messina	Italy	SR
2	Freni, Salvatore	7004444222	8	18	CNR	Italy	SR
3	Noronha, Fabio B.	7004514118	6	42	Natl. Technol. Inst.	Brazil	SR, OSR, PO
4	Frusteri, Francesco	7003418964	6	10	Univ. Messina	Italy	SR
5	Llorca, Jordi	26039349400	5	54	Univ. Barcelona	Spain	SR
6	Mattos, Lisiane V.*	7005199691	5	34	Natl. Technol. Inst.	Brazil	SR, OSR, PO
7	Laborde, Miguel	7003342826	5	22	Ciudad Univ.	Argentina	SR
8	Homs, Narcis	7004247231	5	21	Univ. Barcelona	Spain	SR
9	De la Piscina, Pilar R.*	56819227100	5	11	Univ. Barcelona	Spain	SR
10	Idriss, Hicham	7006768868	5	10	Univ. Auckland	New Zealand	PBR
11	Chiodo, Vitaliano	6603101616	5	9	CNR	Italy	SR
12	Fierro, Jose-Luis G.	7202608150	4	19	CSIC	Spain	SR
13	Jacobs, Gary	7203024020	4	16	Univ. Kentucky	USA	SR, OSR, PO
14	Davis, Burtron H.	36043712200	4	13	Univ. Kentucky	USA	SR, OSR, PO
15	Mirodatos, Claude	57105913200	4	12	Univ. Lyon	France	SR
16	Verykios, Xenophon E.	35551305100	4	12	Univ. Patras	Greece	SR, AR
17	Spadore, Lorenzo	6602740818	4	5	CNR	Italy	SR
18	Cai, Weijie	22033375300	3	27	Chinese Acad. Sci.	China	SR
19	Assabumrungrat, Suttichai	7004487886	3	19	Chulalongkorn Univ.	Thailand	SR
20	Gong, Jinlong	55218059900	3	16	Tianjin Univ.	China	SR
21	Shen, 2	7403601371	3	15	Chinese Acad. Sci.	China	SR
22	Fornasiero, Paolo	7003709305	3	14	Univ. Trieste	Italy	PBR

(Continued)

TABLE 103.2 (*Continued*)
The Most Prolific Authors on Bioethanol-based Biohydrogen Fuels

No.	Author Name	Author Code	Sample Papers	Population Papers	Institution	Country	Res. Front
23	Montini, Tiziano	57208479381	3	13	Univ. Trieste	Italy	PBR
24	Assaf, Jose M.	6603833450	3	11	Univ. Sao Paulo	Thailand	SR
25	Calles, Jose A.	16426457400	3	11	Univ. Rey Juan Carlos	Spain	SR
26	Carrero, Alicia*	16425302700	3	11	Univ. Rey Juan Carlos	Spain	SR
27	Vizcaino, Arturo J.	16426457400	3	11	Univ. Rey Juan Carlos	Spain	SR
28	Navarro, Rufino M.	16234352800	3	11	CSIC	Spain	SR
29	De Lima, Sania M.*	24365720000	3	8	Natl. Technol. Inst.	Brazil	SR, OSR, PO
30	Fierro, Vanessa*	57200997518 57105928700	3	8	CNRS	France	SR
31	Gombac, Valentina*	6506827249	3	7	Univ. Trieste	Italy	PBP
32	Sanchez-Sanchez, Maricruz*	44361828200	3	7	CSIC	Spain	SR
33	Bonura, Giusepppe	6506398547	3	6	Univ. Messina	Italy	SR
34	Kondarides, Dimitris I.	6603982077	3	6	Univ. Patras	Greece	SR

*, Female; Author code, the unique code given by Scopus to the authors; AR, Autothermal reforming; OSR, Oxidative steam reforming; PBR, Photocatalytic biohydrogen production; PO, Partial oxidation; Population papers, the number of papers authored in the population dataset; Sample papers, the number of papers authored in the sample dataset; SR, Steam reforming.

The most prolific authors are Stefano Callaro and Salvatore Freni with nine and eight sample papers, respectively. Fabio B. Noronha and Francesco Frusteri follow these top authors with six sample papers each.

The most prolific institution for the sample dataset is the Superior Council of Scientific Investigations (CSIC) with four authors. The National Institute of Technology, University of Barcelona, the University of Messina, University of Rey Juan Carlos, and University of Trieste follow this top institution with three authors each. In total 17 institutions house these prolific authors.

The most prolific countries for the sample dataset are Spain and Italy with nine authors each. Brazil and China follow these top countries with three authors each. In total, ten countries house these authors.

The most prolific research front is the steam reforming of bioethanol for biohydrogen production with 30 authors. Other prolific research fronts are oxidative steam reforming and partial oxidation of bioethanol fuels for biohydrogen fuel production with five authors each. Additionally, one author each studies the autothermal reforming of bioethanol fuels and photocatalytic biohydrogen production.

On the other hand, there is a significant gender deficit (Beaudry and Lariviere, 2016) for the sample dataset as surprisingly only seven of these top researchers are female with a representation rate of 21%.

103.3.3 The Most Prolific Research Output by Years in Bioethanol-based Biohydrogen Fuels

Information about papers published between 1970 and 2021 is given in Figure 103.1. This figure clearly shows that the bulk of the research papers in the population dataset were published primarily in the 2010s with 64.6% of the population dataset. This was followed by the early 2020s and 2000s with 10.8% and 22.8% of the population papers, respectively. The publication rates for the pre-2000s

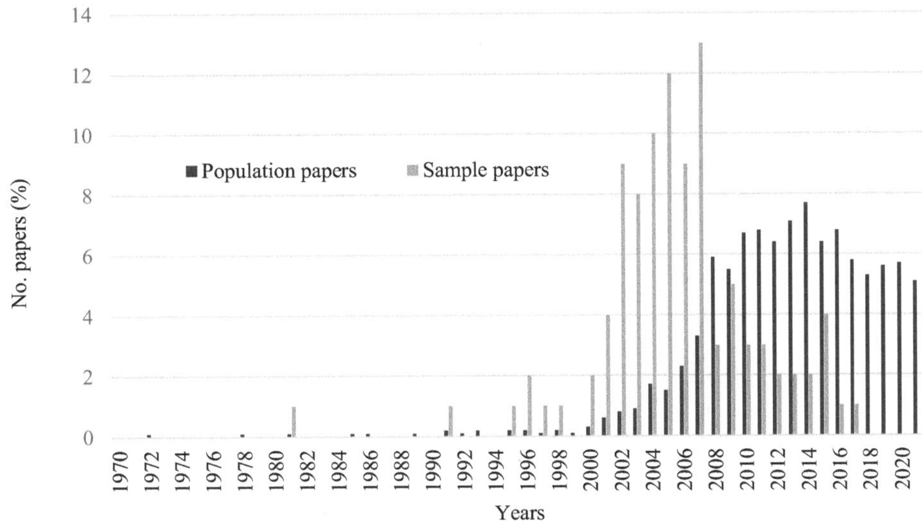

FIGURE 103.1 The research output by years regarding bioethanol-based biohydrogen fuels.

were negligible. It appears that the research output for the population papers rose between 2004 and 2010 and after 2010 it persistently declined to around 7% in 2021.

Similarly, the bulk of the research papers in the sample dataset were published in the 2000s and 2010s with 75% and 18% of the sample dataset, respectively. The publication rates for the 1990s, 1980s, and 1970s were 6%, 4%, and 2% of the sample papers, respectively.

The most prolific publication years for the population dataset were after 2007 with at least 5.1% each of the dataset. Similarly, 61% of the sample papers were published between 2002 and 2007.

103.3.4 The Most Prolific Institutions on Bioethanol-based Biohydrogen Fuels

Information about the 20 most prolific institutions publishing papers on bioethanol-based biohydrogen fuels with at least three sample papers and 0.7% of the population papers each is given in Table 103.3.

The most prolific institutions are the University of Messina and National Research Council (CNR) of Italy with nine papers each. The National Technology Institute, the University of Lyon, University of Barcelona, University of Auckland, CSIC, and University of Patras follow these top institutions with at least five sample papers each.

The top countries for these most prolific institutions are Italy and Brazil with three institutions each. Next, Argentina, China, France, Spain, and Thailand follow these top countries with two institutions each. In total, 11 countries house these top institutions.

On the other hand, the institutions with the most impact are the University of Messina and CNR of Italy with a 7.8% and 5.9% surplus, respectively. The University of Lyon, University of Auckland, and University of Patras follow these top institutions with at least a 3.6% surplus each. Similarly, the institutions with the lowest impact are Tianjin University and CONICET with at least a 1.1% deficit each.

103.3.5 The Most Prolific Funding Bodies on Bioethanol-based Biohydrogen Fuels

Information about the 15 most prolific funding bodies funding with at least two sample papers and 0.7% of the population papers each is given in Table 103.4.

TABLE 103.3
The Most Prolific Institutions on Bioethanol-based Biohydrogen Fuels

No.	Institutions	Country	Sample Papers (%)	Population Papers (%)	Surplus (%)
1	Natl. Res. Counc. (CNR)	Italy	9	3.1	5.9
2	Univ. Messina	Italy	9	1.2	7.8
3	Natl. Technol. Inst.	Brazil	6	2.8	3.2
4	Univ. Lyon	France	6	1.6	4.4
5	Superior Counc. Sci. Inv. (CSIC)	Spain	5	2.0	3
6	Univ. Barcelona	Spain	5	1.7	3.3
7	Univ. Patras	Greece	5	1.4	3.6
8	Univ. Auckland	New Zealand	5	0.7	4.3
9	Chinese Acad. Sci.	China	4	3.0	1
10	Univ. Buenos Aires	Argentina	4	1.3	2.7
11	Fed. Univ. Sao Carlos	Brazil	4	1.2	2.8
12	Univ. Kentucky	USA	4	0.9	3.1
13	Tianjin Univ.	China	3	4.1	−1.1
14	Natl. Counc. Sci. Tech. Res. (CONICET)	Argentina	3	2.6	0.4
15	Sci. Res. Natl. Res. (CNRS)	France	3	1.2	1.8
16	Chulalongkorn Univ.	Thailand	3	1.2	1.8
17	Univ. Trieste	Italy	3	0.9	2.1
18	King Mongkut's Univ. Technol.	Thailand	3	0.8	2.2
19	Univ. Rey Juan Carlos	Brazil	3	0.7	2.3
20	Natl. Inst. Adv. Ind. Sci. Technol.	Japan	3	0.7	2.3

TABLE 103.4
The Most Prolific Funding Bodies on Bioethanol-based Biohydrogen Fuels

No.	Funding bodies	Country	Sample Paper No. (%)	Population Paper No. (%)	Surplus (%)
1	National Natural Science Foundation of China	China	5	10.6	−5.6
2	Catalonia Government	Spain	5	1.0	4
3	National Council for Scientific and Technological Development	Brazil	4	4.1	−0.1
4	European Commission	EU	4	4.1	−0.1
5	Ministry of Science, Technology and Innovation	Spain	4	3.8	0.2
6	University of Buenos Aires	Argentina	4	0.9	3.1
7	Ministry of Science and Technology	China	3	3.1	−0.1
8	Amparo Foundation for Research in the State of Sao Paulo	Brazil	3	2.1	0.9
9	Thailand Research Fund	Thailand	3	1.5	1.5
10	U.S. Department of Energy	USA	2	2.8	−0.8
11	Ministry of Education	Spain	2	2.3	−0.3
12	Financier of Studies and Projects	Brazil	2	1.2	0.8
13	University of Basque Country	Spain	2	0.7	1.3
14	Basque Government	Spain	2	0.7	1.3
15	Ministry of Science and Technology	Spain	2	0.6	1.4

The most prolific funding bodies are the National Natural Science Foundation of China and the Catalonia Government with five sample papers each. The National Council for Scientific and Technological Development (CONICET), the European Commission, Ministry of Science,

Technology and Innovation, and University of Buenos Aires follow these top funding bodies with four sample papers each.

It is notable that 45% and 53.1% of the sample and population papers are funded, respectively.

The most prolific country for these top funding bodies is Spain with six funding bodies. Brazil and China follow this top country with three and two sample papers, respectively. In total, six countries and the European Union house these top funding bodies.

The funding bodies with the highest citation impact are the Catalonia Government and University of Buenos Aires with at least a 3.1% surplus each. Similarly, the funding body with the lowest citation impact is the National Natural Science Foundation of China with a 5.6% deficit. The U.S. Department of Energy and the Ministry of Education of Spain follow this funding body with at least a 0.8% deficit each.

103.3.6 The Most Prolific Source Titles in Bioethanol-based Biohydrogen Fuels

Information about the 15 most prolific source titles publishing at least two sample papers and 1% of the population papers each in bioethanol-based biohydrogen fuels is given in Table 103.5.

The most prolific source titles are the 'International Journal of Hydrogen Energy' and 'Journal of Catalysis' with 19 and 15 sample papers, respectively. 'Applied Catalysis A General' and 'Applied Catalysis B Environmental' follow these top titles with at least ten papers each.

On the other hand, the source title with the highest citation impact is the 'Journal of Catalysis' with a 12.7% surplus. The 'Journal of Power Sources', 'Applied Catalysis A General', and 'Applied Catalysis B Environmental' follow this top title with at least a 5.3% surplus each. Similarly, the source title with the lowest impact is the 'International Journal of Hydrogen Energy' with a 2.3% deficit.

103.3.7 The Most Prolific Countries on Bioethanol-based Biohydrogen Fuels

Information about the 13 most prolific countries publishing at least two sample papers and 2% of the population papers each in bioethanol-based biohydrogen fuels is given in Table 103.6.

The most prolific country is Spain with 18 sample papers. The USA, Italy, China, and Brazil are the other prolific countries with at least 11 sample papers each. It is notable that the five European countries listed in Table 103.6 produce 49% and 33% of the sample and population papers, respectively.

On the other hand, the countries with the highest citation impact are Spain and Greece with a 5.9% and 5% surplus, respectively. Similarly, the country with the lowest citation impact is China with a 6.5% deficit.

TABLE 103.5
The Most Prolific Source Titles in Bioethanol-based Biohydrogen Fuels

No.	Source Titles	Sample Papers (%)	Population Papers (%)	Surplus (%)
1	International Journal of Hydrogen Energy	19	21.3	−2.3
2	Journal of Catalysis	15	2.3	12.7
3	Applied Catalysis A General	11	5.0	6.0
4	Applied Catalysis B Environmental	10	4.7	5.3
5	Journal of Power Sources	9	2.6	6.4
6	Catalysis Today	6	5.9	0.1
7	Catalysis Communications	5	1.7	3.3
8	Chemical Engineering Journal	4	2.7	1.3
9	Catalysis Letters	2	2.2	−0.2
10	Energy and Fuels	2	1.6	0.4
11	ACS Catalysis	2	1.0	1.0

TABLE 103.6
The Most Prolific Countries on Bioethanol-based Biohydrogen Fuels

No.	Countries	Sample Papers (%)	Population Papers (%)	Surplus (%)
1	Spain	18	12.1	5.9
2	USA	14	10.5	3.5
3	Italy	13	10.5	2.5
4	China	12	18.5	−6.5
5	Brazil	11	8.7	2.3
6	France	9	5.5	3.5
7	Japan	7	5.2	1.8
8	Greece	7	2.0	5.0
9	India	6	4.0	2.0
10	Argentina	5	4.4	0.6
11	Thailand	4	2.5	1.5
12	UK	2	2.4	−0.4

TABLE 103.7
The Most Prolific Scopus Subject Categories on Bioethanol-based Biohydrogen Fuels

No.	Scopus Subject Categories	Sample Papers (%)	Population Papers (%)	Surplus (%)
1	Chemical Engineering	64	52.7	11.3
2	Chemistry	54	44.0	10.0
3	Energy	33	39.0	−6.0
4	Physics and Astronomy	22	27.0	−5.0
5	Environmental Science	16	13.7	2.3
6	Engineering	13	16.1	−3.1
7	Materials Science	3	9.4	−6.4

103.3.8 The Most Prolific Scopus Subject Categories on Bioethanol-based Biohydrogen Fuels

Information about the seven most prolific Scopus subject categories indexing at least 3% and 9.4% of the sample and population papers, respectively, is given in Table 103.7.

The most prolific Scopus subject categories on bioethanol-based biohydrogen fuels are ‘Chemistry’, ‘Chemical Engineering’, and ’Energy’ with 64, 54, and 33 sample papers, respectively.

On the other hand, Scopus subject categories with the highest citation impact are ‘Chemical Engineering’ and ‘Chemistry’ with a 11.3% and 10% surplus, respectively. Similarly, Scopus subject categories with the lowest citation impact are ‘Materials Science’, ‘Energy’, and ‘Physics and Astronomy’ with at least a 5% deficit each.

103.3.9 The Most Prolific Keywords on Bioethanol-based Biohydrogen Fuels

Information about the keywords used in at least 5% of the sample and population papers is given in Table 103.8. For this purpose, keywords related to the keyword set given in the appendix are selected from a list of the most prolific keyword set provided by the Scopus database.

These keywords are grouped under four headings: bioethanol fuels, biohydrogen fuels, materials, and catalysis.

TABLE 103.8
The Most Prolific Keywords on Bioethanol-based Biohydrogen Fuels

No.	Keywords	Sample Papers (%)	Population Papers (%)	Surplus (%)
1.	Bioethanol fuels			
	Ethanol	83	86.3	−3.3
	Ethanol steam reforming	22	42.7	−20.7
	Bio-ethanol	8		8
	Steam reforming of ethanol	7	17.9	−10.9
	Ethanol reforming	5	8.8	−3.8
	Ethanol conversion		10.0	−10
	Bioethanol		8.7	−8.7
2.	Biohydrogen fuels			
	Hydrogen	53	35.7	17.3
	Steam reforming	43	55.5	−12.5
	Hydrogen production	43	55.2	−12.2
	Fuel cells	34	8.6	25.4
	Reforming reactions	33	12.4	33
	Steam	25	22.2	2.8
	Oxidation	11	7.1	3.9
	Dehydrogenation	9		9
	Steam engineering	8	13.8	−5.8
	H_2 production	7		7
	Dehydration	5		5
	Reforming	5		5
	Decomposition	5		5
	Catalytic reforming		12.5	−12.5
3.	Materials			
	Nickel	25	21.5	3.5
	Alumina	13	5.6	7.4
	Carbon	12	8.9	3.1
	Cerium compounds	12	6.7	5.3
	Rhodium	11	5.6	5.4
	Cobalt	10	9.9	0.1
	Zinc oxide	8		8
	Ceria	7		7
	Biomass	6		6
	Nickel compounds	6		6
	Titanium dioxide	5		5
	Magnesia	5		5
	Precious metals	5		5
	Aluminum	5		5
	Platinum		7.4	−7.4
4.	Catalysis			
	Catalysts	55	37.8	17.2
	Catalyst activity	19	22.7	−3.7
	Catalysis	18	11.0	7
	Catalyst selectivity	12	11.7	0.3

(Continued)

TABLE 103.8 (*Continued*)
The Most Prolific Keywords on Bioethanol-based Biohydrogen Fuels

No.	Keywords	Sample Papers (%)	Population Papers (%)	Surplus (%)
	Photocatalysis	6		6
	Catalyst supports	5	10.4	−5.4
	Cobalt catalyst	5		5
	Nickel catalysts	5		5
	Catalytic performance		8.5	−8.5
	Catalyst deactivation		5.8	−5.8

TABLE 103.9
The Most Prolific Research Fronts in Bioethanol-based Biohydrogen Fuels

No.	Research Fronts	Sample Papers (%)
1	Biohydrogen production by the steam reforming of bioethanol fuels	73
2	Photocatalytic biohydrogen production from bioethanol fuels	11
3	Biohydrogen production by the oxidative steam reforming of bioethanol fuels	9
4	Biohydrogen production by the partial oxidation of bioethanol fuels	6
5	Autothermal reforming of bioethanol fuels	5
6	Biohydrogen production from bioethanol fuels in general	5

Sample papers, the sample of the 100 most-cited papers.

There are seven keywords used related to bioethanol fuels. The prolific keywords are ethanol and ethanol steam reforming. It is notable that the bioethanol keyword appears in the sample paper keyword list with around 8% of the sample and population papers each.

Prolific keywords related to biohydrogen fuels are hydrogen, steam reforming, hydrogen production, fuel cells, reforming reactions, and steam with at least 25 papers each. The prolific keywords related to materials are nickel, alumina, carbon, and cerium compounds while the prolific keywords related to catalysis are catalysts, catalyst activity, and catalysis.

On the other hand, the keywords with the highest citation impact are reforming reactions, fuel cells, hydrogen, and catalysts. Similarly, keywords with the lowest citation impact are ethanol steam reforming, catalytic reforming, steam reforming, reforming reactions, and hydrogen production.

103.3.10 The Most Prolific Research Fronts in Bioethanol-based Biohydrogen Fuels

Information about the most prolific research fronts for sample papers in bioethanol-based biohydrogen fuels is given in Table 103.9.

There is only one primary research front in this field: biohydrogen production by the steam reforming of bioethanol with 73 sample papers (Haryanto et al., 2005; Ni et al., 2007). Other research fronts are photocatalytic biohydrogen production (Bamwenda et al., 1995; Gombac et al., 2010), biohydrogen production by the oxidative steam reforming of bioethanol (Kugai et al.,

2006; Velu et al., 2005), biohydrogen production by the partial oxidation of bioethanol (de Lima et al., 2008; Rabenstein and Hacker, 2008), autothermal reforming of bioethanol (Deluga et al., 2004; Ni et al., 2007), and biohydrogen production from bioethanol in general (Akande et al., 2005; Mattos et al., 2012).

The focus of these 100 most-cited papers is the development of materials for the most efficient processes for the reforming of bioethanol fuels for the production of biohydrogen fuels for fuel cells.

103.4 DISCUSSION

103.4.1 Introduction

Crude oil-based gasoline fuels have been widely used in the transportation sector since the 1920s. However, there have been great public concerns over the adverse environmental and human impact of these fuels. Hence, biomass-based bioethanol fuels have increasingly been used in blending gasoline fuels and in fuel cells.

In the meantime, research in nanomaterials and nanotechnology has intensified in recent years to become a major research field in scientific research with over one and a half million published papers. In this context, a large number of nanomaterials have been developed for nearly every research field. These materials offer an innovative way to increase the efficiency in the production and utilization of bioethanol fuels as in other scientific fields.

Research in the field of fuel cells has also intensified in recent years with over 50,000 papers in three primary research streams of high-temperature SOFCs, low-temperature PEMFCs, and MFCs. Research in the DEFCs has been a stream of research on low-temperature PEMFCs. The primary focus of the research on fuel cells has been the optimization of the operating conditions and the development of catalysts to maximize fuel cell performance.

Furthermore, research in the use of both biohydrogen fuels and conventional hydrogen fuels in fuel cells has intensified in recent years in the primary research fronts of water-based photocatalytic hydrogen production, water-based electrocatalytic hydrogen production, reforming of bioethanol fuels and other alcoholic fuels for the production of biohydrogen fuels, biohydrogen production, and hydrogen storage. It is also notable that the research in the fields of hydrogen production and storage has focused on nanomaterial development for hydrogen production and storage.

However, it is essential to develop efficient incentive structures for the primary stakeholders to enhance the research in this field. Scientometric analysis has been used in this context to inform the primary stakeholders about the current state of research in this research field.

There has not been any scientometric study on bioethanol-based biohydrogen fuels. Thus, this chapter presents a scientometric study of the research in bioethanol-based biohydrogen fuels. It examines the scientometric characteristics of both the sample and population data presenting scientometric characteristics of both these datasets in the order of documents, authors, publication years, institutions, funding bodies, source titles, countries, Scopus subject categories, keywords, and research fronts.

Search for this study was carried out using the Scopus database in September 2021. As the first step for the search of the relevant literature, keywords were selected using the 200 most-cited papers. The selected keyword list was optimized to obtain a representative sample of papers in this research field. This keyword list was provided in the appendix for future replication studies. Additionally, the information about the most-used keywords was given in Section 103.3.9 to highlight the key research fronts in Section 103.3.10.

As the second step, two sets of data were used for this study. First, a population sample of over 1,600 papers was used to examine the scientometric characteristics of the population data. Second, a sample of 100 most-cited papers was used to examine the scientometric characteristics of these citation classics with over 125 citations each.

Scientometric characteristics of both these sample and population datasets were presented in the order of documents, authors, publication years, institutions, funding bodies, source titles, countries, Scopus subject categories, keywords, and research fronts. Lastly, key scientometric findings for both datasets were discussed to highlight the research landscape for bioethanol-based biohydrogen fuels. Additionally, a number of brief conclusions were drawn and a number of relevant recommendations were made to enhance the future research landscape.

103.4.2 The Most Prolific Documents on Bioethanol-based Biohydrogen Fuels

Articles dominate both the sample and population datasets. Review papers and conference papers have a slight surplus and deficit, respectively.

Scopus differs from the Web of Science database in differentiating and showing articles and conference papers published in journals separately. Hence, it can be said that 93% and 96% of sample and population papers, respectively, are articles by combining the figures for articles and conference papers published in journals. Similarly, Scopus differs from the Web of Science database in introducing short surveys. Thus by combining the figures for review papers and short surveys, it can be said that 93% and 8% of sample papers are articles and reviews, respectively.

It is observed during the search process that there has been inconsistency in the classification of documents in Scopus as well as in other databases such as the Web of Science. This is especially relevant for the classification of papers as reviews or articles as the papers not involving a literature review may be erroneously classified as a review paper. There is also a case of review papers being classified as articles. For example, although there are eight review papers as classified by the Scopus database, nine of the sample papers are review papers.

In this context, it would be helpful to provide a classification note for the published papers in the books and journals at the first instance. It would also be helpful to use the document types listed in Table 103.1 for this purpose. Book chapters may also be classified as articles or reviews as an additional classification to differentiate review chapters from experimental chapters as is done by the Web of Science. It would be further helpful to additionally classify conference papers as articles or review papers as well as is done in the Web of Science database.

103.4.3 The Most Prolific Authors on Bioethanol-based Biohydrogen Fuels

There have been 34 most prolific authors with at least three sample papers and five population papers each as given in Table 103.2. These authors have shaped the development of research in this field.

The most prolific authors are Stefano Callaro, Salvatore Freni, Fabio B. Noronha, and Francesco Frusteri (Table 103.2).

It is important to note the inconsistencies in the indexing of author names in Scopus and other databases. It is especially an issue for names with more than two components such as 'Judge Alex de Camp Sirous'. The probable outcomes are 'Sirous, J.A.D.C.', 'de Camp Sirous, J.A.', or 'Camp Sirous, J.A.D.'. The first choice is the gold standard of the publishing sector as the last word in the name is taken as the last name. In most academic databases such as PubMed and EBSCO databases, this version is used predominantly. The second choice is a strong alternative while the last choice is an undesired outcome as two last words are taken as the last name. It is good practice to combine the words of the last name with a hyphen: 'Camp-Sirous, J.A.D.'. It is notable that inconsistent indexing of author names may cause substantial inefficiencies in the search process for the papers as well as allocating credit to authors as there are different author entries for each outcome in the databases.

There is also a case of shortening Chinese names. For example, 'Yuoyang Wang' is often shortened as 'Wang, Y.', 'Wang, Y.-Y.', and 'Wang Y.Y.' as is done in the Web of Science database as well. However, the gold standard in this case is 'Wang Y' where the last word is taken as the last name and the first word is taken as a single forename. In most academic databases such as PubMed and EBSCO, this first version is used predominantly. Nevertheless, it makes sense to use the third option to differentia Chinese names efficiently: 'Wang Y.Y.'. Therefore, there have been difficulties to locate papers for Chinese authors. In such cases, the use of unique author codes provided for each author by the Scopus database has been helpful.

There is also a difficulty in allowing credit for authors, especially for authors with common names such as 'Wang, Y', 'Huang, Y', or 'Zhu, Y.' in conducting scientometric studies. These difficulties strongly influence the efficiency of scientometric studies as well as allocating credit to authors as there are the same author entries for different authors with the same name, e.g., 'Wang Y.' in the databases.

In this context, the coding of authors in the Scopus database is a welcome innovation compared to other databases such as the Web of Science. In this process, Scopus allocates a unique number to each author in the database. However, there might still be substantial inefficiencies in this coding system, especially for common names. For example, some of the papers by Vanessa Fierro are allocated to another researcher with a different author code. There are two separate entries for 'Fierro' in one or two papers in the Scopus database. This was revised as a single-author entry with three sample papers. Another example is 'de Lima, S' where there are two entries as 'de Lima, S' and 'De Lima, S'. It is possible that Scopus uses a number of software programs to differentiate author names and the program may not be false-proof (Shin et al., 2014).

In this context, it does not help that author names are not given in full in some journals and books. This makes it difficult to differentiate authors with common names and further makes scientometric studies difficult in the author domain. Therefore, author names should be given in all books and journals at the first instance. There is also a cultural issue where some authors do not use their full names in their papers. Instead, they use initials for their forenames: 'Coutancy, A.P.' instead of 'Coutancy, Alas Padras'.

There are also inconsistencies in the naming of authors with more than two components by the authors themselves in journal papers and book chapters. For example, 'Alaspanda, A.P.C.', 'Sakoura, C.E.', and 'Mentaslo, S.J.' might be given as 'Alaspanda, A', 'Sakoura, C', or 'Mentaslo, S' in journals and books, respectively. This also makes scientometric studies difficult in the author domain. Hence, contributing authors should use their names consistently in their publications.

The other critical issue regarding author names is the spelling of the author names in the national spelling (e.g., Gonçalves, Übeiro) rather than in the English spelling (e.g., Goncalves, Ubeiro) in the Scopus database. Scopus differs from the Web of Science database and many other databases in this respect where author names are given only in the English spelling. It is observed that national spellings of author names do not help in conducting scientometric studies as well as in allocating credits to authors as sometimes there are different author entries for the English and national spelling in the Scopus database.

The most prolific institutions for the sample dataset are CSIC, National Institute of Technology, University of Barcelona, the University of Messina, University of Rey Juan Carlos, and University of Trieste. Similarly, the most prolific countries for the sample dataset are Spain, Italy, Brazil, and China. It is not surprising that the authors from these countries dominate this prolific author list.

It is also notable that there is a significant gender deficit for the sample dataset as surprisingly only seven of these top researchers are female. This finding is the most thought-provoking with strong public policy implications. Hence, institutions, funding bodies, and policy makers should take efficient measures to reduce the gender deficit in this field as well as other scientific fields with a strong gender deficit. In this context, it is worth noting the level of representation of researchers from minority groups in science on the basis of race, sexuality, age, and disability, besides gender (Blankenship, 1993; Dirth and Branscombe, 2017; Konur, 2000, 2002a,b,c, 2006a,b, 2007a,b).

103.4.4 The Most Prolific Research Output by Years on Bioethanol-based Biohydrogen Fuels

Research output observed between 1970 and 2021 is illustrated in Figure 103.1. This figure clearly shows that the bulk of the research papers in the population dataset were published primarily in the 2010s. This was followed by the early 2020s and 2000s while the publication rates for the pre-2000s were negligible. Similarly, the bulk of the research papers in the sample dataset were published in the 2000s and to a lesser extent in the 2010s.

These data suggest that the most-cited sample papers were primarily published in the 2000s while population papers were primarily published in the 2010s. These are thought-provoking findings as there has been no significant research in this field in the pre-2000s, but there has been a significant research boom in the past two decades. In this context, the increasing public concerns about climate change (Change, 2007), greenhouse gas emissions (Carlson et al., 2017), and global warming (Kerr, 2007) have been certainly behind the boom in research in this field in the past two decades.

Based on these findings, the size of population papers is likely to more than double in the current decade, provided that public concerns about climate change, greenhouse gas emissions, and global warming are translated efficiently to research funding in this field. However, it is notable that the research output stagnated after 2010 losing its momentum.

103.4.5 The Most Prolific Institutions on Bioethanol-based Biohydrogen Fuels

The 20 most prolific institutions publishing papers on bioethanol-based biohydrogen fuels with at least three sample papers and 0.7% of the population papers each given in Table 103.3 have shaped the development of research in this field.

The most prolific institutions are the University of Messina, CNR, National Technology Institute, University of Lyon, University of Barcelona, University of Auckland, CSIC, and University of Patras. The top countries for these most prolific institutions are Italy, Brazil, Argentina, China, France, Spain, and Thailand.

On the other hand, the institutions with the highest citation impact are the University of Messina, CNR, University of Lyon, University of Auckland, and University of Patras. Similarly, the institutions with the lowest impact are Tianjin University and CONICET.

It appears that Europe dominates the top institution list.

103.4.6 The Most Prolific Funding Bodies on Bioethanol-based Biohydrogen Fuels

The 15 most prolific funding bodies funding at least two sample papers and 0.7% of the population papers each are given in Table 103.4.

The most prolific funding bodies are the National Natural Science Foundation of China, the Catalonia Government, CONICET, the European Commission, Ministry of Science, Technology and Innovation of Spain, and University of Buenos Aires. It is notable that 45% and 53.1% of the sample and population papers are funded, respectively. The most prolific countries for these top funding bodies are Spain, Brazil, and China. In total, six countries and the EU house these top funding bodies. The heavy funding by the Chinese, Brazilian, and Spanish funding bodies is notable as these countries are all major producers of research in this field.

These findings on the funding of the research in this field suggest that the level of funding, mostly in the past two decades, has been largely instrumental in enhancing research in this field (Ebadi and Schiffauerova, 2016) in light of North's institutional framework (North, 1991). However, considering the fall in research output after 2010 and in light of the recent supply shocks, it appears that there is a need for more funding in this research field.

103.4.7 The Most Prolific Source Titles on Bioethanol-based Biohydrogen Fuels

The 15 most prolific source titles publishing at least two sample papers and 1% of the population papers each in bioethanol-based biohydrogen fuels have shaped the development of research in this field (Table 103.5).

The most prolific source titles are 'International Journal of Hydrogen Energy', 'Journal of Catalysis', 'Applied Catalysis A General', and 'Applied Catalysis B Environmental'.

On the other hand, the source titles with the highest impact are 'Journal of Catalysis', 'Journal of Power Sources', 'Applied Catalysis A General', and 'Applied Catalysis B Environmental'. Similarly, the source title with the lowest impact is 'International Journal of Hydrogen Energy'.

It is notable that these top source titles are related to catalysis and a lesser extent to energy. This finding suggests that journals in these fields have significantly shaped the development of research in this field as they focus on the development of nanocatalysts and conventional catalysts for the production and storage of biohydrogen fuels from bioethanol fuels.

The top journal produces 19% and 21% of the sample and population papers. Its scope is highly broad: 'analytical and experimental, covering all aspects of hydrogen energy, including production, storage, transmission, utilization, enabling technologies, environmental impact, economic and international aspects of hydrogen and hydrogen carriers'. Further, 'the utilization includes thermochemical (combustion), photochemical, electrochemical (fuel cells) and nuclear conversion of hydrogen, hydrogen isotopes and/or hydrogen carriers to thermal, mechanical and electrical energies, and their applications in transportation (including aerospace), industrial, commercial and residential sectors' (Journal scope information).

103.4.8 The Most Prolific Countries on Bioethanol-based Biohydrogen Fuels

The 13 most prolific countries publishing at least two papers and 2% of the population papers each have significantly shaped the development of research in this field (Table 103.6).

The most prolific countries are Spain, the USA, Italy, China, and Brazil. It is notable that the five European countries listed in Table 103.6 produce 49% and 33% of the sample and population papers, respectively. On the other hand, the countries with the highest citation impact are Spain and Greece while the country with the lowest citation impact is China.

A close examination of these findings suggests that Europe, the USA, China, and Brazil, are the major producers of research in this field. It is a fact that the USA has been a major player in science (Leydesdorff and Wagner, 2009; Leydesdorff et al., 2014). The USA has further developed a strong research infrastructure to support its corn and grass-based bioethanol industry (Vadas et al., 2008). The USA has also been very active in nanotechnology research (Dong et al., 2016) as well as hydrogen research (Bockris and Veziroglu, 1983).

However, China has been a rising star in scientific research in competition with the USA and Europe (Leydesdorff and Zhou, 2005). China is also a major player in this field as a major producer of bioethanol (Li and Chan-Halbrendt, 2009). China has also been very active in nanotechnology research (Dong et al., 2016) as well as hydrogen research (Ren et al., 2020).

Next, Europe has been a persistent player in scientific research in competition with both the USA and China (Leydesdorff, 2000). Europe has also been a persistent producer of bioethanol along with the USA and Brazil (Gnansounou, 2010). Europe has also been very active in nanotechnology research (Schellekens, 2010) as well as hydrogen research (Contaldi et al., 2008).

Additionally, Brazil has also been a persistent player in scientific research at a moderate level (Glanzel et al., 2006). Brazil has also developed a strong research infrastructure to support its biomass-based bioethanol industry (Soccol et al., 2010). Brazil has also been very active in nanotechnology research (Fonseca and Pereira, 2014) as well as hydrogen research (Dos Santos et al., 2017).

103.4.9 The Most Prolific Scopus Subject Categories on Bioethanol-based Biohydrogen Fuels

The seven most prolific Scopus subject categories indexing at least 3.0% and 9.3% of sample and population papers, respectively, given in Table 103.7 have shaped the development of research in this field.

The most prolific Scopus subject categories on bioethanol-based biohydrogen fuels are 'Chemistry', 'Chemical Engineering', and 'Energy'. On the other hand, Scopus subject categories with the highest citation impact are 'Chemical Engineering' and 'Chemistry'. Similarly, Scopus subject categories with the lowest citation impact are 'Materials Science', 'Energy', and 'Physics and Astronomy'.

These findings are thought-provoking suggesting that the primary subject categories are 'Chemistry', 'Chemical Engineering', and 'Energy'. The other key finding is that social sciences are not well represented both in sample and population papers, unlike the field of evaluative studies in bioethanol fuels. However, environmental and economic issues regarding the production and use of hydrogen fuels are often covered in papers indexed by these prolific subject categories.

These findings are not surprising as the key research fronts in this field are the development of catalysts for the production, storage, and use of hydrogen fuels from bioethanol fuels. All these research fronts are related to hard sciences and engineering.

103.4.10 The Most Prolific Keywords in Bioethanol-based Biohydrogen Fuels

A limited number of keywords have shaped the development of research in this field as shown in Table 103.8 and the Appendix.

These keywords are grouped under four headings: bioethanol fuels, biohydrogen fuels, materials, and catalysis. There are seven keywords used related to bioethanol fuels. However, the most prolific keywords are ethanol and ethanol steam reforming. It is notable that the bioethanol keyword appears in the sample paper list. Prolific keywords related to biohydrogen fuels are hydrogen, steam reforming, hydrogen production, fuel cells, reforming reactions, and steam. Prolific keywords related to materials are nickel, alumina, carbon, and cerium compounds while prolific keywords related to catalysis are catalysts, catalyst activity, and catalysis.

On the other hand, keywords with the highest citation impact are reforming reactions, fuel cells, hydrogen, and catalysts. Similarly, keywords with the lowest citation impact are ethanol steam reforming, catalytic reforming, steam reforming, reforming reactions, and hydrogen production.

These prolific keywords highlight the key research fronts in this field and reflect well the keywords used in sample papers.

103.4.11 The Most Prolific Research Fronts in Bioethanol-based Biohydrogen Fuels

There is only one primary research front in this field: biohydrogen production by the steam reforming of bioethanol fuels. Other research fronts are photocatalytic biohydrogen production, biohydrogen production by oxidative steam reforming of bioethanol, biohydrogen production by the partial oxidation of bioethanol, autothermal reforming of bioethanol, and biohydrogen production from bioethanol in general.

The focus of these 100 most-cited papers is the development of catalysts and catalyst supports based on nanomaterials and conventional materials for the production of biohydrogen fuels.

Research in the field of fuel cells has intensified in recent years with over 50,000 papers in three primary research streams of SOFCs (Adler, 2004), low-temperature PEMFCs (Borup et al., 2007), and MFCs (Logan, 2009). The primary focus of research on fuel cells has been the optimization of the operating conditions to maximize fuel cell performance. Therefore, it is not surprising that there have been no prolific papers on biohydrogen-based biohydrogen fuel cells themselves as most of these prolific papers have focused on catalyst and catalyst support development for reforming bioethanol fuels to produce biohydrogen fuels. This finding suggests that the development of efficient catalysts and catalyst supports is highly critical for the production of biohydrogen fuels.

One innovative way to enhance biohydrogen production has been the use of nanomaterials for both the catalysts themselves and the catalyst supports (Ma et al., 2016; Murdoch et al., 2011).

In parallel with the research on fuel cells, research on nanomaterials and nanotechnology (Geim, 2009; Geim and Novoselov, 2007) has intensified in recent years with over one and half million papers, enriching the material portfolio to be used for fuel cell applications. It is expected that this enriched portfolio of innovative nanomaterials would enhance the development of catalysts and catalyst supports for biohydrogen production through the reforming of bioethanol fuels in the future, benefiting from the superior properties of these nanomaterials.

In the end, these most-cited papers in this field hint that the efficiency of biohydrogen production from the reforming of bioethanol fuels could be optimized using the structure, processing, and property relationships of nanomaterials and conventional materials used as both electrocatalyst materials and catalyst supports (Formela et al., 2016; Konur, 2018a, 2020b, 2021c,d,e,f; Konur and Matthews, 1989).

103.5 CONCLUSION AND FUTURE RESEARCH

Research on bioethanol-based biohydrogen fuels has been mapped through a scientometric study of both sample and population datasets.

The critical issue in this study has been to obtain a representative sample of the research as in any other scientometric study. Therefore, the keyword set has been carefully devised and optimized after a number of runs in the Scopus database.

Another issue has been the selection of a multidisciplinary database to carry out the scientometric study of the research in this field. For this purpose, the Scopus database has been selected. Journal coverage of this database has been wider than that of the Web of Science.

Key scientometric properties of research in this field have been determined and discussed in this chapter. It is evident that a limited number of documents, authors, institutions, publication periods, institutions, funding bodies, source titles, countries, Scopus subject categories, keywords, and research fronts have shaped the development of the research in this field.

There is ample scope to increase the efficiency of scientometric studies in this field in author and document domains by developing consistent policies and practices in both domains across all academic databases. In this respect, authors, journals, and academic databases have a lot to do. Furthermore, the significant gender deficit as in most scientific fields emerges as a public policy issue. Potential deficits on the basis of age, race, disability, and sexuality need also to be explored in this field as in other scientific fields.

Research in this field has boomed in the 2010s possibly promoted by public concerns on global warming, greenhouse gas emissions, and climate change.

The modest funding rate of 45% and 53% for sample and population papers, respectively, suggests that this funding rate significantly enhanced the research in this field primarily in the 2010s, possibly more than doubling in the current decade. However, there is ample room for more funding considering the decline in research output after 2019 and the recent supply shocks.

The most prolific journals have been mostly indexed by the subject categories of chemistry, chemical engineering, and energy as the focus of sample papers has been on the development of catalyst and catalyst supports as well as reforming bioethanol fuels to produce biohydrogen fuels.

Europe, the USA, China, and Brazil have been the major producers of research in this field as major producers of bioethanol fuels from different types of biomass such as corn, sugarcane, and grass as well as other types of biomass. Brazil has also contributed largely to both sample and population papers. These countries have well-developed research infrastructure in bioethanol fuels, bioethanol-based biohydrogen fuels, nanomaterials, and conventional materials. However, it is notable that the research output stagnated after 2010 losing its momentum.

The primary subject categories have been Chemistry, Chemical Engineering, and Energy. Due to the technological emphasis of this field on the development of catalysts and catalyst supports, social sciences have not been fairly represented in both sample and population papers, unlike in evaluative studies in bioethanol fuels.

Ethanol is more popular than bioethanol as a keyword with strong implications for the search strategy. In other words, a search strategy using only the bioethanol keyword would not be helpful. These keywords are grouped under four headings: bioethanol fuels, biohydrogen fuels, materials, and catalysis. These groups of keywords highlight the potential primary research fronts for these fields: The development of catalysts and catalyst supports for bioethanol reforming to produce biohydrogen fuels.

There is only one primary research front in this field: biohydrogen production by the steam reforming of bioethanol. The focus has been on the use of conventional materials for the development of catalyst and catalyst supports with 96% of the sample papers.

These findings are thought-provoking. The focus of these 100 most-cited papers is the development of catalysts and catalyst supports rather than the reforming processes themselves. There are strong structure-processing-property relationships for all of the materials used for these research fronts. In the end, these most-cited papers in this field hint that the efficiency of the materials science and nanotechnology applications could be optimized using the structure, processing, and property relationships of the materials used in these applications.

Thus, scientometric analysis has great potential to gain valuable insights into the evolution of research in this field as in other scientific fields.

It is recommended that further scientometric studies are carried out about the other aspects of both production and utilization of bioethanol fuels. It is further recommended that reviews of the most-cited papers are carried out for each research front to complement these scientometric studies. Next, scientometric studies of the hot papers in these primary fields are carried out.

ACKNOWLEDGMENTS

The contribution of highly cited researchers in the field of bioethanol-based biohydrogen fuels has been gratefully acknowledged.

APPENDIX: THE KEYWORD SET FOR BIOETHANOL-BASED BIOHYDROGEN FUELS

((((TITLE (reform* OR *reforming OR "partial oxidation" OR "re-forming") OR SRCTITLE (reform*)) OR (TITLE ((photoassisted OR photocataly* OR "photo-cataly*") AND (hydrogen OR biohydrogen OR h2 OR "h-2")))) AND ((TITLE (ethanol OR bioethanol OR "bio-ethanol") OR SRCTITLE (ethanol* OR bioethanol* OR "bio-ethanol")))) OR ((TITLE ("h2 production from ethanol*" OR "hydrogen from ethanol*" OR "biohydrogen from ethanol*")) OR (TITLE (("from ethanol*" OR "from neat ethanol*" OR "from water/ethanol" OR "from bioethanol*" OR "ethanol decomposition" OR "coupling of ethanol") AND (hydrogen OR biohydrogen OR h2 OR "h-2"))))) AND NOT (TITLE (ferment* OR emission*)) AND (LIMIT-TO (SRCTYPE, "j") OR LIMIT-TO (SRCTYPE, "b") OR LIMIT-TO (SRCTYPE, "k")) AND (LIMIT-TO (DOCTYPE, "ar") OR LIMIT-TO (DOCTYPE, "cp") OR LIMIT-TO (DOCTYPE, "ch") OR LIMIT-TO (DOCTYPE, "re") OR LIMIT-TO (DOCTYPE, "no") OR LIMIT-TO (DOCTYPE, "sh")) AND (LIMIT-TO (LANGUAGE, "English"))

REFERENCES

Adler, S. B. 2004. Factors governing oxygen reduction in solid oxide fuel cell cathodes. *Chemical Reviews* 104. 4791–4843.

Akande, A. J., R. O. Idem and A. K. Dalai. 2005. Synthesis, characterization and performance evaluation of Ni/Al_2O_3 catalysts for reforming of crude ethanol for hydrogen production. *Applied Catalysis A: General* 287:159–175.

Antolini, E. 2007. Catalysts for direct ethanol fuel cells. *Journal of Power Sources* 170:1–12.

Bamwenda, G. R., S. Tsubota, T. Nakamura and M. Haruta. 1995. Photoassisted hydrogen-production from a water-ethanol solution - a comparison of activities of Au-TiO_2 and Pt-TiO_2. *Journal of Photochemistry and Photobiology A-Chemistry* 89:177–189.

Beaudry, C. and V. Lariviere. 2016. Which gender gap? Factors affecting researchers' scientific impact in science and medicine. *Research Policy* 45:1790–1817.

Blankenship, K. M. 1993. Bringing gender and race in: US employment discrimination policy. *Gender & Society* 7:204–226.

Bockris, J. N. and T. N. Veziroglu. 1983. A solar-hydrogen economy for USA. *International Journal of Hydrogen Energy* 8:323–340.

Borup, R., J. Meyers and B. Pivovar, et al. 2007. Scientific aspects of polymer electrolyte fuel cell durability and degradation. *Chemical Reviews* 107:3904–3951.

Burnham, J. F. 2006. Scopus database: A review. *Biomedical Digital Libraries* 3:1–8.

Carlson, K. M., J. S. Gerber and D. Mueller, et al. 2017. Greenhouse gas emissions intensity of global croplands. *Nature Climate Change* 7:63–68.

Change, C. 2007. Climate change impacts, adaptation and vulnerability. *Science of the Total Environment* 326:95–112.

Chen, X., S. Shen, L. Guo and S. S. Mao. 2010. Semiconductor-based photocatalytic hydrogen generation. *Chemical Reviews* 110:6503–6570.

Contaldi, M., F. Gracceva and A. Mattucci. 2008. Hydrogen perspectives in Italy: Analysis of possible deployment scenarios. *International Journal of Hydrogen Energy* 33:1630–1642.

Das, D. and T.N. Veziroglu. 2001. Hydrogen production by biological processes: A survey of literature. *International Journal of Hydrogen Energy* 26:13–28.

De Lima, S. M., I. O. da Cruz and G. Jacobs, et al. 2008. Steam reforming, partial oxidation, and oxidative steam reforming of ethanol over $Pt/CeZrO_2$ catalyst. *Journal of Catalysis* 257:356–368.

Deluga, G. A., J. R. Salge, L. D. Schmidt and X. E. Verykios. 2004. Renewable hydrogen from ethanol by autothermal reforming. *Science* 303:993–997.

Dirth, T. P. and N. R. Branscombe. 2017. Disability models affect disability policy support through awareness of structural discrimination. *Journal of Social Issues* 73:413–442.

Dong, H., Y. Gao and P. J. Sinko, et al. 2016. The nanotechnology race between China and the United States. *Nano Today* 11:7–12.

Dos Santos, K. G., C. T. Eckert and E. de Rossi, et al. 2017. Hydrogen production in the electrolysis of water in Brazil, a review. *Renewable and Sustainable Energy Reviews* 68:563–571.

Ebadi, A. and A. Schiffauerova. 2016. How to boost scientific production? A statistical analysis of research funding and other influencing factors. *Scientometrics* 106:1093–1116.

Fonseca, P. F. and T. S. Pereira. 2014. The governance of nanotechnology in the Brazilian context: Entangling approaches. *Technology in Society* 37:16–27.

Formela, K., A. Hejna, L. Piszczyk, M. R. Saeb and X. Colom. 2016. Processing and structure-property relationships of natural rubber/wheat bran biocomposites. *Cellulose* 23:3157–3175.

Garfield, E. 1955. Citation indexes for science. *Science* 122:108–111.

Geim, A. K. 2009. Graphene: Status and prospects. *Science* 324:1530–1534.

Geim, A. K. and K. S. Novoselov. 2007. The rise of graphene. *Nature Materials* 6:183–191.

Glanzel, W., J. Leta and B. Thijs. 2006. Science in Brazil. Part 1: A macro-level comparative study. *Scientometrics* 67:67–86.

Gnansounou, E. 2010. Production and use of lignocellulosic bioethanol in Europe: Current situation and perspectives. *Bioresource Technology* 101:4842–4850.

Gombac, V., L. Sordelli and T. Montini. 2010. CuO_x-TiO_2 photocatalysts for H_2 production from ethanol and glycerol solutions. *Journal of Physical Chemistry A* 114:3916–3925.

Greeley, J., T. F. Jaramillo, J. Bonde, I. Chorkendorff and J. K. Norskov. 2006. Computational high-throughput screening of electrocatalytic materials for hydrogen evolution. *Nature Materials* 5:909–913.

Haryanto, A., S. Fernando, N. Murali and S. Adhikari. 2005. Current status of hydrogen production techniques by steam reforming of ethanol: A review. *Energy and Fuels* 19:2098–2106.

Hill, J., E. Nelson, D. Tilman, S. Polasky and D. Tiffany. 2006. Environmental, economic, and energetic costs and benefits of biodiesel and ethanol biofuels. *Proceedings of the National Academy of Sciences of the United States of America* 103:11206–11210.

Hill, J., S. Polasky and E. Nelson, et al. 2009. Climate change and health costs of air emissions from biofuels and gasoline. *Proceedings of the National Academy of Sciences of the United States of America* 106:2077–2082.

Hsieh, W. D., R. H. Chen, T. L. Wu and T. H. Lin. 2002. Engine performance and pollutant emission of an SI engine using ethanol-gasoline blended fuels. *Atmospheric Environment* 36:403–410.

Kamarudin, M. Z. F., S. K. Kamarudin, M. S. Masdar and W. R. W. Daud. 2013. Review: Direct ethanol fuel cells. *International Journal of Hydrogen Energy* 38:9438–9453.

Kerr, R. A. 2007. Global warming is changing the world. *Science* 316:188–190.

Konur, O. 2000. Creating enforceable civil rights for disabled students in higher education: An institutional theory perspective. *Disability & Society* 15:1041–1063.

Konur, O. 2002a. Access to nursing education by disabled students: Rights and duties of nursing programs. *Nurse Education Today* 22:364–374.

Konur, O. 2002b. Assessment of disabled students in higher education: Current public policy issues. *Assessment and Evaluation in Higher Education* 27:131–152.

Konur, O. 2002c. Access to employment by disabled people in the UK: Is the Disability Discrimination Act working? *International Journal of Discrimination and the Law* 5:247–279.

Konur, O. 2006a. Participation of children with dyslexia in compulsory education: Current public policy issues. *Dyslexia* 12:51–67.

Konur, O. 2006b. Teaching disabled students in higher education. *Teaching in Higher Education* 11:351–363.

Konur, O. 2007a. A judicial outcome analysis of the *Disability Discrimination Act*: A windfall for the employers? *Disability & Society* 22:187–204.

Konur, O. 2007b. Computer-assisted teaching and assessment of disabled students in higher education: The interface between academic standards and disability rights. *Journal of Computer Assisted Learning* 23:207–219.

Konur, O. 2011. The scientometric evaluation of the research on the algae and bio-energy. *Applied Energy* 88:3532–3540.

Konur, O. 2012a. Prof. Dr. Ayhan Demirbas' scientometric biography. *Energy Education Science and Technology Part A: Energy Science and Research* 28:727–738.

Konur, O. 2012b. The evaluation of the biogas research: A scientometric approach. *Energy Education Science and Technology Part A: Energy Science and Research* 29:1277–1292.

Konur, O. 2012c. The evaluation of the global energy and fuels research: A scientometric approach. *Energy Education Science and Technology Part A: Energy Science and Research* 30:613–628.

Konur, O. 2012d. The evaluation of the research on the biodiesel: A scientometric approach. *Energy Education Science and Technology Part A: Energy Science and Research* 28:1003–1014.

Konur, O. 2012e. The evaluation of the research on the bioethanol: A scientometric approach. *Energy Education Science and Technology Part A: Energy Science and Research* 28:1051–1064.

Konur, O. 2012f. The evaluation of the research on the biofuels: A scientometric approach. *Energy Education Science and Technology Part A: Energy Science and Research* 28:903–916.

Konur, O. 2012g. The evaluation of the research on the biohydrogen: A scientometric approach. *Energy Education Science and Technology Part A: Energy Science and Research* 29:323–338.

Konur, O. 2012h. The evaluation of the research on the microbial fuel cells: A scientometric approach. *Energy Education Science and Technology Part A: Energy Science and Research* 29:309–322.

Konur, O. 2012i. The scientometric evaluation of the research on the production of bioenergy from biomass. *Biomass and Bioenergy* 47:504–515.

Konur, O. 2015. Current state of research on algal bioethanol. In *Marine Bioenergy: Trends and Developments*, Ed. S. K. Kim and C. G. Lee, pp. 217–244. Boca Raton, FL: CRC Press.

Konur, O. 2016a. Scientometric overview in nanobiodrugs. In *Nanoarchitectonics for Smart Delivery and Drug Targeting*, Ed. A. M. Holban and A. M. Grumezescu, pp. 405–428. Amsterdam: Elsevier.

Konur, O. 2016b. Scientometric overview regarding nanoemulsions used in the food industry. In *Emulsions: Nanotechnology in the Agri-Food Industry*, Ed. A. M. Grumezescu, pp. 689–711. Amsterdam: Elsevier.

Konur, O. 2016c. Scientometric overview regarding the nanobiomaterials in antimicrobial therapy. In *Nanobiomaterials in Antimicrobial Therapy*, Ed. A. M. Grumezescu, pp. 511–535. Amsterdam: Elsevier.

Konur, O. 2016d. Scientometric overview regarding the nanobiomaterials in dentistry. In *Nanobiomaterials in Dentistry*, Ed. A. M. Grumezescu, pp. 425–453. Amsterdam: Elsevier.

Konur, O. 2016e. Scientometric overview regarding the surface chemistry of nanobiomaterials. In *Surface Chemistry of Nanobiomaterials*, Ed. A. M. Grumezescu, pp. 463–486. Amsterdam: Elsevier.

Konur, O. 2016f. The scientometric overview in cancer targeting. In *Nanoarchitectonics for Smart Delivery and Drug Targeting*, Ed. A. M. Holban and A. Grumezescu, pp. 871–895. Amsterdam: Elsevier.

Konur, O. 2017a. Recent citation classics in antimicrobial nanobiomaterials. In *Nanostructures for Antimicrobial Therapy*, Ed. A. Ficai and A. M. Grumezescu, pp. 669–685. Amsterdam: Elsevier.

Konur, O. 2017b. Scientometric overview in nanopesticides. In *New Pesticides and Soil Sensors*, Ed. A. M. Grumezescu, pp. 719–744. Amsterdam: Elsevier.

Konur, O. 2017c. Scientometric overview regarding oral cancer nanomedicine. In *Nanostructures for Oral Medicine*, Ed. E. Andronescu and A. M. Grumezescu, pp. 939–962. Amsterdam: Elsevier;

Konur, O. 2017d. Scientometric overview regarding water nanopurification. In *Water Purification*, Ed. A. M. Grumezescu, pp. 693–716. Amsterdam: Elsevier.

Konur, O. 2017e. Scientometric overview in food nanopreservation. In *Food Preservation*, Ed. A. M. Grumezescu, pp. 703–729. Amsterdam: Elsevier.

Konur, O., Ed. 2018a. *Bioenergy and Biofuels*. Boca Raton, FL: CRC Press.

Konur, O. 2018b. Bioenergy and biofuels science and technology: Scientometric overview and citation classics. In *Bioenergy and Biofuels*, Ed. O. Konur, pp. 3–63. Boca Raton: CRC Press.

Konur, O. 2019a. Cyanobacterial bioenergy and biofuels science and technology: A scientometric overview. In *Cyanobacteria: From Basic Science to Applications*, Ed. A. K. Mishra, D. N. Tiwari and A. N. Rai, pp. 419–442. Amsterdam: Elsevier.

Konur, O. 2019b. Nanotechnology applications in food: A scientometric overview. In *Nanoscience for Sustainable Agriculture*, Ed. R. N. Pudake, N. Chauhan and C. Kole, pp. 683–711. Cham: Springer.

Konur, O. 2020a. The scientometric analysis of the research on the bioethanol production from green macroalgae. In *Handbook of Algal Science, Technology and Medicine*, Ed. O. Konur, pp. 385–401. London: Academic Press.

Konur, O., Ed. 2020b. *Handbook of Algal Science, Technology and Medicine*. London: Academic Press.

Konur, O. 2021a. Nanotechnology applications in diesel fuels and the related research fields: A review of the research. In *Handbook of Biodiesel and Petrodiesel Fuels: Science, Technology, Health, and Environment. Volume 1. Biodiesel Fuels: Science, Technology, Health, and Environment*, Ed. O. Konur, pp. 89–110. Boca Raton, FL: CRC Press.

Konur, O. 2021b. Nanobiosensors in agriculture and foods: A scientometric review. In *Nanobiosensors in Agriculture and Food*, Ed. R. N. Pudake, pp. 365–384. Cham: Springer.

Konur, O., Ed. 2021c. *Handbook of Biodiesel and Petrodiesel Fuels: Science, Technology, Health, and Environment*. Boca Raton, FL: CRC Press.

Konur, O., Ed. 2021d. *Handbook of Biodiesel and Petrodiesel Fuels: Science, Technology, Health, and Environment. Volume 1. Biodiesel Fuels: Science, Technology, Health, and Environment*. Boca Raton, FL: CRC Press.

Konur, O., Ed. 2021e. *Handbook of Biodiesel and Petrodiesel Fuels: Science, Technology, Health, and Environment. Volume 2. Biodiesel Fuels based on the Edible and Nonedible Feedstocks, Wastes, and Algae: Science, Technology, Health, and Environment*. Boca Raton, FL: CRC Press.

Konur, O., Ed. 2021f. *Handbook of Biodiesel and Petrodiesel Fuels: Science, Technology, Health, and Environment. Volume 3. Petrodiesel Fuels: Science, Technology, Health, and Environment*. Boca Raton, FL: CRC Press.

Konur, O. and F. L. Matthews. 1989. Effect of the properties of the constituents on the fatigue performance of composites: A review. *Composites* 20:317–328.

Kugai, J., V. Subramani, C. Song, M. H. Engelhard and Y. H. Chin. 2006. Effects of nanocrystalline CeO_2 supports on the properties and performance of Ni-Rh bimetallic catalyst for oxidative steam reforming of ethanol. *Journal of Catalysis* 238:430–440.

Leydesdorff, L. 2000. Is the European Union becoming a single publication system? *Scientometrics* 47:265–280.

Leydesdorff, L. and C. Wagner. 2009. Is the United States losing ground in science? A global perspective on the world science system. *Scientometrics* 78:23–36.

Leydesdorff, L., C. S. Wagner and L. Bornmann. 2014. The European Union, China, and the United States in the top-1% and top-10% layers of most-frequently cited publications: Competition and collaborations. *Journal of Informetrics* 8:606–617.

Leydesdorff, L. and P. Zhou. 2005. Are the contributions of China and Korea upsetting the world system of science? *Scientometrics* 63:617–630.

Li, S. Z. and C. Chan-Halbrendt. 2009. Ethanol production in (the) People's Republic of China: Potential and technologies. *Applied Energy* 86:S162–S169.

Logan, B. E. 2009. Exoelectrogenic bacteria that power microbial fuel cells. *Nature Reviews Microbiology* 7:375–381.
Ma, H., L. Zeng and H. Tian, et al. 2016. Efficient hydrogen production from ethanol steam reforming over La-modified ordered mesoporous Ni-based catalysts. *Applied Catalysis B: Environmental* 181:321–331.
Ma, X., L. Sun and C. Song. 2002. A new approach to deep desulfurization of gasoline, diesel fuel and jet fuel by selective adsorption for ultra-clean fuels and for fuel cell applications. *Catalysis Today* 77:107–116.
Mattos, L. V., G. Jacobs, B. H. Davis and F. B. Noronha. 2012. Production of hydrogen from ethanol: Review of reaction mechanism and catalyst deactivation. *Chemical Reviews* 112:4094–4123.
Murdoch, M., G. I. N. Waterhouse and M. A. Nadeem, et al. 2011. The effect of gold loading and particle size on photocatalytic hydrogen production from ethanol over Au/TiO_2 nanoparticles. *Nature Chemistry* 3:489–492.
Najafi, G., B. Ghobadian and T. Tavakoli, et al. 2009. Performance and exhaust emissions of a gasoline engine with ethanol blended gasoline fuels using artificial neural network. *Applied Energy* 86:630–639.
Newman, P. W. G. and J. R. Kenworthy. 1989. Gasoline consumption and cities: A comparison of U.S. cities with a global survey. *Journal of the American Planning Association* 55:24–37.
Ni, M., D. Y. C. Leung and M. K. H. Leung. 2007. A review on reforming bio-ethanol for hydrogen production. *International Journal of Hydrogen Energy* 32:3238–3247.
North, D. C. 1991. Institutions. *Journal of Economic Perspectives* 5:97–112.
Orimo, S. I., Y. Nakamori, J. R. Eliseo, A. Zuttel and C. M. Jensen. 2007. Complex hydrides for hydrogen storage. *Chemical Reviews* 107:4111–4132.
Rabenstein, G. and V. Hacker. 2008. Hydrogen for fuel cells from ethanol by steam-reforming, partial-oxidation and combined auto-thermal reforming: A thermodynamic analysis. *Journal of Power Sources* 185:1293–1304.
Ren, X., L. Dong, D. Xu and B. Hu. 2020. Challenges towards hydrogen economy in China. *International Journal of Hydrogen Energy* 45:34326–34345.
Schellekens, M. 2010. Patenting nanotechnology in Europe: Making a good start? An analysis of issues in law and regulation. *Journal of World Intellectual Property*, 13:47–76.
Shin, D., T. Kim, J. Choi and J. Kim. 2014. Author name disambiguation using a graph model with node splitting and merging based on bibliographic information. *Scientometrics* 100:15–50.
Soccol, C. R., L. P. de Souza Vandenberghe and A. B. P. Medeiros, et al. 2010. Bioethanol from lignocelluloses: Status and perspectives in Brazil. *Bioresource Technology* 101:4820–4825.
Vadas, P. A., K. H. Barnett and D. J. Undersander 2008. Economics and energy of ethanol production from alfalfa, corn, and switchgrass in the Upper Midwest, USA. *Bioenergy Research* 1:44–55.
Velu, S., K. Suzuki, M. Vijayaraj, S. Barman and C. S. Gopinath. 2005. *In situ* XPS investigations of $Cu_{1-x}Ni_xZnAl$-mixed metal oxide catalysts used in the oxidative steam reforming of bio-ethanol. *Applied Catalysis B: Environmental* 55:287–299.

104 Bioethanol Fuel-based Biohydrogen Fuels
Review

Ozcan Konur
(Formerly) Ankara Yildirim Beyazit University

104.1 INTRODUCTION

Crude oil-based gasoline fuels (Ma et al., 2002; Newman and Kenworthy, 1989) have been widely used in the transportation sector since the 1920s. However, there have been great public concerns over the adverse environmental and human impacts of these fuels (Hill et al., 2006, 2009). Hence, biomass-based bioethanol fuels (Hill et al., 2006; Konur, 2012, 2015, 2019, 2020a) have increasingly been used in blending gasoline fuels (Hsieh et al., 2002; Najafi et al., 2009) and in fuel cells (Antolini, 2007, 2009).

In the meantime, the research in nanomaterials and nanotechnology has intensified in recent years to become a major research field in scientific research, with over one and a half million published papers (Geim, 2009; Geim and Novoselov, 2007). In this context, large numbers of nanomaterials have been developed nearly for every research field. These materials offer an innovative way to increase efficiency in the production and utilization of bioethanol fuels and bioethanol-based biohydrogen fuels, as well as in other scientific fields, including water-based hydrogen production.

The research in the field of fuel cells has also intensified in recent years, with over 50,000 papers in three primary research streams of high-temperature solid oxide fuel cells (SOFCs) (Adler, 2004), low-temperature polymer electrode membrane fuel cells (PEMFCs) (Borup et al., 2007), and microbial fuel cells (MFCs) (Logan, 2009). The primary focus of the research on fuel cells has been the optimization of the operating conditions and the development of catalysts to maximize fuel cell performance. Bioethanol-based biohydrogen fuels have been used as feedstocks for fuel cells to produce bioelectricity.

Furthermore, the research on the use of both biohydrogen fuels and conventional hydrogen fuels in fuel cells has intensified in recent years on the primary research fronts of water-based photocatalytic hydrogen production (Chen et al., 2010), water-based electrocatalytic hydrogen production (Norskov et al., 2005), reforming of bioethanol fuels and other fuels for the production of biohydrogen fuels (Haryanto et al., 2005; Ni et al., 2007), biohydrogen production (Das and Veziroglu, 2001), and hydrogen storage (Orimo et al., 2007). It is also notable that research in the fields of hydrogen production and storage has focused on nanomaterial and conventional material development for these fields.

Thus, research in the field of the production of biohydrogen fuels from bioethanol fuels has progressed in the research fronts of biohydrogen production by the steam reforming of bioethanol (Haryanto et al., 2005; Ni et al., 2007) and, to a lesser extent, photocatalytic biohydrogen production (Bamwenda et al., 1995; Gombac et al., 2010), biohydrogen production by the oxidative steam reforming of bioethanol (Kugai et al., 2006; Velu et al., 2005), biohydrogen production by the partial oxidation of bioethanol (de Lima et al., 2008; Rabenstein and Hacker, 2008), autothermal reforming of bioethanol (Deluga et al., 2004; Ni et al., 2007), and biohydrogen production from bioethanol in general (Akande et al., 2005; Mattos et al., 2012).

DOI: 10.1201/9781003226574-138

However, it is essential to develop efficient incentive structures (North, 1991) for the primary stakeholders to enhance the research in this field (Konur, 2000, 2002a,b,c, 2006a,b, 2007a,b).

Although there have been a number of review papers on bioethanol-based production of biohydrogen fuels (Haryanto et al., 2005; Mattos et al., 2012; Ni et al., 2007), there has been no review of the 25 most-cited articles in this field.

This chapter presents a review of the 25 most-cited articles in the field of bioethanol-based production of biohydrogen fuels. Then, it discusses the key findings of these highly influential papers and comments on future research priorities in this field.

104.2 MATERIALS AND METHODS

The search for this study was carried out using the Scopus database (Burnham, 2006) in September 2021.

As the first step for the search of the relevant literature, keywords were selected using the 200 most-cited papers. The selected keyword list was optimized to obtain a representative sample of papers for this research field. This keyword list was provided in the appendix of Konur (2023) for future replication studies.

As the second step, a sample dataset was used in this study. The first 25 articles in the sample of 100 most-cited papers with at least 125 citations each were selected for the review study. Key findings from each paper were taken from the abstracts of these papers and were discussed.

Additionally, a number of brief conclusions were drawn and a number of relevant recommendations were made to enhance the future research landscape.

104.3 RESULTS

A brief information about the 25 most-cited papers with at least 222 citations each on the production of biohydrogen fuels from bioethanol fuels is given below under two major headings: The steam reforming of bioethanol fuels to produce biohydrogen fuels and the other methods in the reforming of bioethanol fuels to produce biohydrogen fuels with 21 and four papers, respectively. The latter group includes papers on the photocatalytic biohydrogen fuel production from bioethanol fuels, the autothermal reforming of bioethanol fuels to produce biohydrogen fuels, and the oxidative steam reforming of bioethanol fuels to produce biohydrogen fuels.

104.3.1 The Steam Reforming of Bioethanol Fuels to Produce Biohydrogen Fuels

A brief information about 21 prolific studies with at least 225 citations each on the steam reforming of bioethanol to produce biohydrogen fuels is given in Table 104.1. Furthermore, brief notes on the contents of these studies are also given.

Liguras et al. (2003) evaluates the effect of catalysts and catalyst support on the steam reforming of bioethanol fuels to produce biohydrogen fuels in a paper with 592 citations. They employ rhodium (Rh), ruthenium (Ru), platinum (Pt), palladium (Pd), catalysts and alumina (Al_2O_3), magnesia (MgO), and titania (TiO_2) catalyst supports at the range of 600°C–850°C. They note that for low-loaded catalysts, Rh was significantly more active and selective toward hydrogen formation compared to Ru, Pt, and Pd, which showed a similar behavior. The catalytic performance of Rh and Ru was significantly improved with increasing metal loading, leading to higher ethanol conversions and hydrogen selectivities at given reaction temperatures. Further, the catalytic activity and selectivity of high-loaded Ru catalysts were comparable to those of Rh. Under certain reaction conditions, the 5% Ru/Al_2O_3 catalyst completely converted ethanol with selectivities toward hydrogen above 95%, the only byproduct being methane. Further, the catalyst was acceptably stable.

Fatsikostas and Verykios (2004) evaluates the effect of catalyst supports on the steam reforming of bioethanol fuels to produce biohydrogen fuels for nickel (Ni)-based catalysts in a paper with 532

TABLE 104.1
The Steam Reforming of Bioethanol Fuels to Produce Biohydrogen Fuels

No.	Papers	Catalysts	Catalyst Supports	Issues	Cits
1	Liguras et al. (2003)	Rh, Ru, Pt, Pd	Al_2O_3, MgO, and TiO_2	Effect of catalysts and catalyst supports	592
2	Fatsikostas and Verykios (2004)	Ni	γ-Al_2O_3, La_2O_3, and La_2O_3/γ-Al_2O_3	Effect of catalyst supports	532
3	Llorca et al. (2002)	Co	MgO, γ-Al_2O_3, SiO_2, TiO_2, V_2O_5, ZnO, La_2O_3, CeO_2, and Sm_2O_3	Effect of catalyst supports	489
4	Fatsikostas et al. (2002)	Ni	La_2O_3, Al_2O_3, YSZ, and MgO	Effect of the catalyst supports	432
5	Breen et al. (2002)	Rh, Pd, Ni, Pt	Al_2O_3, CeO_2, and ZrO_2	Effect of catalysts and catalyst supports	396
6	Sanchez-Sanchez et al. (2007)	Ni	La_2O_3, Al_2O_3, MgO, CeO_2, and ZrO_2	Effect of the catalyst supports	367
7	Zhang et al. (2007)	Ir, Co, Ni	CeO_2	Effect of catalysts	365
8	Aupretre et al. (2002)	Rh, Pt, Ni, Cu, Zn, Fe	γ-Al_2O_3, 12%CeO_2–Al_2O_3, CeO_2, and $Ce_{0.63}Zr_{0.37}O_2$	Effect of catalysts and catalyst supports	313
9	Vizcaino et al. (2007)	Cu-Ni	SiO_2, γ-Al_2O_3, MCM-41, SBA-15, and ZSM-5	Effect catalyst supports	309
10	Comas et al. (2004)	Ni	γ-Al_2O_3	Effect of temperature	302
11	Sun et al. (2005)	Ni	Y_2O_3, La_2O_3, and Al_2O_3	Effect of catalyst supports and effect of temperature	277
12	Frusteri et al. (2004)	Pd, Rh, Ni, Co	MgO	Effect catalysts	277
13	Vasudeva et al. (1996)	Na	Na	Thermodynamic analysis	257
14	Diagne et al. (2002)	Rh	CeO_2 and ZrO_2	Effect catalyst supports	248
15	Song and Ozkan (2009)	Co	ZrO_2 and CeO_2	Effect catalyst supports	246
16	Garcia and Laborde (1991)	Na	Na	Thermodynamic analysis	246
17	Haga et al. (1997)	Co	Al_2O_3, SiO_2, MgO, ZrO_2, and C	Effect catalyst supports	241
18	Cavallaro et al. (2003)	Rh	Al_2O_3	Effect of operating conditions	236
19	Llorca et al. (2003)	Co	ZnO	Effect of catalyst properties	228
20	Li et al. (2015)	Ni/SBA-15, Ce-Ni/SBA-15	CeO_2	Effect catalysts	225
21	Marino et al. (2001)	Cu-Ni-K	Al_2O_3	Effect of catalyst	225

Cits., the number of the citations received by each paper; Na, not available

citations. They employ γ-Al_2O_3, lanthana (La_2O_3), and La_2O_3/γ-Al_2O_3 catalyst supports. They note that ethanol interacted strongly with alumina on the surface of which it was dehydrated at low temperatures and less strongly with lanthana on the surface of which it was both dehydrogenated and dehydrated. They also observe cracking reactions on the carriers at intermediate temperatures. In the presence of Ni, catalytic activity was shifted toward lower temperatures. In addition to the above reactions, reforming, water-gas shift, and methanation significantly contributed to product distribution. Carbon deposition was also a significant route. The rate of carbon deposition was a strong function of the carrier, the steam-to-ethanol ratio, and the reaction temperature. The presence of lanthana on the catalyst, the high steam-to-ethanol ratio, and the high temperature offered enhanced resistance toward carbon deposition.

Llorca et al. (2002) evaluates the effect of catalyst supports on the steam reforming of bioethanol fuels to produce biohydrogen fuels for the cobalt (Co) catalysts in a paper with 489 citations. They deposit Co catalysts on MgO, γ-Al_2O_3, silica (SiO_2), TiO_2, vanadia (V_2O_5), zinc oxide (ZnO), La_2O_3, ceria (CeO_2), and samarium oxide (Sm_2O_3) catalyst supports. They note that ethanol steam reforming took place to a large degree over ZnO-, La_2O_3-, Sm_2O_3-, and CeO_2-supported catalysts as Co-free hydrogen was produced. Depending on the support, they identify different cobalt-based phases: Metallic cobalt particles, Co_2C, CoO, and La_2CoO_4. The extent and nature of carbon deposition depended on the sample and the reaction temperature. ZnO-supported samples showed the best catalytic performances. Under 100% ethanol conversion, selectivity up to 73.8% to hydrogen and 24.2% to CO_2 was obtained.

Fatsikostas et al. (2002) evaluates the effect of catalyst supports on the steam reforming of bioethanol fuels to produce biohydrogen fuels in a paper with 432 citations. They use Ni catalysts supported on La_2O_3, Al_2O_3, yttria stabilized zirconia (YSZ), and MgO. They note that the Ni/La_2O_3 catalyst exhibited high activity and selectivity toward hydrogen production and, most importantly, long-term stability for steam reforming of ethanol. They attribute the enhanced stability of this catalyst to the scavenging of coke deposition on the Ni surface by lanthana species, which existed on top of the Ni particles under reaction conditions.

Breen et al. (2002) evaluate the effect of catalysts and catalyst supports on the steam reforming of bioethanol fuels to produce biohydrogen fuels in a paper with 396 citations. They employ the catalysts of Rh, Pd, Ni, and Pt and the catalyst supports of Al_2O_3, CeO_2, and zirconia (ZrO_2). They note that alumina-supported catalysts were very active at lower temperatures for the dehydration of ethanol to ethene, which, at higher temperatures, was converted into H_2, CO, and CO_2 as the major products and CH_4 as a minor product. The order of activity of the metals was Rh >Pd >Ni=Pt. With CeO_2/ZrO_2-supported catalysts, they do not observe the formation of ethene, and the order of activity at higher temperatures was Pt ≥Rh >Pd. By using combinations of a ceria/zirconia-supported metal catalyst with the alumina support, they showed that the formation of ethene did not inhibit the steam reforming reaction at higher temperatures. They conclude that the catalyst support played a significant role in the steam reforming of ethanol.

Sanchez-Sanchez et al. (2007) evaluates the effect of catalyst supports on the steam reforming of bioethanol fuels to produce biohydrogen fuels for the Ni catalysts in a paper with 367 citations. They employ catalyst supports of La_2O_3, Al_2O_3, MgO, ZrO_2, and CeO_2. They modify alumina-supported Ni catalysts with cerium (Ce), magnesium (Mg), zirconium (Zr), and lanthanum (La). They note changes in the acidity, Ni dispersion, and Ni-catalyst support interaction with the type of modifier added to Al_2O_3. They further note that the acidity of catalysts containing Mg, Ce, La, and Zr additives decreased with respect to that supported on bare Al_2O_3. The trend of metal dispersion followed the order: La_2O_3–Al_2O_3 >MgO–Al_2O_3 >CeO_2–Al_2O_3> Al_2O_3 >ZrO_2–Al_2O_3. They observe the development of strong interactions between Ni species and ZrO_2, La_2O_3, and CeO_2 oxides added to catalyst supports. They note the different catalyst functionality that influences on their reforming activity. Thus, they explain the higher reforming activity for Mg-modified catalyst with respect to bare Al_2O_3 in terms of the lower acidity and better dispersion achieved in the former, while for Ce- and Zr-promoted catalysts, they attribute the improvement in intrinsic activity to the enhancement of water adsorption/dissociation on the Ni–Ce and Ni–Zr interfaces developed on these catalysts. On the other hand, they explain the lower intrinsic activity of La-added catalysts in terms of the dilution effect caused by the presence of La on Ni surfaces. La and Ce additives prevented the formation of carbon filaments on Ni surfaces, which was responsible for the changes in product selectivities with reaction time observed on Ni/Al_2O_3, Ni/ZrO_2–Al_2O_3, and Ni/MgO–Al_2O_3.

Zhang et al. (2007) evaluates the effect of catalysts on the steam reforming of bioethanol fuels to produce biohydrogen fuels in a paper with 365 citations. They employ iridium (Ir), Co, and Ni catalysts and ceria catalyst supports. They note that ethanol dehydrogenation to acetaldehyde and ethanol decomposition to methane and carbon monoxide (CO) were the primary reactions at low temperatures, depending on the active metals. At higher temperatures, where all the ethanol and the intermediate compounds, like acetaldehyde and acetone, were completely converted into hydrogen, CO_2, and methane, steam reforming of methane and water gas shift became the major reactions. The Ir/CeO_2 catalyst was significantly more active and selective toward hydrogen production, and they explain the superior catalytic performance in terms of the intimated contact between Ir particles and ceria based on the ceria-mediated redox process.

Aupretre et al. (2002) evaluates the effect of catalysts and catalyst supports on the steam reforming of bioethanol fuels to produce biohydrogen fuels with 313 citations. They employ catalysts of Rh, Pt, Ni, copper (Cu), Zn, and iron (Fe) and the catalyst supports of γ-Al_2O_3, 12%CeO_2–Al_2O_3, CeO_2, and $Ce_{0.63}Zr_{0.37}O_2$. They note that CO_2 was a primary product in the ethanol steam reforming catalytic reaction (SRR) over some supported metal catalysts. They then propose a new strategy for maximizing hydrogen production and minimizing CO formation. They assert that any highly selective catalytic formulation should be free of any promoter in the water gas shift reaction (WGSR), which tends to equilibrate the SRR gas toward higher CO concentrations.

Vizcaino et al. (2007) evaluates the effect of catalyst supports on the steam reforming of bioethanol fuels to produce biohydrogen fuels for the copper (Cu)-Ni catalysts in a paper with 309 citations. They use the catalyst supports of SiO_2 and γ-Al_2O_3, MCM-41 crystalline aluminosilicates, SBA-15 ordered mesoporous silicates, and ZSM-5 aluminosilicate zeolites. They note that aluminum (Al)-containing supports resulted in ethanol dehydration to ethylene in the acid sites, which in turn promoted the coke deactivation process. They obtain the highest hydrogen selectivity with the Cu-Ni/SBA-15 catalyst due to a smaller metallic crystallite size. Nevertheless, the Cu-Ni/SiO_2 catalyst showed the best catalytic performance since a better equilibrium between high hydrogen selectivity and CO_2/CO_x ratio was obtained. Ni was the phase responsible for hydrogen production in a greater grade, although both CO production and coke deposition decreased when Cu was added to the catalyst

Comas et al. (2004) evaluates the effect of temperature on the steam reforming of bioethanol fuels to produce biohydrogen fuels for the Ni/γAl_2O_3 catalyst system in a paper with 302 citations. They note that a higher water/ethanol ratio (6:1) and a higher temperature (773 K) promoted hydrogen production (91% selectivity). Over Ni-based catalyst, there would be no evidence that WGSR occurred. The presence of oxygen in the feed produced a favorable effect on carbon deposition. Nevertheless, CO production was not reduced.

Sun et al. (2005) evaluates the effect of temperature and catalyst supports on the steam reforming of bioethanol fuels to produce biohydrogen fuels for in a paper with 277 citations. They use Ni catalysts and the catalyst supports of yttria (Y_2O_3), La_2O_3, and Al_2O_3. They observe a first-order reaction with respect to ethanol. The catalysts of Ni/Y_2O_3 and Ni/La_2O_3 exhibited relative high activity for ethanol steam reforming at 250°C, with a conversion of ethanol of 81.9% and 80.7%, and a selectivity of hydrogen of 43.1% and 49.5%, respectively. When the temperature reached 320°C, the conversion of ethanol increased to 93.1% and 99.5%, and the selectivity of hydrogen was 53.2% and 48.5%, respectively. The catalyst Ni/Al_2O_3 exhibited relative lower activity for ethanol steam reforming and hydrogen selectivity. However, the three catalysts all had long-term stability for ethanol steam reforming.

Frusteri et al. (2004) evaluates the effect of temperature and catalyst supports on the steam reforming of bioethanol fuels to produce biohydrogen fuels at 650°C in a paper with 277 citations. They employ Pd, Rh, Ni, and Co catalysts with MgO catalyst support. They note that Rh/MgO showed the best performance both in terms of activity and stability, however, it was not so selective toward hydrogen production. Ni, Co, and Pd catalysts were affected by deactivation mainly due to metal sintering. Ni/MgO displayed the best performance in terms of hydrogen selectivity (>95%). They observe a very low coke formation rate on Rh/MgO catalyst; however, even on Ni/MgO catalyst coke formation took place at modest rate. A significant difference in terms of metal-specific activity existed as Rh sites were 2.2, 3.7, and 5.8 times more active than Pd, Co, and Ni, respectively.

Vasudeva et al. (1996) carry out the thermodynamic analysis of the steam reforming of bioethanol to produce biohydrogen fuels under conditions conducive to carbon formation in a paper with 257 citations. They extend the computations to high water-to-ethanol ratios (and higher temperatures) as applicable to the dilute stream produced during the fermentation of molasses. They note that equilibrium hydrogen yields as high as 5.5 moles per mole of ethanol in the feed were obtainable as against the stoichiometric value of 6.0.

Diagne et al. (2002) evaluates the effect of catalyst supports on the steam reforming of bioethanol fuels to produce biohydrogen fuels for the Rh catalyst with the CeO_2 and ZrO_2 catalyst supports in excess of water (1–8 molar ratio) in a paper with 248 citations. They use the catalysts Rh/CeO_2, Rh/ZrO_2, and Rh/CeO_2-ZrO_2 (Ce/Zr=4, 2, and 1). At 400°C–500°C, they note that all catalysts showed high activity and selectivity toward hydrogen production (between 5 and 5.7 mol of H_2 per mol of ethanol inlet) despite the considerable textural differences of the oxides (fluorite, monoclinic, and tetragonal). The large variations of Rh dispersion between all catalysts had a small effect on H_2 production. Although the reaction was not sensitive to either the oxide or the metal structure, Rh/CeO_2 (the most basic catalyst) was the least reactive.

Song and Ozkan (2009) evaluates the effect of catalyst supports on the steam reforming of bioethanol fuels to produce biohydrogen fuels for the Co catalysts with CeO_2 and ZrO_2 catalyst supports in a paper with 246 citations. They focus on the effect of oxygen mobility on the ethanol steam reforming of ZrO_2- and CeO_2-supported Co catalysts. They note that the catalyst deactivation was due mostly to the deposition of various types of carbon on the surface although Co-sintering could also be contributing to the deactivation. They assert that the addition of ceria improved the catalytic stability as well as activity, primarily due to the higher oxygen mobility of ceria.

Garcia and Laborde (1991) carry out the thermodynamic analysis of the steam reforming of bioethanol to produce biohydrogen fuels in a paper with 246 citations. They consider the parameters of the pressure 1–9 atm, temperature 400–800 K, and water-to-ethanol feed ratio 0:1–10:1. They note that the best conditions for hydrogen production occur at T>650 K, atmospheric pressure, and water in excess in the feed. In this condition, methane production was minimized and carbon formation was thermodynamically inhibited. The behavior of the system was very similar to that of steam reforming of methanol, except that in the case of ethanol, higher temperatures and higher water-to-ethanol feed ratios were needed in order to obtain the best hydrogen production.

Haga et al. (1997) evaluates the effect of catalyst supports on the steam reforming of bioethanol fuels to produce biohydrogen fuels for the Co catalysts in a paper with 241 citations. They employ catalyst supports of Al_2O_3, SiO_2, MgO, ZrO_2, and C. They note that the properties of these Co catalysts were greatly affected by the supports. However, Co/Al_2O_3 exhibited the highest selectivity for steam reforming of ethanol by suppression of methanation of CO and decomposition of ethanol.

Cavallaro et al. (2003) evaluates the effect of operating conditions on the steam reforming of bioethanol fuels to produce biohydrogen fuels in a paper with 230 citations. They employ a 5% Rh/Al_2O_3 catalyst system. They focus on the influence of reaction temperature, H_2O/ethanol molar ratio, gas hourly space velocities, and O_2. They note an extensive formation of encapsulated carbon, while there were great benefits, both in terms of catalyst stability and coke formation, from adding 0.4vol.% of oxygen in the reaction stream. Oxygen, however, promoted metal sintering.

Llorca et al. (2003) evaluates the effect of catalyst properties on the steam reforming of bioethanol fuels to produce biohydrogen fuels in a paper with 228 citations. They employ Co/ZnO catalysts. They note that the use of $Co_2(CO)_8$ as a precursor produced a catalyst that was highly stable and selective for the production of CO-free hydrogen at reaction temperatures as low as 623 K. The only byproduct was methane, and selectivity of 73% to H_2 and 25% to CO_2 was obtained. Finally, under reaction conditions, the catalyst showed 92% of reduced Co, mainly as small particles.

Li et al. (2015) evaluates the effect of catalyst supports on the steam reforming of bioethanol fuels to produce biohydrogen fuels in a paper with 225 citations. They use Ce-Ni/SBA-15 and Ni/SBA-15 catalysts and CeO_2 catalyst support. They note that the incorporation of CeO_2 could effectively control the particle size of Ni via strong metal-support interaction and promote the homogeneous distribution of Ni and Ce to achieve a large Ni-CeO_2 interface. Consequently, the

CeNi/SBA-15 catalysts exhibited superior activity with respect to the reference Ni/SBA-15 catalyst without ceria addition. The catalyst with a Ce/Ni atomic ratio of 1 was the optimized composition owing to its strong metal-support interaction and high nickel active surface area. The ceria-promoted catalysts showed enhanced long-term stability in ethanol steam reforming. The strong interaction between Ni and CeO_2, as well as the confinement of SBA-15 support, restricted the Ni particle growth under harsh reaction conditions. Additionally, the ceria promoter contributed to suppressing coke deposition effectively, which led to the enhanced coking-resistance as observed on Ce-Ni/SBA-15 catalysts.

Marino et al. (2001) evaluates the effect of catalyst supports on the steam reforming of bioethanol fuels to produce biohydrogen fuels in a paper with 225 citations. They use a Cu-Ni-K/γ-Al_2O_3 catalyst system. They note that these catalysts produced acceptable amounts of hydrogen working at atmospheric pressure and a temperature of 300°C. Nickel addition enhanced ethanol gasification, increasing the gas yield, and reducing acetaldehyde and acetic acid production.

104.3.2 The Other Methods in the Reforming of Bioethanol Fuels to Produce Biohydrogen Fuels

A brief information about four prolific studies with at least 222 citations each on the other methods in the reforming of bioethanol fuels to produce biohydrogen fuels is given in Table 104.2. Furthermore, brief notes on the contents of these studies are also given. These include the photocatalytic biohydrogen fuel production from bioethanol fuels, the autothermal reforming of bioethanol fuels, and the oxidative steam reforming of bioethanol fuels.

104.3.2.1 The Photocatalytic Biohydrogen Fuel Production from Bioethanol Fuels

Murdoch et al. (2011) evaluates the effect of gold (Au) loading and particle size on photocatalytic hydrogen production from ethanol over Au/TiO_2 nanoparticles (NPs) in a paper with 932 citations. They note that Au particles in the size range of 3–30 nm on TiO_2 were very active in hydrogen production from ethanol. Au particles of similar size on anatase NPs delivered a rate two orders of magnitude higher than that recorded for Au on rutile TiO_2 nanoparticles. Surprisingly, Au particle size did not affect the photoreaction rate over the 3–12 nm range.

Bamwenda et al. (1995) evaluates the effect of the catalyst on the photoassisted hydrogen production from a water-ethanol solution in a paper with 531 citations. They deposit Au and Pt on the

TABLE 104.2
The Other Methods in the Reforming of Bioethanol Fuels to Produce Biohydrogen Fuels

No.	Papers	Catalysts	Catalyst Supports	Issues	Cits.
		Photocatalytic biohydrogen production from bioethanol			
1	Murdoch et al. (2011)	Au	TiO_2	Effect of nanostructure	932
2	Bamwenda et al. (1995)	Au, Pt	TiO_2	Effect of catalyst	531
		Autothermal reforming of bioethanol			
3	Deluga et al. (2004)	Rh	CeO_2	Effect of operating conditions	883
		Oxidative steam reforming of bioethanol			
4	Kugai et al. (2006)	Ni-Rh	CeO_2	Effect of catalyst support properties	222

Cits., the number of the citations received by each paper.

TiO_2 catalyst support. They note that the main reaction products were hydrogen, methane, CO_2, and acetaldehyde. The overall activity of Au samples was generally about 30% lower than that of Pt samples. The activity of Au samples strongly depended on the method of preparation. The activities of the Pt samples were less sensitive to the preparation method. Au and Pt precursors calcined in air at 573 K showed the highest activity toward hydrogen generation, followed by a decline in activity with increasing calcination temperature. The hydrogen yield was dependent on the metal content of TiO_2 and showed a maximum in the ranges of 0.3–1 wt.% Pt and 1–2 wt.% Au. The exposed surface area of Au had only a small influence on the rate of hydrogen generation. On the other hand, the rate of H_2 production was strongly dependent on the initial pH of the suspension. pH values in the range of 4–7 gave better yields, whereas highly acidic and basic suspensions resulted in a considerable decrease in the H_2 yield.

104.3.2.2 The Autothermal Reforming of Bioethanol Fuels to Produce Biohydrogen Fuels

Deluga et al. (2004) evaluates the autothermal reforming of bioethanol fuels to produce biohydrogen fuels in a paper with 883 citations. They convert ethanol and ethanol-water mixtures directly into hydrogen with ~100% selectivity and >95% conversion by catalytic partial oxidation, with a residence time on Rh-CeO_2 catalysts of <10 ms. They perform rapid vaporization and mixing with air with an automotive fuel injector at temperatures sufficiently low and times sufficiently fast that homogeneous reactions producing carbon, acetaldehyde, ethylene, and total combustion products can be minimized.

104.3.2.3 The Oxidative Steam Reforming of Bioethanol Fuels to Produce Biohydrogen Fuels

Kugai et al. (2006) evaluates the oxidative steam reforming (OSR) of bioethanol fuels to produce biohydrogen fuels in a paper with 222 citations. They employ Ni-Rh/CeO_2 catalysts containing 5 wt% Ni and 1 wt% Rh with CeO_2 supports with different crystal sizes and surface areas. They note that the surface areas of these supports increase in the order of CeO_2-I (74 m^2/g) <CeO_2-II (92 m^2/g) <CeO_2-III (154 m^2/g), but their crystallite sizes are about 10.2, 29.3, and 6.5 nm, respectively. They observe that the Rh metal dispersion increased while the Ni metal dispersion decreased with decreasing crystallite sizes of CeO_2. There was an Rh–CeO_2 metal–support interaction as well as Ni–Rh interaction in the Ni–Rh bimetallic catalyst supported on CeO_2-III with a crystallite size of about 6.5 nm. The reduced Ni and Rh species were reversibly oxidized, suggesting the existence of Ni–Rh redox species rather than NiRh surface alloy in the present catalyst system. The Rh species became highly dispersed when the crystallite size of the CeO_2 support was smaller. Both ethanol conversion and H_2 selectivity increased, and the selectivity for undesirable byproducts decreased with increasing Rh metal dispersion. The best catalytic performance for OSR was achieved by supporting Ni–Rh bimetallic catalysts on nanocrystalline CeO_2-III. The Ni–Rh/CeO_2-III catalyst exhibited stable activity and selectivity during on-stream operations at 450°C as well as at 600°C.

104.4 DISCUSSION

104.4.1 Introduction

Crude oil-based gasoline fuels have been widely used in the transportation sector since the 1920s. However, there have been great public concerns over the adverse environmental and human impact of these fuels. Hence, biomass-based bioethanol fuels have increasingly been used in blending gasoline fuels and in fuel cells.

In the meantime, the research in nanomaterials and nanotechnology has intensified in recent years to become a major research field in scientific research, with over one and a half million published papers. In this context, a large number of nanomaterials have been developed for nearly every research field. These materials offer an innovative way to increase efficiency in the production and utilization of bioethanol fuels, as in other scientific fields.

The research in the field of fuel cells has also intensified in recent years, with over 50,000 papers in the three primary research streams of high-temperature SOFCs, low-temperature PEMFCs, and MFCs. The research in the direct ethanol fuel cells (DEFCs) has been a stream of the research on the low-temperature PEMFCs. The primary focus of the research on fuel cells has been the optimization of the operating conditions and the development of catalysts to maximize fuel cell performance. Bioethanol-based biohydrogen fuels have been used as feedstocks in these fuel cells.

Furthermore, research in the use of both biohydrogen fuels and conventional hydrogen fuels in fuel cells has intensified in recent years in the primary research fronts of water-based photocatalytic hydrogen production, water-based electrocatalytic hydrogen production, reforming of bioethanol fuels and other fuels for the production of biohydrogen fuels, biohydrogen production, and hydrogen storage. It is also notable that the research in the fields of hydrogen production and storage has focused on nanomaterial and conventional material development for these fields.

Thus, the research in the field of the production of biohydrogen fuels from bioethanol fuels has progressed in the research fronts of biohydrogen production by the steam reforming of bioethanol and, to a lesser extent, photocatalytic biohydrogen production, biohydrogen production by the oxidative steam reforming of bioethanol, biohydrogen production by the partial oxidation of bioethanol, autothermal reforming of bioethanol, and biohydrogen production from bioethanol in general.

However, it is essential to develop efficient incentive structures for the primary stakeholders to enhance research in this field. Although there have been a number of review papers on bioethanol-based production of biohydrogen fuels, there has been no review of the 25 most-cited articles in this field.

As the first step for the search of the relevant literature, the keywords were selected using the 200 most-cited papers. The selected keyword list was optimized to obtain a representative sample of papers for this research field. This keyword list was provided in the appendix of Konur (2023) for future replication studies.

As the second step, a sample dataset was used in this study. The first 25 articles in the sample of 100 most-cited papers with at least 125 citations each were selected for the review study. Key findings from each paper were taken from the abstracts of these papers and discussed. Additionally, a number of brief conclusions were drawn, and a number of relevant recommendations were made to enhance the future research landscape.

A brief information about the 25 most-cited papers with at least 222 citations each on the production of biohydrogen fuels from bioethanol fuels is given below under two major headings: The steam reforming of bioethanol fuels to produce biohydrogen fuels and the other methods in the reforming of bioethanol fuels to produce biohydrogen fuels with 21 and four papers, respectively. The latter group includes papers on the photocatalytic biohydrogen fuel production from bioethanol fuels, the autothermal reforming of bioethanol fuels to produce biohydrogen fuels, and the oxidative steam reforming of bioethanol fuels to produce biohydrogen fuels.

104.4.2 The Steam Reforming of Bioethanol Fuels to Produce Biohydrogen Fuels

A brief information about 21 prolific studies with at least 225 citations each on the steam reforming of bioethanol to produce biohydrogen fuels is given in Table 104.1. Furthermore, brief notes on the contents of these studies are also given.

The prolific catalysts used in these studies are Co (Frusteri et al., 2004; Haga et al., 1997; Llorca et al., 2002, 2003; Song and Ozkan, 2009; Zhang et al., 2007), Ni (Aupretre et al., 2002; Breen et al., 2002; Comas et al., 2004; Fatsikostas and Verykios, 2004; Fatsikostas et al., 2002; Frusteri et al., 2004; Li et al., 2015; Marino et al., 2001; Sanchez-Sanchez et al., 2007; Sun et al., 2005; Vizcaino et al., 2007; Zhang et al., 2007), Rh (Aupretre et al., 2002; Breen et al., 2002; Cavallaro et al., 2003; Diagne et al., 2002; Frusteri et al., 2004; Liguras et al., 2003), Pt (Aupretre et al., 2002; Breen et al., 2002; Liguras et al., 2003), and Pd (Breen et al., 2002; Frusteri et al., 2004; Liguras et al., 2003). The other catalysts used are Ir (Zhang et al., 2007), Ru (Liguras et al., 2003), Zn (Aupretre et al., 2002), Cu (Aupretre et al., 2002; Vizcaino et al., 2007), and Ce (Li et al., 2015).

The prolific catalyst supports used in these studies are Al_2O_3 (Aupretre et al., 2002; Breen et al., 2002; Cavallaro et al., 2003; Haga et al., 1997; Liguras et al., 2003; Sanchez-Sanchez et al., 2007; Sun et al., 2005), γ-Al_2O_3 (Aupretre et al., 2002; Comas et al., 2004; Fatsikostas and Verykios, 2004; Fatsikostas et al., 2002; Llorca et al., 2002, Marino et al., 2001; Vizcaino et al., 2007), MgO (Frusteri et al., 2004; Haga et al., 1997; Liguras et al., 2003; Llorca et al., 2002), La_2O_3 (Fatsikostas and Verykios, 2004; Fatsikostas et al., 2002; Llorca et al., 2002; Sanchez-Sanchez et al., 2007; Sun et al., 2005), CeO_2 (Aupretre et al., 2002; Breen et al., 2002; Diagne et al., 2002; Li et al., 2015; Llorca et al., 2002; Sanchez-Sanchez et al., 2007; Song and Ozkan, 2009; Zhang et al., 2007), and ZrO_2 (Breen et al., 2002; Diagne et al., 2002; Haga et al., 1997; Sanchez-Sanchez et al., 2007; Song and Ozkan, 2009). The other catalyst supports used are SiO_2 (Haga et al., 1997; Llorca et al., 2002; Vizcaino et al., 2007), TiO_2 (Liguras et al., 2003; Llorca et al., 2002), ZnO (Llorca et al., 2002, 2003), V_2O_5 (Llorca et al., 2002), Sm_2O_3 (Llorca et al., 2002), yttria stabilized Zr (Fatsikostas et al, 2002), MCM-41, SBA-15, ZSM-5 (Vizcaino et al., 2007), and C (Haga et al., 1997).

Twelve of these prolific studies evaluate the effect of catalyst supports on bioethanol-based biohydrogen fuel production, while seven studies evaluate the effect of catalysts and their properties on bioethanol-based biohydrogen fuel production. Two studies carry out the thermodynamic analysis of bioethanol-based biohydrogen fuel production (Garcia and Laborde, 1991; Vasudeva et al., 1996). Additionally, three papers evaluate the effect of operating conditions such as temperature on bioethanol-based biohydrogen fuel production (Cavallaro et al., 2003; Comas et al., 2004; Sun et al, 2005).

These findings are thought-provoking in seeking ways to optimize bioethanol-based biohydrogen fuel production. The types and properties of the catalysts and the catalyst supports emerge as the key variables impacting on bioethanol-based biohydrogen fuel production.

104.4.3 The Other Methods in the Reforming of Bioethanol Fuels to Produce Biohydrogen Fuels

A brief information about four prolific studies with at least 222 citations each on the other methods in the reforming of bioethanol fuels to produce biohydrogen fuels is given in Table 104.2. Furthermore, brief notes on the contents of these studies are also given. These include the photocatalytic biohydrogen fuel production from bioethanol fuels, the autothermal reforming of bioethanol fuels, and the oxidative steam reforming of bioethanol fuels.

The studied catalysts in these prolific studies are Au (Bamwenda et al., 1995; Murdoch et al., 2011), Rh (Deluga et al., 2004; Kugai et al., 2006), Pt (Bamwenda et al., 1995), and Ni (Kugai et al., 2006). Similarly, the studied catalyst supports are TiO_2 (Bamwenda et al., 1995; Murdoch et al., 2011) and CeO_2 (Deluga et al., 2004; Kugai et al., 2006).

These studies evaluate the effect of the catalyst nanostructure (Murdoch et al., 2011), catalyst type (Bamwenda et al., 1995), catalyst support properties (Kugai et al., 2006), and operating conditions (Deluga et al., 2004).

These findings are thought-provoking in seeking ways to optimize bioethanol-based biohydrogen fuel production through autothermal, partial oxidation, and oxidative steam reforming of bioethanol fuels. The types and properties of the catalysts and the catalyst supports emerge as the key variables impacting bioethanol-based biohydrogen fuel production. Additionally, one study highlights the importance of nanomaterials in catalyst and catalyst support development (Murdoch et al., 2011).

104.4.4 The Overall Remarks

A number of the prolific studies reviewed in the preceding sections hint that the primary catalysts used and researched for bioethanol-based biohydrogen fuel production are Co (Jongsomjit et al., 2001), Ni (Sehested, 2006), Rh (Biloen et al., 1980), Pt (Chen et al., 2016), and Pd (Toebes et al., 2001). As these studies suggest, these catalysts have a wide range of applications besides bioethanol-based biohydrogen fuel production with the satisfactory results.

Additionally, a number of the prolific studies reviewed in the preceding sections hint that the primary catalyst supports used and researched for bioethanol-based biohydrogen fuel production are Al_2O_3 (Zielinski, 1982), γ-Al_2O_3 (Ravenelle et al., 2011), MgO (Diez et al., 2000), La_2O_3 (Cornaglia et al., 2004), CeO_2 (Kim et al., 2012), and ZrO_2 (Heuer et al., 1982). As these studies suggest, these catalyst supports have a wide range of applications besides bioethanol-based biohydrogen fuel production with the satisfactory results.

The research in the field of fuel cells has intensified in recent years, with over 50,000 papers in three primary research streams of high-temperature SOFCs (Adler, 2004), low-temperature PEMFCs (Borup et al., 2007), and MFCs (Logan, 2009). The primary focus of the research on fuel cells has been the optimization of the operating conditions to maximize fuel cell performance. Biohydrogen fuels have been used in these fuel cells to produce bioelectricity.

Therefore, it is surprising that there have been a limited number of prolific papers among the most cited 25 papers on the reforming processes themselves (Haryanto et al., 2005). This finding suggests that the development of efficient catalysts and catalyst supports is highly critical for the development of efficient reforming processes for bioethanol-based biohydrogen fuel production.

The findings narrated in the previous sections are thought-provoking in seeking ways to optimize bioethanol-based biohydrogen fuel production using a number of prolific catalysts and catalyst supports. The nanostructure, catalyst supports, catalysts, and operating conditions of the biohydrogen production emerge as the key variables impacting bioethanol-based biohydrogen fuel production through the reforming of bioethanol fuels.

One innovative way to enhance hydrogen production has been the use of nanomaterials for both the catalysts themselves and the catalyst supports (Jaramillo et al., 2007). In parallel with the research on fuel cells, the research on nanomaterials and nanotechnology (Geim, 2009; Geim and Novoselov, 2007) has intensified in recent years with over one and a half million papers, enriching the material portfolio to be used for fuel cell applications. It is expected that this enriched portfolio of innovative nanomaterials would enhance bioethanol-based biohydrogen fuel production as both catalyst materials and catalyst supports in the future, benefiting from the superior properties of these nanomaterials.

In the end, these most-cited papers in this field hint that the efficiency of bioethanol-based biohydrogen fuel production could be optimized using the structure, processing, and property relationships of the nanomaterials and conventional materials used as both catalyst materials and catalyst supports (Formela et al., 2016; Konur, 2018, 2020b, 2021a,b,c,d; Konur and Matthews, 1989).

104.5 CONCLUSION AND FUTURE RESEARCH

A brief information about the key research fronts covered by the 25 most-cited papers with at least 222 citations each in the field of bioethanol-based biohydrogen fuels is given in Table 104.3.

There are two major research fronts for this field: the steam reforming of bioethanol fuels to produce biohydrogen fuels and the other methods in the reforming of bioethanol fuels to produce biohydrogen fuels, with 21 and four papers, respectively. The latter group includes papers on the photocatalytic biohydrogen fuel production from bioethanol fuels, the autothermal reforming of bioethanol fuels to produce biohydrogen fuels, and the oxidative steam reforming of bioethanol fuels to produce biohydrogen fuels. These finding hints that steam reforming of bioethanol fuels has been the gold standard for the conversion of bioethanol fuel to biohydrogen fuels through ethanol reforming.

It is notable that there is similarity of these research fronts in the sample of reviewed papers and the sample of the 100 most-cited papers in this field. This finding suggests that the set of these 25 most prolific papers is a representative sample of the wider research on bioethanol-based biohydrogen fuels. However, there is a slight overrepresentation and underrepresentation in the fields of steam reforming and other reforming methods, respectively.

The research in the field of fuel cells has intensified in recent years in three primary research streams of high-temperature SOFCs, low-temperature PEMFCs, and MFCs. The research in the

TABLE 104.3
The Most Prolific Research Fronts in Bioethanol-based Biohydrogen Fuel Production

No.	Research Fronts	Sample Papers (%)	Reviewed Papers (%)
1	Biohydrogen production by the steam reforming of bioethanol	73	84
2	Photocatalytic biohydrogen production	11	8
3	Biohydrogen production by the oxidative steam reforming of bioethanol	9	4
4	Biohydrogen production by the partial oxidation of bioethanol	6	0
5	Autothermal reforming of bioethanol	5	4
6	Biohydrogen production from bioethanol in general	5	0
	Sample	100	25

Reviewed papers, the total number of reviewed papers = 25; Sample papers, the sample of the 100 most-cited papers.

DEFCs has been a stream of the research on the low-temperature PEMFCs. The primary focus of the research on fuel cells has been the optimization of the operating conditions to maximize fuel cell performance. Biohydrogen fuels produced through ethanol reforming have been one of the most used feedstocks for these fuel cells.

Furthermore, the research in the use of both biohydrogen fuels and conventional hydrogen fuels in fuel cells has intensified in recent years in the primary research fronts of water-based photocatalytic hydrogen production, water-based electrocatalytic hydrogen production, reforming of bioethanol fuels and other alcoholic fuels for the production of biohydrogen fuels, biohydrogen production, and hydrogen storage. It is also notable that the research in the fields of hydrogen production and storage has focused on the nanomaterial development for the catalysts and catalyst supports for these fields.

Therefore, it is not surprising that there have been a limited number of prolific papers among the 25 most-cited papers on the reforming processes themselves. This finding suggests that the development of efficient catalysts and catalyst supports is highly critical for the development of efficient reforming processes for bioethanol-based biohydrogen fuel production as for the development of materials for the catalysts and catalyst supports for water-based biohydrogen production and storage.

The findings narrated in the previous sections are thought-provoking in seeking ways to optimize bioethanol-based biohydrogen fuel production using a number of prolific catalysts and catalyst supports. The nanostructure, catalyst supports, catalysts, and operating conditions of the biohydrogen production emerge as the key variables impacting bioethanol-based biohydrogen fuel production through the reforming of bioethanol fuels.

One innovative way to enhance hydrogen production has been the use of nanomaterials for both the catalysts themselves and the catalyst supports. In parallel with the research on fuel cells, the research on nanomaterials and nanotechnology has intensified in recent years, enriching the material portfolio to be used for bioethanol-based hydrogen fuel production applications. It is expected that this enriched portfolio of innovative nanomaterials would enhance bioethanol-based biohydrogen fuel production as both catalyst materials and catalyst support in the future, benefiting from the superior properties of these nanomaterials as for the development of materials for the catalysts and catalyst supports for water-based biohydrogen production and storage.

In the end, these most-cited papers in this field hint that the efficiency of bioethanol-based biohydrogen fuel production could be optimized using the structure, processing, and property relationships of the nanomaterials and conventional materials used as both catalyst materials and catalyst supports.

It is recommended that such review studies should be performed for the other research fronts on both the production and utilization of bioethanol fuels, complementing the corresponding scientometric studies.

ACKNOWLEDGMENTS

The contribution of the highly cited researchers in the field of bioethanol-based biohydrogen fuels has been gratefully acknowledged.

REFERENCES

Adler, S. B. 2004. Factors governing oxygen reduction in solid oxide fuel cell cathodes. *Chemical Reviews* 104. 4791–4843.

Akande, A. J., R. O. Idem and A. K. Dalai. 2005. Synthesis, characterization and performance evaluation of Ni/Al_2O_3 catalysts for reforming of crude ethanol for hydrogen production. Applied Catalysis A: General 287:159–175.

Antolini, E. 2007. Catalysts for direct ethanol fuel cells. *Journal of Power Sources* 170:1–12.

Antolini, E. 2009. Palladium in fuel cell catalysis. *Energy and Environmental Science* 2:915–931.

Aupretre, F., C. Descorme and P. Duprez. 2002. Bio-ethanol catalytic steam reforming over supported metal catalysts. *Catalysis Communications* 3:263–267.

Bamwenda, G. R., S. Tsubota, T. Nakamura and M. Haruta. 1995. Photoassisted hydrogen-production from a water-ethanol solution - a comparison of activities of Au-TiO_2 and Pt- TiO_2. *Journal of Photochemistry and Photobiology A-Chemistry* 89:177–189

Biloen, P., J. N. Helle, H. Verbeek, F. M. Dautzenberg and W. M. H. Sachtler. 1980. The role of rhenium and sulfur in platinum-based hydrocarbon-conversion catalysts. *Journal of Catalysis* 63:112–118.

Borup, R., J. Meyers and B. Pivovar, et al. 2007. Scientific aspects of polymer electrolyte fuel cell durability and degradation. *Chemical Reviews* 107:3904–3951.

Breen, J. P., R. Burch and H. M. Coleman. 2002. Metal-catalysed steam reforming of ethanol in the production of hydrogen for fuel cell applications. *Applied Catalysis B: Environmental* 39:65–74.

Burnham, J. F. 2006. Scopus database: A review. *Biomedical Digital Libraries* 3:1–8.

Cavallaro, S., V. Chiodo, S. Freni, N. Mondello and F. Frusteri. 2003. Performance of Rh/Al_2O_3 catalyst in the steam reforming of ethanol: H_2 production for MCFC. *Applied Catalysis A: General* 249:119–128.

Chen, G., C. Xu and X. Huang, et al. 2016. Interfacial electronic effects control the reaction selectivity of platinum catalysts. *Nature Materials* 15:564–569.

Chen, X., S. Shen, L. Guo and S. S. Mao. 2010. Semiconductor-based photocatalytic hydrogen generation. *Chemical Reviews* 110:6503–6570.

Comas, J., F. Marino, M. Laborde and N. Amadeo. 2004. Bio-ethanol steam reforming on Ni/Al_2O_3 catalyst. *Chemical Engineering Journal* 98:61–68.

Cornaglia, L. M., J. Munera, S. Irusta and E. A. Lombardo. 2004. Raman studies of Rh and Pt on La_2O_3 catalysts used in a membrane reactor for hydrogen production. *Applied Catalysis A: General* 263:91–101.

Das, D. and T.N. Veziroglu. 2001. Hydrogen production by biological processes: A survey of literature. *International Journal of Hydrogen Energy* 26:13–28.

De Lima, S. M., I. O. da Cruz and G. Jacobs, et al. 2008. Steam reforming, partial oxidation, and oxidative steam reforming of ethanol over $Pt/CeZrO_2$ catalyst. Journal of Catalysis 257:356–368.

Deluga, G. A., J. R. Salge, L. D. Schmidt and X. E. Verykios. 2004. Renewable hydrogen from ethanol by autothermal reforming. *Science* 303:993–997.

Diagne, C., H. Idriss and A. Kiennemann. 2002. Hydrogen production by ethanol reforming over Rh/CeO_2-ZrO_2 catalysts. *Catalysis Communications* 3:565–571.

Diez, V. K., C. R. Apesteguia and J. I. di Cosimo. 2000. Acid-base properties and active site requirements for elimination reactions on alkali-promoted MgO catalysts. *Catalysis Today* 63:53–62.

Fatsikostas, A. N., D. I. Kondarides and X. E. Verykios. 2002. Production of hydrogen for fuel cells by reformation of biomass-derived ethanol. *Catalysis Today* 75:145–155.

Fatsikostas, A. N. and X. E. Verykios. 2004. Reaction network of steam reforming of ethanol over Ni-based catalysts. *Journal of Catalysis* 225:439–452.

Formela, K., A. Hejna, L. Piszczyk, M. R. Saeb and X. Colom. 2016. Processing and structure-property relationships of natural rubber/wheat bran biocomposites. *Cellulose* 23:3157–3175.

Frusteri, F., S. Freni and L. Spadaro, et al. 2004. H_2 production for MC fuel cell by steam reforming of ethanol over MgO supported Pd, Rh, Ni and Co catalysts. *Catalysis Communications* 5:611–615.

Garcia, E. Y. and M. A. Laborde. 1991. Hydrogen production by the steam reforming of ethanol: Thermodynamic analysis. *International Journal of Hydrogen Energy* 16:307–312.

Geim, A. K. 2009. Graphene: Status and prospects. *Science* 324:1530–1534.

Geim, A. K. and K. S. Novoselov. 2007. The rise of graphene. *Nature Materials* 6:183–191.

Gombac, V., L. Sordelli and T. Montini. 2010. CuOx-TiO_2 photocatalysts for H_2 production from ethanol and glycerol solutions. *Journal of Physical Chemistry A* 114:3916–3925.

Haga, F., T. Nakajima, H. Miya and S. Mishima. 1997. Catalytic properties of supported cobalt catalysts for steam reforming of ethanol. *Catalysis Letters* 48:223–227.

Haryanto, A., S. Fernando, M. Murali and S. Adhikari. 2005. Current status of hydrogen production techniques by steam reforming of ethanol: A review. *Energy and Fuels* 19:2098–2106.

Heuer, A. H., N. Claussen, W. M. Kriven and M. Ruhle. 1982. Stability of tetragonal ZrO_2 particles in ceramic matrices. *Journal of the American Ceramic Society* 65:642–650.

Hill, J., E. Nelson, D. Tilman, S. Polasky and D. Tiffany. 2006. Environmental, economic, and energetic costs and benefits of biodiesel and ethanol biofuels. *Proceedings of the National Academy of Sciences of the United States of America* 103:11206–11210.

Hill, J., S. Polasky and E. Nelson, et al. 2009. Climate change and health costs of air emissions from biofuels and gasoline. *Proceedings of the National Academy of Sciences of the United States of America* 106:2077–2082.

Hsieh, W. D., R. H. Chen, T. L. Wu and T. H. Lin. 2002. Engine performance and pollutant emission of an SI engine using ethanol-gasoline blended fuels. *Atmospheric Environment* 36:403–410.

Jaramillo, T. F., K. P. Jorgensen and J. Bonde, et al. 2007. Identification of active edge sites for electrochemical H_2 evolution from MoS_2 nanocatalysts. *Science* 317:100–102.

Jongsomjit, B., J. Panpranot and J. G. Goodwin. 2001. Co-support compound formation in alumina-supported cobalt catalysts. *Journal of Catalysis* 204:98–109.

Kim, H. Y., H. M. Lee and G. Henkelman. 2012. CO oxidation mechanism on CeO_2-supported Au nanoparticles. *Journal of the American Chemical Society* 134:1560–1570.

Konur, O. 2000. Creating enforceable civil rights for disabled students in higher education: An institutional theory perspective. *Disability & Society* 15:1041–1063.

Konur, O. 2002a. Access to nursing education by disabled students: Rights and duties of nursing programs. *Nurse Education Today* 22:364–374.

Konur, O. 2002b. Assessment of disabled students in higher education: Current public policy issues. *Assessment and Evaluation in Higher Education* 27:131–152.

Konur, O. 2002c. Access to employment by disabled people in the UK: Is the Disability Discrimination Act working? *International Journal of Discrimination and the Law* 5:247–279.

Konur, O. 2006a. Participation of children with dyslexia in compulsory education: Current public policy issues. *Dyslexia* 12:51–67.

Konur, O. 2006b. Teaching disabled students in higher education. *Teaching in Higher Education* 11:351–363.

Konur, O. 2007a. A judicial outcome analysis of the *Disability Discrimination Act*: A windfall for the employers? *Disability & Society* 22:187–204.

Konur, O. 2007b. Computer-assisted teaching and assessment of disabled students in higher education: The interface between academic standards and disability rights. *Journal of Computer Assisted Learning* 23:207–219.

Konur, O. 2012. The evaluation of the research on the bioethanol: A scientometric approach. *Energy Education Science and Technology Part A: Energy Science and Research* 28:1051–1064.

Konur, O. 2015. Current state of research on algal bioethanol. In *Marine Bioenergy: Trends and Developments*, Ed. S. K. Kim and C. G. Lee, pp. 217–244. Boca Raton, FL: CRC Press.

Konur, O., Ed. 2018. *Bioenergy and Biofuels*. Boca Raton, FL: CRC Press.

Konur, O. 2019. Cyanobacterial bioenergy and biofuels science and technology: A scientometric overview. In *Cyanobacteria: From Basic Science to Applications*, Ed. A. K. Mishra, D. N. Tiwari and A. N. Rai, pp. 419–442. Amsterdam: Elsevier.

Konur, O. 2020a. The scientometric analysis of the research on the bioethanol production from green macroalgae. In *Handbook of Algal Science, Technology and Medicine*, Ed. O. Konur, pp. 385–401. London: Academic Press.

Konur, O., Ed. 2020b. *Handbook of Algal Science, Technology and Medicine.* London: Academic Press.

Konur, O., Ed. 2021a. *Handbook of Biodiesel and Petrodiesel Fuels: Science, Technology, Health, and Environment*. Boca Raton, FL: CRC Press.

Konur, O., Ed. 2021b. *Handbook of Biodiesel and Petrodiesel Fuels: Science, Technology, Health, and Environment. Volume 1. Biodiesel Fuels: Science, Technology, Health, and Environment.* Boca Raton, FL: CRC Press.

Konur, O., Ed. 2021c. *Handbook of Biodiesel and Petrodiesel Fuels: Science, Technology, Health, and Environment. Volume 2. Biodiesel Fuels based on the Edible and Nonedible Feedstocks, Wastes, and Algae: Science, Technology, Health, and Environment.* Boca Raton, FL: CRC Press.

Konur, O., Ed. 2021d. *Handbook of Biodiesel and Petrodiesel Fuels: Science, Technology, Health, and Environment. Volume 3. Petrodiesel Fuels: Science, Technology, Health, and Environment.* Boca Raton, FL: CRC Press.

Konur, O. 2023. Bioethanol fuel-based biohydrogen fuels: Scientometric study. In *Evaluation and Utilization of Bioethanol Fuels. II.: Biohydrogen Fuels, Fuel Cells, Biochemicals, and Country Experiences. Handbook of Bioethanol Fuels Volume 6*, Ed. O. Konur, pp. 215–236. Boca Raton, FL: CRC Press.

Konur, O. and F. L. Matthews. 1989. Effect of the properties of the constituents on the fatigue performance of composites: A review. *Composites* 20:317–328.

Kugai, J., V. Subramani, C. Song, M. H. Engelhard and Y. H. Chin. 2006. Effects of nanocrystalline CeO_2 supports on the properties and performance of Ni-Rh bimetallic catalyst for oxidative steam reforming of ethanol. *Journal of Catalysis* 238:430–440.

Li, D., L. Zeng and X. Li, et al. 2015. Ceria-promoted Ni/SBA-15 catalysts for ethanol steam reforming with enhanced activity and resistance to deactivation. *Applied Catalysis B: Environmental* 176–177:532–541.

Liguras, D. K., D. I. Kondarides and X. E. Verykios. 2003. Production of hydrogen for fuel cells by steam reforming of ethanol over supported noble metal catalysts. *Applied Catalysis B: Environmental* 43:345–354.

Llorca, J., P. R. De la Piscina, J. A. Dalmon, J. Sales and N. Homs. 2003. Co-free hydrogen from steam-reforming of bioethanol over ZnO-supported cobalt catalysts: Effect of the metallic precursor. *Applied Catalysis B: Environmental* 43:355–369.

Llorca, J., N. Homs, J. Sales and P. R. De la Piscina. 2002. Efficient production of hydrogen over supported cobalt catalysts from ethanol steam reforming. *Journal of Catalysis* 209:306–317.

Logan, B. E. 2009. Exoelectrogenic bacteria that power microbial fuel cells. *Nature Reviews Microbiology* 7:375–381.

Ma, X., L. Sun and C. Song. 2002. A new approach to deep desulfurization of gasoline, diesel fuel and jet fuel by selective adsorption for ultra-clean fuels and for fuel cell applications. *Catalysis Today* 77:107–116.

Marino, F., M. Boveri, G. Baronetti and M. Laborde. 2001. Hydrogen production from steam reforming of bioethanol using Cu/Ni/K/γ-Al_2O_3 catalysts. Effect of Ni. *International Journal of Hydrogen Energy* 26:665–668.

Mattos, L. V., G. Jacobs, B. H. Davis and F. B. Noronha. 2012. Production of hydrogen from ethanol: Review of reaction mechanism and catalyst deactivation. *Chemical Reviews* 112:4094–4123.

Murdoch, M., G. I. N. Waterhouse and M. A. Nadeem, et al. 2011. The effect of gold loading and particle size on photocatalytic hydrogen production from ethanol over Au/TiO_2 nanoparticles. *Nature Chemistry* 3:489–492.

Najafi, G., B. Ghobadian and T. Tavakoli, et al. 2009. Performance and exhaust emissions of a gasoline engine with ethanol blended gasoline fuels using artificial neural network. *Applied Energy* 86:630–639.

Newman, P. W. G. and J. R. Kenworthy. 1989. Gasoline consumption and cities: A comparison of U.S. cities with a global survey. *Journal of the American Planning Association* 55:24–37.

Ni, M., D. Y. C. Leung and M. K. H. Leung. 2007. A review on reforming bio-ethanol for hydrogen production. *International Journal of Hydrogen Energy* 32:3238–3247.

Norskov, J. K., T. Bligaard and A. Logadottir, et al. 2005. Trends in the exchange current for hydrogen evolution. *Journal of the Electrochemical Society* 152:J23–J26.

North, D. C. 1991. Institutions. *Journal of Economic Perspectives* 5:97–112.

Orimo, S. I., Y. Nakamori, J. R. Eliseo, A. Zuttel and C. M. Jensen. 2007. Complex hydrides for hydrogen storage. *Chemical Reviews* 107:4111–4132.

Rabenstein, G. and V. Hacker. 2008. Hydrogen for fuel cells from ethanol by steam-reforming, partialoxidation and combined auto-thermal reforming: A thermodynamic analysis. Journal of Power Sources 185:1293–1304.

Ravenelle, R. M., J. R. Copeland, W. G. Kim, J. C. Crittenden and C. Sievers. 2011. Structural changes of γ-Al_2O_3-supported catalysts in hot liquid water. *ACS Catalysis* 1:552–561.

Sanchez-Sanchez, M. C., R. M. Navarro and J. L. G. Fierro. 2007. Ethanol steam reforming over Ni/M_xO_y-Al_2O_3 (M=Ce, La, Zr and Mg) catalysts: Influence of support on the hydrogen production. *International Journal of Hydrogen Energy* 32:1462–1471.

Sehested, J. 2006. Four challenges for nickel steam-reforming catalysts. *Catalysis Today* 111:103–110.

Song, H. and U. S. Ozkan. 2009. Ethanol steam reforming over Co-based catalysts: Role of oxygen mobility. *Journal of Catalysis* 261:66–74.

Sun, J., X. P. Qiu, F. Wu and W. T. Zhu. 2005. H_2 from steam reforming of ethanol at low temperature over Ni/Y_2O_3 and Ni/La_2O_3 catalysts for fuel-cell application. *International Journal of Hydrogen Energy* 30:437–445.

Toebes, M. L., J. A. van Dillen and K. P. de Jong. 2001. Synthesis of supported palladium catalysts. *Journal of Molecular Catalysis A: Chemical* 173:75–98.

Vasudeva, K., N. Mitra, P. Umasankar and S. C. Dhingra. 1996. Steam reforming of ethanol for hydrogen production: Thermodynamic analysis. *International Journal of Hydrogen Energy* 21:13–18.

Velu, S., K. Suzuki, M. Vijayaraj, S. Barman and C. S. Gopinath. 2005. In situ XPS investigations of Cu1-xNixZnAl-mixed metal oxide catalysts used in the oxidative steam reforming of bio-ethanol. Applied Catalysis B: Environmental 55:287–299.

Vizcaino, A. J., A. Carrero and J. A. Calles. 2007. Hydrogen production by ethanol steam reforming over Cu-Ni supported catalysts. *International Journal of Hydrogen Energy* 32:1450–1461.

Zhang, B., X. Tang, Y. Li, Y. Xu and W. Shen. 2007. Hydrogen production from steam reforming of ethanol and glycerol over ceria-supported metal catalysts. *International Journal of Hydrogen Energy* 32:2367–2373.

Zielinski, J. 1982. Morphology of nickel/alumina catalysts. *Journal of Catalysis* 76:157–163.

105 Recent Advances in the Steam Reforming of Bioethanol for the Production of Biohydrogen Fuels

Thanh Khoa Phung and Khanh B. Vu
International University
Vietnam National University

105.1 INTRODUCTION

Hydrogen, a feedstock for fuel cells, electricity generation, and transportation, is a highly efficient energy carrier (143 MJ/kg) and is considered one of the promising clean energy carriers for the future (Muradov and Veziroglu, 2008; Phung and Busca, 2020). Hydrogen is also utilized in the production of ammonia, fertilizers, and the petroleum industry (Nazir et al., 2020). The demand for hydrogen is expected to increase due to the requirement of using renewable energy instead of fossil energy to reduce the greenhouse gas (GHG) effect, in particular in using hydrogen in fuel cells (Roychowdhury et al. 2021). Currently, hydrogen is mainly produced by steam reforming of methane and oil-derived hydrocarbons with nearly 95% hydrogen production worldwide (Rakib et al., 2010; Sliwa and Samson, 2021). Using natural gas- or oil-derived hydrocarbons, on the other hand, is a dirty process that contributes to CO_2 emissions. Therefore, hydrogen needs to be produced from green processes using green resources.

Besides hydrogen production from steam reforming of gas-derived fossil resources, hydrogen can be obtained by photocatalytic methods (Ganguly et al., 2019), water electrolysis (Kovac et al., 2019), and steam reforming of several renewable organic compounds such as methanol, ethanol, glycerol, acetic acid, and acetol (Campos et al., 2019; Cerda-Moreno et al., 2019; Ogo and Sekine, 2020). Among mentioned methods, the ethanol steam reforming (ESR) process is green and has been receiving much attention because of hydrogen production from green ethanol arising from biomass via the conventional fermentation process (Olsson and Hahn-Hagerdal, 1996). According to Sliwa and Samson (2021), ethanol has low toxicity and volatility, high solubility in water, is easy to transport and store, and provides a high hydrogen yield in comparison with steam reforming of other chemicals.

Hydrogen (six moles) can be produced from ethanol (one mole) through the ESR reaction $C_2H_5OH+3H_2O \rightarrow 2CO_2+6H_2$ ($\Delta H=+173.5$ kJ/mol) (Ogo and Sekine, 2020). However, the amount of hydrogen decreases if other byproducts are produced during the ESR process. In fact, along with the steam reforming of ethanol, there are several reactions including hydrogenation of ethanol, dehydrogenation, acetaldehyde decomposition, steam reforming of acetaldehyde, water–gas shift reaction, and dehydration of ethanol to ethylene can take place.

$$C_2H_5OH+3H_2O \rightarrow 2CO_2+6H_2 \qquad (105.1)$$

DOI: 10.1201/9781003226574-139

$$C_2H_5OH+H_2O \rightarrow 2CO+4H_2 \quad (105.2)$$

$$C_2H_5OH+2H_2 \rightarrow 2CH_4+H_2O \quad (105.3)$$

$$C_2H_5OH \rightarrow CH_3CHO+H_2 \quad (105.4)$$

$$CH_3CHO \rightarrow CH_4+CO \quad (105.5)$$

$$CH_3CHO+H_2O \rightarrow 2CO+3H_2 \quad (105.6)$$

$$CO+H_2O \rightarrow CO_2+H_2 \quad (105.7)$$

$$C_2H_5OH \rightarrow C_2H_4+H_2O \quad (105.8)$$

ESR is an endothermic reaction; therefore, hydrogen production via ESR has to be performed at high temperatures. In conditions of thermodynamic control of a starting ethanol/H_2O mole ratio of 1/3 using the Soave–Redlich–Kwong (SRK) equation at atmospheric pressure (Reid et al., 1987) and HSC Chemistry 5.11 program. Figure 105.1 shows that the amount of hydrogen increases with an increase in reaction temperature from 0°C to 300°C and approaches a maximum yield above 400°C (Figure 105.1a). In the case of the presence of other byproducts such as CO, CO_2, and CH_4 along with H_2 (Figure 105.1b), the amount of hydrogen increases with an increase in reaction temperature but decreases after 800°C due to the reverse water–gas shift reaction. High reaction temperatures favor side reactions such as ethanol dehydration and subsequent hydrogen reactions (such as the reverse water–gas shift reaction and hydrogenation), which diminish overall hydrogen production (Chen et al., 2018).

Many catalysts, including noble metals and non-noble metals, have been discovered and used to lessen the effect of side reactions and increase the rate of ESR. (Abello et al., 2013; Aguero et al., 2015; Aupretre et al., 2002; Chiu et al., 2013; Diagne et al., 2004; Erdohelyi et al., 2006; Frusteri et al., 2004; Liguras et al., 2003). Noble catalysts, including Ru, Rh, Pd, and Pt, were reported as excellent catalysts for ESR, which can give a hydrogen yield of 96% (Aupretre et al., 2002;

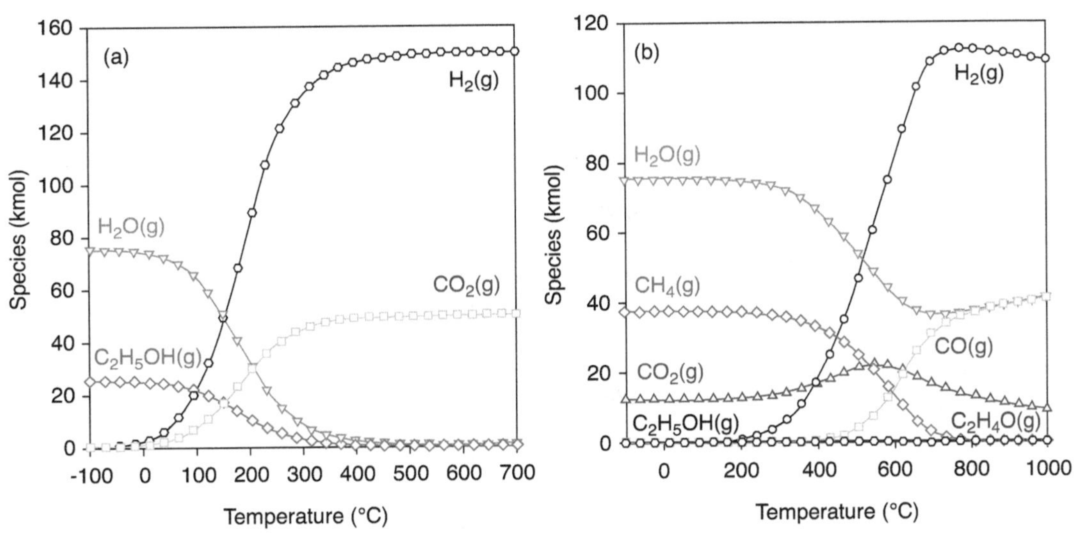

FIGURE 105.1 Thermodynamic equilibrium curves as a function of temperature in the gas phase of (a) ethanol, water, hydrogen and carbon dioxide; and (b) ethanol, water, acetaldehyde, hydrogen, CO and CO_2. (Reproduced from Ref. Phung and Busca (2020).)

Chiu et al., 2013; Diagne et al., 2004; Erdohelyi et al., 2006; Frusteri et al., 2004; Liguras et al., 2003). Non-noble metals such as Co, Cu, and Ni are also employed in the ESR process (Akande et al., 2005; Comas et al., 2004; Fatsikostas and Verykios, 2004). In this chapter, we will summarize the active metals for the ESR process, in particular, focusing on noble metals,- Ni-and Co-based catalysts, the support, promoters, and reaction conditions that affect the yield of hydrogen and the reaction mechanism.

105.2 NOBLE METAL CATALYSTS

Noble metal catalysts, such as Ru, Rh, Pd, and Pt, are well-known and very active catalysts for ESR (Aupretre et al., 2002; Chiu et al., 2013; Diagne et al., 2004; Erdohelyi et al., 2006; Frusteri et al., 2004; Liguras et al., 2003). Among them, Rh and Ru are excellent catalysts for ESR. However, their activities depend on the catalyst loading (Chiu et al., 2013; Diagne et al., 2004; Frusteri et al., 2004). Rh shows high activity at low catalyst loading (0–5 wt.%). In contrast, Ru exhibits high catalytic activity at high catalyst loading (Chiu et al., 2013; Diagne et al., 2004; Frusteri et al., 2004). In Table 105.1, Rh is the best catalyst with 100% ethanol conversion and ≥95% hydrogen selectivity (Liguras et al., 2003). In fact, Rh is an excellent catalyst for ESR and has a good capacity for the

TABLE 105.1
A List of Noble Catalysts that are Utilized in Ethanol Steam Reforming

Catalyst	Metal Loading (wt.%)	Temp. (°C)	$C_2H_5OH/H_2O/O_2$ (molar ratio)	Ethanol Conv. (%)	Hydrogen Selectivity (%)	Ref.
Rh/CeO_2	2	300	8/1	58.5	59.7	Diagne et al. (2004)
		400		100	66.3	
		450		100	69.1	
Rh/ZrO_2		300		100	57.4	
		450		100	70.3	
Ru/CeO_2	1	450	-	>90	57 (20 min)	Erdohelyi et al. (2006)
					25 (100 min)	
Rh/CeO_2	1				82 (20 min)	
					56 (80 min)	
Ru/Al_2O_3	6	630	1/1	100	67	Chiu et al. (2013)
Rh/Al_2O_3	6			100	72	
Pd/Al_2O_3	6			100	60	
Pt/Al_2O_3	6			100	58	
Rh/MgO	3	650	8/5/1	99% (10 h)	91	Frusteri et al. (2004)
Pd/MgO	3			10 (10 h)	70	
Ru/Al_2O_3	0.7	700	1/3	100	46	Aupretre et al. (2002)
Rh/Al_2O_3	1			100	72	
Pd/Al_2O_3	0.8			100	55	
Pt/Al_2O_3	1			100	38	
Ru/Al_2O_3	1	800	3/1	42	~55	Liguras et al. (2003)
	5			100	~96	
Rh/Al_2O_3	1			100	~95	
	2			100	~96	
Pd/Al_2O_3	1			55	~50	
Pt/Al_2O_3	1			60	~65	

Source: Reproduced from Ref. Phung and Busca (2020).

C–C bond cleavage in comparison with other noble metals (Hou et al., 2015a). Rh has a strong interaction with γ-Al_2O_3, leading to the high stability of metal particles. As a result, the sintering of Rh metals at high temperatures is lessened (Silberova et al., 2005).

The CeO_2 support containing oxygen vacancies can avoid the formation of carbon deposits on the surface of metals (Hou et al., 2015a). Therefore, the combination of CeO_2 and γ-Al_2O_3 increases the stability of the catalyst and catalytic activity (Peela et al., 2011; Roychowdhury et al., 2021). In fact, the combination of 2 wt%Rh/20 wt%CeO_2/γ-Al_2O_3 shows a higher catalytic activity for ESR in comparison with Rh/CeO_2 only (Peela et al., 2011). Moreover, Li et al. (2016) confirmed that Rh on the interfacial region of CeO_2–Al_2O_3 is more active than that on CeO_2 or Al_2O_3 alone. There is not only a strong interaction between Rh and γ-Al_2O_3 but also a strong interfacial interaction between Rh and ceria, leading to the easier reduction of RhO_x on CeO_2. The presence of CeO_x-Al_2O_3 interfacial region makes a better Rh dispersion and stronger interaction of metals with support. This bi-oxide support provides sites for steam adsorption and activation while the C–C bond cleavage occurs on the Rh interface (Roychowdhury et al., 2021). Moreover, basic supports such as MgO and MgAl-based spinel oxides can be used to reduce the coke formation rate (Frusteri et al., 2004). This is due to a decrease in the surface acidity of those supports, which lessens the reactions of coke formation. Rh dispersed on the basic support (MgO) also exhibits a full conversion of ethanol, a higher hydrogen selectivity (91%) (Table 105.1), and also better stability at 650°C than Pd/MgO catalyst (Frusteri et al., 2004).

105.3 Ni METAL CATALYSTS

Compared to noble metals, Ni shows a lower catalytic activity. However, it is commonly used for ESR because of its low cost, promising activity, and efficiency (Abello et al., 2013; Aguero et al., 2015; Akande et al., 2005; Comas et al., 2004; Fatsikostas and Verykios, 2004). Several supports including Al_2O_3, La_2O_3, and CeO_2-ZrO_2 are used for Ni-based catalysts for steam reforming reaction (Table 105.2) (Akande et al., 2005; Comas et al., 2004; Fatsikostas and Verykios, 2004). Among them, Al_2O_3 is a good support with a full conversion of ethanol (100%) and an excellent selectivity of hydrogen (96% at 800°C) (Table 105.2) (Fatsikostas and Verykios, 2004).

In order to upgrade Ni-based catalysts to achieve a high yield of hydrogen production and high Ni dispersion, and to enhance their stability, many synthetic methods have been applied such as sol-gel driven by single epoxide step, nanoparticles encapsulated with multilayered graphene, single-step evaporation-induced self-assembly (EISA), and co-impregnation of metallic precursors on xerogel support (Chen et al., 2019a; Han et al., 2014; He et al., 2017; Song et al., 2016a,b,c, 2017). In addition, many supports and promoters have been used to improve the activity and stability of Ni-based catalysts in the ESR process (Calles et al., 2015; Chen et al., 2018; Chiou et al., 2014; Han et al., 2014; Song et al., 2016a,b,c; Song et al., 2017). Among them, nickel citrate precursor used as a nickel precursor improved the interaction of NiO with the support and also formed the small Ni particles dispersed on SBA-15 support (He et al., 2017). In addition, nickel citrate precursor can enhance the catalyst stability because of a small amount of carbon deposit formation on the catalyst's surface (He et al., 2017). The nanoparticles encapsulated with multilayered graphene and sol-gel methods can enhance the catalytic activity and stability of Ni-based catalysts by providing a good dispersion of Ni. Moreover, this Ni protected in the graphene shell avoids the sintering of Ni catalysts at high temperatures and the poisoning of Ni surface by carbonaceous deposition (Chen et al., 2019a). The high dispersion of Ni is also obtained by the sol-gel method, resulting in higher ethanol conversion and hydrogen yield as well as a better stability of catalysts (Song et al., 2016b,c).

Both supports and promoters are very important factors for Ni-based catalysts in the ESR process. The effect of support and promoter has been summarized in the recent review paper (Phung and Busca, 2020). For supports of Ni-based catalysts, both acid and basic supports have been used. Acid supports showed high ethanol conversion but also enhanced ethylene formation via ethanol dehydration (Akande et al., 2005). Basic supports can suppress carbon deposits by strengthening

TABLE 105.2
The Summary of Ni Metals Used in Ethanol Steam Reforming

Catalyst	Promoter	Support	Metal Loading (wt.%)	Temp. (°C)	H_2O/C_2H_5OH (molar ratio)	Ethanol conv. (%)	Hydrogen Selectivity (%)	Ref.
1.5Fe1.5Cu10Ni-Beta	Fe-Cu	Beta zeolite	10	450	8/1	100	70	Zheng et al. (2017)
Ni/Al_2O_3		Al_2O_3	16.1	250	3/1	76	44	Sun et al. (2005)
Ni/$Al_{S.G600}$		Al_2O_3	6	400	6/1	98	37	Yaakob et al. (2013)
Ni/Al_2O_3		Al_2O_3	78	400	8/1	92	35	Wang et al. (2016)
Ni/γ-Al_2O_3		γ-Al_2O_3	35	500	6/1	100	91	Comas et al. (2004)
Ni/Al_2O_3		Al_2O_3 (pre-treatment at 550°C)	3.8	650	3/1	100	89.0	Comas et al. (2004)
Ni/Al_2O_3		Al_2O_3 (pre-treatment at 700°C)		650		100	87.4	Comas et al. (2004)
Ni/γ-Al_2O_3		γ-Al_2O_3	10	650	8/1	100	78.2	Akande et al. (2005)
Ni/Al_2O_3		Al_2O_3	20	800	3/1	100	96	Fatsikostas and Verykios (2004)
Ni-5Ca/Al_2O_3	Ca	Al_2O_3	5	650	3/1	~100	57	Elias et al. (2013)
Ni-Co/Al_2O_3	Co	Al_2O_3	7.5	550	13/1	100	97.2	Zhao and Lu (2016)
3LaNiAl	La	Al_2O_3	7	600	4/1	100	~78	Ma et al. (2016)
Ni-Ti		TiO_2	10	500	3/1	10	4	Menegazzo et al. (2017)
12 wt.% Ni/ 20 wt.% MMT–TiO_2		MMT-TiO_2	12	500	10/1	89	55	Mulewa et al. (2017)
Ni-Zr		ZrO_2	10	500	3/1	100	54	Menegazzo et al. (2017)
Ni/Z		ZrO_2	6.8	600	3.7/1	95.0	80.7	Calles et al. (2015)
CaNi-Zr	Ca (9.0)	ZrO_2	10	500	3/1	100	75	Menegazzo et al. (2017)
Ni/La_2O_3–ZrO_2	La	ZrO_2	8	350	9/1	100	~60	Dan et al. (2015)
Ni/La_2O_3		La_2O_3	15.3	250	3/1	80.7	49.5	Sun et al. (2005)
Ni/ La_2O_3		La_2O_3	10	650	8/1	100	89.3	Akande et al. (2005)
Ni/La_2O_3		La_2O_3	20	800	3/1	~100	95	Fatsikostas and Verykios (2004)

(*Continued*)

TABLE 105.2 (*Continued*)
The Summary of Ni Metals Used in Ethanol Steam Reforming

Catalyst	Promoter	Support	Metal Loading (wt.%)	Temp. (°C)	H_2O/C_2H_5OH (molar ratio)	Ethanol conv. (%)	Hydrogen Selectivity (%)	Ref.
$CeNi_1O_Y$		CeO_2		450	3/1	100	~ 63	Pirez et al. (2016)
Ni-Ce		CeO_2	10	500	3/1	100	50	Menegazzo et al. (2017)
Ni/C		CeO_2	6.7	600	3.7/1	99.1	82.5	Calles et al. (2015)
CaNi-Ce	Ca(8.2)	CeO_2	10	500	3/1	100	70	Menegazzo et al. (2017)
AuNi-Ce	Au	CeO_2	10	500	3/1	100	70	Menegazzo et al. (2017)
AuCaNi-Ce	Au-Ca	CeO_2	10	500	3/1	100	~80	Menegazzo et al. (2017)
10Ni/SBA-15		SBA-15		500	3.7/1	69.3	59.4	Rodriguez-Gomez and Caballero (2018)
8Ni²Co/SBA-15	Co	SBA-15		500	3.7/1	85.6	62.4	Rodriguez-Gomez and Caballero (2018)
Ni/S		SiO_2 (S)	6.9	600	3.7/1	97.1	84.2	Calles et al. (2015)
Ni/ZrS	Zr(9.9)	SiO_2 (S)	6.9	600	3.7/1	100	88.3	Calles et al. (2015)
Ni/CeS	Ce(9.2)	SiO_2 (S)	6.8	600	3.7/1	100	84.3	Calles et al. (2015)
Ni/CeZrS-3	Ce (2.4)-Zr(6.6)	SiO_2 (S)	6.6	600	3.7/1	100	85.8	Calles et al. (2015)
Ni/Al-0La		Al_2O_3-La_2O_3	15	450	6/1	100	100.1	Song et al. (2017)
0-NAZ (0–24: P123 concentration (mM) (N= Ni)		Al_2O_3-ZrO_2 (AZ)	15	500	6/1	100	94	Han et al. (2014)
Ni/AZ		Al_2O_3-ZrO_2 (AZ)	15	450	6/1	100	73.4	Song et al. (2016c)
15Ni/AZ		Al_2O_3-ZrO_2	15	450	6/1	100	87.9	Song et al. (2016a)
Ni-Mg/AZ	Mg		15	450	6/1	100	81.1	Song et al. (2016c)
Ni-Ca/AZ	Ca		15	450	6/1	100	76.8	Song et al. (2016c)
Ni-Sr/AZ	Sr		15	450	6/1	100	87.9	Song et al. (2016c)
Ni-Ba/AZ	Ba		15	450	6/1	100	80.4	Song et al. (2016c)

(*Continued*)

TABLE 105.2 (*Continued*)
The Summary of Ni Metals Used in Ethanol Steam Reforming

Catalyst	Promoter	Support	Metal Loading (wt.%)	Temp. (°C)	H_2O/C_2H_5OH (molar ratio)	Ethanol conv. (%)	Hydrogen Selectivity (%)	Ref.
15Ni-2Sr/AZ	Sr	Al_2O_3-ZrO_2		450	6/1	100	94.2	Song et al. (2016a)
15Ni-4Sr/AZ	Sr	Al_2O_3-ZrO_2		450	6/1	100	96.2	Song et al. (2016a)
15Ni-6Sr/AZ	Sr	Al_2O_3-ZrO_2		450	6/1	100	99.9	Song et al. (2016a)
15Ni-8Sr/AZ	Sr	Al_2O_3-ZrO_2		450	6/1	100	92.4	Song et al. (2016a)
15Ni-10Sr/AZ	Sr	Al_2O_3-ZrO_2		450	6/1	100	86.2	Song et al. (2016a)
Ni/CeZr(N)		Ce0.5Zr0.5O2	5	450	13/1	100	96.7	Chiou et al. (2014)
NiCeZr(C) (C: coprecipitation)	-	$Ce_{0.5}Zr_{0.5}O_2$	10	450	37/3	100	89	Wu et al. (2019)
NiFe/CeZr(N)	Fe (5 wt%)		5	450	13/1	100	98.3	Chiou et al. (2014)
NiCo/CeZr(N)	Co (5 wt%)		5	450	13/1	100	85.0	Chiou et al. (2014)
NiRh/CeZr(N)	Rh (1 wt%)		5	450	13/1	100	93.3	Chiou et al. (2014)
BNiCeZr(C)	B	$Ce_{0.5}Zr_{0.5}O_2$	10	450	37/3	100	92	Wu et al. (2019)
BNiCeZr(I)	B	$Ce_{0.5}Zr_{0.5}O_2$	10	450	37/3	100	68	Wu et al. (2019)
Ni/$Ce_{0.9}Sm_{0.1}O_{2-\delta}$		$Ce_{0.9}Sm_{0.1}O_{2-\delta}$	1	550	3/1	100	~60	Rodrigues et al. (2019)
NiMgAl-r		MgAl	8.3	600	3.7/1	100	78.2	Vizcaino et al. (2012)
NiCaAl-r		CaAl	6.9	600	3.7/1	100	87.3	Vizcaino et al. (2012)
NiZnAl-r		ZnAl	7.4	600	3.7/1	100	79.4	Vizcaino et al. (2012)
NiMgLa-r		MgLa	5.8	600	3.7/1	~82	77.6	Vizcaino et al. (2012)
NiMgCe-r		MgCe	8.9	600	3.7/1	~60	82.2	Vizcaino et al. (2012)
Ni/ZnO		ZnO	10	650	8/1	100	89.1	Akande et al. (2005)
Ni/MgO		MgO	10	650	8/1	100	82.2	Akande et al. (2005)
Ni/MgO		MgO	21	650	8/5	42 (10 h)	97	Frusteri et al. (2004)
Ni/Mg-Al		Mg-Al	10	650	8.4/1	16.9	69	Coleman et al. (2009)
Ni/10Mg-ATP	Mg	ATP (Attapulgite)	20	700	3/1	100	96.7	Chen et al. (2018)

Source: Reproduced from Ref. Phung and Busca (2020).

the metal-support interaction (Coleman et al., 2009). Additionally, the mixture of acidic-basic oxides (Han et al., 2014; Sanchez-Sanchez et al., 2007; Song et al., 2016b,c) and the new preparation method (Han et al., 2014; Rodrigues et al., 2019; Song et al., 2016a,b,c, 2017) have been applied to enhance the stability and hydrogen yield.

For acid supports, several oxides have been used such as Al_2O_3, TiO_2, SiO_2, and β zeolite (Table 105.2). Al_2O_3 is commonly used as support for ESR because its acidic property enhances ethanol conversion but also favors ethanol dehydration into ethylene, which further converts into coke (Hou et al., 2015b; Phung et al., 2015). For Al_2O_3, it is reported that the phase of alumina strongly affects the hydrogen selectivity and the stability of the catalyst. Indeed, γ-Al_2O_3 shows an excellent support in comparison to other phases of alumina, and the highest hydrogen yield is obtained at the calcination temperature of 800°C for catalyst (Figure 105.2) (Yaakob et al., 2013). The calcination temperature of 800°C exhibiting the highest hydrogen selectivity is due to the increase in the strength of NiO-Al_2O_3 interaction, Ni dispersion, and Ni surface area (Figure 105.3) (Yaakob et al., 2013). In the case of basic supports, several supports are commonly used such as La_2O_3, MgO, ZnO, and CeO_2 (Table 105.2). They can avoid the drawback of acid supports, i.e., suppress coke formation and enhance the stability of catalysts (Frusteri et al., 2004).

Moreover, the mixed oxides are also considered promising supports due to possessing both acid and basic characters (Han et al., 2014; Song et al., 2016a,b,c, 2017). For instance, the mixture of Al_2O_3 and La_2O_3 with different La/Al ratios influences the hydrogen yield (Figure 105.4) (Song et al., 2017). The change of the La/Al ratio changed the acidity and basicity of catalysts as well as adjusted the Ni dispersion and Ni surface area, resulting in the change in hydrogen selectivity (Figure 105.4). Besides, some modern preparation techniques for catalyst supports have been developed to achieve a high hydrogen yield (Han et al., 2014; Rodrigues et al., 2019; Song et al., 2016a,b,c, 2017). The sol-gel method can enhance the mass-transfer efficiency and diminish the deactivation rate (Wang et al., 2018). This method also tailors the Ni particle size, promoting the dispersion of Ni on the surface of the support. Additionally, the combination of the epoxide-driven sol-gel method and subsequent supercritical CO_2 drying method created the small size of Ni particles and led to high Ni dispersion, resulting in high hydrogen selectivity and suppression of catalyst

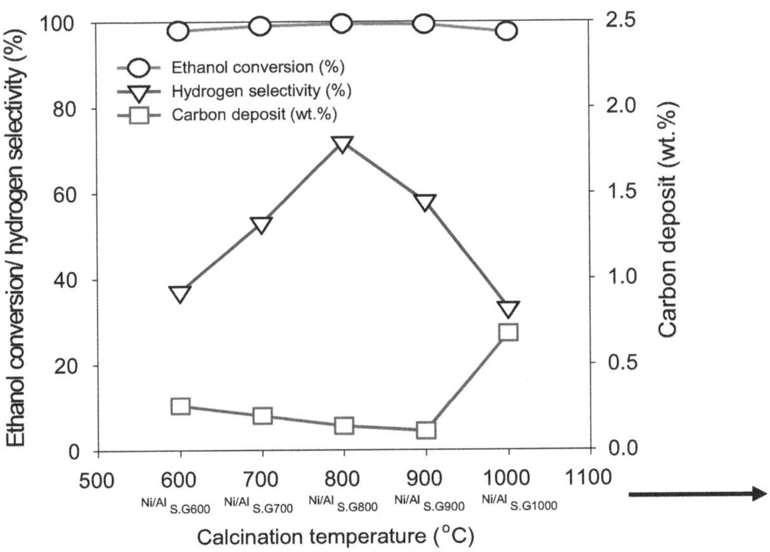

FIGURE 105.2 Ethanol conversion, hydrogen selectivity, and carbon deposits over Ni/sol gel made alumina catalysts as a function of calcination temperature of the alumina supports. Reaction condition: 400°C, steam/ethanol=6/1, 8h. (Reproduced from open-access Ref. Phung and Busca (2020); Yaakob et al. (2013).)

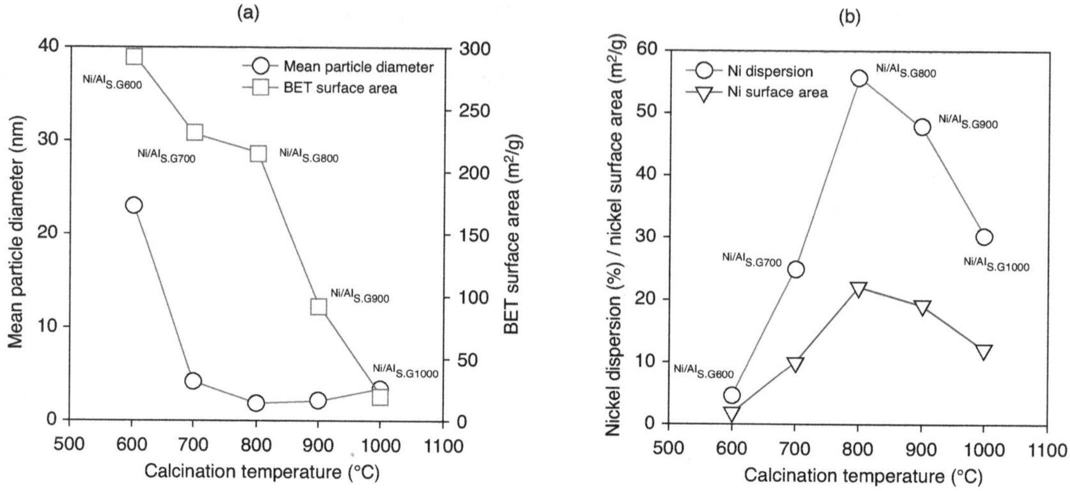

FIGURE 105.3 (a) BET surface area of the sol-gel made alumina support (m^2/g) and mean particle diameter of Ni (nm), and (b) nickel dispersion (%) and nickel surface area (m^2/g) as a function of calcination temperature of nickel supported on sol-gel made alumina. (Reproduced from open-access Ref. Phung and Busca (2020); Yaakob et al. (2013).)

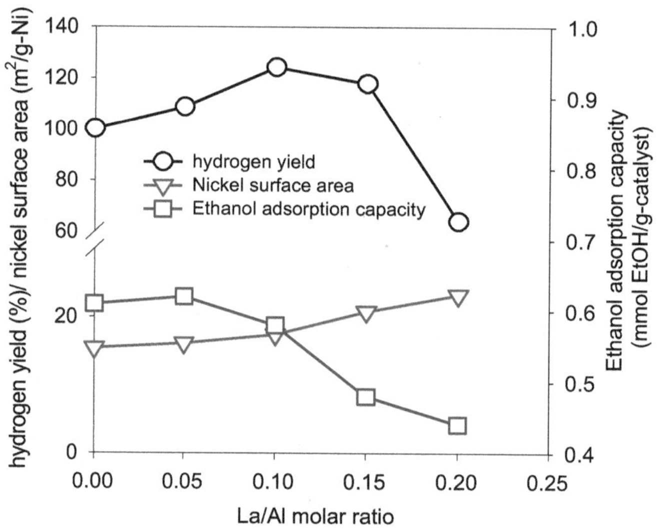

FIGURE 105.4 Hydrogen yield (15 h time-on-stream, 450°C), nickel surface area and ethanol adsorption capacity as a function of La/Al molar ratio. (Reproduced from Ref. Phung and Busca (2020); Song et al. (2017).)

deactivation (Song et al., 2016b). Another development of catalyst prepared by sol-gel method is the modification of pore size by a mesoporous-maker polymer (Han et al., 2014). The utilization of different concentrations of polymer P123 shows different yields of hydrogen. The highest hydrogen yield is obtained at the lowest Ni particle size and highest Ni surface area (Figure 105.5). Moreover, several methods including dealumination (Gac et al., 2018a), coprecipitation (Vizcaino et al., 2012), nanowire structure $Ni/Ce_{0.9}Sm_{0.1}O_{2-\delta}$ (Rodrigues et al., 2019), perovskite-type oxides (Aguero et al., 2015), the three-dimensional pore network material (KIT-6) (Parlett et al., 2017), and using different impregnation media (ethanol or water) (Olivares et al., 2018) have been developed to improve the catalytic performance in the ESR reaction.

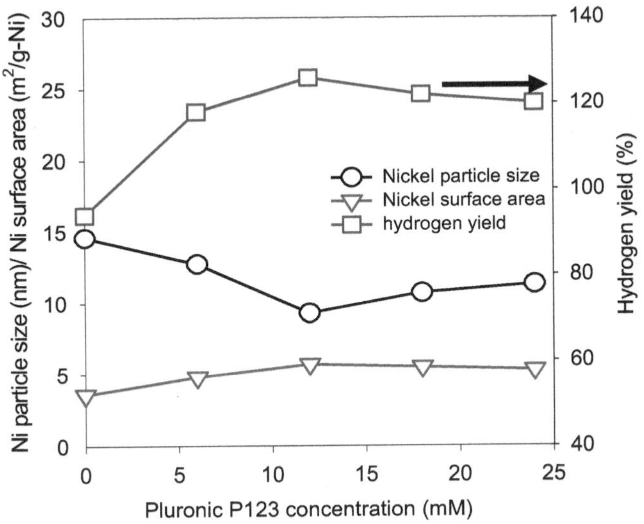

FIGURE 105.5 Hydrogen yield (500°C, 1,000 min), nickel particle size and nickel surface area as a function of the concentration of preparation-assisted pluronic P123 copolymer. (Reproduced from Ref. Han et al. (2014); Phung and Busca (2020).)

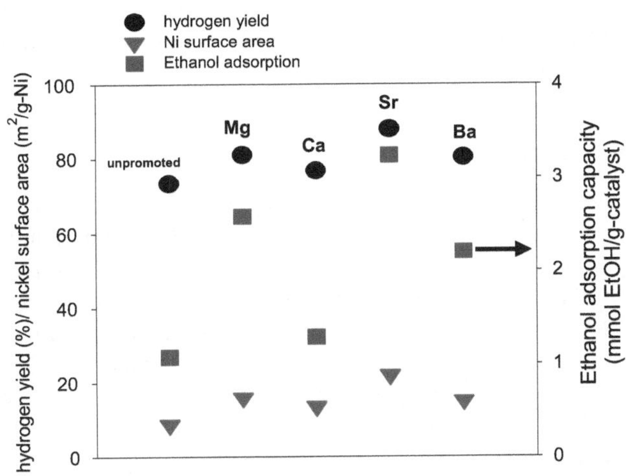

FIGURE 105.6 Hydrogen yield (450°C, 1,000 min), nickel surface area, and ethanol adsorption capacity of Ni–X/Al_2O_3–ZrO_2 (X=Mg, Ca, Sr, and Ba) and Ni/Al_2O_3–ZrO_2 catalysts reduced at 650°C for 3 h. (Reproduced from Ref. Phung and Busca (2020); Song et al. (2016c).)

The hydrogen yield, catalytic activity, and stability of catalysts are enhanced by both supports and promoters (Calles et al., 2015; Rodriguez-Gomez and Caballero, 2018). Figure 105.6 shows that the presence of an alkali promoter enhances the hydrogen yield by increasing Ni surface area and ethanol adsorption. That is due to the fact that the addition of promoter improves the metal-support interaction, leading to a good dispersion and surface area of Ni (Song et al., 2016c). The addition of different alkali metals exhibits different catalytic activities because the basic character of the alkali promoter modifies the acid/basic sites of catalysts (Phung and Busca, 2020). Additionally, the addition of different percentages of promoters can vary the hydrogen yield (Figure 105.7) (Song et al., 2016a). Figure 105.7 shows that the highest hydrogen yield is obtained at 6 wt.% Sr loading. That is due to the fact that the Sr loading modified the Ni surface area and ethanol adsorption capacity of the pristine catalyst. Moreover, another explanation for

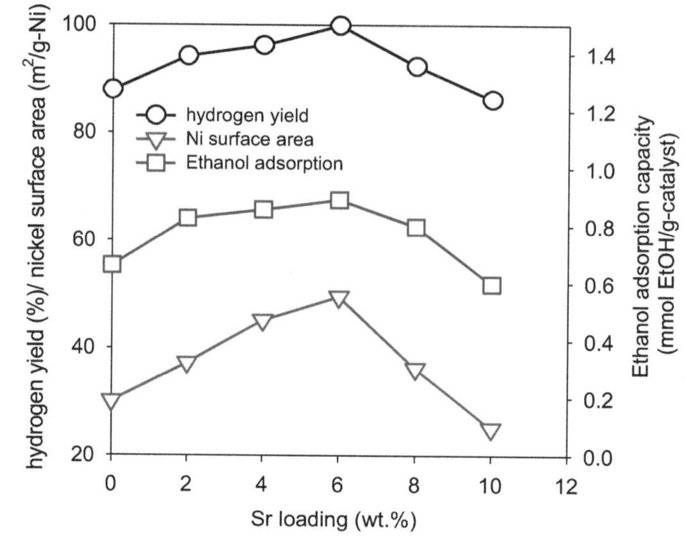

FIGURE 105.7 Hydrogen yield (15h time-on-stream, 450°C), nickel surface area, and ethanol adsorption capacity as a function of Sr loading. (Reproduced from Ref. Phung and Busca (2020); Song et al. (2016a).)

the highest hydrogen yield is due to the acidity of the catalyst after the addition of Sr promoter with the acidity of 1.005, 0.878, 0.844, 0.763 and 0.624 mmol-NH_3/g-catalyst for 2, 4, 6, 8 and 10 wt.% Sr loading, respectively.

105.4 Co METAL CATALYSTS

Besides Ni-based catalysts, Co-based catalysts are also excellent candidates for ESR due to their high activity and lower formation of CH_4 (Ogo and Sekine, 2020). For example, a full conversion of ethanol and high hydrogen selectivity (93%) were obtained at 420°C over Co/CeO_2-nano catalyst (Machocki et al., 2010). This is comparable to Ni/Al_2O_3 catalyst at 800°C catalyst with 100% ethanol conversion and 95% hydrogen selectivity over (Fatsikostas and Verykios, 2004).

For cobalt-based catalysts, the Co oxidation state is the crucial point in the ESR (Ogo et al., 2015). However, there is a debate on the active sites of Co in the ESR. Generally, the metallic Co species are active sites (Hou et al., 2015); however, many other reports confirmed that the metallic Co^0 species are active centers for coke formation (Espinal et al., 2012; Li et al., 2019; Ogo et al., 2015; Sharma et al., 2017). Ogo et al. (2015) and Espinal et al. (2012) concluded that Co^{2+} species are highly active sites and stable with only a small amount of coke formation. Busca et al. (2010) found that the presence of cobalt oxide species enhances the desired products (hydrogen and carbon dioxide). However, the presence of both metallic Co^0 and Co^{2+} species improves the catalytic activity and stability of Co-based catalysts (Chen et al., 2019b; Ogo et al., 2015; Sohn et al., 2017; Zanchet et al., 2015). Regarding the reaction mechanism of this reaction on the surface of Co-based catalysts, C–C bond scission occurs on the surface of Co^0 species while water activation and acetate species formation occur on the surface of Co^{2+} species (Ogo and Sekine, 2020). Therefore, the presence of Co^{2+} favors the acetate pathway to form acetaldehyde followed by reacting with water to produce CO_x and H_2, without CH_4 formation, which can reduce the coke formation through the decomposition of acetaldehyde (see details in Section 105.6). Also, the presence of Co^0 as mentioned favors the cleavage of the C–C bond, supporting for steam reforming of acetaldehyde to form CO_x and H_2. In short, the presence of both Co^0 and Co^{2+} with a proper Co^0/Co^{2+} ratio can enhance the hydrogen yield in the ESR process and avoid catalyst deactivation by carbon deposition (Ogo and Sekine, 2020).

For supports of Co-based catalysts, there are many supports used including Al_2O_3 (Ogo et al., 2017), SiO_2 (Kim et al., 2017), Sc-ZnO (Liang et al., 2017), Y-ZrO_2 (Gaudillere et al., 2017), CeO_2 (Sohn et al., 2017; Turczyniak et al., 2017), MnO_x, (Gac et al., 2018b), zeolite (Li et al., 2019), mesoporous silica (Rodriguez-Gomez et al., 2017), sepiolite (Chen et al., 2019b; Hernandez-Soto et al., 2019), hydroxyapatite (Ogo et al., 2017), hydrotalcite (Hernandez-Soto et al., 2019), carbon (Augusto et al., 2020). $LaAlO_3$ (Ohno et al., 2018) and bismuth vanadate ($BiVO_4$) (Sharma et al., 2018) (Table 105.3). Among them, two categories of supports can be divided including acid supports and basic supports. The acid supports such as Al_2O_3 usually favor the formation of ethylene on acid sites via ethanol dehydration. The basic supports can reduce coke and byproduct formation (Sharma et al., 2017). Hence, doping CaO to Al_2O_3 support leads to a decrease in acidity and an increase in basicity of catalyst, resulting in a decrease in ethylene formation and enhancing the stability of catalysts (Han et al., 2017).

Other basic supports such as Sr-hydroxyapatite (Ogo et al., 2017) and $SiO_2@Co_{0.75}Mg_{0.25}O$ (Kim et al., 2017) showed a low ethylene selectivity and high coke resistance, in which $SiO_2@Co_{0.75}Mg_{0.25}O$ support shows high catalytic activity during 100h. In addition, using the redox support such as CeO_x can be an excellent way to avoid catalyst deactivation by carbon deposits. Indeed, the redox supports like CeO_x with the reversible redox reaction of Ce^{4+}/Ce^{3+} having high oxygen mobility and oxygen storage capacity help oxidize coke and/or coke precursors (adsorbed CH_x species) on the surface of catalysts, resulting in lower coke formation (Ogo and Sekine, 2020). Moreover, CeO_x also affects the Co oxidation state, making a strong impact on the ESR reaction as above discussion via the acetate route. For example, 10wt%Co-7.1wt%Ce/Sepiolite shows high activity during 100 h on stream (Chen et al., 2019b).

TABLE 105.3
Results of ESR Activity Over Various Co-based Catalysts

Catalyst	Temp. (°C)	H_2O/C_2H_5OH (molar ratio)	Ethanol Conv. (%)	H_2 Yield (%)	Ref.
5Co/α-Al_2O_3	550	5	43	28	Ogo et al. (2017)
5Co/$Sr_{10}(PO_4)_6(OH)_2$	550	5	44	30	Ogo et al. (2017)
10Co/CeO_2	450	5	100	93	Sohn et al. (2017)
8.5Co-2K/CeO_2	420	1.5–6	100	93	Turczyniak et al. (2017)
$Co_{1.25}Zn_1Sc_{0.3}O_x$	450	1.5	17	18	Liang et al. (2017)
10Co/SBA-15	500	1.85	89	49	Rodriguez-Gomez et al. (2017)
10Co/Sepiolite	600	3	54	34	Chen et al. (2019b)
10Co-7.1Ce/Sepiolite	600	3	91	69	Chen et al. (2019b)
$SiO_2@30(Co_{0.75}Mg_{0.25}O)$	600	1.5	100	75	Kim et al. (2017)
$SiO_2@30(Co_{0.75}Mg_{0.25}O)$	650	1.5	100	87	Kim et al. (2017)
17Co/MnO_x	420	6	90	98	Gac et al. (2018b)
	450	6	100	98	Gac et al. (2018b)
$Bi_4(V_{0.9}Co_{0.1})_2O_{11}$-δ	400	11.5	88	63	Sharma et al. (2018)
20Co/Sepiolite	400	6.5	100	74	Hernandez-Soto et al. (2019)
20Co/Zn_5-Al hydrotalcite	400	6.5	100	75	Hernandez-Soto et al. (2019)
20Co-1La/Zn_5-Al hydrotalcite	400	5	100	75	Cerda-Moreno et al. (2019)
20Co-1La/Zn_5-Al hydrotalcite	500	5	100	75	Cerda-Moreno et al. (2019)
1Co/CNF	500	1.5	60	62	Augusto et al. (2020)
2.5Ni-2.5Cu/BEA@10Co/MFI	500	1.5	100	75	Li et al. (2019)
3Co/Y_2O_3-ZrO_2	400	6.5	99	68	Gaudillere et al. (2017)
3Co-1La/Y_2O_3-ZrO_2	400	6.5	99	72	Gaudillere et al. (2017)

Source: Reproduced from Ref. Ogo and Sekine (2020).

Although a proper support can enhance the stability and tailor the Co oxidation of Co-based catalysts as aforementioned. Co-based catalysts can be suffered by different factors including carbonaceous deposition, the active phase sintering, changes in the Co oxidation state, and poisoning by impurities in the raw bioethanol (Hou et al., 2015b). Therefore, Co-based catalysts need promoters to enhance the stability and activity. The addition of La promoter to Co/hydrotalcite enhanced the stability and decreased the ethylene formation in the case of even for using industrial alcoholic waste as feedstock Cerda-Moreno et al., 2019). An addition of a small amount of K to Co/CeO_2 suppressed the coke formation to one-tenth (Turczyniak et al., 2017). In fact, potassium is a promising promoter to solve the problem of Co-based catalysts for ESR (Hou et al., 2015b). The addition of alkali promoter such as potassium reduces carbon deposits on the surface of the catalyst resulting in an increase in stability and catalytic activity (Espinal et al., 2012; Frusteri et al., 2004). In addition, the addition of alkali promoters can modify the catalyst acidity by adding basic character to supports, leading to the suppression of ethylene formation via ethanol dehydration, which can further convert into coke (Sanchez-Sanchez et al., 2007). Moreover, the presence of potassium induces the presence of Co^{2+} (Ogo et al., 2015), which is an active site for ESR. Indeed, Grzybek et al. (2020) have recently confirmed that the potassium slightly inhibited the cobalt reducibility, resulting in remaining the active sites for ESR. Furthermore, throughout the activation and ESR procedures, potassium moved from the cobalt nanograins to the alumina support, stabilizing the high dispersion of cobalt crystallites (Grzybek et al., 2020). Potassium also keeps them from detaching from the surface of the support and being subsequently encapsulated by the growing carbon deposits Grzybek et al. (2020).

105.5 EFFECTS OF OPERATING CONDITIONS

105.5.1 Effect of Reaction Conditions

It is well-known that there are many factors including catalysts, reaction conditions (e.g., reaction temperature, feed composition, and residence time), feed impurities, and membrane affect the yield and selectivity of hydrogen in ESR. In the case of reaction temperature, a higher reaction temperature will increase the operation cost but suppress the hydrogenation of ethanol into methane (Ni et al., 2007), resulting in high hydrogen yield. In addition, a higher molar ratio of steam/ethanol and a longer residence time improve the production of hydrogen (De la Piscina and Homs, 2008). The higher water-to-ethanol ratio (>3) also enhances hydrogen production due to favoring the forward water–gas shift reaction and avoiding the formation of coke (Ozkan et al., 2019).

105.5.2 Effect of Impurities on the ESR

The real ethanol contains different impurities due to ethanol production processes including methanol, 2-methyl-1-propanol, 1-propanol, ethyl acetate, 3-methyl-1-butanol, acetaldehyde, acetic acid, and sulfur components (Habe et al., 2013). Those impurities may have impacted the ESR, resulting in reducing the H_2 yield (Sanchez et al., 2020). Figure 105.8 shows that methanol has a positive effect during ESR while fusel alcohols can show both effects during ESR. The longer carbon chain length of fused alcohols leads to the dominantly negative effect because alcohols favor decomposing to olefins via dehydration reaction, which is further converted to coke that covers the active sites of catalysts (Bilal and Jackson, 2017). Amines show a positive effect during ESR because amines possess basic character which can adsorb on the surface of acid sites, resulting in a decrease in the dehydration of ethanol toward ethylene and further isomerization into higher hydrocarbons and coke (Sanchez et al., 2020). Additionally, the basic character of amines also enhances the dehydrogenation reaction of ethanol to produce hydrogen (Devianto et al., 2011; Le Valant et al., 2008).

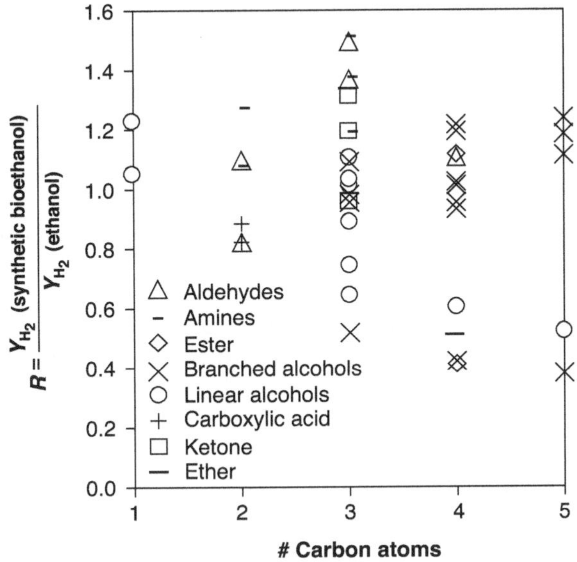

FIGURE 105.8 Effect of various contaminants on H_2 yield during ESR of synthetic bioethanol, categorized by carbon atom number and functional group. (Reproduced from Ref. Sanchez et al. (2020).)

For the effect of amines to metal surface, amines may have impacted the electronic properties of the metal; that may be a positive effect. However, in some cases, amines can poison the meal phase by strong adsorption, causing hinder the reaction of ethanol on metal surface (Le Valant et al., 2008) Aldehydes exhibit a positive effect during ESR; however, acetic acid exhibits a negative effect and enhances the ethanol dehydration due to an increase in acidity of catalysts (Sanchez et al., 2020). Acetic acid can also deactivate the active sites of catalysts by carbon deposits arising from the ketonization reaction of acids and oxidizing metal surface of the active metal surface. In the case of ketone and ethyl acetate, they reduce the hydrogen yield by strong interaction with the catalyst surface (Bilal and Jackson, 2017; Le Valant et al., 2008); however, their effects need to study more due to the limited number of investigations. Moreover, sulfur usually poisons the catalyst through sulfidation. For example, the formation of NiS-like species in the presence of sulfur compounds leads to a decrease in hydrogen production on ESR (Garbarino et al., 2013).

The effect of impurities on the ESR process can be described in Figure 105.9, showing both effects of the metal surface and support. Figure 105.9a shows the deactivation mechanisms on the surface of metals due to the changes in the metal properties, competitive adsorption, and active sites covered by carbon deposits. Figure 105.9b presents the impurities that modify the acidity, basicity and hydrophobicity of support on ESR.

Recently, Sanchez et al. (Sanchez et al., 2020) summarized the effect of impurities during ESR and provided the source of those impurities as well as the strategy to control the impurities as illustrated in Figure 105.10. The impurities of ethanol mainly come from biomass resources and are produced during the fermentation process. Therefore, pretreatment of biomass is necessary to reduce the amount of impurities during the fermentation process. Additionally, the fermentation culture design strategies and yeast metabolism modifications are also important to reduce the production of harmful impurities and increase the harmless impurities. Moreover, the purification process after fermentation needs to be installed to remove the harmful impurities produced during the fermentation process. Besides, the use of bimetallic catalysts supported on oxides with oxygen vacancy capacity can help to burn carbon deposits on the surface of catalysts, which reduces the catalyst deactivation (Sanchez et al., 2020).

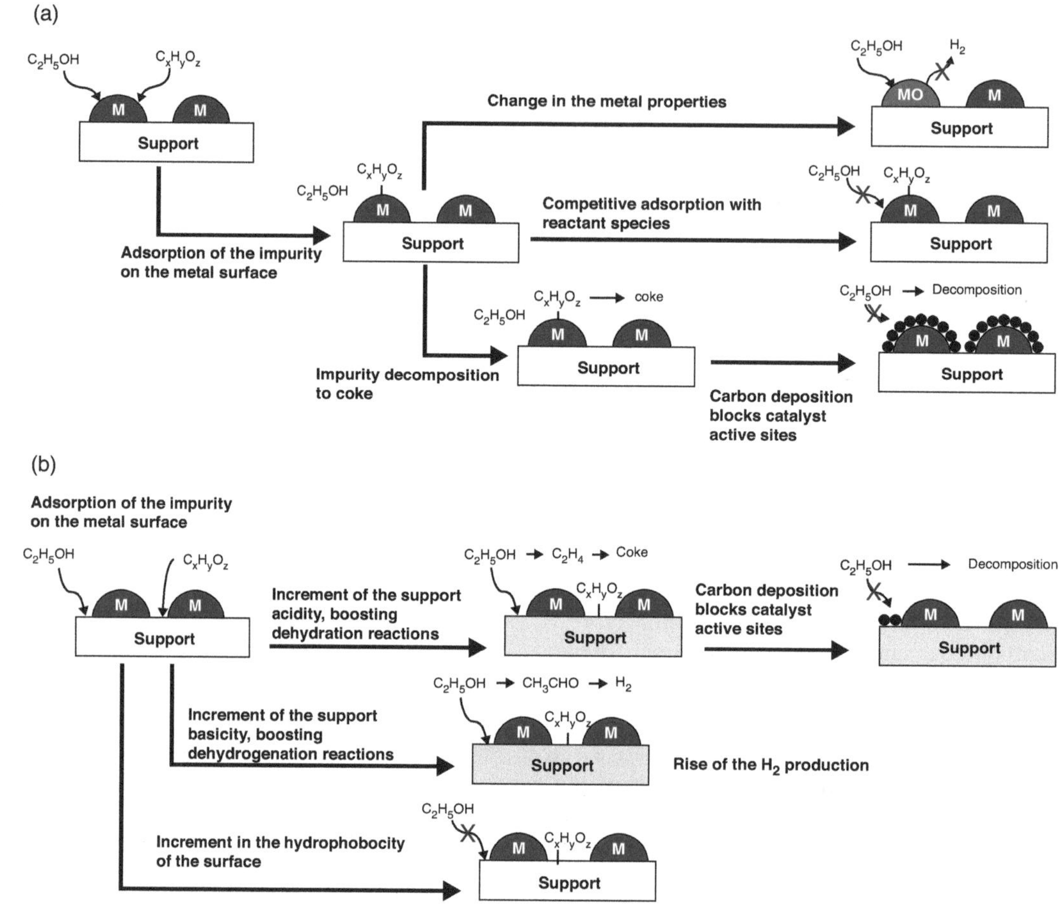

FIGURE 105.9 Possible effects of impurities on the active sites (a) and on the support (b) during ESR of the simulated bioethanol samples (Sanchez et al., 2020).

105.5.3 Membrane-Assisted ESR

A barrier function in the membrane allows for the continuous movement of selective species driven by a gradient such as partial pressure or concentration gradient. Due to the continuous selective species, the membrane can shift the equilibrium of the reaction to form more products based on the Le Chatelier's principle. For example, in the case of ESR, if H_2 is continuously removed by a membrane, hydrogenation of ethanol and reverse water–gas shift reaction are declined, resulting in an increase in hydrogen production. Additionally, membranes can also be used to control the dose of the reactant, allowing it to control the reaction system.

In parallel with the membrane, adsorbents can be used to assist the reaction to form more products by shifting toward the equilibrium of the reaction. Based on this manner, adsorbents can be used to adsorb CO_2 from ESR to shift toward water–gas shift and ESR reactions, leading to the increase of H_2 yield. However, all adsorbents have a maximum capacity to adsorb materials when they get a breakthrough point or saturated adsorbate, such as saturated CO_2. At that point, H_2 production will be reduced through ESR due to its reaction with CO_2 in a reverse water–gas shift reaction. Therefore, adsorbents need to be regenerated before the breakthrough point region. That means that adsorbents need to remove the saturated adsorbate by a regeneration process at higher temperatures and/or lower pressures, in contrast with the continuous membrane separation process.

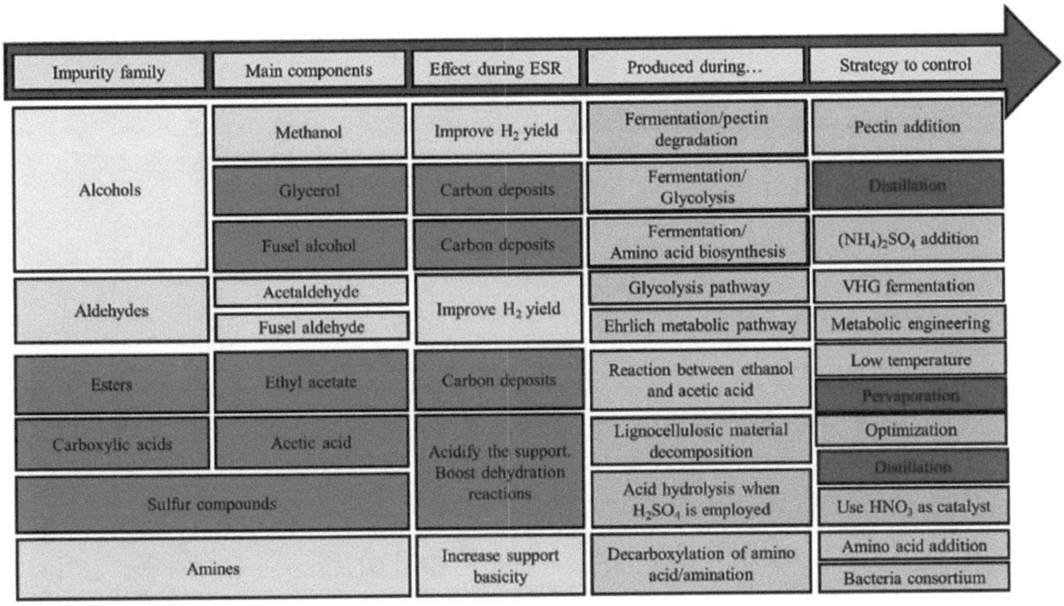

FIGURE 105.10 The effect of bioethanol impurities during ESR: The origin of impurities and strategies to control their formation. (Reproduced from Ref. Sanchez et al. (2020).)

According to the manner discussed above, membranes can be installed in the ESR process inside the reactor (Figure 105.11). Membranes can be made of a variety of materials, including polymers, metals, ceramics, zeolites, and metal–organic frameworks (Bac et al., 2020). Metal–organic frameworks are promising materials for H_2 purification via adsorption and membranes (Bac et al., 2020).

105.5.4 Reaction Mechanisms of ESR

Recently, the reaction mechanism of ESR has been summarized in detail by Zanchet et al. (2015) and Ogo and Sekine (2020). Figure 105.12 depicts the reaction pathway for ESR, which includes ethanol dehydrogenation, acetaldehyde steam reforming, and the water–gas shift reaction. Ethanol is mainly dehydrogenated into acetaldehyde and hydrogen, followed by steam reforming of acetaldehyde to CO and hydrogen, and then water–gas shift reaction converts CO and H_2O into CO_2 and H_2 (Figure 105.12). Besides, other side reactions also occur such as dehydration of ethanol, decomposition of acetaldehyde, acetone formation, CO disproportionation, CO methanation, and CO_2 methanation, followed by decomposition of hydrocarbons to form coke (Figure 105.12).

Acetone formation from acetaldehyde will reduce hydrogen yield in comparison with steam reforming reaction. Additionally, methanation of CO and CO_2 consumed hydrogen proceeded slightly over Co-based catalysts (Ogo et al., 2015; Sekine et al., 2014). To avoid byproduct formation, the catalysts have to develop the desired reaction routes with high selectivity to obtain the highest hydrogen. For example, the adsorption of acetate species on the surface of the catalyst can prevent acetaldehyde decomposition and favor promotion of the acetaldehyde steam reforming. This also decreases the acetaldehyde decomposition to enhance hydrogen yield and suppress the formation of methane, which converts to coke at high temperatures. Moreover, reaction condition is also an important factor to reduce coke formation. The higher molar ratios of steam to ethanol favor ESR

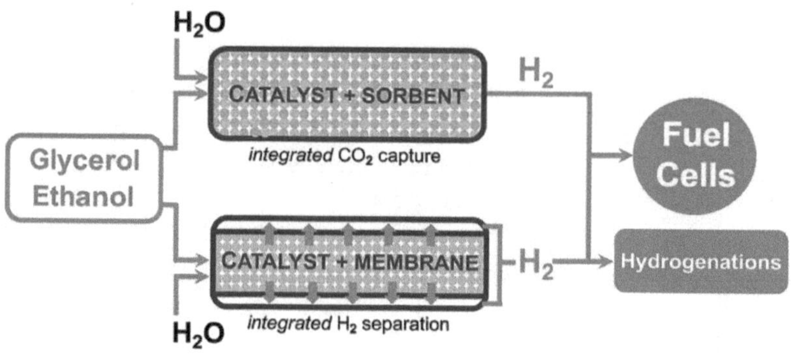

FIGURE 105.11 Adsorption and membrane assistance for hydrogen production in ethanol/glycerol steam reforming. (Reproduced from Ref. Bac et al. (2020).)

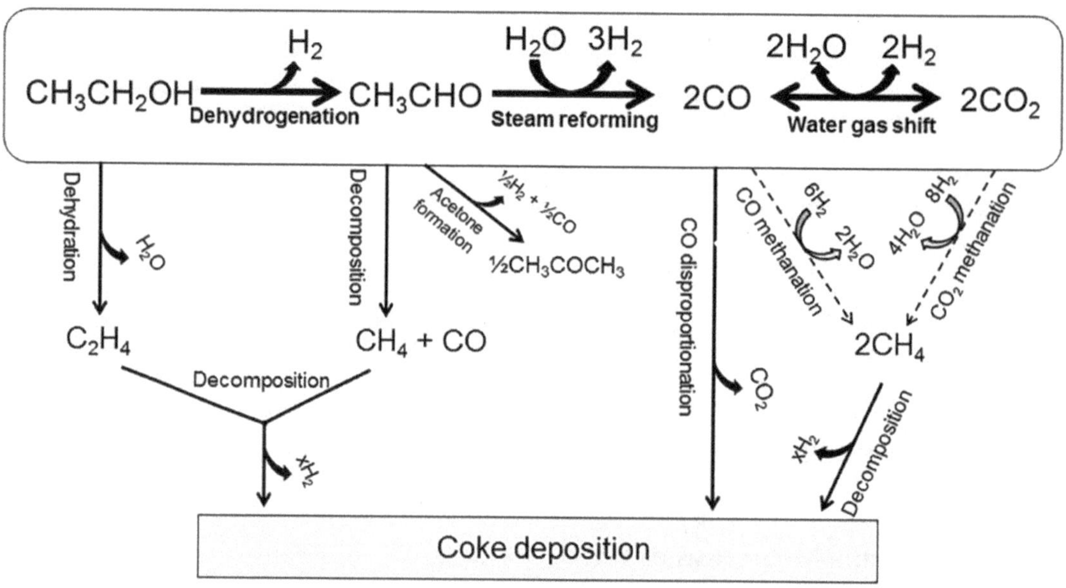

FIGURE 105.12 Reaction pathway of ESR *via* acetaldehyde steam reforming. (Reproduced from Ref. Ogo and Sekine (2020).)

and the forward water–gas shift reaction and avoid the formation of coke (James et al., 2011); however, an introduction of excessive steam to ESR increases heating costs and lowers energy efficiency.

In the case of Ni catalysts, the mechanism of ethanol decomposition into methane has been proposed (Figure 105.13) (Sharma et al., 2017; Zanchet et al., 2015). Nickel catalysts possess a property of cleavage C–C bond, leading to the decomposition of ethanol into methane or the decomposition of acetaldehyde into CO (Ogo and Sekine, 2020). Further reaction can convert methane into CO through steam reforming, followed by water–gas shift reaction into CO_2 and H_2 (Figure 105.13).

Regarding the adsorption species on the surface of the catalyst, Zanchet et al. (2015) reported that the ESR occurs via three main pathways over different catalysts (Figure 105.14). The first reaction pathway ('red' pathway) is through the cleavage of the OH bond followed by the dehydrogenation forming acetaldehyde (CH_3CHO^*), acetyl (CH_3C^*O), ketene ($^*CH_2C^*O$) and ketenyl ($^*CHC^*O$) as sequence intermediates. For the second pathway ('blue' pathway), the carbon of the alcohol group is adsorbed on the surface of metal catalysts followed by successive dehydrogenation

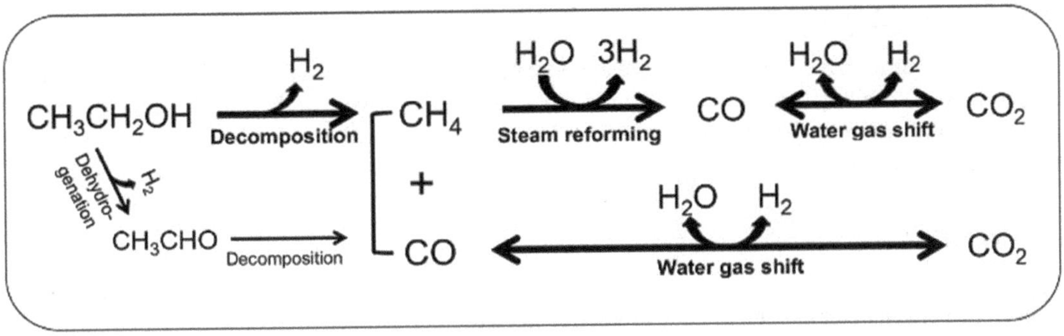

FIGURE 105.13 ESR reaction pathway involving ethanol breakdown and subsequent methane steam reforming. (Reproduced from Ref. Ogo and Sekine (2020).)

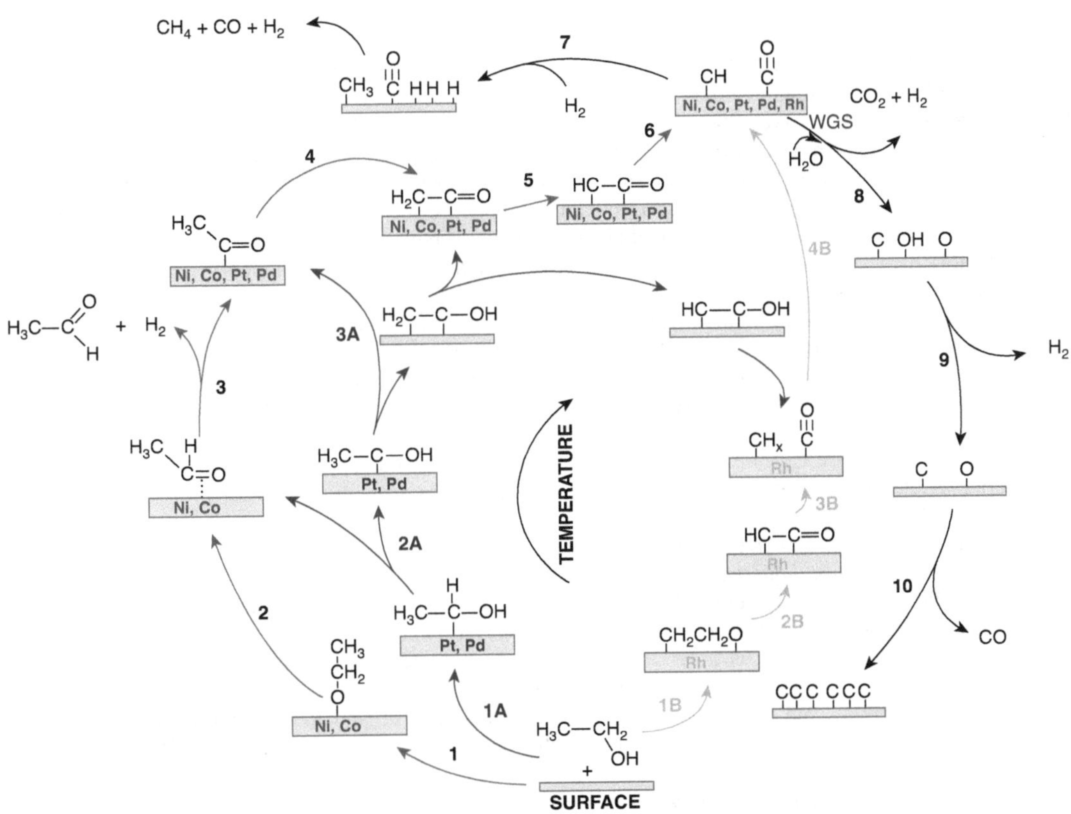

FIGURE 105.14 SRE reaction pathways as a function of temperature for various metal surfaces. This Figure shows the primary routes on Ni or Co, Pt or Pd, and Rh. Secondary routes are indicated by dashed lines. (Reproduced from Ref. Zanchet et al. (2015).)

to form hydrogen while preserving O–H bond. The third pathway ('green' pathway) breaks O–H and C–H bonds to form an oxametallacycle intermediate ($^*CH_2CH_2O^*$) on the surface of Rh catalyst, followed by cleavage of the C–C bond of the intermediates, hydrogenation/dehydrogenation of CH_{x^*}, water activation and oxidation of C* species. Overall, the steam reforming reaction pathway depends on the nature of the catalyst and reaction temperature.

105.6 CONCLUSIONS

This chapter summarized all recently advanced catalysts in the steam reforming of bioethanol for the production of biohydrogen fuels. The main active metals as well as the supports and promoters for these catalysts have been summarized. Additionally, the reaction conditions and mechanism have been discussed.

REFERENCES

Abello, S., E. Bolshak and D. Montan. 2013. Ni-Fe catalysts derived from hydrotalcite-like precursors for hydrogen production by ethanol steam reforming. *Applied Catalysis A* 450:261–274.

Aguero, F. N., M. R. Morales and S. Larregola, et al. 2015. $La_{1-x}Ca_xAl_{1-y}NiyO_3$ perovskites used as precursors of nickel based catalysts for ethanol steam reforming. *International Journal of Hydrogen Energy* 40:15510–15520.

Akande, A. J., R. O. Idem and A. K. Dalai. 2005. Synthesis, characterization and performance evaluation of Ni/Al_2O_3 catalysts for reforming of crude ethanol for hydrogen production. *Applied Catalysis A* 287:159–175.

Augusto, B. L., M. C. Ribeiro, F. J. C. S. Aires, V. T. da Silva and F. B. Noronha. 2020. Hydrogen production by the steam reforming of ethanol over cobalt catalysts supported on different carbon nanostructures. *Catalysis Today* 344:66–74.

Aupretre, F., C. Descorme and D. Duprez. 2002. Bio-ethanol catalytic steam reforming over supported metal catalysts. Catalysis Communications 3:263–267.

Bac, S., S. Keskin and A. K. Avci. 2020. Recent advances in materials for high purity H_2 production by ethanol and glycerol steam reforming. *International Journal of Hydrogen Energy* 45:34888–34917.

Bilal, M. and D. M. Jackson. 2017. Ethanol steam reforming over Pt/Al_2O_3 and Rh/Al_2O_3 catalysts: The effect of impurities on selectivity and catalyst deactivation. *Applied Catalysis A: General* 529:98–107.

Busca, G., U. Costantino and T. Montanari, et al. 2010. Nickel versus cobalt catalysts for hydrogen production by ethanol steam reforming: Ni-Co-Zn-Al catalysts from hydrotalcite-like precursors. *International Journal of Hydrogen Energy* 35:5356–5366.

Calles, J., A. Carrero, A. Vizcaino and M. Lindo. 2015. Effect of Ce and Zr addition to Ni/SiO_2 catalysts for hydrogen production through ethanol steam reforming. *Catalysis* 5:58–76.

Campos, C. H., G. Pecchi, J. L. G. Fierro and P. Osorio-Vargas. 2019. Enhanced bimetallic Rh-Ni supported catalysts on alumina doped with mixed lanthanum-cerium oxides for ethanol steam reforming. *Molecular Catalysis* 469:87–97.

Cerda-Moreno, C., J. F. A. Costa-Serra and A. Chica. 2019. Co and La supported on Zn-hydrotalcite-derived material as efficient catalyst for ethanol steam reforming. *International Journal of Hydrogen Energy* 44:12685–12692.

Chen, D., W. Wang and C. Liu 2019a. Ni-encapsulated graphene chainmail catalyst for ethanol steam reforming. *International Journal of Hydrogen Energy* 44:6560–6572.

Chen, M., C. Wang and Y. Wang, et al. 2019b. Hydrogen production from ethanol steam reforming: Effect of Ce content on catalytic performance of Co/Sepiolite catalyst. *Fuel* 247:344–355.

Chen, M., Y. Wang and Z. Yang, et al. 2018. Effect of Mg-modified mesoporous Ni/attapulgite catalysts on catalytic performance and resistance to carbon deposition for ethanol steam reforming. *Fuel* 220:32–46.

Chiou, J. Y.Z., C. L. Lai and S. W. Yu, et al. 2014. Effect of Co, Fe and Rh addition on coke deposition over Ni/$Ce_{0.5}Zr_{0.5}O_2$ catalysts for steam reforming of ethanol. *International Journal of Hydrogen Energy* 39:20689–20699.

Chiu, W. C., R. F. Horng and H. M. Chou. 2013. Hydrogen production from an ethanol reformer with energy saving approaches over various catalysts. *International Journal of Hydrogen Energy* 38:2760–2769.

Coleman, L. J. I., W. Epling, R. R. Hudgins and E. Croiset. 2009. Ni/Mg-Al mixed oxide catalyst for the steam reforming of ethanol. *Applied Catalysis A* 363:52–63.

Comas, J., F. Marino, M. Laborde and N. Amadeo. 2004. Bio-ethanol steam reforming on Ni/Al_2O_3 catalyst. *Chemical Engineering Journal* 98:61–68.

Dan, M., M. Mihet and Z. Tasnadi-Asztalos, et al. 2015. Hydrogen production by ethanol steam reforming on nickel catalysts: Effect of support modification by CeO_2 and La_2O_3. *Fuel* 147:260–268.

De la Piscina, P. R. and N. Homs. 2008. Use of biofuels to produce hydrogen (reformation processes). *Chemical Society Reviews* 37:24592467.

Devianto, H., J. Han and S. P. Yoon, et al. 2011. The effect of impurities on the performance of bioethanol-used internal reforming molten carbonate fuel cell. *International Journal of Hydrogen Energy* 36: 10346–10354.

Diagne, C., H. Idriss, K. Pearson, M. A. Gomez-Garcia and A. Kiennemann. 2004. Efficient hydrogen production by ethanol reforming over Rh catalysts. Effect of addition of Zr on CeO_2 for the oxidation of CO to CO_2. *Comptes Rendus Chimie* 7:617–622.

Elias, K. F. M., A. F. Lucredio and E. M. Assaf. 2013. Effect of CaO addition on acid properties of Ni-Ca/Al_2O_3 catalysts applied to ethanol steam reforming. *International Journal of Hydrogen Energy* 38:4407–4417.

Erdohelyi, A., J. Rasko and T. Kecskes, et al. 2006. Hydrogen formation in ethanol reforming on supported noble metal catalysts. *Catalysis Today* 116:367–376.

Espinal, R., E. Taboada and E. Molins, et al. 2012. Cobalt hydrotalcites as catalysts for bioethanol steam reforming. The promoting effect of potassium on catalyst activity and long-term stability. *Applied Catalysis B: Environmental* 127:59–67.

Fatsikostas, A. N. and X. E. Verykios. 2004. Reaction network of steam reforming of ethanol over Ni-based catalysts. *Journal of Catalysis* 225:439–452.

Frusteri, F., S. Freni and L. Spadaro, et al. 2004. H_2 production for MC fuel cell by steam reforming of ethanol over MgO supported Pd, Rh, Ni and Co catalysts. Catalysis Communications 5:611–615.

Gac, W., M. Greluk and G. Słowik, et al. 2018a. Effects of dealumination on the performance of Ni-containing BEA catalysts in bioethanol steam reforming. *Applied Catalysis B: Environmental* 237:94–109.

Gac, W., M. Greluk, G. Słowik and S. Turczyniak-Surdacka. 2018b. Structural and surface changes of cobalt modified manganese oxide during activation and ethanol steam reforming reaction. *Applied Surface Science* 440:1047–1062.

Ganguly, P., M. Harb and Z. Cao, et al. 2019. 2D nanomaterials for photocatalytic hydrogen production. *ACS Energy Letters* 4:1687–1709.

Garbarino, G., A. R. Perez, E. Finocchio and G. Busca. 2013. A study of the deactivation of low loading Ni/Al_2O_3 steam reforming catalyst by tetrahydrothiophene. *Catalysis Communications* 38:67–73.

Gaudillere, C., J. J. Gonzalez, A. Chica and J. M. Serra. 2017. YSZ monoliths promoted with Co as catalysts for the production of H_2 by steam reforming of ethanol. *Applied Catalysis A: General* 538:165–173.

Grzybek, G., M. Greluk and P. Indyka, et al. 2020. Cobalt catalyst for steam reforming of ethanol-Insights into the promotional role of potassium. *International Journal of Hydrogen Energy* 45:22658–22673.

Habe, H., T. Shinbo and T. Yamamoto, et al. 2013. Chemical analysis of impurities in diverse bioethanol samples. *Journal of the Japan Petroleum Institute* 56414–422.

Han, S. J., Y. Bang and J. Yoo, et al. 2014. Hydrogen production by steam reforming of ethanol over P123-assisted mesoporous Ni-Al_2O_3-ZrO_2 xerogel catalysts. *International Journal of Hydrogen Energy* 39:10445–10453.

Han, X., Y. Wang, Y. Zhang, Y. Yu and H. He. 2017. Hydrogen production from oxidative steam reforming of ethanol over Ir catalysts supported on Ce-La solid solution. *International Journal of Hydrogen Energy* 42:11177–11186.

He, S., Z. Mei and N. Liu, et al. 2017. Ni/SBA-15 catalysts for hydrogen production by ethanol steam reforming: Effect of nickel precursor. *International Journal of Hydrogen Energy* 42:14429–14438.

Hernandez-Soto, M. C., J. F. Da Costa-Serra, J. Carratala, R. Beneito and A. Chica. 2019. Valorization of alcoholic wastes from the vinery industry to produce H_2. *International Journal of Hydrogen Energy* 44:9763–9770.

Hou, T., S. Zhang, Y. Chen, D. Wang and W. Cai. 2015. Hydrogen production from ethanol reforming: Catalysts and reaction mechanism. *Renewable and Sustainable Energy Reviews* 44:132–148.

Hou, T., B. Yu and S. Zhang, et al. 2015a. Hydrogen production from ethanol steam reforming over Rh/CeO_2 catalyst. *Catalysis Communications* 58137–140.

Hou, T., S. Zhang, Y. Chen, D. Wang and W. Cai. 2015b. Hydrogen production from ethanol reforming: Catalysts and reaction mechanism. *Renewable and Sustainable Energy Reviews* 44:132–148.

James, O. O., S. Maity and M. A. Mesubi, et al. 2011. Towards reforming technologies for production of hydrogen exclusively from renewable resources. *Green Chemistry* 13:2272.

Kim, K. M., B. S. Kwak and Y. Im, et al. 2017. Effective hydrogen production from ethanol steam reforming using CoMg co-doped SiO2@ Co1−xMgxO catalyst. *Journal of Industrial and Engineering Chemistry* 51:140–152.

Kovac, A., D. Marcius and L. Budin. 2019. Solar hydrogen production via alkaline water electrolysis. *International Journal of Hydrogen Energy* 44:9841–9848.

Le Valant, A., A. Garron, N. Bion, F. Epron and D. Duprez. 2008. Hydrogen production from raw bioethanol over Rh/$MgAl_2O_4$ catalyst: Impact of impurities: Heavy alcohol, aldehyde, ester, acid and amine. *Catalysis Today* 138:169–174.

Li, X., Z. Zheng and S. Wang, et al. 2019. Preparation and characterization of core-shell composite zeolite BEA@ MFI and their catalytic properties in ESR. *Catalysis Letters* 149:766–777.

Li, Y., X. Wang and C. Song. 2016. Spectroscopic characterization and catalytic activity of Rh supported on CeO_2-modified Al_2O_3 for low-temperature steam reforming of propane. Catalysis Today 263:22–34.

Liang, X., X. Shi and F. Zhang, et al. 2017. Improved H_2 production by ethanol steam reforming over Sc_2O_3-doped Co-ZnO catalysts. *Catalysts* 7:241.

Liguras, D. K., D. I. Kondarides and X. E. Verykios. 2003. Production of hydrogen for fuel cells by steam reforming of ethanol over supported noble metal catalysts. *Applied Catalysis B: Environmental* 43:345–354.

Ma, H., L. Zeng and H. Tian, et al. 2016. Efficient hydrogen production from ethanol steam reforming over La-modified ordered mesoporous Ni-based catalysts. *Applied Catalysis B: Environmental* 181:321–331.

Machocki, A., A. Denis, W. Grzegorczyk and W. Gac. 2010. Nano- and micro-powder of zirconia and ceria-supported cobalt catalysts for the steam reforming of bio-ethanol. *Applied Surface Science* 256:5551–5558.

Menegazzo, F., C. Pizzolitto and D. Zanardo, et al. 2017. Hydrogen production by ethanol steam reforming on Ni-based catalysts: Effect of the support and of CaO and Au doping. *ChemistrySelect* 2:9523–9531.

Mulewa, W., M. Tahir and N. A. S. Amin. 2017. MMT-supported Ni/TiO_2 nanocomposite for low temperature ethanol steam reforming toward hydrogen production. *Chemical Engineering Journal* 326:956–969.

Muradov, N. Z. and T. N. Vezirolu. 2008. "Green" path from fossil-based to hydrogen economy: An overview of carbon-neutral technologies. *International Journal of Hydrogen Energy* 33:6804–6839.

Nazir, H., N. Muthuswamy and C. Louis, et al. 2020. Is the H_2 economy realizable in the foreseeable future? Part III: H_2 usage technologies, applications, and challenges and opportunities. *International Journal of Hydrogen Energy* 45:28217–28239.

Ni, M., D. Y. C. Leung and M. K. H. Leung. 2007. A review on reforming bio-ethanol for hydrogen production. *International Journal of Hydrogen Energy* 32:3238–3247.

Ogo, S., S. Maeda and Y. Sekine. 2017. Coke resistance of Sr-hydroxyapatite supported Co catalyst for ethanol steam reforming. *Chemistry Letters* 46:729–732.

Ogo, S. and Y. Sekine. 2020. Recent progress in ethanol steam reforming using non-noble transition metal catalysts: A review. *Fuel Processing Technology* 199:106238.

Ogo, S., T. Shimizu, Y. Nakazawa, K. Mukawa, D. Mukai and Y. Sekine. 2015. Steam reforming of ethanol over K promoted Co catalyst. *Applied Catalysis A: General* 495:30–38.

Ohno, T., S. Ochibe and H. Wachi, et al. 2018. Preparation of metal catalyst component doped perovskite catalyst particle for steam reforming process by chemical solution deposition with partial reduction. *Advanced Powder Technology* 29:584–589.

Olivares, A. C. V., M. F. Gomez, M. N. Barroso and M. C. Abello. 2018. Ni-supported catalysts for ethanol steam reforming: Effect of the solvent and metallic precursor in catalyst preparation. *International Journal of Industrial Chemistry* 9:61–73.

Olsson, L. and B. Hahn-Hagerdal. 1996. Fermentation of lignocellulosic hydrolysates for ethanol production. *Enzyme and Microbial Technology* 18:312–331.

Ozkan, G., B. Sahbudak and G. Ozkan. 2019. Effect of molar ratio of water/ethanol on hydrogen selectivity in catalytic production of hydrogen using steam reforming of ethanol. *International Journal of Hydrogen Energy* 44:9823–9829.

Parlett, C. M. A., A. Aydin and L. J. Durndell, et al. 2017. Tailored mesoporous silica supports for Ni catalysed hydrogen production from ethanol steam reforming. *Catalysis Communications* 91:76–79.

Peela, N. R., A. Mubayi and D. Kunzru. 2011. Steam reforming of ethanol over Rh/CeO_2/Al_2O_3 catalysts in a microchannel reactor. Chemical Engineering Journal 167:578–587.

Phung, T. K. and G. Busca. 2020. Selective bioethanol conversion to chemicals and fuels via advanced catalytic approaches. In *Biorefinery of Alternative Resources: Targeting Green Fuels and Platform Chemicals,* Ed. S. Nanda, D. V. N. Vo and P. K. Sarangi, pp. 75–103. Singapore: Singapore.

Phung, T. K., L. P. Hernandez and G. Busca. 2015. Conversion of ethanol over transition metal oxide catalysts: Effect of tungsta addition on catalytic behaviour of titania and zirconia. *Applied Catalysis A* 489:180–187.

Rakib, M. A., J. R. Grace, C. J. Lim, S. S. Elnashaie and B. Ghiasi. 2010. Steam reforming of propane in a fluidized bed membrane reactor for hydrogen production. *International Journal of Hydrogen Energy* 35:6276–6290.

Reid, R. C., J. M. Prausnitz and B. E. Poling 1987. *The Properties of Gases and Liquids.* New York: McGraw Hill.

Rodrigues, T. S., A. B. L. de Moura and F. A. e Silva, et al. 2019. Ni supported $Ce_{0.9}Sm_{0.1}O_{2-\delta}$ nanowires: An efficient catalyst for ethanol steam reforming for hydrogen production. *Fuel* 237:1244–1253.

Rodriguez-Gomez, A. and A. Caballero. 2018. Bimetallic Ni-Co/SBA-15 catalysts for reforming of ethanol: How cobalt modifies the nickel metal phase and product distribution. *Molecular Catalysis* 449:122–130.

Rodriguez-Gomez, A., J. P. Holgado and A. Caballero. 2017. Cobalt carbide identified as catalytic site for the dehydrogenation of ethanol to acetaldehyde. *ACS Catalysis* 7:5243–5247.

Roychowdhury, S., M. M. Ali, S. Dhua, T. Sundararajan and G. R. Rao. 2021. Thermochemical hydrogen production using Rh/CeO_2/γ-Al_2O_3 catalyst by steam reforming of ethanol and water splitting in a packed bed reactor. *International Journal of Hydrogen Energy* 46:19254–19269.

Sanchez, N., R. Ruiz, V. Hacker and M. Cobo. 2020. Impact of bioethanol impurities on steam reforming for hydrogen production: A review. *International Journal of Hydrogen Energy* 45:11923–11942.

Sanchez-Sanchez, M. C., R. M. Navarro and J. L. G Fierro. 2007. Ethanol steam reforming over Ni/La-Al2O3 catalysts: Influence of lanthanum loading. *Catalysis Today* 129:336–345.

Sekine, Y., Y. Nakazawa, K. Oyama, T. Shimizu and S. Ogo. 2014. Effect of small amount of Fe addition on ethanol steam reforming over Co/Al_2O_3 catalyst. *Applied Catalysis A: General* 472:113–122.

Sharma, S., B. Patil and A. Pathak, et al. 2018. Application of BICOVOX catalyst for hydrogen production from ethanol. *Clean Technologies and Environmental Policy* 20:695–701.

Sharma, Y. C., A. Kumar, R. Prasad and S. N. Upadhyay. 2017. Ethanol steam reforming for hydrogen production: Latest and effective catalyst modification strategies to minimize carbonaceous deactivation. *Renewable and Sustainable Energy Reviews* 74:89–103.

Silberova, B., H. J. Venvik, J. C. Walmsley and A. Holmen. 2005. Small-scale hydrogen production from propane. Catalysis Today 100:457–462.

Sliwa, M. and K. Samson. 2021. Steam reforming of ethanol over copper-zirconia based catalysts doped with Mn, Ni, Ga. *International Journal of Hydrogen Energy* 46:555–564.

Sohn, H., G. Celik and S. Gunduz, et al. 2017. Oxygen mobility in pre-reduced nano-and macro-ceria with Co loading: An AP-XPS, in-situ DRIFTS and TPR study. *Catalysis Letters* 147:2863–2876.

Song, J. H., S. J. Han and J. Yoo, et al. 2016a. Effect of Sr content on hydrogen production by steam reforming of ethanol over Ni-Sr/Al_2O_3-ZrO_2 xerogel catalysts. *Journal of Molecular Catalysis A: Chemical* 418–419:68–77.

Song, J. H., S. J. Han and J. Yoo, et al. 2016b. Hydrogen production by steam reforming of ethanol over Ni-Sr-Al_2O_3-ZrO_2 aerogel catalyst. *Journal of Molecular Catalysis A: Chemical* 424:342–350.

Song, J. H., S. J. Han and J. Yoo, et al. 2016c. Hydrogen production by steam reforming of ethanol over Ni-X/Al_2O_3-ZrO_2 (X=Mg, Ca, Sr, and Ba) xerogel catalysts: Effect of alkaline earth metal addition. *Journal of Molecular Catalysis A: Chemical* 415:151–159.

Song, J. H., S. Yoo and J. Yoo, et al. 2017. Hydrogen production by steam reforming of ethanol over Ni/Al_2O_3-La_2O_3 xerogel catalysts. *Molecular Catalysis* 434:123–133.

Sun, J., X. P. Qiu, F. Wu and W. T. Zhu. 2005. H_2 from steam reforming of ethanol at low temperature over Ni/Y_2O_3, Ni/La_2O_3 and Ni/Al_2O_3 catalysts for fuel-cell application. *International Journal of Hydrogen Energy* 30:437–445.

Turczyniak, S., M. Greluk and G. Słowik, et al. 2017. Surface state and catalytic performance of ceria-supported cobalt catalysts in the steam reforming of ethanol. *ChemCatChem* 9:782–797.

Vizcaino, A. J., M. Lindo, A. Carrero and J. A. Calles. 2012. Hydrogen production by steam reforming of ethanol using Ni catalysts based on ternary mixed oxides prepared by coprecipitation. *International Journal of Hydrogen Energy* 37:1985–1992.

Wang, S., B. He and R. Tian, et al. 2018. Ni-hierarchical Beta zeolite catalysts were applied to ethanol steam reforming: Effect of sol gel method on loading Ni and the role of hierarchical structure. *Molecular Catalysis* 453:64–73.

Wang, T., H. Ma and L. Zeng, et al. 2016. Highly loaded Ni-based catalysts for low temperature ethanol steam reforming. *Nanoscale* 8:10177–10187.

Wu, R. C., C. W. Tang and H. H. Huang, et al. 2019. Effect of boron doping and preparation method of Ni/$Ce_{0.5}Zr_{0.5}O_2$ catalysts on the performance for steam reforming of ethanol. *International Journal of Hydrogen Energy* 44:14279–14289.

Yaakob, Z., A. Bshish, A. Ebshish, S. M. Tasirin and F. H. Alhasan. 2013. Hydrogen production by steam reforming of ethanol over nickel catalysts supported on sol gel made alumina: Influence of calcination temperature on supports. *Materials* 6:2229–2239.

Zanchet, D., J. B. O. Santos, S. Damyanova, J. M.R. Gallo and J. M. C. Bueno. 2015. Toward understanding metal-catalyzed ethanol reforming. *ACS Catalysis* 5:3841–3863.

Zhao, X. and G. Lu. 2016. Modulating and controlling active species dispersion over Ni-Co bimetallic catalysts for enhancement of hydrogen production of ethanol steam reforming. *International Journal of Hydrogen Energy* 41:3349–3362.

Zheng, Z., C. Sun, R. Dai and S. Wang, et al. 2017. Ethanol steam reforming on Ni-based catalysts: Effect of Cu and Fe addition on the catalytic activity and resistance to deactivation. *Energy and Fuels* 31:3091–3100.

Part 31

Bioethanol Fuel Cells

106 Bioethanol Fuel Cells
Scientometric Study

Ozcan Konur
(Formerly) Ankara Yildirim Beyazit University

106.1 INTRODUCTION

Crude oil-based gasoline fuels (Ma et al., 2002; Newman and Kenworthy, 1989) have been widely used in the transportation sector since the 1920s. However, there have been great public concerns over the adverse environmental impact of these fuels (Hill et al., 2006, 2009). Hence, biomass-based bioethanol fuels (Hill et al., 2006; Konur, 2012e, 2015, 2019a, 2020a) have increasingly been used in blending gasoline fuels (Hsieh et al., 2002; Najafi et al., 2009) and in fuel cells (Antolini, 2007, 2009).

In the meantime, the research in nanomaterials and nanotechnology has intensified in recent years to become a major research field in scientific research with over one and a half million published papers (Geim, 2009; Geim and Novoselov, 2007). In this context, a large number of nanomaterials have been developed nearly for every research field. These materials offer an innovative way to increase the efficiency in the production and utilization of bioethanol fuels as in the other scientific fields (Konur 2016a,b,c,d,e,f, 2017a,b,c,d,e, 2019b, 2021a,b).

The research in the field of fuel cells has also intensified in recent years with over 50,000 papers in three primary research streams of high-temperature solid oxide fuel cells (SOFCs) (Adler, 2004), low-temperature polymer electrode membrane fuel cells (PEMFCs) (Borup et al., 2007), and microbial fuel cells (MFCs) (Logan, 2009). The research in the direct ethanol fuel cells (DEFCs) has been a stream of the research on the low-temperature PEMFCs. The primary focus of the research on fuel cells has been the optimization of the operating conditions and the development of the catalysts to maximize the fuel cell performance.

On the other hand, the research on the direct utilization of bioethanol fuels in these DEFCs has intensified in recent years, primarily in the research fronts of the ethanol electrooxidation on the platinum (Pt) electrocatalysts (Kowal et al., 2009; Lamy et al., 2004, Zhou et al., 2003) and on the palladium (Pd) electrocatalysts for the DEFCs (Liang et al., 2009; Shen et al., 2010; Tian et al., 2010). The other research fronts are the ethanol electrooxidation on both Pt and Pd electrocatalysts for the DEFCs (Xu et al., 2007a,b), the ethanol electrooxidation on the other electrocatalysts for the DEFCs (Christensen et al., 2006; Larsson and Anderson, 1998), the other issues in the ethanol electrooxidation (Camara and Iwasita, 2005; Egolfopoulos et al., 1992), and the DEFCs themselves (Kamarudin et al., 2013; Song and Tsiakaras, 2006).

However, it is essential to develop efficient incentive structures (North, 1991) for the primary stakeholders to enhance the research in this field (Konur, 2000, 2002a,b,c, 2006a,b, 2007a,b).

The scientometric analysis has been used in this context to inform the primary stakeholders about the current state of the research in a selected research field (Garfield, 1955; Konur, 2011, 2012a,b,c,d,e,f,g,h,i, 2015, 2016a,b,c,d,e,f, 2017a,b,c,d,e,f, 2018a,b, 2019a,b, 2020a, 2021a,b).

As there has not been any scientometric study on bioethanol fuel cells, this book chapter presents a scientometric study of the research in bioethanol fuel cells. It examines the scientometric characteristics of both the sample and population data presenting the scientometric characteristics of these both datasets in the order of documents, authors, publication years, institutions, funding bodies, source titles, countries, Scopus subject categories, keywords, and research fronts.

DOI: 10.1201/9781003226574-141

106.2 MATERIALS AND METHODS

The search for this study was carried out using the Scopus database (Burnham, 2006) in September 2021.

As a first step for the search of the relevant literature, the keywords were selected using the first most-cited 200 papers. The selected keyword list was optimized to obtain a representative sample of papers for the searched research field. This keyword list was provided in the appendix for future replication studies. Additionally, the information about the most-used keywords was given in the Section 106.3.9 to highlight the key research fronts in the Section 106.3.10.

As a second step, two sets of data were used for this study. First, a population sample of over 3,100 papers was used to examine the scientometric characteristics of the population data. Secondly, a sample of 100 most-cited papers was used to examine the scientometric characteristics of these citation classics with over 95 citations each.

The scientometric characteristics of these both sample and population datasets were presented in the order of documents, authors, publication years, institutions, funding bodies, source titles, countries, Scopus subject categories, keywords, and research fronts.

Lastly, the key scientometric findings for both datasets were discussed to highlight the research landscape for the bioethanol fuel cells. Additionally, a number of brief conclusions were drawn and a number of relevant recommendations were made to enhance the future research landscape.

106.3 RESULTS

106.3.1 The Most Prolific Documents in Bioethanol Fuel Cells

The information on the types of documents for both datasets is given in Table 106.1. The articles dominate both the sample and population datasets. The review papers and articles have a slight surplus and deficit, respectively.

It is also interesting to note that all of the papers in the sample dataset were published in the journals, while only 97.9% of the papers were published in the journals for the population dataset. Furthermore, 1.6% and 0.7% of the population papers were published in book series and books, respectively.

106.3.2 The Most Prolific Authors in Bioethanol Fuel Cells

The information about the most-prolific 33 authors with at least three sample papers and six population papers each is given in Table 106.2.

TABLE 106.1
Documents in Bioethanol Fuel Cells

Documents	Sample Dataset (%)	Population Dataset (%)	Surplus (%)
Article	90	94.1	–4.1
Conference paper	5	3.9	1.1
Review	5	0.9	4.1
Book chapter	0	0.7	–0.7
Letter	0	0.2	–0.2
Note	0	0.1	–0.1
Editorial	0	0.1	–0.1
Book	0	0.0	0.0
Short Survey	0	0.0	0.0
Sample size	100	3,166	

TABLE 106.2
Most-Prolific Authors in Bioethanol Fuel Cells

No.	Author Name	Author Code	Sample Papers	Population Papers	Institution	Country	Res. Front
1	Sun, Gongquan	7402760735	9	30	Chinese Acad. Sci.	China	Pt catalysts
2	Xin, Quin	28167866000	8	24	Chinese Acad. Sci.	China	Pt catalysts
3	Leger, Jean Michel	7201980020 13105762400	7	29	Univ. Poitiers	France	Pt catalysts
4	Tsiakaras, Panagiotis	7003948427	7	28	Univ. Thessalia	Greece	Pt catalysts
5	Jiang, Luhua	57209054216	7	14	Chinese Acad. Sci.	China	Pt catalysts
6	Coutanceau, Christophe	8714035200	7	11	Univ. Poitiers	France	Pt catalysts
7	Lamy, Claude	7007017658	7	8	Univ. Poitiers	France	Pt catalysts
8	Zhao, Tianshou	13004121800	5	30	Hong Kong Univ. Sci. Technol.	China	Pd catalysts
9	Shen, Peikang	7201767641	5	24	Sun Yat-Sen Univ.	China	Pt, Pd catalysts
10	Song, Shuqin	7403349881	5	16	Chinese Acad. Sci.	China	Pt catalysts
11	Xu, Changwei	9248835900	5	13	Jinan Univ.	China	Pt, Pd catalysts
12	Li, Wenzhen	8922678800	5	8	Chinese Acad. Sci.	China	Pt catalysts
13	Belgsir, el Mustapha	6701740559	5	5	Univ. Poitiers	France	Pt catalysts
14	Gonzalez, Ernesto R	35596037000 57199756260	4	36	Univ. Sao Paulo	Brazil	Pt catalysts
15	Sun, Shi-Gang	7404510197	4	22	Xiamen Univ.	China	Pt, Pd catalysts
16	Behm, Rolf Jurgen	36885065400	4	10	Univ. Ulm	Germany	Pt catalysts
17	Xu, Jianbo	22636447700	4	10	Hong Kong Univ. Sci. Technol.	China	Pd catalysts
18	Zhou, Zhenhua	7406094713	4	9	Chinese Acad. Sci.	China	Pt catalysts
19	Jusys, Zenonas	7003450828	4	9	Univ. Ulm	Germany	Pt catalysts
20	Iwasita, Teresa	7004504751	4	8	Univ. Sao Paulo	Brazil	Pt catalysts
21	Zhou, Weijiang	56003143300	4	7	Chinese Acad. Sci.	China	Pt catalysts
22	Tremiliosi-Filho, Germano	6701503483	3	36	Univ. Sao Paulo	Brazil	Pt catalysts
23	Kamarudin, Siti Kartoom	6506009910	3	16	Malaysia Natl. Univ.	Malaysia	DEFCs
24	Li, Yinshi	15762935100	3	14	Hong Kong Univ. Sci. Technol.	China	Pd catalysts
25	Liu, Ping	36065603800	3	14	Brookhaven Natl. Lab.	USA	Pt catalysts
26	Su, Dong	55154198900	3	13	Brookhaven Natl. Lab.	USA	Pd catalysts
27	Adzic, Radoslav R	7006804065	3	11	Brookhaven Natl. Lab.	USA	Pt catalysts
28	Leger, Jean Marc	13105762400	3	10	Chinese Acad. Sci.	China	Pt catalysts
29	Liang, Zheng Xing	7402178316	3	10	Hong Kong Univ. Sci. Technol.	China	Pd catalysts
30	Zhou, Zhi-You	7406098551	3	10	Xiamen Univ.	China	Pt, Pd catalysts
31	Hahn, Francoise	7202561708	3	8	Univ. Poitiers	France	Pt catalysts
32	Tian, Na	57203214786	3	7	Xiamen Univ.	China	Pt, Pd catalysts
33	Rousseau, Severine	8714035100	3	6	Univ. Poitiers	France	Pt catalysts

Author code, the unique code given by Scopus to the authors; Population papers, the number of papers authored in the population dataset; Sample papers, the number of papers authored in the sample dataset.

The most prolific authors are Gongquan Sun and Quin Xin with nine and eight sample papers, respectively. Jean Michel Leger, Panagiotis Tsiakaras, Luhua Jiang, Christophe Coutanceau, and Claude Lamy follow these top authors with seven sample papers each.

The most prolific institution for the sample dataset is the Chinese Academy of Sciences with eight authors. The University of Poitiers, Hong Kong University of Science and Technology, Brookhaven National Laboratory, University of Sao Paulo, and Xiamen University follow this top institution with at least three papers each.

The most prolific country for the sample dataset is China with 17 authors. France, Brazil, and the USA follow this top country with at least three authors each. In total, seven countries house these authors.

The most-prolific research front is the electrooxidation of ethanol on the Pt electrocatalysts for the DEFCs with 27 papers. The other prolific research front is the electrooxidation of ethanol on the Pd electrocatalysts for the DEFCs with 10 papers. Additionally, one author focuses on the DEFCs themselves.

On the other hand, there is a significant gender deficit (Beaudry and Lariviere, 2016) for the sample dataset as surprisingly nearly all of these top researchers are male.

106.3.3 The Most Prolific Research Output by Years in Bioethanol Fuel Cells

Information about papers published between 1970 and 2021 is given in Figure 106.1. This figure clearly shows that the bulk of the research papers in the population dataset were published primarily in the 2010s with 64% of the population dataset. This was followed by the early 2020s and 2000s with 11.7% and 15.6% of the population papers, respectively. The publication rates for the pre-2000s were negligible. There was a rising trend for the population papers between 2004 and 2014, and after that, the research output steadied around 7% each year.

Similarly, the bulk of the research papers in the sample dataset were published in the 2000s and 2010s with 54% and 31% of the sample dataset, respectively. The publication rates for the 1990s, 1980s, and 1970s were 13%, 4%, and 2% of the sample papers, respectively.

The most prolific publication years for the population dataset were after 2012 with at least 6.1% each of the dataset. Similarly, 45% of the sample papers were published between 2004 and 2010.

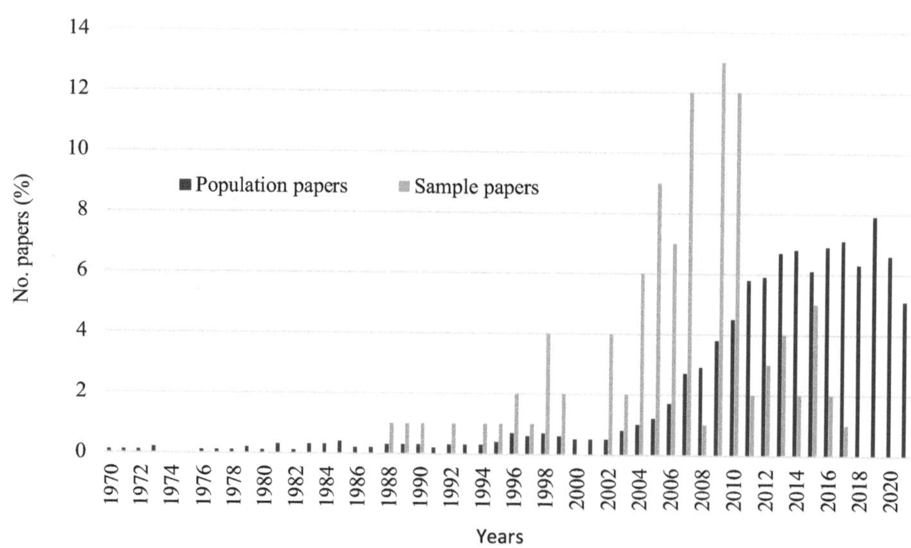

FIGURE 106.1 The research output by years regarding the bioethanol fuel cells.

106.3.4 The Most Prolific Institutions in Bioethanol Fuel Cells

Information about the most prolific 18 institutions publishing papers on the bioethanol fuel cells with at least two sample papers and 0.3% of the population papers each is given in Table 106.3.

The most prolific institution is the Chinese Academy of Sciences with nine papers. The University of Sao Paulo, University of Poitiers, and Sun Yat-Sen University follow this top institution with eight sample papers each.

The top country for these most prolific institutions is China with seven institutions. Next, the USA follows China with two institutions. In total, ten countries house these top institutions.

On the other hand, the institutions with the most citation impact are the Sun Yat-Sen University and the University of Poitiers with a 6.8% surplus each. The Chinese Academy of Sciences, University of Thessaly, and Brookhaven National Laboratory follow these top institutions with at least 4.0% surplus each. Similarly, the institutions with the least impact are Soochow University and Tsinghua University with at least 0.1% surplus each.

106.3.5 The Most Prolific Funding Bodies in Bioethanol Fuel Cells

Information about the most prolific seven funding bodies funding at least two sample papers and 0.3% of the population papers each is given in Table 106.4.

The most prolific funding body is the National Natural Science Foundation of China with 16 sample papers. Sao Paulo State Research Support Foundation and the US Department of Energy follow this top funding body with six and five sample papers, respectively.

It is notable that 50% and 54.9% of the sample and population papers are funded, respectively.

The most prolific country for these top funding bodies is Brazil with four funding bodies. China and the USA follow this top funding body with three sample papers each. In total, four countries and the European Union house these top funding bodies.

The funding bodies with the most citation impact are the US Department of Energy and the German Research Foundation with at least a 3.4% surplus. Similarly, the funding bodies with

TABLE 106.3
The Most Prolific Institutions in Bioethanol Fuel Cells

No.	Institutions	Country	Sample Papers (%)	Population Papers (%)	Surplus (%)
1	Chinese Acad. Sci.	China	9	4.4	4.6
2	Univ. Sao Paulo	Brazil	8	5.5	2.5
3	Univ. Poitiers	France	8	1.2	6.8
4	Sun Yat-Sen Univ.	China	8	1.2	6.8
5	Hong Kong Univ. Sci. Technol.	China	5	1.1	3.9
6	Univ. Thessaly	Greece	5	1.0	4.0
7	Brookhaven Natl. Lab.	USA	5	1.0	4.0
8	Xiamen Univ.	China	4	2.3	1.7
9	Univ. Ulm	Germany	4	0.3	3.7
10	Nanyang Technol. Univ.	Singapore	3	1.1	1.9
11	Soochow Univ.	China	2	1.9	0.1
12	Tsinghua Univ.	China	2	1.4	0.6
13	Malaysia Natl. Univ.	Malaysia	2	0.6	1.4
14	Fudan Univ.	China	2	0.4	1.6
15	US Army	USA	2	0.4	1.6
16	Denmark Tech. Univ.	Denmark	2	0.3	1.7
17	Leiden Univ.	Netherlands	2	0.3	1.7

TABLE 106.4
The Most Prolific Funding Bodies in Bioethanol Fuel Cells

No.	Funding Bodies	Country	Sample Paper No. (%)	Population Paper No. (%)	Surplus (%)
1	National Natural Science Foundation of China	China	16	17.2	−1.2
2	Sao Paulo State Research Support Foundation	Brazil	6	5.3	0.7
3	US Department of Energy	USA	5	1.5	3.5
4	National Council for Scientific and Technological Development	Brazil	4	6.0	−2.0
5	Ministry of Science, Technology and Innovation	Brazil	4	5.3	−1.3
6	Coordination for the Improvement of Higher Education Personnel	Brazil	4	4.2	−0.2
7	European Commission	EU	4	1.7	2.3
8	German Research Foundation	Germany	4	0.6	3.4
9	National Science foundation of Quangdong Province	China	3	0.7	2.3
10	US Army	USA	3	0.4	2.6
11	Guangdong Science and Technology Department	China	2	0.3	1.7
12	American Chemical Society	USA	2	0.3	1.7

the least citation impact are the National Council for Scientific and Technological Development, the Ministry of Science, Technology and Innovation of Brazil and the National Natural Science Foundation of China with at least a 2% deficit each.

106.3.6 The Most Prolific Source Titles in Bioethanol Fuel Cells

Information about the most prolific 17 source titles publishing at least two sample papers and 0.3% of the population papers each in bioethanol fuel cells is given in Table 106.5.

The most prolific source titles are 'Journal of Power Sources' and 'Electrochimica Acta' with 19 and 13 sample papers, respectively. 'International Journal of Hydrogen Energy', 'Applied Catalysis B Environmental', 'Journal of Electroanalytical Chemistry', and 'Journal of Catalysis' follow these top titles with at least five papers each.

On the other hand, the source titles with the most citation impact are 'Journal of Power Sources' and 'Electrochimica Acta' with 11.7% and 6.5% surplus, respectively. Similarly, the source titles with the least impact are the 'International Journal of Hydrogen Energy', 'Journal of the Electrochemical Society', and 'Journal of Physical Chemistry C' with at least 1% deficit each.

106.3.7 The Most Prolific Countries in Bioethanol Fuel Cells

Information about the most prolific 12 countries publishing at least two sample papers and 1.8% of the population papers each in bioethanol fuel cells is given in Table 106.6.

The most prolific countries are China and the USA with 35 and 25 sample papers, respectively. France, Brazil, Germany, and Greece follow these top countries with at least seven sample papers each. Further, six European countries listed in Table 106.6 produce 33% and 15% of the sample and population papers, respectively. It is notable that China is also the largest producer of population papers with a 32.2% publication rate.

On the other hand, the countries with the most citation impact are the USA and France with 13.5% and 7.9% surplus, respectively. Similarly, the countries with the least citation impact are India, Brazil and Italy with at least a 3.5% deficit each.

TABLE 106.5
The Most Prolific Source Titles in Bioethanol Fuel Cells

No.	Source Titles	Sample Papers (%)	Population Papers (%)	Surplus (%)
1	Journal of Power Sources	19	7.3	11.7
2	Electrochimica Acta	13	6.5	6.5
3	International Journal of Hydrogen Energy	7	8.0	−1,0
4	Applied Catalysis B Environmental	7	2.5	4.5
5	Journal of Electroanalytical Chemistry	5	2.6	2.4
6	Journal of Catalysis	5	1.4	3.6
7	Journal of the American Chemical Society	4	0.7	3.3
8	Angewandte Chemie International Edition	4	0.4	3.6
9	Journal of Physical Chemistry B	4	0.2	3.8
10	Electrochemistry Communications	3	1.7	1.3
11	Journal of the Electrochemical Society	2	2.5	−0.5
12	Journal of Physical Chemistry C	2	1.7	0.3
13	Journal of Applied Electrochemistry	2	1.1	0.9
14	Catalysis Today	2	0.9	1.1
15	Physical Chemistry Chemical Physics	2	0.9	1.1
16	Advanced Materials	2	0.4	1.6
17	Carbon	2	0.3	1.7

TABLE 106.6
The Most Prolific Countries in Bioethanol Fuel Cells

No.	Countries	Sample Papers (%)	Population Papers (%)	Surplus (%)
1	China	35	32.2	2.8
2	USA	25	11.5	13.5
3	France	11	3.1	7.9
4	Brazil	9	11.1	−2.1
5	Germany	7	3.8	3.2
6	Greece	7	1.5	5.5
7	India	3	6.3	−3.3
8	Italy	3	3.5	−0.5
9	Singapore	3	1.6	1.4
10	Netherlands	3	1.1	1.9
11	Poland	2	1.9	0.1
12	Argentina	2	1.8	0.2

106.3.8 The Most Prolific Scopus Subject Categories in Bioethanol Fuel Cells

Information about the most prolific eight Scopus subject categories indexing at least 6% and 4.5% of the sample and population papers, respectively, is given in Table 106.7.

The most prolific Scopus subject category in bioethanol fuel cells is ‘Chemistry’ with 80 sample papers. This top category is followed by ‘Chemical Engineering’, ‘Energy’, and ‘Engineering’ with at least 26 sample papers each.

On the other hand, the Scopus subject categories with the most citation impact are ‘Chemistry’ and ‘Chemical Engineering’ with 13.8% and 8.1% surplus, respectively. Similarly, the Scopus subject categories with the least citation impact are ‘Materials Science’ and ‘Physics and Astronomy’ with at least a 9% deficit each.

TABLE 106.7
The Most Prolific Scopus Subject Categories in Bioethanol Fuel Cells

No.	Scopus Subject Categories	Sample Papers (%)	Population Papers (%)	Surplus (%)
1	Chemistry	80	66.2	13.8
2	Chemical Engineering	47	38.9	8.1
3	Energy	34	29.0	5.0
4	Engineering	26	21.0	5.0
5	Materials Science	18	27.5	−9.5
6	Physics and Astronomy	11	20.0	−9
7	Environmental Science	11	6.7	4.3
8	Biochemistry, Genetics and Molecular Biology	6	4.5	1.5

106.3.9 The Most Prolific Keywords in Bioethanol Fuel Cells

Information about the keywords used in at least 5% of the sample and population papers is given in Table 106.8. For this purpose, keywords related to the keyword set given in the appendix are selected from a list of the most prolific keyword set provided by the Scopus database.

These keywords are grouped under six headings: Bioethanol fuels, nanomaterials, oxidation, fuel cells, materials, and catalysis.

There are five keywords used related to bioethanol fuels. These are ethanol, ethanol oxidation, and ethanol electrooxidation. It is notable that the bioethanol keyword does not appear in this list.

The only prolific keyword related to nanomaterials is nanoparticles. The prolific keywords related to the oxidation are oxidation, electrooxidation, electrochemistry, and cyclic voltammetry while the prolific keywords related to fuel cells are fuel cells, direct ethanol fuel cells (DEFCs), cyclic voltammetry, and alkaline media.

The prolific keywords related to the materials are platinum and palladium while the prolific keywords related to the catalysis are catalysts, electrocatalysts, and catalyst activity.

On the other hand, the keywords with the most citation impact are oxidation, fuel cells, cyclic voltammetry, catalysts, and alkaline media. Similarly, the keywords with the least citation impact are ethanol oxidation reaction, catalysis, and binary alloys.

106.3.10 The Most Prolific Research Fronts in Bioethanol Fuel Cells

Information about the most prolific research fronts for the sample papers in bioethanol fuel cells is given in Table 106.9.

There are two primary research fronts in this field: The ethanol electrooxidation on the Pt electrocatalysts for the DEFCs (Kowal et al., 2009; Lamy et al., 2004, Zhou et al., 2003) and the ethanol electrooxidation on the Pd electrocatalysts for the DEFCs (Liang et al., 2009; Shen et al., 2010; Tian et al., 2010) with 46 and 23 sample papers, respectively.

The other research fronts are the ethanol electrooxidation on both Pt and Pd electrocatalysts for the DEFCs (Xu et al., 2007a,b), the ethanol electrooxidation on the other electrocatalysts for the DEFCs (Christensen et al., 2006; Larsson and Anderson, 1998), the other issues in the ethanol electrooxidation (Camara and Iwasita, 2005; Egolfopoulos et al., 1992), and the DEFCs (Kamarudin et al., 2013; Song and Tsiakaras, 2006) with eight, 13, four, and seven sample papers, respectively.

The focus of these most-cited 100 papers is the development of nanomaterials and conventional materials for the most efficient processes for ethanol electrooxidation and the DEFCs.

TABLE 106.8
The Most Prolific Keywords in Bioethanol Fuel Cells

No.	Keywords	Sample Papers (%)	Population Papers (%)	Surplus (%)
1.	Bioethanol fuels			
	Ethanol	83	74.7	8.3
	Ethanol oxidation	33	26.8	6.2
	Ethanol electrooxidation	23	27.2	−4.2
	Ethanol fuels	15	19.5	−4.5
	Ethanol oxidation reaction	6	17.8	−11.8
2.	Nanomaterials			
	Nanoparticles	12	14.6	−2.6
	Nanocatalysts		5.0	−5
	Graphene		5.6	−5.6
3.	Oxidation			
	Oxidation	60	37.5	22.5
	Electrooxidation	25	26.4	−1.4
	Acetic acid	11		11
	Carbon dioxide	9	6.2	2.8
	Acetaldehyde	9	5.3	3.7
	Catalytic oxidation	7	11.5	−4.5
	Aldehydes	6		6
4	Fuel cells			
	Fuel cells	49	26.6	22.4
	Direct ethanol fuel cells	38	26.9	11.1
	Electrolysis	7	5.4	1.6
	Electrodes	6	8.5	−2.5
	Current density	5		5
5	Materials			
	Platinum	34	23.6	10.4
	Palladium	24	13.1	10.9
	Carbon	10	8.5	1.5
	Platinum alloys	9	8.8	0.2
	Platinum compounds	8		8.0
	Gold	8		8.0
	Ruthenium	5		5.0
	Binary alloys		7.8	−7.8
	Tin		6.0	−6
	Nickel		5.4	−5.4
6.	Catalysis			
	Catalysts	34	17.0	17
	Electrocatalysts	33	26.2	6.8
	Catalyst activity	24	21.6	2.4
	Electrocatalytic activity	11	13.5	−2.5
	Electrocatalysis	11	11.4	−0.4
	Catalysis		9.2	−9.2
	Catalyst supports		5.3	−5.3

TABLE 106.9
The Most Prolific Research Fronts in Bioethanol Fuel Cells

No.	Research Fronts	Sample Papers (%)
1.	Pd and Pt electrocatalysts	8
	Conventional electrocatalysts	4
	Nanomaterial-based electrocatalysts	4
2.	Pt electrocatalysts	43
	Conventional electrocatalysts	41
	Nanomaterial-based electrocatalysts	2
3.	Pd electrocatalysts	26
	Conventional electrocatalysts	12
	Nanomaterial-based electrocatalysts	14
4.	Other electrocatalysts	13
	Conventional electrocatalysts	9
	Nanomaterial-based electrocatalysts	4
5.	Other issues in ethanol electrooxidation	4
6.	Ethanol fuel cells	7

Sample papers, the sample of the most-cited 100 papers.

106.4 DISCUSSION

106.4.1 Introduction

Crude oil-based gasoline fuels have been widely used in the transportation sector since the 1920s. However, there have been great public concerns over the adverse environmental impact of these fuels. Hence, biomass-based bioethanol fuels have increasingly been used in blending gasoline fuels and in fuel cells. In the meantime, the research in nanotechnology has intensified in recent years.

The research in fuel cells has also intensified in recent years in three primary research streams of SOFCs, PEMFCs, and MFCs. The research on the DEFCs has been a stream of the research on the PEMFCs. On the other hand, the research on the direct utilization of bioethanol fuels in these DEFCs has intensified in recent years, primarily in the research fronts of the ethanol electrooxidation on the Pt electrocatalysts and on the Pd electrocatalysts for the DEFCs. The other research fronts are the ethanol electrooxidation on both Pt and Pd electrocatalysts and on the other electrocatalysts, the other issues in the ethanol electrooxidation, and the DEFCs themselves. However, it is essential to develop efficient incentive structures for the primary stakeholders to enhance the research in this field.

The scientometric analysis has been used in this context to inform the primary stakeholders about the current state of the research in a selected research field. As there has not been any scientometric study on bioethanol fuel cells, this book chapter presents a scientometric study of the research in bioethanol fuel cells. It examines the scientometric characteristics of both the sample and population data presenting the scientometric characteristics of these both datasets in the order of documents, authors, publication years, institutions, funding bodies, source titles, countries, Scopus subject categories, keywords, and research fronts.

The search for this study was carried out using the Scopus database. As a first step for the search of the relevant literature, the keywords were selected using the first most-cited 200 papers. The selected keyword list was optimized to obtain a representative sample of papers for the searched research field. This keyword list was provided in the appendix for future replication studies. Additionally, the information about the most-used keywords was given in Section 106.3.9 to highlight the key research fronts in Section 106.3.10.

As a second step, two sets of data were used for this study. First, a population sample of over 3,100 papers was used to examine the scientometric characteristics of the population data. Secondly, a sample of 100 most-cited papers was used to examine the scientometric characteristics of these citation classics with over 95 citations each.

The scientometric characteristics of these both sample and population datasets were presented in the order of documents, authors, publication years, institutions, funding bodies, source titles, countries, Scopus subject categories, keywords, and research fronts. Lastly, the key scientometric findings for both datasets were discussed to highlight the research landscape for bioethanol fuel cells. Additionally, a number of brief conclusions were drawn and a number of relevant recommendations were made to enhance the future research landscape.

106.4.2 The Most Prolific Documents in Bioethanol Fuel Cells

The articles dominate both the sample and population datasets. The review papers and articles have a slight surplus and deficit, respectively. Scopus differs from the Web of Science database in differentiating and showing articles and conference papers published in the journals separately. Hence, it can be said that 95% and 98% of the sample and population papers, respectively, are articles by combining the figures for articles and conference papers published in the journals.

It is observed during the search process that there has been inconsistency in the classification of the documents in Scopus as well as in other databases such as Web of Science. This is especially relevant for the classification of papers as reviews or articles as the papers not involving a literature review may be erroneously classified as a review paper. There is also a case of review papers being classified as articles. For example, although there are five review papers as classified by the Scopus database, eight of the sample papers are review papers.

In this context, it would be helpful to provide a classification note for the published papers in the books and journals at the first instance. It would also be helpful to use the document types listed in Table 106.1 for this purpose. Book chapters may also be classified as articles or reviews as an additional classification to differentiate review chapters from experimental chapters as it is done by the Web of Science. It would be further helpful to additionally classify the conference papers as articles or review papers as well as it is done in the Web of Science database.

106.4.3 The Most Prolific Authors in Bioethanol Fuel Cells

There have been most-prolific 33 authors with at least two sample papers and six population papers each as given in Table 106.2. These authors have shaped the development of research in this field.

The most prolific authors are Gongquan Sun, Quin Xin, Jean Michel Leger, Panagiotis Tsiakaras, Luhua Jiang, Christophe Coutanceau, and Claude Lamy (Table 106.2).

It is important to note the inconsistencies in the indexing of the author names in Scopus and other databases. It is especially an issue for names with more than two components such as 'Judge Alex de Camp Sirous'. The probable outcomes are 'Sirous, J.A.D.C.', 'de Camp Sirous, J.A.' or 'Camp Sirous, J.A.D.'. The first choice is the gold standard of the publishing sector as the last word in the name is taken as the last name. In most of the academic databases such as PubMed, EBSCO databases, this version is used predominantly. The second choice is a strong alternative while the last choice is an undesired outcome as two last words are taken as the last name. It is good practice to combine the words of the last name with a hyphen: 'Camp-Sirous, J.A.D.'. It is notable that inconsistent indexing of the author names may cause substantial inefficiencies in the search process for the papers as well as allocating credit to the authors as there are different author entries for each outcome in the databases.

There is also a case of shortening Chinese names. For example, 'Yuoyang Wang' is often shortened as 'Wang, Y.', 'Wang, Y.-Y.', and 'Wang Y.Y.' as it is done in the Web of Science database as well. However, the gold stand in this case is 'Wang Y' where the last word is taken as the last name and the first word is taken as a single forename. In most of the academic databases such as PubMed and EBSCO, this first version is used predominantly. However, it makes sense to use the third option to differentiate Chinese names efficiently: 'Wang Y.Y... Therefore, there have been difficulties to locate papers for Chinese authors for example by Gongquan Sun, Quin Xin, and Luhua Zhang where there is inconsistency in the shortening of these names.

In such cases, the use of the unique author codes provided for each author by the Scopus database has been helpful. For example, there are two separate entries for 'Sun, G.' and 'Sun, G.Q.' with five and four papers, separately. This entry was revised as 'Sun, Gongquan' with nine papers. Similarly, there are two separate entries for 'Jiang, L.' and 'Jiang, L.H.' with four and three sample papers, respectively. This entry was revised as 'Jiang, Luhua' with seven papers. The other example is 'Li, Wenzhen' with separate entries of 'Li, Z.' and 'Li, W.Z." in the Scopus database.

There is also a difficulty in allowing credit for the authors, especially for the authors with common names such as 'Wang, Y.', or 'Huang, Y.' or 'Zhu, Y.' in conducting scientometric studies. These difficulties strongly influence the efficiency of the scientometric studies as well as allocating credit to the authors as there are the same author entries for different authors with the same name, e.g., 'Wang Y.' in the databases.

In this context, the coding of authors in the Scopus database is a welcome innovation compared to other databases such as Web of Science. In this process, Scopus allocates a unique number to each author in the database. However, there might still be substantial inefficiencies in this coding system, especially for common names. For example, some of the papers for Jean Michel Leger are allocated to another researcher with a different author code. There are two separate entries for 'Léger, J.M' with four and three papers in the Scopus database. This was revised as single-author entry with seven sample papers. It is possible that Scopus uses a number of software programs to differentiate the author names (Shin et al., 2014).

In this context, it does not help that author names are not given in full in some journals. This makes it difficult to differentiate authors with common names and makes the scientometric studies further difficult in the author domain. Therefore, the author names should be given in all books and journals at the first instance. There is also a cultural issue where some authors do not use their full names in their papers. Instead, they use initials for their forenames: 'Coutancy, A.P.' instead of 'Coutancy, Alas Padras'.

There are also inconsistencies in the naming of the authors with more than two components by the authors themselves in journal papers and book chapters. For example, 'Faaij, A.P.C.', 'Wyman, C.E.', and 'Shuai, S.J.' might be given as 'Faaij, A', 'Wyman, C', or 'Shuai, S' in the journals and books. This also makes the scientometric studies difficult in the author domain. Hence, contributing authors should use their names consistently in their publications.

The other critical issue regarding the author names is the spelling of the author names in the national spellings (e.g., Gonçalves, Übeiro) rather than in the English spellings (e.g., Goncalves, Ubeiro) in the Scopus database. Scopus differs from the Web of Science database in this respect where the author names are given only in English spellings. It is observed that national spellings of the author names do not help in conducting scientometric studies as well as in allocating credits to the authors as sometimes there are different author entries for the English and National spellings in the Scopus database.

The most prolific institutions for the sample dataset are the Chinese Academy of Sciences, the University of Poitiers, the Hong Kong University of Science and Technology, Brookhaven National Laboratory, the University of Sao Paulo, and Xiamen University. Similarly, the most prolific countries for the sample dataset are China, France, Brazil, and the USA. It is not surprising that the Chinese authors dominate this prolific author list as China is a major producer of research in this field.

It is also notable that there is a significant gender deficit for the sample dataset as surprisingly nearly all of these top researchers are male. This finding is the most thought-provoking with strong public policy implications. Hence, institutions, funding bodies, and policy makers should take efficient measures to reduce the gender deficit in this field as well as other scientific fields with strong gender deficits. In this context, it is worth to note the level of representation of the researchers from minority groups in science on the basis of race, sexuality, age, and disability, besides gender (Blankenship, 1993; Dirth and Branscombe, 2017; Konur, 2000, 2002a,b,c, 2006a,b, 2007a,b).

106.4.4 The Most Prolific Research Output by Years in Bioethanol Fuel Cells

The research output observed between 1970 and 2021 is illustrated in Figure 106.1. This figure clearly shows that the bulk of the research papers in the population dataset were published primarily in the 2010s, followed by the early 2020s and 2000s. Similarly, the bulk of the research papers in the sample dataset were published in the 2000s and 2010s.

These data suggest that the most-cited sample papers were primarily published in the 2000s and 2010s while the population papers were primarily published in the 2010s. However, it is notable that the momentum for the rise in the research output has been lost after 2014, suggesting the consolidation in the research output for unknown reasons.

These are thought-provoking findings as there has been no significant research in this field in the pre-2000s, but there has been a significant research boom in the last two decades. In this context, the increasing public concerns about climate change (Change, 2007), greenhouse gas emissions (Carlson et al., 2017), and global warming (Kerr, 2007) have been certainly behind the boom in the research in this field in the last two decades.

The data in Figure 106.1 also suggest that the research in this field has boomed in the last two decades and the size of the population papers likely to more than double in the current decade, provided that the public concerns about climate change, greenhouse gas emissions, and global warming are translated efficiently to the research funding in this field.

106.4.5 The Most Prolific Institutions in Bioethanol Fuel Cells

The most prolific 18 institutions publishing papers on bioethanol fuel cells with at least two sample papers and 0.3% of the population papers each given in Table 106.3 have shaped the development of the research in this field.

The most prolific institutions are the Chinese Academy of Sciences, the University of Sao Paulo, the University of Poitiers, and Sun Yat-Sen University. The top countries for these most prolific institutions are China and the USA. It is not surprising that the Chinese institutions dominate this prolific institution list as China is a major producer of research in this field.

On the other hand, the institutions with the most citation impact are Sun Yat-Sen University, the University of Poitiers, the Chinese Academy of Sciences, the University of Thessaly, and Brookhaven National Laboratory. Similarly, the institutions with the least impact are Soochow University and Tsinghua University.

106.4.6 The Most Prolific Funding Bodies in Bioethanol Fuel Cells

The most prolific seven funding bodies funding at least two sample papers and 0.3% of the population papers each are given in Table 106.4.

The most prolific funding bodies are the National Natural Science Foundation of China, the Sao Paulo State Research Support Foundation, and the US Department of Energy. It is notable that 50% and 54.9% of the sample and population papers are funded, respectively. The most prolific countries for these top funding bodies are Brazil, China and the USA. In total, four countries and the

European Union house these top funding bodies. The heavy funding by the Chinese funding bodies is notable as China is a major producer of research in this field.

These findings on the funding of the research in this field suggest that the level of funding, mostly in the 2010s been largely instrumental in enhancing the research in this field (Ebadi and Schiffauerova, 2016) in the light of North's institutional framework (North, 1991). However, the consolidation in the research output after 2014 should be carefully considered in terms of funding rates and policies.

106.4.7 The Most Prolific Source Titles in Bioethanol Fuel Cells

The most prolific 17 source titles publishing at least two sample papers and 0.3% of the population papers each in bioethanol fuel cells have shaped the development of the research in this field (Table 106.5).

The most prolific source titles are 'Journal of Power Sources', 'Electrochimica Acta', 'International Journal of Hydrogen Energy', 'Applied Catalysis B Environmental', 'Journal of Electroanalytical Chemistry', and 'Journal of Catalysis'.

On the other hand, the source titles with the most citation impact are 'Journal of Power Sources' and 'Electrochimica Acta'. Similarly, the source titles with the least impact are the 'International Journal of Hydrogen Energy', 'Journal of the Electrochemical Society', and 'Journal of Physical Chemistry C'.

It is notable that these top source titles are related to materials science, material chemistry, energy, and chemical engineering. This finding suggests that the journals in these fields have significantly shaped the development of the research in this field as they focus on the development of nanomaterials and conventional materials for efficient ethanol electrooxidation and the DEFCs.

106.4.8 The Most Prolific Countries in Bioethanol Fuel Cells

The most prolific 12 countries publishing at least two papers and 1.8% of the population papers each have significantly shaped the development of the research in this field (Table 106.6).

The most prolific countries are China, the USA, France, Brazil, Germany, and Greece. On the other hand, the countries with the most citation impact are the USA and France while the countries with the least citation impact are India, Brazil, and Italy. Further, six European countries listed in Table 106.6 produce 33% and 15% of the sample and population papers, respectively. This makes Europe as the second largest prolific country after China.

Further, the countries with the most citation impact are the USA and France with 13.5% and 7.9% surplus, respectively. Similarly, the countries with the least citation impact are India, Brazil and Italy with at least a 3.5% deficit each. Six European countries listed in Table 106.6 have an 18% surplus, making Europe the most influential producer of research in this field. These European countries, France, Greece, Germany, and Netherlands are the most influential countries.

A close examination of these findings suggests that the USA, China, Europe and to a lesser extent Brazil, India, and Singapore are the major producers of research in this field. It is a fact that the USA has been a major player in science (Leydesdorff and Wagner, 2009; Leydesdorff et al., 2014). The USA has further developed a strong research infrastructure to support its corn and grass-based bioethanol industry (Vadas et al., 2008). The USA has also been very active in nanotechnology research (Dong et al., 2016) as well as bioethanol fuel cell research (Lee et al., 2019).

However, China has been a rising star in scientific research in competition with the USA and Europe (Leydesdorff and Zhou, 2005). China is also a major player in this field as a major producer of bioethanol (Li and Chan-Halbrendt, 2009). China has also been very active in nanotechnology research (Dong et al., 2016) as well as bioethanol fuel cell research (Liu et al., 2018).

Next, Europe has been a persistent player in scientific research in competition with both the USA and China (Leydesdorff, 2000). Europe has also been a persistent producer of bioethanol along with the USA and Brazil (Gnansounou, 2010). The European Union has also been very active in nanotechnology research (Scheufele et al., 2009) as well as bioethanol fuel cell research (Borthwick, 2000).

Additionally, Brazil has also been a persistent player in scientific research at a moderate level (Glanzel et al., 2006). Brazil has also developed a strong research infrastructure to support its biomass-based bioethanol industry (Soccol et al., 2010). Brazil has also been very active in nanotechnology research (Fonseca and Pereira, 2004) as well as bioethanol fuel cell research (Silva et al., 2013).

106.4.9 The Most Prolific Scopus Subject Categories in Bioethanol Fuel Cells

The most prolific eight Scopus subject categories indexing at least 6% and 4.5% of the sample and population papers, respectively, given in Table 106.7 have shaped the development of the research in this field.

The most prolific Scopus subject categories in bioethanol fuel cells are 'Chemistry', 'Chemical Engineering', 'Energy', and 'Engineering'.

On the other hand, the Scopus subject categories with the most citation impact are 'Chemistry' and 'Chemical Engineering'. Similarly, the Scopus subject categories with the least citation impact are 'Materials Science' and 'Physics and Astronomy'.

These findings are thought-provoking suggesting that the primary subject categories are 'Chemistry', 'Chemical Engineering', 'Energy', and 'Engineering'. The other key finding is that social sciences are not well represented in both the sample and population papers, unlike the field of evaluative studies in bioethanol fuels.

These findings are not surprising as the key research fronts in this field are the performance and development of innovative nanomaterials and conventional materials for ethanol electrooxidation for the DEFCs in this field. All these research fronts are related to the hard sciences and engineering.

106.4.10 The Most Prolific Keywords in Bioethanol Fuel Cells

A limited number of keywords have shaped the development of the research in this field as shown in Table 106.8 and the appendix.

These keywords are grouped under six headings: Bioethanol fuels, nanomaterials, oxidation, fuel cells, materials, and catalysis.

The most prolific keywords across all the headings are ethanol, oxidation, fuel cells, direct ethanol fuel cells (DEFCs), platinum, catalysts, ethanol oxidation, electrocatalysts, electrooxidation, palladium, catalyst activity, and ethanol electrooxidation. Similarly, the most influential keywords are oxidation, fuel cells, cyclic voltammetry, catalysts, alkaline media, electrochemistry, direct ethanol fuel cells (DEFCs), acetic acid, palladium, and platinum.

These prolific keywords highlight the key research fronts in this field and reflect well the keywords used in the sample papers.

106.4.11 The Most Prolific Research Fronts in Bioethanol Fuel Cells

There are two primary research fronts in this field: The ethanol electrooxidation on the Pt electrocatalysts for the DEFCs (Kowal et al., 2009; Lamy et al., 2004, Zhou et al., 2003) and the ethanol electrooxidation on the Pd electrocatalysts for the DEFCs (Liang et al., 2009; Shen et al., 2010; Tian et al., 2010) with 46 and 23 sample papers, respectively.

The other research fronts are the ethanol electrooxidation on the Pt and Pd electrocatalysts for the DEFCs (Xu et al., 2007a,b), the ethanol electrooxidation on the other electrocatalysts for the DEFCs (Christensen et al., 2006; Larsson and Anderson, 1998), the other issues in the ethanol electrooxidation (Camara and Iwasita, 2005; Egolfopoulos et al., 1992), and the DEFCs (Kamarudin et al., 2013; Song and Tsiakaras, 2006) with eight, 13, four, and seven sample papers, respectively.

The focus of these most-cited 100 papers is the development of the nanomaterials and conventional materials for ethanol electrooxidation for the DEFCs. The primary electrocatalysts used and researched for ethanol electrooxidation are Pt (Adzic et al. 2007) and Pd electrocatalysts (Antolini, 2009). The Pt electrooxidation has been a gold standard to use for ethanol electrooxidation in the DEFCs. However, these electrocatalysts have been used with a number of additives in the DEFCs to increase the efficiency of ethanol electrooxidation. On the other hand, Pd electrocatalysts have been a rising star for ethanol electrooxidation in the DEFCs.

The research in the field of fuel cells has intensified in recent years with over 50,000 papers in three primary research streams of high-temperature SOFCs (Adler, 2004), low-temperature PEMFCs (Borup et al., 2007), and MFCs (Logan, 2009). The research in the DEFCs (Song and Tsiakaras, 2006) has been a stream of research on the low-temperature PEMFCs. The primary focus of research on fuel cells has been the optimization of the operating conditions to maximize the fuel cell performance.

Therefore, it is not surprising that there has been a limited number of prolific papers on the bioethanol fuel cells themselves (Kamarudin et al., 2013; Song and Tsiakaras, 2006) as most of these prolific papers have focused on the electrocatalyst development for ethanol electrooxidation in the DEFCs. This finding suggests that the development of efficient electrocatalysts and catalyst supports is highly critical for the development of efficient DEFCs.

One innovative way to enhance the fuel cell performance has been, as a number of studies for ethanol electrooxidation hint, the use of nanomaterials for both the electrocatalysts themselves and the catalyst supports (Seger and Kamat, 2009).

In parallel with the research onfuel cells, the research onnanomaterials and nanotechnology (Geim, 2009; Geim and Novoselov, 2007) has intensified in recent years with over one and half million papers, enriching the material portfolio to be used for the fuel cell applications. It is expected that this enriched portfolio of innovative nanomaterials would enhance the ethanol electrooxidation in the DEFCs as both electrocatalyst materials and electrocatalyst supports in the future, benefiting from the superior properties of these nanomaterials.

In the end, these most-cited papers in this field hint that the efficiency of ethanol electrooxidation in the DEFCs could be optimized using the structure, processing, and property relationships of the nanomaterials and conventional materials used as both electrocatalyst materials and catalyst supports (Formela et al., 2016; Konur, 2018c, 2020b, 2021c,d,e,f; Konur and Matthews, 1989).

106.5 CONCLUSION AND FUTURE RESEARCH

The research on bioethanol fuel cells has been mapped through a scientometric study of both sample and population datasets.

The critical issue in this study has been to obtain a representative sample of the research as in any other scientometric study. Therefore, the keyword set has been carefully devised and optimized after a number of runs in the Scopus database.

The other issue has been the selection of a multidisciplinary database to carry out the scientometric study of the research in this field. For this purpose, the Scopus database has been selected. The journal coverage of this database has been wider than that of the Web of Science.

The key scientometric properties of the research in this field have been determined and discussed in this book chapter. It is evident that a limited number of documents, authors, institutions, publication periods, institutions, funding bodies, source titles, countries, Scopus subject categories, keywords, and research fronts have shaped the development of the research in this field.

There is ample scope to increase the efficiency of the scientometric studies in this field in the author and document domains by developing consistent policies and practices in both domains across all academic databases. In this respect, authors, journals, and academic databases have a lot to do. Furthermore, the significant gender deficit as in most scientific fields emerges as a public policy issue. The potential deficits on the basis of age, race, disability, and sexuality need also to be explored in this field as in other scientific fields.

The research in this field has boomed in the late 2010s possibly promoted by the public concerns on global warming, greenhouse gas emissions, and climate change. Institutions from the USA, China, and Europe have mostly shaped the research in this field. However, it is notable that there was consolidation in the research output after 2014 raising questions about its source.

The moderate funding rate of 50% and 54.9% for sample and population papers, respectively, suggests that this funding rate significantly enhanced the research in this field primarily in the 2010s, possibly more than doubling in the current decade. However, there is ample room for more funding, especially considering the consolidation in research output after 2014 and the recent supply shocks caused by the COVID-19 pandemic and the Ukrainian war.

The most-prolific source titles have been mostly indexed by the subject categories of materials science, material chemistry, energy, and chemical engineering as the focus of the research in this field has been on the development of efficient catalyst systems.

The USA, China, and Europe have been the major producers of the research in this field as the major producers of bioethanol fuels from different types of biomass such as corn, sugarcane, and grass as well as other types of biomass. Brazil has also contributed largely to both sample and population papers. These countries have well-developed research infrastructure in bioethanol fuels, bioethanol fuel cells, nanomaterials, and conventional materials.

The primary subject categories have been 'Chemistry', 'Chemical Engineering', 'Energy', and 'Engineering'. Due to the technological emphasis of this field on the development of ethanol electrocatalysts for the DEFCs, social sciences have not been fairly represented in both the sample and population papers, unlike the evaluative studies in bioethanol fuels.

Ethanol is more popular than bioethanol as a keyword with strong implications for the search strategy. In other words, the search strategy using only the bioethanol keyword would not be much helpful. However, it is recommended that the term "bioethanol fuels' is used instead of ethanol or bio-ethanol or bioethanol.

These keywords are grouped under six headings: Bioethanol fuels, nanomaterials, oxidation, fuel cells, materials, and catalysis. These groups of keywords highlight the potential primary research fronts for these fields: The development of ethanol electrocatalysts for ethanol electrooxidation in the DEFCs, mainly Pt and Pd-based electrocatalysts.

There are two primary research fronts in this field: Ethanol electrooxidation on the Pt and Pd electrocatalysts for the DEFCs. Furthermore, there are further research fronts within each of these research fronts: Nanomaterial-based catalysts and conventional material-based catalysts. In total, there are brute 24 and 66 papers for the nanomaterial-based fuel cells and conventional material-based bioethanol fuel cells, respectively. Therefore, there is ample room for the use of nanomaterials for the DEFCs.

These findings are thought-provoking. The focus of these most-cited 100 papers is the development of nanomaterials and conventional materials for ethanol electrocatalysts. There are strong structure-processing-property relationships for all of the materials used for these research fronts. In the end, these most-cited papers in this field hint that the efficiency of the nanotechnology and materials science applications could be optimized using the structure, processing, and property relationships of the materials used in these applications.

Thus, the scientometric analysis has a great potential to gain valuable insights into the evolution of the research in this field as in other scientific fields.

It is recommended that further scientometric studies are carried out about the other aspects of both production and utilization of bioethanol fuels. It is further recommended that reviews of the most-cited papers are carried out for each research front to complement these scientometric studies. Next, the scientometric studies of the hot papers in these primary fields are carried out.

ACKNOWLEDGMENTS

The contribution of the highly cited researchers in the field of bioethanol fuel cells has been gratefully acknowledged.

APPENDIX: THE KEYWORD SET FOR BIOETHANOL FUEL CELLS

((TITLE (ethanol OR bioethanol OR c2h5oh OR defc*) OR SRCTITLE (ethanol* OR bioethanol*)) AND (TITLE ("fuel cell*" OR defc* OR adefc* OR "biofuel cell*" OR electrooxidation OR "electro-oxidation" OR *oxidation OR *oxidizing OR pemfc* OR aemfc* OR dafcs) OR SRCTITLE ("fuel cell*" OR oxidation OR electrooxidation))) AND NOT (TITLE (hepatic OR *diesel OR dichloride* OR pretreatment* OR ethane OR immun* OR microsom* OR cytochrome* OR lipid* OR bisethanol OR catalase OR extract* OR yeast* OR periodate OR reforming OR "partial oxidation") OR SRCTITLE (biochem* OR biol*) OR SUBJAREA (medi OR neur OR immu OR phar OR heal OR dent OR nurs OR vete OR psyc)) AND (LIMIT-TO (DOCTYPE, "ar") OR LIMIT-TO (DOCTYPE, "cp") OR LIMIT-TO (DOCTYPE, "re") OR LIMIT-TO (DOCTYPE, "ch") OR LIMIT-TO (DOCTYPE, "no") OR LIMIT-TO (DOCTYPE, "le") OR LIMIT-TO (DOCTYPE, "ed") OR LIMIT-TO (DOCTYPE, "cr")) AND (LIMIT-TO (LANGUAGE, "English")) AND (LIMIT-TO (SRCTYPE, "j") OR LIMIT-TO (SRCTYPE, "k") OR LIMIT-TO (SRCTYPE, "b"))

REFERENCES

Adler, S. B. 2004. Factors governing oxygen reduction in solid oxide fuel cell cathodes. *Chemical Reviews* 104. 4791–4843.

Adzic, R. R., J. Zhang, J. and K. Sasaki, et al. 2007. Platinum monolayer fuel cell electrocatalysts. Topics in Catalysis 46:249–262.

Antolini, E. 2007. Catalysts for direct ethanol fuel cells. *Journal of Power Sources* 170:1–12.

Antolini, E. 2009. Palladium in fuel cell catalysis. *Energy and Environmental Science* 2:915–931.

Beaudry, C. and V. Lariviere. 2016. Which gender gap? Factors affecting researchers' scientific impact in science and medicine. *Research Policy* 45:1790–1817.

Blankenship, K. M. 1993. Bringing gender and race in: US employment discrimination policy. *Gender & Society* 7:204–226.

Borthwick, W. K. 2000. The European Union approach to fuel cell development. *Journal of Power Sources* 86:52–56.

Borup, R., J. Meyers and B. Pivovar, et al. 2007. Scientific aspects of polymer electrolyte fuel cell durability and degradation. *Chemical Reviews* 107:3904–3951.

Burnham, J. F. 2006. Scopus database: A review. *Biomedical Digital Libraries* 3:1–8.

Camara, G. A. and T. Iwasita. 2005. Parallel pathways of ethanol oxidation: The effect of ethanol concentration. *Journal of Electroanalytical Chemistry* 578:315–321.

Carlson, K. M., J. S. Gerber and D. Mueller, et al. 2017. Greenhouse gas emissions intensity of global croplands. *Nature Climate Change* 7:63–68.

Change, C. 2007. Climate change impacts, adaptation and vulnerability. *Science of the Total Environment* 326:95–112.

Christensen, C. H., B. Jorgensen and J. Rass-Hansen, et al. 2006. Formation of acetic acid by aqueous-phase oxidation of ethanol with air in the presence of a heterogeneous gold catalyst. *Angewandte Chemie - International Edition* 45:4648–4651.

Dirth, T. P. and N. R. Branscombe. 2017. Disability models affect disability policy support through awareness of structural discrimination. *Journal of Social Issues* 73:413–442.

Dong, H., Y. Gao and P. J. Sinko, et al. 2016. The nanotechnology race between China and the United States. *Nano Today* 11:7–12.

Ebadi, A. and A. Schiffauerova. 2016. How to boost scientific production? A statistical analysis of research funding and other influencing factors. *Scientometrics* 106:1093–1116.

Egolfopoulos, F. N., D. X. Du and C. K. Law. 1992. A study on ethanol oxidation kinetics in laminar premixed flames, flow reactors, and shock tubes. *Symposium (International) on Combustion* 24:833–841.

Fonseca, P. F. and T. S. Pereira. 2014. The governance of nanotechnology in the Brazilian context: Entangling approaches. *Technology in Society* 37:16–27.

Formela, K., A. Hejna, L. Piszczyk, M. R. Saeb and X. Colom. 2016. Processing and structure-property relationships of natural rubber/wheat bran biocomposites. *Cellulose* 23:3157–3175.

Garfield, E. 1955. Citation indexes for science. *Science* 122:108–111.

Geim, A. K. 2009. Graphene: Status and prospects. *Science* 324:1530–1534.

Geim, A. K. and K. S. Novoselov. 2007. The rise of graphene. *Nature Materials* 6:183–191.

Glanzel, W., J. Leta and B. Thijs. 2006. Science in Brazil. Part 1: A macro-level comparative study. *Scientometrics* 67:67–86.

Gnansounou, E. 2010. Production and use of lignocellulosic bioethanol in Europe: Current situation and perspectives. *Bioresource Technology* 101:4842–4850.

Hill, J., E. Nelson, D. Tilman, S. Polasky and D. Tiffany. 2006. Environmental, economic, and energetic costs and benefits of biodiesel and ethanol biofuels. *Proceedings of the National Academy of Sciences of the United States of America* 103:11206–11210.

Hill, J., S. Polasky and E. Nelson, et al. 2009. Climate change and health costs of air emissions from biofuels and gasoline. *Proceedings of the National Academy of Sciences of the United States of America* 106:2077–2082.

Hsieh, W. D., R. H. Chen, T. L. Wu and T. H. Lin. 2002. Engine performance and pollutant emission of an SI engine using ethanol-gasoline blended fuels. *Atmospheric Environment* 36:403–410.

Kamarudin, M. Z. F., S. K. Kamarudin, M. S. Masdar and W. R. W. Daud. 2013. Review: Direct ethanol fuel cells. *International Journal of Hydrogen Energy* 38:9438–9453.

Kerr, R. A. 2007. Global warming is changing the world. *Science* 316:188–190.

Konur, O. 2000. Creating enforceable civil rights for disabled students in higher education: An institutional theory perspective. *Disability & Society* 15:1041–1063.

Konur, O. 2002a. Access to nursing education by disabled students: Rights and duties of nursing programs. *Nurse Education Today* 22:364–374.

Konur, O. 2002b. Assessment of disabled students in higher education: Current public policy issues. *Assessment and Evaluation in Higher Education* 27:131–152.

Konur, O. 2002c. Access to employment by disabled people in the UK: Is the Disability Discrimination Act working? *International Journal of Discrimination and the Law* 5:247–279.

Konur, O. 2006a. Participation of children with dyslexia in compulsory education: Current public policy issues. *Dyslexia* 12:51–67.

Konur, O. 2006b. Teaching disabled students in higher education. *Teaching in Higher Education* 11:351–363.

Konur, O. 2007a. A judicial outcome analysis of the *Disability Discrimination Act*: A windfall for the employers? *Disability & Society* 22:187–204.

Konur, O. 2007b. Computer-assisted teaching and assessment of disabled students in higher education: The interface between academic standards and disability rights. *Journal of Computer Assisted Learning* 23:207–219.

Konur, O. 2011. The scientometric evaluation of the research on the algae and bio-energy. *Applied Energy* 88:3532–3540.

Konur, O. 2012a. Prof. Dr. Ayhan Demirbas' scientometric biography. *Energy Education Science and Technology Part A: Energy Science and Research* 28:727–738.

Konur, O. 2012b. The evaluation of the biogas research: A scientometric approach. *Energy Education Science and Technology Part A: Energy Science and Research* 29:1277–1292.

Konur, O. 2012c. The evaluation of the global energy and fuels research: A scientometric approach. *Energy Education Science and Technology Part A: Energy Science and Research* 30:613–628.

Konur, O. 2012d. The evaluation of the research on the biodiesel: A scientometric approach. *Energy Education Science and Technology Part A: Energy Science and Research* 28:1003–1014.

Konur, O. 2012e. The evaluation of the research on the bioethanol: A scientometric approach. *Energy Education Science and Technology Part A: Energy Science and Research* 28:1051–1064.

Konur, O. 2012f. The evaluation of the research on the biofuels: A scientometric approach. *Energy Education Science and Technology Part A: Energy Science and Research* 28:903–916.

Konur, O. 2012g. The evaluation of the research on the biohydrogen: A scientometric approach. *Energy Education Science and Technology Part A: Energy Science and Research* 29:323–338.

Konur, O. 2012h. The evaluation of the research on the microbial fuel cells: A scientometric approach. *Energy Education Science and Technology Part A: Energy Science and Research* 29:309–322.

Konur, O. 2012i. The scientometric evaluation of the research on the production of bioenergy from biomass. *Biomass and Bioenergy* 47:504–515.

Konur, O. 2015. Current state of research on algal bioethanol. In *Marine Bioenergy: Trends and Developments*, Ed. S. K. Kim and C. G. Lee, pp. 217–244. Boca Raton, FL: CRC Press.

Konur, O. 2016a. Scientometric overview in nanobiodrugs. In *Nanoarchitectonics for Smart Delivery and Drug Targeting*, Ed. A. M. Holban and A. M. Grumezescu, pp. 405–428. Amsterdam: Elsevier.

Konur, O. 2016b. Scientometric overview regarding nanoemulsions used in the food industry. In *Emulsions: Nanotechnology in the Agri-Food Industry*, Ed. A. M. Grumezescu, pp. 689–711. Amsterdam: Elsevier.

Konur, O. 2016c. Scientometric overview regarding the nanobiomaterials in antimicrobial therapy. In *Nanobiomaterials in Antimicrobial Therapy*, Ed. A. M. Grumezescu, pp. 511–535. Amsterdam: Elsevier.

Konur, O. 2016d. Scientometric overview regarding the nanobiomaterials in dentistry. In *Nanobiomaterials in Dentistry*, Ed. A. M. Grumezescu, pp. 425–453. Amsterdam: Elsevier.

Konur, O. 2016e. Scientometric overview regarding the surface chemistry of nanobiomaterials. In *Surface Chemistry of Nanobiomaterials*, Ed. A. M. Grumezescu, pp. 463–486. Amsterdam: Elsevier.

Konur, O. 2016f. The scientometric overview in cancer targeting. In *Nanoarchitectonics for Smart Delivery and Drug Targeting*, Ed. A. M. Holban and A. Grumezescu, pp. 871–895. Amsterdam: Elsevier.

Konur, O. 2017a. Recent citation classics in antimicrobial nanobiomaterials. In *Nanostructures for Antimicrobial Therapy*, Ed. A. Ficai and A. M. Grumezescu, pp. 669–685. Amsterdam: Elsevier.

Konur, O. 2017b. Scientometric overview in nanopesticides. In *New Pesticides and Soil Sensors*, Ed. A. M. Grumezescu, pp. 719–744. Amsterdam: Elsevier.

Konur, O. 2017c. Scientometric overview regarding oral cancer nanomedicine. In *Nanostructures for Oral Medicine*, Ed. E. Andronescu, A. M. Grumezescu, pp. 939–962. Amsterdam: Elsevier;

Konur, O. 2017d. Scientometric overview regarding water nanopurification. In *Water Purification*, Ed. A. M. Grumezescu, pp. 693–716. Amsterdam: Elsevier.

Konur, O. 2017e. Scientometric overview in food nanopreservation. In *Food Preservation*, Ed. A. M. Grumezescu, pp. 703–729. Amsterdam: Elsevier.

Konur, O. 2017f. The top citation classics in alginates for biomedicine. In *Seaweed Polysaccharides: Isolation, Biological and Biomedical Applications*, Ed. J. Venkatesan, S. Anil and S. K. Kim, pp. 223–249. Amsterdam: Elsevier.

Konur, O. 2018a. Scientometric evaluation of the global research in spine: An update on the pioneering study by Wei et al. *European Spine Journal* 27:525–529.

Konur, O. 2018b. Bioenergy and biofuels science and technology: Scientometric overview and citation classics. In *Bioenergy and Biofuels*, Ed. O. Konur, pp. 3–63. Boca Raton: CRC Press.

Konur, O., Ed. 2018c. *Bioenergy and Biofuels*. Boca Raton, FL: CRC Press.

Konur, O. 2019a. Cyanobacterial bioenergy and biofuels science and technology: A scientometric overview. In *Cyanobacteria: From Basic Science to Applications*, Ed. A. K. Mishra, D. N. Tiwari and A. N. Rai, pp. 419–442. Amsterdam: Elsevier.

Konur, O. 2019b. Nanotechnology applications in food: A scientometric overview. In *Nanoscience for Sustainable Agriculture*, Ed. R. N. Pudake, N. Chauhan and C. Kole, pp. 683–711. Cham: Springer.

Konur, O. 2020a. The scientometric analysis of the research on the bioethanol production from green macroalgae. In *Handbook of Algal Science, Technology and Medicine*, Ed. O. Konur, pp. 385–401. London: Academic Press.

Konur, O., Ed. 2020b. *Handbook of Algal Science, Technology and Medicine*. London: Academic Press.

Konur, O. 2021a. Nanotechnology applications in diesel fuels and the related research fields: A review of the research. In *Handbook of Biodiesel and Petrodiesel Fuels: Science, Technology, Health, and Environment. Volume 1. Biodiesel Fuels: Science, Technology, Health, and Environment*, Ed. O. Konur, pp. 89–110. Boca Raton, FL: CRC Press.

Konur, O. 2021b. Nanobiosensors in agriculture and foods: A scientometric review. In *Nanobiosensors in Agriculture and Food*, Ed. R. N. Pudake, pp. 365–384. Cham: Springer.

Konur, O., Ed. 2021c. *Handbook of Biodiesel and Petrodiesel Fuels: Science, Technology, Health, and Environment*. Boca Raton, FL: CRC Press.

Konur, O., Ed. 2021d. *Handbook of Biodiesel and Petrodiesel Fuels: Science, Technology, Health, and Environment. Volume 1. Biodiesel Fuels: Science, Technology, Health, and Environment*. Boca Raton, FL: CRC Press.

Konur, O., Ed. 2021e. *Handbook of Biodiesel and Petrodiesel Fuels: Science, Technology, Health, and Environment. Volume 2. Biodiesel Fuels based on the Edible and Nonedible Feedstocks, Wastes, and Algae: Science, Technology, Health, and Environment*. Boca Raton, FL: CRC Press.

Konur, O., Ed. 2021f. *Handbook of Biodiesel and Petrodiesel Fuels: Science, Technology, Health, and Environment. Volume 3. Petrodiesel Fuels: Science, Technology, Health, and Environment*. Boca Raton, FL: CRC Press.

Konur, O. and F. L. Matthews. 1989. Effect of the properties of the constituents on the fatigue performance of composites: A review. *Composites* 20:317–328.

Kowal, A., M. Li and M. Shao, et al. 2009. Ternary Pt/Rh/SnO_2 electrocatalysts for oxidizing ethanol to CO_2. *Nature Materials* 8:325–330.

Lamy, C., S. Rousseau, E. M. Belgsir, C. Coutanceau and J. M. Leger. 2004. Recent progress in the direct ethanol fuel cell: Development of new platinum-tin electrocatalysts. *Electrochimica Acta* 49:3901–3908.

Larsson, P. O. and A. Andersson. 1998. Complete oxidation of CO, ethanol, and ethyl acetate over copper oxide supported on titania and ceria modified titania. *Journal of Catalysis* 179:72–89.

Lee, D. Y., A. Elgowainy and R. Vijayagopal. 2019. Well-to-wheel environmental implications of fuel economy targets for hydrogen fuel cell electric buses in the United States. *Energy Policy* 128:565–583.

Leydesdorff, L. 2000. Is the European Union becoming a single publication system? *Scientometrics* 47:265–280.

Leydesdorff, L. and C. Wagner. 2009. Is the United States losing ground in science? A global perspective on the world science system. *Scientometrics* 78:23–36.

Leydesdorff, L., C. S. Wagner and L. Bornmann. 2014. The European Union, China, and the United States in the top-1% and top-10% layers of most-frequently cited publications: Competition and collaborations. *Journal of Informetrics* 8:606–617.

Leydesdorff, L. and P. Zhou. 2005. Are the contributions of China and Korea upsetting the world system of science? *Scientometrics* 63:617–630.

Li, S. Z. and C. Chan-Halbrendt. 2009. Ethanol production in (the) People's Republic of China: Potential and technologies. *Applied Energy* 86:S162–S169.

Liang, Z. X., T. S. Zhao, J. B. Xu and L. D. Zhu. 2009. Mechanism study of the ethanol oxidation reaction on palladium in alkaline media. *Electrochimica Acta* 54:2203–2208.

Liu, F., F. Zhao, Z. Liu and H. Hao, H. 2018. The impact of fuel cell vehicle deployment on road transport greenhouse gas emissions: The China case. *International Journal of Hydrogen Energy* 43:22604–22621.

Logan, B. E. 2009. Exoelectrogenic bacteria that power microbial fuel cells. *Nature Reviews Microbiology* 7:375–381.

Ma, X., L. Sun and C. Song. 2002. A new approach to deep desulfurization of gasoline, diesel fuel and jet fuel by selective adsorption for ultra-clean fuels and for fuel cell applications. *Catalysis Today* 77:107–116.

Najafi, G., B. Ghobadian and T. Tavakoli, et al. 2009. Performance and exhaust emissions of a gasoline engine with ethanol blended gasoline fuels using artificial neural network. *Applied Energy* 86:630–639.

Newman, P. W. G. and J. R. Kenworthy. 1989. Gasoline consumption and cities: A comparison of U.S. cities with a global survey. *Journal of the American Planning Association* 55:24–37.

North, D. C. 1991. Institutions. *Journal of Economic Perspectives* 5:97–112.

Scheufele, D. A., E. A. Corley, T. J. Shih, K. E. Dalrymple and S. S. Ho. 2009. Religious beliefs and public attitudes toward nanotechnology in Europe and the United States. *Nature Nanotechnology* 4:91–94.

Seger, B. and P. V. Kamat. 2009. Electrocatalytically active graphene-platinum nanocomposites. Role of 2-D carbon support in PEM fuel cells. *Journal of Physical Chemistry C* 113:7990–7995.

Shen, S.Y., T. S. Zhao, J. B. Xu and Y. S. Li. 2010. Synthesis of PdNi catalysts for the oxidation of ethanol in alkaline direct ethanol fuel cells. *Journal of Power Sources* 195:1001–1006.

Shin, D., T. Kim, J. Choi and J. Kim. 2014. Author name disambiguation using a graph model with node splitting and merging based on bibliographic information. *Scientometrics* 100:15–50.

Silva, S. B., M. M. Severino and M. A. G. de Oliveira. 2013. A stand-alone hybrid photovoltaic, fuel cell and battery system: A case study of Tocantins, Brazil. *Renewable Energy* 57:384–389.

Soccol, C. R., L. P. de Souza Vandenberghe and A. B. P. Medeiros, et al. 2010. Bioethanol from lignocelluloses: Status and perspectives in Brazil. *Bioresource Technology* 101:4820–4825.

Song, S. and P. Tsiakaras. 2006. Recent progress in direct ethanol proton exchange membrane fuel cells (DE-PEMFCs). *Applied Catalysis B: Environmental* 63:187–193.

Tian, N., Z. Y. Zhou, N. F. Yu, L. Y. Wang and S. G. Sun. 2010. Direct electrodeposition of tetrahexahedral Pd nanocrystals with high-index facets and high catalytic activity for ethanol electrooxidation. *Journal of the American Chemical Society* 132:7580–7581.

Vadas, P. A., K. H. Barnett and D. J. Undersander 2008. Economics and energy of ethanol production from alfalfa, corn, and switchgrass in the Upper Midwest, USA. *Bioenergy Research* 1:44–55.

Xu, C., L. Cheng, P. Shen and Y. Liu. 2007a. Methanol and ethanol electrooxidation on Pt and Pd supported on carbon microspheres in alkaline media. *Electrochemistry Communications* 9:997–1001.

Xu, C., P. K. Shen and Y. Liu. 2007b. Ethanol electrooxidation on Pt/C and Pd/C catalysts promoted with oxide. *Journal of Power Sources* 164:527–531.

Zhou, W., Z. Zhou and S. Song, et al. 2003. Pt based anode catalysts for direct ethanol fuel cells. *Applied Catalysis B: Environmental* 46:273–285.

107 Bioethanol Fuel Cells
Review

Ozcan Konur
(Formerly) Ankara Yildirim Beyazit University

107.1 INTRODUCTION

Crude oil-based gasoline fuels (Ma et al., 2002; Newman and Kenworthy, 1989) have been widely used in the transportation sector since the 1920s. However, there have been great public concerns over the adverse environmental impact of these fuels (Hill et al., 2006, 2009). Hence, biomass-based bioethanol fuels (Hill et al., 2006; Konur, 2012, 2015, 2019a, 2020a) have increasingly been used in blending gasoline fuels (Hsieh et al., 2002; Najafi et al., 2009) and in fuel cells (Antolini, 2007, 2009).

In the meantime, research in nanomaterials and nanotechnology has intensified in recent years to become a major research field in scientific research with over one and a half million published papers (Geim, 2009; Geim and Novoselov, 2007). In this context, large numbers of nanomaterials have been developed nearly for every research field. These materials offer an innovative way to increase the efficiency in the production and utilization of bioethanol fuels as in other scientific fields (Konur 2016a,b,c,d,e,f, 2017a,b,c,d,e, 2019b, 2021a,b).

Research in the field of fuel cells has also intensified in recent years with over 50,000 papers in three primary research streams of high-temperature solid oxide fuel cells (SOFCs) (Adler, 2004), low-temperature polymer electrode membrane fuel cells (PEMFCs) (Borup et al., 2007), and microbial fuel cells (MFCs) (Logan, 2009). Research in DEFCs has been a stream of research on low-temperature PEMFCs. The primary focus of research on fuel cells has been the optimization of the operating conditions to maximize fuel cell performance.

On the other hand, research on the direct utilization of bioethanol fuels in these DEFCs has intensified in recent years, primarily in research fronts of ethanol electrooxidation on platinum (Pt) electrocatalysts (Kowal et al., 2009; Lamy et al., 2004, Zhou et al., 2003) and on palladium (Pd) electrocatalysts for DEFCs (Liang et al., 2009; Shen et al., 2010; Tian et al., 2010). Other research fronts are ethanol electrooxidation on both Pt and Pd electrocatalysts for DEFCs (Xu et al., 2007a,b), ethanol electrooxidation on other electrocatalysts for DEFCs (Christensen et al., 2006; Larsson and Anderson, 1998), other issues in ethanol electrooxidation (Camara and Iwasita, 2005; Egolfopoulos et al., 1992), and the DEFCs themselves (Kamarudin et al., 2013; Song and Tsiakaras, 2006).

However, it is essential to develop efficient incentive structures (North, 1991) for the primary stakeholders to enhance research in this field (Konur, 2000, 2002a,b,c, 2006a,b, 2007a,b).

Although there have been a number of review papers on bioethanol fuel cells (Antolini, 2007; Kamarudin et al., 2013; Song and Tsiakaras, 2006), there has been no review of the 25 most-cited articles in this field.

This chapter presents a review of the 25 most-cited articles in the field of bioethanol fuel cells. Then, it discusses the key findings of these highly influential papers and comments on future research priorities in this field.

 DOI: 10.1201/9781003226574-142

107.2 MATERIALS AND METHODS

Search for this study was carried out using the Scopus database (Burnham, 2006) in September 2021.

As the first step for the search of the relevant literature, keywords were selected using the 200 most-cited papers. The selected keyword list was optimized to obtain a representative sample of papers for the searched research field. This keyword list was provided in the appendix of Konur (2023) for future replication studies.

As the second step, a sample dataset was used for this study. The first 25 articles in the sample of the 100 most-cited papers with at least 186 citations each were selected for the review study. Key findings from each paper were taken from the abstracts of these papers and were discussed.

Additionally, a number of brief conclusions were drawn and a number of relevant recommendations were made to enhance the future research landscape.

107.3 RESULTS

Brief information about the 25 most-cited papers with at least 186 citations each on ethanol electrooxidation in DEFCs is given below under five headings: the ethanol electrooxidation on both Pt- and Pd-based electrocatalysts, the ethanol electrooxidation on Pt-based electrocatalysts, the ethanol electrooxidation on Pd-based electrocatalysts, the ethanol electrooxidation on other electrocatalysts, and the other issues on ethanol oxidation with five, eleven, six, two, and one papers, respectively.

107.3.1 The Platinum and Palladium-based Bioethanol Electrooxidation in DEFCs

Brief information about five prolific studies with at least 278 citations each on both Pt and Pd-based bioethanol electrooxidation is given in Table 107.1. Furthermore, brief notes on the contents of these studies are also given.

Zhou et al. (2003) evaluate the effect of the additives of ruthenium (Ru), Pd, tin (Sn), and tungsten (W) on Pt-based anode catalysts supported on Vulcan XC-72R carbon black in a paper with 580 citations. They note that all of these catalysts had uniform nanoparticles with sharp distribution and

TABLE 107.1
The Pt-based Bioethanol Electrooxidation

No.	Papers	Catalysts	Catalyst supports	Issues	Cits
1	Zhou et al. (2003)	Pt-Sn/C, Pt-Ru/C, Pt-W/C, Pt-Pd/C, Pt/C	C black	Electrocatalyst development: Effect of Sn, Ru, W, and Pd additives	580
2	Xu et al. (2007a)	Pt/C, Pd/C	C black, C microspheres	Electrocatalyst development: Effect of catalyst support; effect of electrocatalyst	462
3	Bambagioni et al. (2009)	Pd NP/MWCNT, Pt-Ru NP/ MWCNT	MWCNTs	Electrocatalyst development: Effect of nanostructure; effect of electrocatalyst; effect of Ru additives	378
4	Xu et al. (2007c)	Pt/C, Pd/C, Pt-CeO_2/C, Pt-NiO/C, Pd-CeO_2/C, Pd-Nio/C	C	Electrocatalyst development: Effect of electrocatalyst; effect of CeO_2, NiO additives	355
5	Hu et al. (2012)	Pt-Pd-Cu nanoboxes/Gr, Pt/C	3D Graphene, C	Electrocatalyst development: Effect of Pd, Cu additives; effect of nanostructure	278

Cits., the number of citations received by each paper.

the Pt lattice parameter decreased with the addition of Ru or Pd and increased with the addition of Sn or W. The presence of Sn, Ru, and W enhanced the activity of Pt toward ethanol electrooxidation. A single DEFC having Pt_1-Sn_1/C, Pt_3-Sn_2/C, or Pt_2-Sn_1/C as an anode catalyst showed better performances than those with Pt_3-Sn_1/C or Pt_4-Sn_1/C. The latter two DEFCs exhibited higher performances than the single cell using Pt_1-Ru_1/C. They attribute this distinct difference in DEFC performance between these catalysts to the bifunctional mechanism and to the electronic interaction between Pt and the additives. An amount of -OH_{ads}, an amount of surface Pt active sites, and the conductivity effect of Pt-Sn/C catalysts determine the activity of Pt-Sn/C with different Pt/Sn ratios. At 75°C, the single DEFC with Pt_3-Sn_2/C as the anode catalyst showed comparable performance to that with Pt_2-Sn1/C, but at a higher temperature of 90°C, the latter presented a much better performance. They assert that Pt_2-Sn1/C, supplying sufficient -OH_{ads} and having adequate active Pt sites and acceptable ohmic effects, could be the appropriate anode catalyst for DEFC.

Xu et al. (2007a) evaluate the ethanol electrooxidation on Pt and Pd supported on carbon microspheres (CMs) and carbon black in an alkaline media in a paper with 462 citations. They note that these noble metal electrocatalysts supported on CMs gave better performances compared to the carbon black supports. Pd showed excellent higher activity and better steady-state electrolysis than Pt for ethanol electrooxidation in an alkaline media. They showed a synergistic effect with the interaction between Pd and CMs. Pd supported on CMs possessed excellent electrocatalytic properties and had great potential in DEFCs.

Bambagioni et al. (2009) evaluate the ethanol electrooxidation on Pd and Pt/Ru nanoparticle (NP) anode electrocatalysts supported on multiwalled carbon nanotubes (MWCNTs) and their use in passive and active direct alcohol fuel cells (DAFCs) with an anion-exchange membrane in a paper with 378 citations. They note that Pd NP/MWCNT exhibited unrivaled activity as an anode electrocatalyst for ethanol oxidation. Ethanol was selectively oxidized to acetic acid (AA) and detected as an acetate ion in the alkaline media of the reaction. On the other hand, Pt-Ru NP/MWCNT anodes in acid media had activity inferior as anode electrocatalysts for alcohol oxidation compared to Pd NP/MWCNT electrodes in alkaline media.

Xu et al. (2007c) evaluate the ethanol electrooxidation on Pt/C and Pd/C catalysts with CeO_2 and NiO additives in alkaline media in a paper with 355 citations. They note that the Pd/C electrocatalyst had higher catalytic activity and better steady-state behavior for ethanol oxidation compared to the Pt/C electrocatalyst. They further observed that the electrocatalysts with a weight ratio of noble metal (Pt, Pd) to CeO_2 of 2:1 and a noble metal to NiO ratio of 6:1 showed the highest catalytic activity for ethanol electrooxidation. The Pt/C and Pd/C electrocatalysts with these oxide additives showed higher activity compared to the commercial E-TEK Pt-Ru/C electrocatalyst for ethanol oxidation in alkaline media.

Hu et al. (2012) evaluate the ethanol electrooxidation on ternary Pt-Pd-Cu nanoboxes anchored on a three-dimensional graphene framework in a paper with 278 citations. They note that this structurally well-defined electrocatalyst possessed an approximately 4-fold improvement in catalytic activity for ethanol oxidation in alkaline media over the commercial 20% Pt/C catalyst as normalized by the total mass of active metals. They assert that this electrocatalyst has great potential for DEFCs.

107.3.2 The Platinum-based Bioethanol Electrooxidation in DEFCs

Brief information about 11 prolific studies with at least 288 citations each on Pt-based bioethanol electrooxidation is given in Table 107.2. Furthermore, brief notes on the contents of these studies are also given.

Kowal et al. (2009) evaluate the ethanol electrooxidation on a ternary Pt-Rh-SnO_2/C electrocatalyst in a paper with 615 citations. They deposit Pt and rhodium (Rh) atoms on carbon-supported tin dioxide (SnO_2) nanoparticles (NPs). They note that this electrocatalyst effectively split the C–C bond in ethanol at room temperature in acid solutions, facilitating its oxidation at low potentials to CO_2. They assert that the electrocatalytic activity of this electrocatalyst was due to the specific property of each of its constituents, induced by their interactions.

TABLE 107.2
The Pt-based Bioethanol Electrooxidation

No.	Papers	Catalysts	Catalyst Supports	Issues	Cits
1	Kowal et al. (2009)	Pt-Rh-SnO_2/C	C	Electrocatalyst development: Effect of Rh and SnO2 additives	615
2	Dong et al. (2010)	Pt NP/C, Pt-Ru NP/C	Graphene sheets and C black	Electrocatalyst development: Effect of Ru additives; effect of catalyst support; effect of nanostructure	524
3	Lamy et al. (2004)	Pt-Sn/C	C	Electrocatalyst development: Effect of Sn additives	547
4	Vigier et al. (2004a)	Pt-Sn/C	C	Electrocatalyst development: Effect of Sn additives	466
5	Rousseau et al. (2006)	Pt/C, Pt-Sn/C, Pt-Sn-Ru/C	C	Electrocatalyst development: Effect of Sn, Ru additives	445
6	Vigier et al. (2004b)	Pt/C, Pt-Sn/C, Pt-Re/C	C powder	Electrocatalyst development: Effect of Sn, Re additives	337
7	Wang and Liu (2008)	Pt		Electrocatalyst development: Effect of Pt surfaces	327
8	De Souza et al. (2002)	Pt, Rh, Pt-Rh		Electrocatalyst development: Effect of Rh additives	327
9	Wang et al. (1995)	Pt, Pt-Ru		Electrocatalyst development: Effect of Ru additives	319
10	Wang et al. (2004a)	Pt NP/C, Pt/C	C	Electrocatalyst development: Effect of nanostructure, effect of ethanol concentration, and reaction temperature	292
11	Zhou et al. (2010)	Pt NC/C, Pt/C	C black	Electrocatalyst development: Effect of nanostructure,	288

Cits., the number of citations received by each paper.

Dong et al. (2010) evaluate the ethanol electrooxidation on Pt and Pt-Ru NPs on graphene sheets and carbon black in a paper with 524 citations. They note that in comparison to the Vulcan XC-72R carbon black catalyst supports, graphene-supported Pt and Pt-Ru NPs demonstrated enhanced efficiency for ethanol electrooxidation with regard to diffusion efficiency, oxidation potential, forward oxidation peak current density, and the ratio of the forward peak current density to the reverse peak current density. For example, the forward peak current density of ethanol oxidation for graphene- and carbon black-supported Pt NPs is 16.2 and 13.8 mA/cm^2, respectively and the ratio is 0.90. They recommend the use of graphene sheets as catalyst supports for ethanol fuel cells.

Lamy et al. (2004) evaluate the ethanol electrooxidation on Pt-Sn electrocatalysts in DEFCs in a paper with 547 citations. They note that the overall electrocatalytic activity was greatly enhanced at low potentials. The optimum composition in Sn was in the range of 10–20 at.%. With this composition, poisoning by adsorbed CO coming from ethanol dissociative chemisorption was greatly reduced leading to a significant enhancement of electrode activity. However, the oxidation of ethanol was not complete leading to the formation of C_2 products.

Vigier et al. (2004a) evaluate the electrooxidation of Pt and Pt-Sn catalysts in a paper with 466 citations. They note the beneficial effect of Sn for ethanol electrooxidation by cyclic voltammetry as the Pt/Sn catalyst activity was almost double compared to pure Pt. They detect adsorbed CO, adsorbed CH_3CO, CH_3CHO, CH_3COOH, and CO_2. They assert that there were two effects involved in ethanol electrooxidation on the Pt/Sn electrocatalyst: the bifunctional mechanism and the ligand effect. The presence of Sn allowed ethanol to adsorb dissociatively, then to break the C–C bond, at lower potentials and with a higher selectivity compared to pure Pt. It allowed the formation of AA at lower potentials compared to pure Pt.

Rousseau et al. (2006) evaluate the ethanol electrooxidation on Pt-based anodes with the additives of Sn and Ru in a DEFC focusing on the reaction product distribution in a paper with 445 citations. They detect acetaldehyde (AAL), AA, and CO_2 in the reaction products. The addition of Sn to Pt increased the activity of the catalyst by several orders of magnitude compared to pure Pt while the electrical performance of DEFC was greatly enhanced from a few mW/cm^2 to $30\,mW/cm^2$ at 80°C, with Pt/C and Pt/Sn/C catalysts, respectively. Moreover, at Pt-Sn-C and Pt-Sn-Ru/C the formation of CO_2 and AAL was lowered whereas the formation of AA was increased compared to a Pt/C catalyst. The addition of Ru to Pt-Sn only enhanced the electrical performance of the DEFC, i.e., the activity of the electrocatalyst, but did not modify the product distribution. They observe very good stability in the open circuit voltage of the DEFC at around 0.75 V over a period of 2 weeks at 90°C, with the DEFC undergoing start-run-stop cycles each day. They also observe good stability under operating conditions at a given current density over 6 h.

Vigier et al. (2004b) evaluate the ethanol electrooxidation on Pt, Pt-Sn, and Pt-Rhenium (Re) electrocatalysts dispersed on a high surface area carbon powder for DEFCs in a paper with 337 citations. They varied the atomic composition of the bimetallic catalyst and obtained the best results with an atomic ratio Pt:X close to 100:20. Under voltammetric conditions and in a single DEFC, they observe that Pt-Sn/C was the most active catalyst. During electrolysis, ethanol was oxidized to AAL, AA, and carbon dioxide (CO_2). On Pt-Sn/C and Pt-Re/C, the ratio AA/AAL was always lower than unity. Otherwise, Pt-Sn/C electrocatalysts were the most selective toward the production of CO_2 compared to Pt/C and Pt-Re/ electrodes.

Wang and Liu (2008) explore the reaction network of ethanol electrooxidation on Pt electrocatalysts with different surfaces in a paper with 327 citations. They employ different Pt surfaces, including close-packed Pt{111}, stepped Pt{211}, and open Pt{100}, with an efficient reaction path searching method. They note that the selectivity of ethanol oxidation on Pt depended markedly on the surface structure, which can be attributed to the structure-sensitivity of two key reaction steps: The initial dehydrogenation of ethanol and the oxidation of acetyl (CH_3CO). On open surface sites, ethanol prefers C–C bond cleavage via strongly adsorbed intermediates (CH_2CO or $CHCO$), which leads to complete oxidation to CO_2. However, only partial oxidizations to CH_3CHO and CH_3COOH occur on Pt{111}. The open surface Pt{100} was the best facet to fully oxidize ethanol at low coverages. They identify two fundamental quantities that dictate the selectivity of ethanol oxidation: The ability of surface metal atoms to bond with unsaturated C-containing fragments and the relative stability of hydroxyl at surface atop sites with respect to other sites. They assert that the ethanol oxidation on Pt is a typical multistep and multiselectivity heterogeneous catalytic process.

De Souza et al. (2002) evaluate the ethanol electrooxidation on Pt, Rh, and Pt-Rh electrocatalysts using online differential electrochemical mass spectrometry (DEMS) and *in situ* infrared spectroscopy (FTIR) in a paper with 327 citations. They detect three products: CO_2 and AAL (detected by DEMS) and AA (detected by *in situ* FTIR). They find that Rh was the far less active electrocatalyst for ethanol electrooxidation. Pure Pt and $Pt_{90}Rh_{10}$ presented similar overall normalized current density, but $Pt_{90}Rh_{10}$ presented a better CO_2 yield compared to pure Pt. The best CO_2 yield was for $Pt_{73}Rh_{27}$ electrodes. The AAL yield decreased as Rh is added to the electrode. The ratio CO_2/CH_3CH_2O increased when Rh is added to the electrode.

Wang et al. (1995) evaluate ethanol electrooxidation in a direct methanol fuel cell (DMFC) in a paper with 317 citations. They note that for water/ethanol mole ratios between 5 and 2, ethanol was the main product, while CO_2 was a minor product. However, an increase in the water/ethanol mole ratio increased the relative product distribution of CO_2 slightly. Between 150°C and 190°C, the product distributions for the electrooxidation of ethanol did not depend significantly on the temperature. There were no differences in the product selectivities of Pt-Ru and Pt-black. They assert that ethanol is a promising alternative fuel for DMFCs with an electrochemical activity comparable to that of methanol.

Wang et al. (2004a) evaluate ethanol electrooxidation on the carbon-supported Pt nanoparticle (NP) catalyst focusing on the reaction kinetics and product yields in a paper with 292 citations.

They study the ethanol oxidation reaction (EOR) by cyclic voltammetry and potential-step measurements as a function of ethanol concentration and reaction temperature (23°C–60°C). They note that incomplete ethanol oxidation to AAL and AA prevailed over complete oxidation to CO_2 under all conditions. The dominant products were AA at low (1 mM) and AAL at high (0.5 M) ethanol concentration or low catalyst loading/electrode roughness. The current efficiency and product yield for CO_2 formation were of the order of a few percent. The reaction orders for ethanol on Pt/C were 0.3, 0.6, and 0.9 for CO_2, AA, and AAL formation and 0. 6 for the total Faradaic current, respectively. The temperature dependence in this temperature range resulted in an apparent activation energy for the total reaction (Faradaic current) of 32 kJ/mol. The respective values for the partial reactions for CO_2, AA, and AAL formation were 20, 28, and 43 kJ/mol, respectively.

Zhou et al. (2010) evaluate the ethanol electrooxidation on high-index faceted Pt nanocrystals (NCs) supported on carbon black in a paper with 288 citations. They note that these nanocrystals had high-index facets and a high density of atomic steps. These catalysts exhibited at least twice the activity and selectivity of commercial Pt/C catalysts for ethanol electrooxidation into CO_2 due to their high density,

107.3.3 The Palladium-based Bioethanol Electrooxidation in DEFCs

Brief information about six prolific studies with at least 283 citations each on Pd-based ethanol electrooxidation is given in Table 107.3. Furthermore, brief notes on the contents of these studies are also given.

Liang et al. (2009) study the mechanism of EOR on a Pd electrode in alkaline media using the cyclic voltammetry method in a paper with 655 citations. They find that the dissociative adsorption of ethanol proceeded rather quickly and the rate-determining step was the removal of the adsorbed ethoxy by the adsorbed hydroxyl on the Pd electrode. The adsorption of OH^- ions followed the Temkin-type isotherm on the Pd electrode. They assert that at higher potentials, kinetics was not only affected by the adsorption of OH^- ions but also by the formation of the inactive oxide layer on the Pd electrode.

Xu et al. (2007b) evaluate ethanol electrooxidation on the highly ordered Pd nanowire arrays (NWAs) in DAFCs in a paper with 450 citations. They note that these Pd nanowires were highly ordered, with uniform diameter and length. These Pd NWAs exhibited a face-centered cubic (FCC)

TABLE 107.3
The Pd-based Bioethanol Oxidation in DEFCs

No.	Papers	Catalysts	Catalyst Supports	Issues	Cits
1	Liang et al. (2009)	Pd		Ethanol oxidation reaction mechanism	655
2	Xu et al. (2007b)	Pd nanowire arrays		Electrocatalyst development: Effect of nanostructure	450
3	Tian et al. (2010)	Tetrahexahedral Pd NC/C, Pd/C	Glassy C	Electrocatalyst development: Effect of nanostructure	396
4	Shen et al. (2010)	Pd-Ni/C, Pd/C	C	Electrocatalyst development: Effect of Ni additives	374
5	Wang et al. (2013)	Pd-PANI-Pd sandwich-structured nanotube array (SNTA), Pd/C, Pd NTAs	C	Electrocatalyst development: Effect of PANI additives; effect of nanostructure	296
6	Nguyen et al. (2009)	Pd-Ag/C, Pd/C, Pt/C	C	Electrocatalyst development: Effect of Ag additives; effect of electrocatalyst	283

lattice structure. They observe that these Pd NWAs had high electrocatalytic activity and a resultant superior performance for the electrooxidation reaction of ethanol in DAFC.

Tian et al. (2010) evaluate ethanol electrooxidation on the tetrahexahedral Pd nanocrystals (THH Pd NCs) with high-index facets and high catalytic activity in a paper with 396 citations. They fabricate these nanocrystals with {730} high-index facets on a glassy carbon substrate in a dilute $PdCl_2$ solution by a programmed electrodeposition method. They observe that these nanocatalysts exhibited 4–6 times higher catalytic activity than commercial Pd black catalysts toward ethanol electrooxidation in alkaline solutions due to their high density of surface atomic steps.

Shen et al. (2010) evaluate ethanol electrooxidation on the Pd-Ni catalysts on the C support in alkaline DEFCs in a paper with 374 citations. They confirm the formation of FCC crystalline Pd and $Ni(OH)_2$ on the carbon powder for Pd-Ni/C catalysts. They observe that the metal particles were well-dispersed on carbon powder with a uniform distribution of Ni around Pd. They note that the Pd_2-Ni_3/C catalyst exhibited higher activity and stability for EOR in an alkaline medium compared to the Pd/C catalyst. Application of Pd_2-Ni_3/C as the anode catalyst of an alkaline DEFC with an anion-exchange membrane yielded a maximum power density of 90 mW/cm^2 at 60°C.

Wang et al. (2013) evaluate ethanol electrooxidation on the Pd-polyaniline (PANI)-Pd sandwich-structured nanotube array (SNTA) catalysts compared to the Pd SNTAs and Pd/C electrocatalysts to exploit shape effects and synergistic effects of Pd-PANI composites in a paper with 296 citations. They note that these SNTAs exhibited significantly improved electrocatalytic activity and durability compared with Pd NTAs and commercial Pd/C catalysts. Unique SNTAs provide fast transport and short diffusion paths for electroactive species and a high utilization rate of catalysts. Besides the merits of nanotube arrays, they attribute the improved electrocatalytic activity and durability to the special Pd/PANI/Pd sandwich-like nanostructures, which results in electron delocalization between Pd d orbitals and PANI π-conjugated ligands and in electron transfer from Pd to PANI. They assert that these SNTAs have high activity and long-term durability at a low cost as viable alternative non-Pt electrocatalysts.

Nguyen et al. (2009) evaluate ethanol electrooxidation on the carbon-supported Pd-Ag/C catalyst in alkaline media in a DEFC in a paper with 283 citations. They note those alloy NPs with FCC structures were successfully formed. These electrocatalysts exhibited excellent activity, enhanced CO tolerance, and better stability than Pt/C and Pd/C. They assert that this electrocatalyst is suitable for a DEFC.

107.3.4 The Other Catalyst-based Bioethanol Oxidation in DEFCs

Brief information about two prolific studies with at least 281 citations each on the ethanol electrooxidation on other electrocatalysts is given in Table 107.4. Furthermore, brief notes on the contents of these studies are also given.

Klosek and Raftery (2002) evaluate ethanol photoelectrooxidation on visible light-driven vanadium (V)-doped titanium dioxide (TiO_2) photocatalyst in a paper with 534 citations. They note that this catalyst photooxidized ethanol produced mostly CO_2 with small amounts of AAL, formic

TABLE 107.4
The Other Catalyst-based Bioethanol Electrooxidation in DEFCs

No.	Papers	Catalysts	Catalyst Supports	Issues	Cits
1	Klosek and Raftery (2002)	TiO_2/V		Photocatalyst development: effect of V additives; effect of irradiation	534
2	Liu and Hensen (2013)	Au NP	$MgCuCr_2O_4$	Electrocatalyst development: Effect of the nanostructure	281

TABLE 107.5
The Other Issues in Bioethanol Oxidation in DEFCs

No.	Papers	Catalysts	Issues	Cits
1	Marinov (1999)	None	Chemical kinetic model development	667

acid, and carbon monoxide (CO) under visible irradiation. Under ultraviolet (UV) irradiation, the catalyst had comparable activity and product distribution as a similarly prepared TiO_2 thin-film monolayer catalyst.

Liu and Hensen (2013) evaluate the ethanol electrooxidation to AAL on gold (Au) NPs on $MgCuCr_2O_4$ catalyst supports in a paper with 281 citations. They note that these NPs were highly active and selective for the aerobic oxidation of ethanol to AAL (conversion 100%; yield ~95%). This catalyst was stable for at least 500 h. They reason that the unprecedented catalytic performance was due to a strong synergy between metallic Au NPs and surface Cu^+ species. Cu^+ species were already formed during catalyst preparation and become more dominant at the surface during ethanol oxidation. These Cu^+ species were stabilized at the surface of the ternary $MgCuCr_2O_4$ catalyst support. The Cu^+ species acted as sites for O_2 activation.

107.3.5 The Other Issues in Bioethanol Oxidation in DEFCs

Brief information about a study with 667 citations each on the other issues of ethanol oxidation is given in Table 107.5. Furthermore, brief notes on the contents of these studies are also given.

Marinov (1999) develops a chemical kinetic model for high-temperature bioethanol oxidation in a paper with 667 citations. He finds that high-temperature ethanol oxidation exhibits strong sensitivity to the fall-off kinetics of ethanol decomposition, branching ratio selection for $C_2H_5OH+OH \leftrightarrow$ products, and reactions involving the hydroperoxyl (HO_2) radical. He defines the $C_2H_5OH(+M) \leftrightarrow CH_3+C_2H_5OH(+M)$ rate expression and the $C_2H_5OH(+M) \leftrightarrow C_2H_4+H_2O(+M)$ rate expression.

107.4 DISCUSSION

107.4.1 Introduction

Crude oil-based gasoline fuels have been widely used in the transportation sector since the 1920s. However, there have been great public concerns over the adverse environmental impact of these fuels. Hence, biomass-based bioethanol fuels have increasingly been used in blending gasoline fuels and in fuel cells. In the meantime, research in nanotechnology has intensified in recent years. Research in fuel cells has also intensified in recent years in three primary research streams of SOFCs, PEMFCs, and MFCs. Research in DEFCs has been a stream of the research on PEMFCs.

On the other hand, research on the direct utilization of bioethanol fuels in these DEFCs has intensified in recent years, primarily in the research fronts of the ethanol electrooxidation on the Pt electrocatalysts and on the Pd electrocatalysts for DEFCs. Other research fronts are ethanol electrooxidation on both Pt and Pd electrocatalysts and on other electrocatalysts, other issues in the ethanol electrooxidation, and the DEFCs themselves.

However, it is essential to develop efficient incentive structures for the primary stakeholders to enhance the research in this field. Although there have been a number of review papers on bioethanol fuel cells, there has been no review of the 25 most-cited articles in this field. This chapter presents a review of the 25 most-cited articles in the field of bioethanol fuel cells. Then, it discusses the key findings of these highly influential papers and comments on future research priorities in this field.

Brief information about the 25 most-cited papers with at least 186 citations each on ethanol electrooxidation in DEFCs is given in this chapter under five headings: The ethanol electrooxidation on Pt and Pd-based electrocatalysts, the ethanol electrooxidation on Pt-based electrocatalysts, the ethanol electrooxidation on Pd-based electrocatalysts, ethanol electrooxidation on the other electrocatalysts, and the other issues on ethanol oxidation with five, eleven, six, two, and one papers, respectively.

107.4.2 The Platinum and Palladium-based Bioethanol Electrooxidation in DEFCs

Brief information about five prolific studies with at least 278 citations each on Pt and Pd-based bioethanol electrooxidation is given in Table 107.1. Furthermore, brief notes on the contents of these studies are also given (Bambagioni et al., 2009; Hu et al., 2012; Xu et al., 2007a,c; Zhou et al., 2003).

Two of these studies develop nanomaterial-based electrocatalysts for ethanol electrooxidation in DEFCs (Bambagioni et al., 2009; Hu et al., 2012). The other three studies develop conventional material-based electrocatalysts for ethanol electrooxidation (Xu et al., 2007a,c; Zhou et al., 2003).

These studies also use catalyst supports to support these used electrocatalysts. These are carbon black in four studies, carbon microspheres in one study (Xu et al., 2007a), MWCNT (Bambagioni et al., 2009), and graphene (Hu et al., 2012) in two studies.

These studies evaluate the effect of the additives on ethanol electrooxidation: Ru, Sn, and W (Zhou et al., 2003), CeO_2 and NiO (Xu et al., 2007c), Ru (Bambagioni et al., 2009), and Cu (Hu et al., 2012).

Three studies compare ethanol electrooxidation on Pt and Pd electrocatalysts (Bambagioni et al., 2009; Xu et al., 207a,c) while two studies evaluate ethanol electrooxidation on both Pt and Pd electrocatalysts where Pd catalysts are considered as additives for the Pt electrocatalysts (Hu et al., 2012; Zhou et al., 2003). The studies in the first group showed that Pd/C electrocatalysts had better electrocatalytic performance for ethanol electrooxidation compared to Pt/C electrocatalysts. The second group of studies also showed that the addition of Pd enhanced the electrocatalytic performance of Pt electrocatalysts for ethanol electrooxidation. The ternary Pt-Pd-Cu nanoboxes anchored on a three-dimensional graphene framework had a four-fold improvement compared to pure 20% Pt/C electrocatalysts (Hu et al., 2012).

These findings are thought-provoking in seeking ways to optimize ethanol electrooxidation performance using Pd and Pt electrocatalysts in DEFCs. The nanostructure, additives, catalyst supports, and Pd addition emerges as the key variables impacting ethanol electrooxidation performance in DEFCs.

107.4.3 The Platinum-based Bioethanol Electrooxidation in DEFCs

Brief information about 11 prolific studies with at least 288 citations each on Pt-based bioethanol electrooxidation is given in Table 107.2. Furthermore, brief notes on the contents of these studies are also given.

Three of these studies focus on the use of nanomaterial-based electrocatalysts for ethanol electrooxidation in DEFCs (Dong et al., 2010; Wang et al., 2004a; Zhou et al., 2010). These studies show that the nanostructured Pt electrocatalysts had better electrocatalytic activity for ethanol electrooxidation compared to conventional Pt/C electrocatalysts.

Nine of these studies focus on the effect of additives on ethanol electrooxidation performance: Re (Vigier et al., 2004b), Rh (de Souza et al., 2002; Kowal et al., 2009), Ru (Dong et al., 2010; Rousseau et al., 2006; Wang et al., 1995), and Sn (Kowal et al., 2009; Lamy et al., 2004; Vigier et al., 2004a). All of these studies show that all of these additives had a beneficial effect on ethanol electrooxidation performance.

Only three of these studies specify electrocatalyst supports for ethanol electrooxidation: graphene sheets and carbon black (Dong et al., 2010), carbon powder (Vigier et al., 2004b), and carbon black (Zhou et al., 2010). Only Dong et al. (2010) compared the effect of the electrocatalyst support and found that graphene sheet-supported electrocatalysts had better electrocatalytic activity for ethanol electrooxidation.

Additionally, Wang et al. (2004a) evaluated the effect of ethanol concentration and reaction temperature on ethanol electrooxidation and found that these parameters had a strong effect on both reaction kinetics and product yields. Wang and Liu (2008) evaluated the effect of Pt surfaces on ethanol electrooxidation and found that these surfaces had a strong effect on ethanol electrooxidation.

These findings are thought-provoking in seeking ways to optimize ethanol electrooxidation using Pt electrocatalysts in DEFCs. The nanostructure, additives, catalyst supports, Pt surfaces, ethanol concentration, and reaction temperature emerge as the key variables impacting ethanol electrooxidation performance in DEFCs.

107.4.4 The Palladium-based Bioethanol Electrooxidation in the DEFCs

Brief information about six prolific studies with at least 283 citations each on the Pd-based ethanol electrooxidation is given in Table 107.3. Furthermore, brief notes on the contents of these studies are also given.

Three of these studies use nanomaterial-based electrocatalysts for ethanol electrooxidation in DEFCs: Pd nanowire arrays (Xu et al., 2007b), tetrahedral PD nanocrystals (Tian et al., 2010), Pd-PANI-Pd sandwich-structured nanotube array (SNTA) and Pd NTAs (Wang et al., 2013). All of these studies show that these nanomaterial-based electrocatalysts significantly improved electrocatalytic activity for ethanol electrooxidation compared to the conventional Pd electrocatalysts.

On the other hand, three studies evaluate the effect of additives on the Pd catalytic activity for ethanol electrooxidation: Ni (Shen et al., 2010), PANI (Wang et al., 2013), and Ag (Nguyen et al., 2009). All of these studies show that all of these additives had a beneficial effect on electrocatalytic activity of Pd catalysts ethanol electrooxidation.

Additionally, Liang et al. (2009) evaluate the determinants of EOR in Pd electrocatalysts.

These findings are thought-provoking in seeking ways to optimize ethanol electrooxidation using Pd electrocatalysts in DEFCs. The nanostructure, additives, and catalyst supports emerge as the key variables impacting on ethanol electrooxidation performance in DEFCs.

107.4.5 The Other Catalyst-based Bioethanol Oxidation in DEFCs

Brief information about two prolific studies with at least 281 citations each on ethanol electrooxidation on other electrocatalysts is given in Table 107.4. Furthermore, brief notes on the contents of these studies are also given.

Klosek and Raftery (2002) evaluate the effect of V addition to TiO_2 photocatalysts and the effect of irradiation and Liu and Hensen (2013) use gold (Au) NPs on $MgCuCr_2O_4$ catalyst supports for ethanol electrooxidation.

These prolific papers hint that TiO_2 photocatalysts (Klosek and Raftery, 2002; Larsson and Andersson, 1998) and Au electrocatalysts (Christensen et al., 2006, Liu and Hensen, 2013) are important research streams for ethanol electrooxidation.

107.4.6 The Other Issues in Bioethanol Oxidation in DEFCs

Brief information about a study with 667 citations each on other issues on ethanol oxidation is given in Table 107.5. Furthermore, brief notes on the contents of these studies are also given. Marinov (1999) developed a chemical kinetic model for high-temperature bioethanol oxidation.

107.4.7 Overall Remarks

A number of prolific studies reviewed in the preceding sections hint that the primary electrocatalysts used and researched for ethanol electrooxidation are Pt (Adzic et al. 2007) and Pd electrocatalysts (Antolini, 2009). Pt electrooxidation has been a gold standard to use for ethanol electrooxidation in DEFCs. However, these electrocatalysts have been used with a number of additives in DEFCs to increase the efficiency of ethanol electrooxidation. On the other hand, Pd electrocatalysts have been a rising star for ethanol electrooxidation in DEFCs.

Research in the field of fuel cells has intensified in recent years with over 50,000 papers in three primary research streams of high-temperature SOFCs (Adler, 2004), low-temperature PEMFCs (Borup et al., 2007), and MFCs (Logan, 2009). Research in DEFCs (Badwal et al., 2015) has been a stream of research on low-temperature PEMFCs. The primary focus of the research on fuel cells has been the optimization of the operating conditions to maximize fuel cell performance.

Therefore, it is surprising that there have been no prolific papers among the 25 most-cited papers on bioethanol fuel cells themselves (Badwal et al., 2015) as all of these prolific papers have focused on the electrocatalyst development for ethanol electrooxidation in DEFCs. This finding suggests that the development of efficient electrocatalysts and catalyst supports is highly critical for the development of efficient DEFCs.

One innovative way to enhance fuel cell performance has been, as a number of studies for ethanol electrooxidation hint, the use of nanomaterials for both the electrocatalysts themselves and the catalyst supports (Seger and Kamat, 2009).

The findings narrated in the previous sections are thought-provoking in seeking ways to optimize ethanol electrooxidation performance using primarily Pd and Pt electrocatalysts in DEFCs. The nanostructure, additives, catalyst supports, Pd addition, Pt surfaces, ethanol concentration, and reaction temperature emerge as the key variables impacting ethanol electrooxidation performance using both Pd and Pt electrocatalysts as well as other electrocatalysts in DEFCs.

In parallel with the research on fuel cells, research on nanomaterials and nanotechnology (Geim and Novoselov, 2007) has intensified in recent years with over one and a half million papers, enriching the material portfolio to be used for fuel cell applications. It is expected that this enriched portfolio of innovative nanomaterials would enhance ethanol electrooxidation in DEFCs as both electrocatalyst materials and electrocatalyst supports in the future, benefiting from the superior properties of these nanomaterials.

In the end, these most-cited papers in this field hint that the efficiency of ethanol electrooxidation in DEFCs could be optimized using the structure, processing, and property relationships of nanomaterials and conventional materials used as both electrocatalyst materials and catalyst supports (Formela et al., 2016; Konur, 2018, 2020b, 2021c,d,e,f; Konur and Matthews, 1989).

107.5 CONCLUSION AND FUTURE RESEARCH

Brief information about the key research fronts covered by the 25 most-cited papers with at least 186 citations each in the field of bioethanol fuel cells is given in Table 107.6.

There are three major research fronts for this field: ethanol electrooxidation on both Pt and Pd-based electrocatalysts, Pt-based electrocatalysts, and Pd-based electrocatalysts. This finding hints that Pt and Pd electrocatalysts are the primary research fronts for the field of ethanol electrooxidation. Pt electrocatalysts have been a gold standard to use for ethanol electrooxidation in DEFCs. However, these electrocatalysts have been used with a number of additives in DEFCs while Pd electrocatalysts have been a rising star for ethanol electrooxidation in the DEFCs.

It is notable that there is a similarity in these research fronts in the sample of reviewed papers and the sample of the 100 most-cited papers (the first column data in Table 107.6) in this field. This finding suggests that these 25 most prolific papers are a representative sample of the wider research on bioethanol fuel cells. There is an overrepresentation and underrepresentation in the fields of Pd and Pt electrocatalysts and the field of bioethanol fuel cells, respectively.

Research in the field of fuel cells has intensified in recent years with over 50,000 papers in three primary research streams of high-temperature SOFCs, low-temperature PEMFCs, and MFCs. Research in DEFCs has been a stream of the research on low-temperature PEMFCs. The primary

TABLE 107.6
The Most Prolific Research Fronts in Bioethanol Fuel Cells

No.	Research Fronts	Sample Papers (%)	Reviewed Papers (%)
1.	Pd and Pt electrocatalysts	8	20
	Conventional electrocatalysts	4	12
	Nanomaterial-based electrocatalysts	4	8
2.	Pt electrocatalysts	43	44
	Conventional electrocatalysts	41	32
	Nanomaterial-based electrocatalysts	2	12
3.	Pd electrocatalysts	26	24
	Conventional electrocatalysts	12	12
	Nanomaterial-based electrocatalysts	14	12
4.	Other electrocatalysts	13	8
	Conventional electrocatalysts	9	4
	Nanomaterial-based electrocatalysts	4	4
5.	Other issues in ethanol electrooxidation	4	4
6.	Ethanol fuel cells	7	0
	Sample	100	25

Reviewed papers, The total number of reviewed papers=25. Sample papers, the sample of the 100 most-cited papers.

focus of the research on fuel cells has been the optimization of operating conditions to maximize fuel cell performance.

Therefore, it is surprising that there have been no prolific papers among the 25 most cited papers on bioethanol fuel cells themselves as all of these prolific papers have focused on the electrocatalyst development for ethanol electrooxidation in DEFCs. This finding suggests that the development of efficient electrocatalysts and catalyst supports is highly critical for the development of efficient DEFCs.

One innovative way to enhance fuel cell performance has been, as a number of studies for ethanol electrooxidation hint, the use of nanomaterials for both the electrocatalysts themselves and the catalyst supports.

The findings narrated in Section 107.3 and discussed in Section 107.4 are thought-provoking in seeking ways to optimize ethanol electrooxidation performance using primarily Pd and Pt electrocatalysts in DEFCs. The nanostructure, additives, catalyst supports, Pd addition, Pt surfaces, ethanol concentration, and reaction temperature emerge as the key variables impacting ethanol electrooxidation performance using both Pd and Pt electrocatalysts as well as other electrocatalysts in DEFCs.

In parallel with the research on fuel cells, the research on nanomaterials and nanotechnology has intensified in recent years with over one and a half million papers, enriching the material portfolio to be used for fuel cell applications. The findings from the 25 most-cited papers confirm that the application of nanomaterials and nanotechnology for ethanol electrooxidation in DEFCs significantly improves the efficiency of the electrooxidation process as well as of the DEFCs processes through the enhancement of the structure-processing-property relationships. This would make bioethanol fuels a viable alternative to hydrogen fuels as well as other fuels in fuel cells.

It is expected that this enriched portfolio of innovative nanomaterials would enhance ethanol electrooxidation in DEFCs as both electrocatalyst materials and electrocatalyst supports in the future, benefiting from the superior properties of these nanomaterials.

In the end, these most-cited papers in this field hint that the efficiency of ethanol electrooxidation in DEFCs could be optimized using the structure, processing, and property relationships of the nanomaterials and conventional materials used as both electrocatalyst materials and catalyst supports.

It is recommended that such review studies should be performed for the other research fronts on both the production and utilization of bioethanol fuels complementing the corresponding scientometric studies.

ACKNOWLEDGMENTS

The contribution of highly cited researchers in the field of bioethanol fuel cells has been gratefully acknowledged.

REFERENCES

Adler, S. B. 2004. Factors governing oxygen reduction in solid oxide fuel cell cathodes. *Chemical Reviews* 104:4791–4843.

Adzic, R. R., J. Zhang and K. Sasaki, et al. 2007. Platinum monolayer fuel cell electrocatalysts. *Topics in Catalysis* 46:249–262.

Antolini, E. 2007. Catalysts for direct ethanol fuel cells. *Journal of Power Sources* 170:1–12.

Antolini, E. 2009. Palladium in fuel cell catalysis. *Energy and Environmental Science* 2:915–931.

Badwal, S. P. S., S. Giddey, A. Kulkarni, J. Goel and S. Basu. 2015. Direct ethanol fuel cells for transport and stationary applications - A comprehensive review. *Applied Energy* 145:80–103.

Bambagioni, V., C. Bianchini and A. Marchionni, et al. 2009. Pd and Pt-Ru anode electrocatalysts supported on multi-walled carbon nanotubes and their use in passive and active direct alcohol fuel cells with an anion-exchange membrane (alcohol=methanol, ethanol, glycerol). *Journal of Power Sources* 190:241–251.

Borup, R., J. Meyers and B. Pivovar, et al. 2007. Scientific aspects of polymer electrolyte fuel cell durability and degradation. *Chemical Reviews* 107:3904–3951.

Burnham, J. F. 2006. Scopus database: A review. *Biomedical Digital Libraries* 3:1–8.

Camara, G. A. and T. Iwasita. 2005. Parallel pathways of ethanol oxidation: The effect of ethanol concentration. *Journal of Electroanalytical Chemistry* 578:315–321.

Christensen, C. H., B. Jorgensen and J. Rass-Hansen, et al. 2006. Formation of acetic acid by aqueous-phase oxidation of ethanol with air in the presence of a heterogeneous gold catalyst. *Angewandte Chemie - International Edition* 45:4648–4651.

De Souza, J. P. I., S. L. Queiroz, K. Bergamaski, E. R. Gonzalez and F. C. Nart. 2002. Electro-oxidation of ethanol on Pt, Rh, and PtRh electrodes. A study using DEMS and *in-situ* FTIR techniques. *Journal of Physical Chemistry B* 106:9825–9830.

Dong, L., R. R. S. Gari, Z. Li, M. M. Craig and S. Hou. 2010. Graphene-supported platinum and platinum-ruthenium nanoparticles with high electrocatalytic activity for methanol and ethanol oxidation. *Carbon* 48:781–787.

Egolfopoulos, F. N., D. X. Du and C. K. Law. 1992. A study on ethanol oxidation kinetics in laminar premixed flames, flow reactors, and shock tubes. *Symposium (International) on Combustion* 24:833–841.

Formela, K., A. Hejna, L. Piszczyk, M. R. Saeb and X. Colom. 2016. Processing and structure-property relationships of natural rubber/wheat bran biocomposites. *Cellulose* 23:3157–3175.

Geim, A. K. 2009. Graphene: Status and prospects. *Science* 324:1530–1534.

Geim, A. K. and K. S. Novoselov. 2007. The rise of graphene. *Nature Materials* 6:183–191.

Hill, J., E. Nelson, D. Tilman, S. Polasky and D. Tiffany. 2006. Environmental, economic, and energetic costs and benefits of biodiesel and ethanol biofuels. *Proceedings of the National Academy of Sciences of the United States of America* 103:11206–11210.

Hill, J., S. Polasky and E. Nelson, et al. 2009. Climate change and health costs of air emissions from biofuels and gasoline. *Proceedings of the National Academy of Sciences of the United States of America* 106:2077–2082.

Hsieh, W. D., R. H. Chen, T. L. Wu and T. H. Lin. 2002. Engine performance and pollutant emission of an SI engine using ethanol-gasoline blended fuels. *Atmospheric Environment* 36:403–410.

Hu, C., H. Cheng and Y. Zhao, et al. 2012. Newly-designed complex ternary Pt/PdCu nanoboxes anchored on three-dimensional graphene framework for highly efficient ethanol oxidation. *Advanced Materials* 24:5493–5498.

Kamarudin, M. Z. F., S. K. Kamarudin, M. S. Masdar and W. R. W. Daud. 2013. Review: Direct ethanol fuel cells. *International Journal of Hydrogen Energy* 38:9438–9453.

Klosek, S. and D. Raftery. 2002. Visible light driven V-doped TiO_2 photocatalyst and its photooxidation of ethanol. *Journal of Physical Chemistry B* 105:2815–2819.

Konur, O. 2000. Creating enforceable civil rights for disabled students in higher education: An institutional theory perspective. *Disability & Society* 15:1041–1063.

Konur, O. 2002a. Access to nursing education by disabled students: Rights and duties of nursing programs. *Nurse Education Today* 22:364–374.

Konur, O. 2002b. Assessment of disabled students in higher education: Current public policy issues. *Assessment and Evaluation in Higher Education* 27:131–152.

Konur, O. 2002c. Access to employment by disabled people in the UK: Is the Disability Discrimination Act working? *International Journal of Discrimination and the Law* 5:247–279.

Konur, O. 2006a. Participation of children with dyslexia in compulsory education: Current public policy issues. *Dyslexia* 12:51–67.

Konur, O. 2006b. Teaching disabled students in higher education. *Teaching in Higher Education* 11:351–363.

Konur, O. 2007a. A judicial outcome analysis of the *Disability Discrimination Act*: A windfall for the employers? *Disability & Society* 22:187–204.

Konur, O. 2007b. Computer-assisted teaching and assessment of disabled students in higher education: The interface between academic standards and disability rights. *Journal of Computer Assisted Learning* 23:207–219.

Konur, O. 2012. The evaluation of the research on the bioethanol: A scientometric approach. *Energy Education Science and Technology Part A: Energy Science and Research* 28:1051–1064.

Konur, O. 2015. Current state of research on algal bioethanol. In *Marine Bioenergy: Trends and Developments*, Ed. S. K. Kim and C. G. Lee, pp. 217–244. Boca Raton, FL: CRC Press.

Konur, O. 2016a. Scientometric overview in nanobiodrugs. In *Nanoarchitectonics for Smart Delivery and Drug Targeting*, Ed. A. M. Holban and A. M. Grumezescu, pp. 405–428. Amsterdam: Elsevier.

Konur, O. 2016b. Scientometric overview regarding nanoemulsions used in the food industry. In *Emulsions: Nanotechnology in the Agri-Food Industry*, Ed. A. M. Grumezescu, pp. 689–711. Amsterdam: Elsevier.

Konur, O. 2016c. Scientometric overview regarding the nanobiomaterials in antimicrobial therapy. In *Nanobiomaterials in Antimicrobial Therapy*, Ed. A. M. Grumezescu, pp. 511–535. Amsterdam: Elsevier.

Konur, O. 2016d. Scientometric overview regarding the nanobiomaterials in dentistry. In *Nanobiomaterials in Dentistry*, Ed. A. M. Grumezescu, pp. 425–453. Amsterdam: Elsevier.

Konur, O. 2016e. Scientometric overview regarding the surface chemistry of nanobiomaterials. In *Surface Chemistry of Nanobiomaterials*, Ed. A. M. Grumezescu, pp. 463–486. Amsterdam: Elsevier.

Konur, O. 2016f. The scientometric overview in cancer targeting. In *Nanoarchitectonics for Smart Delivery and Drug Targeting*, Ed. A. M. Holban and A. Grumezescu, pp. 871–895. Amsterdam: Elsevier.

Konur, O. 2017a. Recent citation classics in antimicrobial nanobiomaterials. In *Nanostructures for Antimicrobial Therapy*, Ed. A. Ficai and A. M. Grumezescu, pp. 669–685. Amsterdam: Elsevier.

Konur, O. 2017b. Scientometric overview in nanopesticides. In *New Pesticides and Soil Sensors*, Ed. A. M. Grumezescu, pp. 719–744. Amsterdam: Elsevier.

Konur, O. 2017c. Scientometric overview regarding oral cancer nanomedicine. In *Nanostructures for Oral Medicine*, Ed. E. Andronescu and A. M. Grumezescu, pp. 939–962. Amsterdam: Elsevier.

Konur, O. 2017d. Scientometric overview regarding water nanopurification. In *Water Purification*, Ed. A. M. Grumezescu, pp. 693–716. Amsterdam: Elsevier.

Konur, O. 2017e. Scientometric overview in food nanopreservation. In *Food Preservation*, Ed. A. M. Grumezescu, pp. 703–729. Amsterdam: Elsevier.

Konur, O., Ed. 2018. *Bioenergy and Biofuels*. Boca Raton, FL: CRC Press.

Konur, O. 2019a. Cyanobacterial bioenergy and biofuels science and technology: A scientometric overview. In *Cyanobacteria: From Basic Science to Applications*, Ed. A. K. Mishra, D. N. Tiwari and A. N. Rai, pp. 419–442. Amsterdam: Elsevier.

Konur, O. 2019b. Nanotechnology applications in food: A scientometric overview. In *Nanoscience for Sustainable Agriculture*, Ed. R. N. Pudake, N. Chauhan and C. Kole, pp. 683–711. Cham: Springer.

Konur, O. 2020a. The scientometric analysis of the research on the bioethanol production from green macroalgae. In *Handbook of Algal Science, Technology and Medicine*, Ed. O. Konur, pp. 385–401. London: Academic Press.

Konur, O., Ed. 2020b. *Handbook of Algal Science, Technology and Medicine.* London: Academic Press.

Konur, O. 2021a. Nanotechnology applications in diesel fuels and the related research fields: A review of the research. In *Handbook of Biodiesel and Petrodiesel Fuels: Science, Technology, Health, and Environment. Volume 1. Biodiesel Fuels: Science, Technology, Health, and Environment*, Ed. O. Konur, pp. 89–110. Boca Raton, FL: CRC Press.

Konur, O. 2021b. Nanobiosensors in agriculture and foods: A scientometric review. In *Nanobiosensors in Agriculture and Food*, Ed. R. N. Pudake, pp. 365–384. Cham: Springer.

Konur, O., Ed. 2021c. *Handbook of Biodiesel and Petrodiesel Fuels: Science, Technology, Health, and Environment.* Boca Raton, FL: CRC Press.

Konur, O., Ed. 2021d. *Handbook of Biodiesel and Petrodiesel Fuels: Science, Technology, Health, and Environment. Volume 1. Biodiesel Fuels: Science, Technology, Health, and Environment.* Boca Raton, FL: CRC Press.

Konur, O., Ed. 2021e. *Handbook of Biodiesel and Petrodiesel Fuels: Science, Technology, Health, and Environment. Volume 2. Biodiesel Fuels based on the Edible and Nonedible Feedstocks, Wastes, and Algae: Science, Technology, Health, and Environment.* Boca Raton, FL: CRC Press.

Konur, O., Ed. 2021f. *Handbook of Biodiesel and Petrodiesel Fuels: Science, Technology, Health, and Environment. Volume 3. Petrodiesel Fuels: Science, Technology, Health, and Environment.* Boca Raton, FL: CRC Press.

Konur, O. 2023. Bioethanol fuel cells: Scientometric study. In *Evaluation and Utilization of Bioethanol Fuels. II.: Biohydrogen Fuels, Fuel Cells, Biochemicals, and Country Experiences. Handbook of Bioethanol Fuels Volume 6*, Ed. O. Konur, pp. 277–297. Boca Raton, FL: CRC Press.

Konur, O. and F. L. Matthews. 1989. Effect of the properties of the constituents on the fatigue performance of composites: A review. *Composites* 20:317–328.

Kowal, A., M. Li and M. Shao, et al. 2009. Ternary Pt/Rh/SnO_2 electrocatalysts for oxidizing ethanol to CO_2. Nature Materials 8:325–330.

Lamy, C., S. Rousseau, E. M. Belgsir, C. Coutanceau and J. M. Leger. 2004. Recent progress in the direct ethanol fuel cell: Development of new platinum-tin electrocatalysts. *Electrochimica Acta* 49:3901–3908

Larsson, P. O. and A. Andersson. 1998. Complete oxidation of CO, ethanol, and ethyl acetate over copper oxide supported on titania and ceria modified titania. *Journal of Catalysis* 179:72–89.

Liang, Z. X., T. S. Zhao, J. B. Xu and L. D. Zhu. 2009. Mechanism study of the ethanol oxidation reaction on palladium in alkaline media. *Electrochimica Acta* 54:2203–2208.

Liu, P. and E. J. M. Hensen. 2013. Highly efficient and robust Au/$MgCuCr_2O_4$ catalyst for gas-phase oxidation of ethanol to acetaldehyde. *Journal of the American Chemical Society* 135:14032–14035.

Logan, B. E. 2009. Exoelectrogenic bacteria that power microbial fuel cells. *Nature Reviews Microbiology* 7:375–381.

Ma, X., L. Sun and C. Song. 2002. A new approach to deep desulfurization of gasoline, diesel fuel and jet fuel by selective adsorption for ultra-clean fuels and for fuel cell applications. *Catalysis Today* 77:107–116.

Marinov, N. M. 1999. A detailed chemical kinetic model for high temperature ethanol oxidation. *International Journal of Chemical Kinetics* 31:183–220.

Najafi, G., B. Ghobadian and T. Tavakoli, et al. 2009. Performance and exhaust emissions of a gasoline engine with ethanol blended gasoline fuels using artificial neural network. *Applied Energy* 86:630–639.

Newman, P. W. G. and J. R. Kenworthy. 1989. Gasoline consumption and cities: A comparison of U.S. cities with a global survey. *Journal of the American Planning Association* 55:24–37.

Nguyen, S. T., H. M. Law and H. T. Nguyen, et al. 2009. Enhancement effect of Ag for Pd/C towards the ethanol electro-oxidation in alkaline media. *Applied Catalysis B: Environmental* 91:507–515.

North, D. C. 1991. Institutions. *Journal of Economic Perspectives* 5:97–112.

Rousseau, S., C. Coutanceau, C. Lamy and J. M. Leger. 2006. Direct ethanol fuel cell (DEFC): Electrical performances and reaction products distribution under operating conditions with different platinum-based anodes. *Journal of Power Sources* 158:18–24.

Seger, B. and P. V. Kamat. 2009. Electrocatalytically active graphene-platinum nanocomposites. Role of 2-D carbon support in PEM fuel cells. *Journal of Physical Chemistry C* 113:7990–7995.

Shen, S. Y., T. S. Zhao, J. B. Xu and Y. S. Li. 2010. Synthesis of PdNi catalysts for the oxidation of ethanol in alkaline direct ethanol fuel cells. *Journal of Power Sources* 195:1001–1006.

Song, S. and P. Tsiakaras. 2006. Recent progress in direct ethanol proton exchange membrane fuel cells (DE-PEMFCs). *Applied Catalysis B: Environmental* 63:187–193.

Tian, N., Z. Y. Zhou, N. F. Yu, L. Y. Wang and S. G. Sun. 2010. Direct electrodeposition of tetrahexahedral Pd nanocrystals with high-index facets and high catalytic activity for ethanol electrooxidation. *Journal of the American Chemical Society* 132:7580–7581.

Vigier, F., C. Coutanceau, F. Hahn, E. M. Belgsir and C. Lamy. 2004a. On the mechanism of ethanol electro-oxidation on Pt and PtSn catalysts: Electrochemical and *in situ* IR reflectance spectroscopy studies. *Journal of Electroanalytical Chemistry* 563:81–89.

Vigier, F., C. Coutanceau, A. Perrard, E. M. Belgsir and C. Lamy. 2004b. Development of anode catalyst for a direct ethanol fuel cell. *Journal of Applied Electrochemistry* 34:439–446.

Wang, A. L., H. Xu and J. X. Feng, et al. 2013. Design of Pd/PANI/Pd sandwich-structured nanotube array catalysts with special shape effects and synergistic effects for ethanol electrooxidation. *Journal of the American Chemical Society* 135:10703–10709.

Wang, H., Z. Jusys and R. J. Behm. 2004a. Ethanol electrooxidation on a carbon-supported Pt catalyst: Reaction kinetics and product yields. *Journal of Physical Chemistry B* 108:19413–19424.

Wang, H. F. and Z. P. Liu. 2008. Comprehensive mechanism and structure-sensitivity of ethanol oxidation on platinum: New transition-state searching method for resolving the complex reaction network. *Journal of the American Chemical Society* 130:10996–11004.

Wang, J., S. Wasmus and R. F. Savinell. 1995. Evaluation of ethanol, 1-propanol, and 2-propanol in a direct oxidation polymer-electrolyte fuel cell: A real-time mass spectrometry study. *Journal of the Electrochemical Society* 142:4218–4224.

Xu, C., L. Cheng, P. Shen and Y. Liu. 2007a. Methanol and ethanol electrooxidation on Pt and Pd supported on carbon microspheres in alkaline media. *Electrochemistry Communications* 9:997–1001.

Xu, C., P. K. Shen and Y. Liu. 2007c. Ethanol electrooxidation on Pt/C and Pd/C catalysts promoted with oxide. *Journal of Power Sources* 164:527–531.

Xu, C., H. Wang, P. K. Shen and S. P. Jiang. 2007b. Highly ordered Pd nanowire arrays as effective electrocatalysts for ethanol oxidation in direct alcohol fuel cells. *Advanced Materials* 19:4256–4259.

Zhou, W., Z. Zhou and S. Song, et al. 2003. Pt based anode catalysts for direct ethanol fuel cells. *Applied Catalysis B: Environmental* 46:273–285.

Zhou, Z. Y., Z. Z. Huang and D. J. Chen, et al. 2010. High-index faceted platinum nanocrystals supported on carbon black as highly efficient catalysts for ethanol electrooxidation. *Angewandte Chemie - International Edition* 49:411–414.

Part 32

Bioethanol Fuel-based Biochemicals

108 Bioethanol Fuel-based Biochemicals
Scientometric Study

Ozcan Konur
(Formerly) Ankara Yildirim Beyazit University

108.1 INTRODUCTION

Crude oil-based gasoline fuels (Ma et al., 2002; Newman and Kenworthy, 1989) have been widely used in the transportation sector since the 1920s. However, there have been great public concerns over the adverse environmental and human impacts of these fuels (Hill et al., 2006, 2009). Hence, biomass-based bioethanol fuels (Konur, 2012e, 2015, 2019a, 2020a) have increasingly been used in blending gasoline fuels (Hsieh et al., 2002; Najafi et al., 2009) and in fuel cells (Antolini, 2007). Additionally, bioethanol fuels have been used to produce valuable biochemicals (Nikolau et al., 2008; Zhang et al., 2016) in a biorefinery (Cherubini et al., 2009; Maity, 2015) context.

In the meantime, research in the nanomaterials and nanotechnology has intensified in recent years to become a major research field in scientific research, with over one and a half million published papers (Geim, 2009; Geim and Novoselov, 2007). In this context, a large number of nanomaterials have been developed for nearly every research field. These materials offer an innovative way to increase the efficiency in the production and utilization of bioethanol fuels, as in other scientific fields (Konur 2016a,b,c,d,e,f, 2017a,b,c,d,e, 2019b, 2021a,b).

There are four primary research fronts in this field: Bioethylenes (Morschbacker, 2009; Takahara et al., 2005; Zhang and Yu, 2013), biobutadienes (Angelici et al., 2013; Makshina et al., 2012; Sushkevich et al., 2014), biobutanols (Fu et al., 2017; Ndou et al., 2003; Tsuchida et al., 2006), and bioacetaldehydes (Gong and Mullins, 2008; Jackson et al., 1991; Santacesaria et al., 2012). The other research fronts are bioethyl acetates (Inui et al., 2002), biopropylenes (Goto et al., 2010), bioacetic acids (Christensen et al., 2006), biodiethyl ethers (Christiansen et al., 2013), and bioisobutenes (Liu et al., 2013).

However, it is essential to develop efficient incentive structures (North, 1991) for the primary stakeholders to enhance the research in this field (Konur, 2000, 2002a,b,c, 2006a,b, 2007a,b). Scientometric analysis has been used in this context to inform the primary stakeholders about the current state of research in this research field (Garfield, 1955; Konur, 2011, 2012a,b,c,d,e,f,g,h,i, 2015, 2016a,b,c,d,e,f, 2017a,b,c,d,e, 2018a, 2019a,b, 2020a, 2021a,b).

As there have been no scientometric studies on biochemical production in general and on biochemical production from bioethanol fuels, this chapter presents a scientometric study of research on bioethanol fuel-based biochemical production. It examines the scientometric characteristics of both the sample and population data and presents them in the order of documents, authors, publication years, institutions, funding bodies, source titles, countries, Scopus subject categories, keywords, and research fronts.

108.2 MATERIALS AND METHODS

The search for this study was carried out using the Scopus database (Burnham, 2006) in September 2021.

DOI: 10.1201/9781003226574-144

As the first step for the search of the relevant literature, keywords were selected using the 200 most-cited papers. The selected keyword list was optimized to obtain a representative sample of papers for this research field. This keyword list was provided in the appendix for future replication studies. Additionally, the information about the most-used keywords was given in Section 108.3.9 to highlight the key research fronts in Section 108.3.10.

As the second step, two sets of data were used in this study. First, a population sample of over 1100 papers was used to examine the scientometric characteristics of the population data. Second, a sample of 100 most-cited papers was used to examine the scientometric characteristics of these citation classics with over 46 citations each.

The scientometric characteristics of both these sample and population datasets were presented in the order of documents, authors, publication years, institutions, funding bodies, source titles, countries, Scopus subject categories, keywords, and research fronts.

Lastly, the key scientometric findings for both datasets were discussed to highlight the research landscape for bioethanol fuel-based biochemical production. Additionally, a number of brief conclusions were drawn and a number of relevant recommendations were made to enhance the future research landscape.

108.3 RESULTS

108.3.1 The Most Prolific Documents on Bioethanol-based Biochemical Production

Information on the types of documents for both datasets is given in Table 108.1. Articles dominate both the sample and population datasets, while the review papers and articles have a slight surplus and deficit, respectively.

It is also interesting to note that all the papers in the sample dataset were published in journals, while only 99.3% of the papers were published in the journals for the population dataset. Furthermore, 0.4% and 0.4% of the population papers were published in book series and books, respectively.

108.3.2 The Most Prolific Authors on Bioethanol Fuel-based Biochemical Production

Information about the 32 most prolific authors with at least two sample papers and four population papers each is given in Table 108.2.

TABLE 108.1
Documents in Bioethanol Fuel-based Biochemical Production

Documents	Sample Dataset (%)	Population Dataset (%)	Surplus (%)
Article	93	97.1	–4.1
Review	6	1.5	4.5
Letter	1	0.5	0.5
Note	0	0.5	–0.5
Book chapter	0	0.3	–0.3
Book	0	0.1	–0.1
Conference paper	0	0.0	0
Short Survey	0	0.0	0
Editorial	0	0.0	0
Sample size	100	1100	

TABLE 108.2
Most Prolific Authors in Bioethanol Fuel-based Biochemical Production

No.	Author Name	Author Code	Sample Papers	Population Papers	Institution	Country	Res. Front
1	Fujitani, Tadahiro	7005367660	5	17	AIST	Japan	Propylene
2	Takahashi, Atsushi	55728349100	5	12	AIST	Japan	Propylene
3	Ivanova, Irina I.	7201988013	4	7	Lomonosov Moscov State Univ.	Russia	Butadiene
4	Sushkevich, Vitaly L.	25634689900	4	5	Lomonosov Moscov State Univ.	Russia	Butadiene
5	Angelici, Carlo	6508075661	4	5	Utrecht Univ.	The Netherlands	Butadiene
6	Bruijnincx, Pieter C. A.	8283902100	4	5	Utrecht Univ.	The Netherlands	Butadiene
7	Weckhuysen, Bert M.	7005917472	4	5	Utrecht Univ.	The Netherlands	Butadiene
8	Sun, Junming	55632741500	3	5	Washington State Univ.	USA	Isobutene
9	Hensen, Emiel J. M.	35597384500	3	4	Eindhoven Univ. Technol.	The Netherlands	Acetaldehyde
10	Wass, Duncan F.	6701707900	3	4	Univ. Bristol	UK	Butanol
11	Wingad, Richard L.	8415058400	3	4	Univ. Bristol	UK	Butanol
12	Kyriienko, Pavlo I.	55314336200	2	24	Natl. Acad. Sci.	Ukraine	Butadiene
13	Soloviev, Sergiy O.	15048785900	2	24	Natl. Acad. Sci.	Ukraine	Butadiene
14	Larina, Olga V.	56911170000	2	21	Natl. Acad. Sci.	Ukraine	Butadiene
15	Appel, Lucia G.	7102658832	2	11	Natl. Technol. Inst.	Venezuela	Butanol
16	Nakamura, Isao	7202458542	2	9	AIST	Japan	Propylene
17	Inaba, Megumu	7202428788	2	7	AIST	Japan	Propylene
18	Murata, Kazuhisa	56419254200	2	7	AIST	Japan	Propylene
19	Takahara, Isao	6701457373	2	7	AIST	Japan	Propylene
20	Jones, Matthew D.	35230030600	2	6	Univ. Bath	UK	Butadiene
21	Sano, Tsuneji	55566058400	2	5	Hiroshima Univ.	Japan	propylene
22	Di Serio, Martino	57218579597	2	5	Univ. Naples	Italy	Ethyl acetate
23	Santacesaria, Elio	7004435446	2	5	Univ. Naples	Italy	Acetate
24	Guan, Naijia	55608312600	2	4	AIST	Japan	Butadiene
25	Furumoto, Yoshiyashu	7005082305	2	4	Hiroshima Univ.	Japan	Propylene
26	Sadakane, Masahiro	6602196986	2	4	Hiroshima Univ.	Japan	propylene
27	Dai, Weili	56683883800	2	4	Nankai Univ.	China	Butadiene
28	Li, Landong	7501449617	2	4	Nankai Univ.	China	Butadiene
29	Wu, Guangjun	57190295874	2	4	Nankai Univ.	China	Butadiene
30	Bell, Alexis T.	36062590400	2	4	Univ. Calif. Berkeley	USA	Butadiene
31	Tesser, Riccardo	7004607875	2	4	Univ. Naples	Italy	Acetaldehyde
32	Wang, Yong	47962677000 55992345500	2	4	Washington State Univ.	USA	Isobutene

Author code, the unique code given by Scopus to the authors; Population papers, The number of papers authored in the population dataset; Sample papers, the number of papers authored in the sample dataset.

The most prolific authors are Tadahiro Fujitani and Atsushi Takahashi, with five sample papers each. Irina I. Ivanova, Vitaly L. Sushkevich, Carlo Angelici, Pieter C. A. Bruijnincx, and Bert M. Weckhuysen follow these top authors with four sample papers each.

The most prolific institution for the sample dataset is the National Institute of Advanced Industrial Science and Technology (AIST), with seven authors. Hiroshima University, Nankai University, the National Academy of Sciences of Ukraine, the University of Naples, and Utrecht University follow this top institution with three authors each. In total, 13 institutions house these prolific authors.

The most prolific country in the sample dataset is Japan, with ten authors. The Netherlands follows Japan with four authors. The other prolific countries with three authors each are China, Italy, the UK, Ukraine, and the USA. In total, nine countries house these authors.

The most prolific research front is bioethanol-based butadiene and propylene production, with 14 and nine authors, respectively. Other prolific research fronts are bioethanol-based butanol and acetaldehyde production, with three and two papers, respectively.

On the other hand, there is a significant gender deficit (Beaudry and Lariviere, 2016) for the sample dataset, as surprisingly, nearly all of these top researchers are male.

108.3.3 The Most Prolific Research Output by Years in Bioethanol-based Biochemical Production

Information about papers published between 1970 and 2021 is given in Figure 108.1. This figure clearly shows that the bulk of the research papers in the population dataset was published primarily in the 2010s with 50.1% of the population dataset. The publication rates for the 2020s, 2000s, 1990s, 1980s, 1970s, and pre-1970s were 14.6%, 12.1%, 7.2%, 7.1%, 2.9%, and 4.5%, respectively. There was a rising trend for the population papers between 2007 and 2017, and thereafter it steadied around 7% each year after a sharp rise in 2017.

Similarly, the bulk of the research papers in the sample dataset were published in the 2010s with 69% of the sample dataset and, to a lesser extent, in the 2000s with 20% of the sample papers. The publication rates for the 1990s, 1980s, and 1970s were 7%, 4%, and 2% of the sample papers, respectively.

The most prolific publication years for the population dataset were after 2013, with at least 4.8% each of the dataset. Similarly, 60% of the sample papers were published between 2011 and 2017.

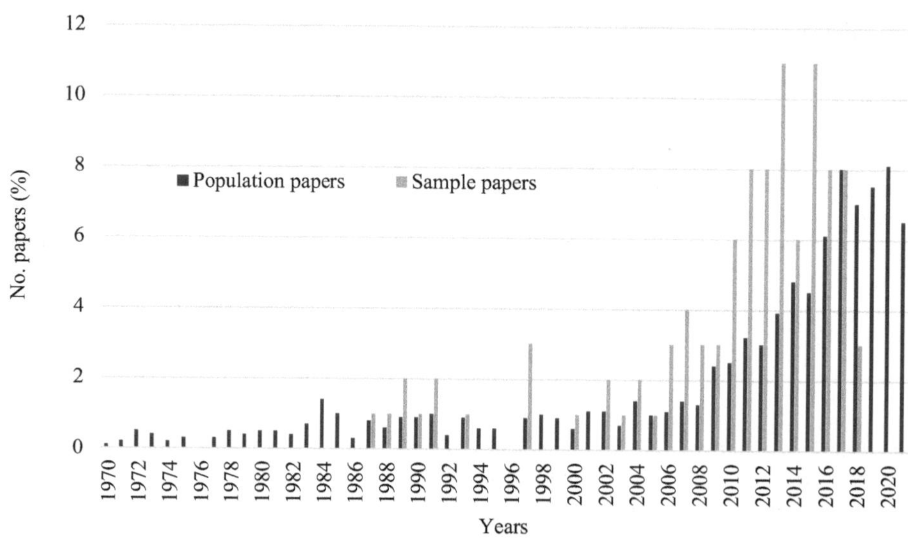

FIGURE 108.1 The research output by years regarding bioethanol fuel-based biochemical production.

108.3.4 The Most Prolific Institutions on Bioethanol-based Biochemical Production

Information about the 22 most prolific institutions publishing papers on bioethanol fuel-based biochemical production with at least two sample papers and 0.5% of the population papers each is given in Table 108.3.

The most prolific institution is the AIST, with seven sample papers. The National Scientific Research Centre (CNRS), Lomonosov Moscow State University, and Utrecht University follow the AIST with four sample papers each.

The top countries for these most prolific institutions are China, the USA, and Japan, with five, four, and three institutions, respectively. Next, the Netherlands and Russia follow these top countries with two institutions each. In total, 11 countries house these top institutions.

On the other hand, institution with the highest citation impact is the AIST, with a 4.7% surplus. Utrecht University and Lomonosov Moscow State University follow these top institutions with 3.2% surpluses each. Similarly, the institutions with the lowest impact are Tianjin University and the National Academy of Sciences of Ukraine, with at least a 1% deficit each.

108.3.5 The Most Prolific Funding Bodies on Bioethanol-based Biochemical Production

Information about the most prolific ten funding bodies, each funding at least two sample papers and 1% of the population papers is given in Table 108.4.

TABLE 108.3
The Most Prolific Institutions in Bioethanol Fuel-based Biochemical Production

No.	Institutions	Country	Sample Papers (%)	Population Papers (%)	Surplus (%)
1	AIST	Japan	7	2.3	4.7
2	CNRS	France	4	1.4	2.6
3	Lomonosov Moscow State Univ.	Russia	4	0.8	3.2
4	Utrecht Univ.	The Netherlands	4	0.8	3.2
5	Tianjin Univ.	China	3	4.0	−1
6	Chinese Acad. Sci.	China	3	2.4	0.6
7	Russ. Acad. Sci.	Russia	3	1.4	1.6
8	Pacific NW Natl. Lab.	USA	3	0.9	2.1
9	Univ. Calif. Berkeley	USA	3	0.6	2.4
10	Eindhoven Tech. Univ.	The Netherlands	3	0.5	2.5
11	Natl. Acad. Sci. Ukraine	Ukraine	2	2.6	−0.6
12	Nat. Technol. Inst.	Brazil	2	1.2	0.8
13	Dalian Univ. Technol.	China	2	0.9	1.1
14	Hokkaido Univ.	Japan	2	0.7	1.3
15	Univ. Naples	Italy	2	0.6	1.4
16	Nankai Univ.	China	2	0.6	1.4
17	Denmark Tech. Univ.	Denmark	2	0.5	1.5
18	Tsinghua Univ.	China	2	0.5	1.5
19	Univ. Bath	UK	2	0.5	1.5
20	Hiroshima Univ.	Japan	2	0.5	1.5
21	Lawrence Berkeley Natl. Lab.	USA	2	0.5	1.5
22	Washington State Univ.	USA	2	0.5	1.5

TABLE 108.4
The Most Prolific Funding Bodies in Bioethanol Fuel-based Biochemical Production

No.	Funding Bodies	Country	Sample Paper No. (%)	Population Paper No. (%)	Surplus (%)
1	National Natural Science Foundation of China	China	7	9.8	−2.8
2	US Department of Energy	USA	7	3.1	3.9
3	National Science Foundation	USA	5	3.0	2
4	Ministry of Science and Technology	China	3	0.9	2.1
5	Engineering and Physical Sciences Research Council	UK	3	0.6	2.4
6	New Energy and Industrial Technology Development Organization	Japan	3	0.5	2.5
7	Chinese Academy of Sciences	China	2	2.4	−0.4
8	Ministry of Education	China	2	1.8	0.2
9	European Commission	EU	2	1.5	0.5
10	Los Alamos National Laboratory	USA	2	1.0	1

The most prolific funding bodies are the National Natural Science Foundation of China and the US Department of Energy, with seven sample papers each. The National Science Foundation follows these top funding bodies with five sample papers.

It is notable that 39% and 40.1% of the sample and population papers are funded, respectively.

The most prolific countries for these top funding bodies are China and the USA, with five and four funding bodies, respectively. In total, four countries and the European Union house these top funding bodies.

The funding body with the highest citation impact is the US Department of Energy, with a 3.9% surplus. Similarly, the funding body with the lowest citation impact is the National Natural Science Foundation of China, with a 2.8% deficit.

108.3.6 The Most Prolific Source Titles on Bioethanol-based Biochemical Production

Information about the 17 most prolific source titles publishing at least two sample papers and 0.5% of the population papers each in bioethanol fuel-based biochemical production is given in Table 108.5.

The most prolific source titles are 'ACS Catalysis', 'Applied Catalysis A General', and 'Journal of Catalysis', with 12, 10, and 10 sample papers, respectively. The 'Catalysis Letters', 'Journal of the American Chemical Society', and 'ChemSusChem' follow these top titles with seven sample papers each.

On the other hand, the source title with the highest citation impact is 'ACS Catalysis', with a 10.2% surplus. The 'Journal of Catalysis', 'Applied Catalysis A General', and 'ChemSusChem' follow this top title with at least 5.7% surplus each. Similarly, the source titles with the lowest impact are 'Chemical Engineering Journal', 'ACS Sustainable Chemistry and Engineering', and 'Catalysts', with at least a 1% deficit each.

108.3.7 The Most Prolific Countries on Bioethanol-based Biochemical Production

Information about the 16 most prolific countries publishing at least two sample papers and 1% of the population papers each in bioethanol fuel-based biochemical production is given in Table 108.6.

The most prolific countries are the USA and China and Japan with 24, 20, and 15 sample papers, respectively. The Netherlands, Russia, and the UK are the other prolific countries, with at least six

TABLE 108.5
The Most Prolific Source Titles in Bioethanol Fuel-based Biochemical Production

No.	Source Titles	Sample Papers (%)	Population Papers (%)	Surplus (%)
1	ACS Catalysis	12	1.8	10.2
2	Applied Catalysis A General	10	3.5	6.5
3	Journal of Catalysis	10	3.3	6.7
4	Catalysis Letters	7	2.6	4.4
5	Journal of the American Chemical Society	7	2.0	5
6	Chemsuschem	7	1.3	5.7
7	Industrial and Engineering Chemistry Research	6	2.9	3.1
8	Catalysis Today	4	2.0	2
9	Catalysis Communications	3	1.9	1.1
10	Applied Catalysis B Environmental	3	1.4	1.6
11	Catalysis Science and Technology	3	1.0	2
12	Applied Catalysis	3	0.5	2.5
13	Angewandte Chemie International Edition	3	0.5	2.5
14	Chemical Engineering Journal	2	1.3	0.7
15	ACS Sustainable Chemistry and Engineering	2	1.2	0.8
16	Catalysts	2	1.0	1
17	Journal of Molecular Catalysis A Chemical	2	0.6	1.4

TABLE 108.6
The Most Prolific Countries in Bioethanol Fuel-based Biochemical Production

No.	Countries	Sample Papers (%)	Population Papers (%)	Surplus (%)
1	USA	24	15.4	8.6
2	China	20	22.0	−2
3	Japan	15	9.8	5.2
4	The Netherlands	7	2.0	5
5	Russia	6	4.1	1.9
6	UK	6	3.7	2.3
7	Italy	4	3.3	0.7
8	France	4	2.0	2
9	Denmark	4	1.0	3
10	Brazil	3	4.2	−1.2
11	Canada	3	2.5	0.5
12	Germany	3	1.9	1.1
13	Belgium	3	1.4	1.6
14	Spain	2	4.7	−2.7
15	Ukraine	2	2.7	−0.7
16	Sweden	2	1.4	0.6

sample papers each. On the other hand, 11 European countries listed in Table 18.6 produce 43% and 28% of the sample and population papers, respectively.

On the other hand, the countries with the highest citation impact are the USA, Japan, and the Netherlands, with at least 5% surplus each. Similarly, the countries with the lowest citation impact are Spain, China, and Brazil, with at least a 2.7% deficit each.

108.3.8 The Most Prolific Scopus Subject Categories on Bioethanol-based Biochemical Production

Information about the eight most prolific Scopus subject categories indexing at least six and 10% of the sample and population papers, respectively, is given in Table 108.7.

The most prolific Scopus subject categories in bioethanol fuel-based biochemical production are 'Chemical Engineering' and 'Chemistry' with 90 and 69 sample papers, respectively.

On the other hand, the Scopus subject category with the highest citation impact is 'Chemical Engineering', with a 37.2% surplus. Similarly, the Scopus subject categories with the lowest citation impact are 'Physics and Astronomy', 'Materials Science', and 'Biochemistry', with at least a 3.5% deficit each.

108.3.9 The Most Prolific Keywords on Bioethanol-based Biochemical Production

Information about the keywords used with at least 5% of the sample and population papers each is given in Table 108.8. For this purpose, keywords related to the keyword set given in the appendix are selected from a list of the most prolific keyword sets provided by the Scopus database.

These keywords are grouped under the five headings: Bioethanol fuels, biochemicals, processes, materials, and catalysis related to the biochemical production from bioethanol fuels.

There are seven keywords used related to bioethanol fuels. These are ethanol, alcohols, ethanol dehydration, ethanol conversion, and bioethanol. It is notable that the bioethanol keyword appears in the sample paper keyword list, with around 13% and 5% of the sample papers for the keywords of bioethanol and bio-ethanol, respectively.

The prolific keywords related to the biochemicals are ethylene, acetaldehyde, butadiene, and 1,3-butadiene, with at least 16 papers each, while the prolific keywords related to the processes are dehydration, dehydrogenation, and catalytic conversion.

The prolific keywords related to the materials are H-ZSM 5, zeolites, silica, magnesia, and alumina, while the prolific keywords related to the catalysis are catalysts, catalyst selectivity, catalysis, and catalyst activity.

On the other hand, the keywords with the highest citation impact are catalyst activity, catalytic conversion, 1,3-butadiene, ethanol conversion, dehydrogenation, and butenes. Similarly, the keywords with the lowest citation impact are acetic acid, zirconium, and ethyl acetate.

Further, the most prolific keywords across all the subject headings are ethanol, catalysts, ethylene, catalyst selectivity, catalysis, dehydration, catalyst activity, acetaldehyde, alcohols, butadiene, ethanol dehydration, 1,3-butadiene, ethanol conversion, dehydrogenation, and H-ZSM 5, with 15 to 87% of the sample papers, respectively.

TABLE 108.7
The Most Prolific Scopus Subject Categories in Bioethanol Fuel-based Biochemical Production

No.	Scopus Subject Categories	Sample Papers (%)	Population Papers (%)	Surplus (%)
1	Chemical Engineering	90	52.8	37.2
2	Chemistry	69	64.3	4.7
3	Engineering	15	13.4	1.6
4	Environmental Science	15	10.7	4.3
5	Materials Science	13	17.3	–4.3
6	Energy	13	11.7	1.3
7	Biochemistry	8	11.5	–3.5
8	Physics and Astronomy	6	12.4	–6.4

TABLE 108.8
The Most Prolific Keywords in Bioethanol Fuel-based Biochemical Production

No.	Keywords	Sample Papers (%)	Population Papers (%)	Surplus (%)
1.	Bioethanol fuels			
	Ethanol	87	67.5	19.5
	Alcohols	19	14.2	4.8
	Ethanol dehydration	18	7.5	10.5
	Ethanol conversion	15	6.8	8.2
	Bioethanol	13	7.0	6
	Ethanol dehydrogenation	12	4.9	7.1
	Bio-ethanol	5	1.3	3.7
2.	Biochemicals			
	Ethylene	25	18.8	6.2
	Acetaldehyde	21	18.1	2.9
	Butadiene	19	5.1	13.9
	1,3-Butadiene	16	7.8	8.2
	Butanol	13	2.7	10.3
	Butenes	11	3.1	7.9
	Ethyl acetate	10	9.0	1
	Diethyl ethers	9	3.9	5.1
	1-butanol	8	1.5	6.5
	Acetic acid	6	8.2	–2.2
	Propylene	6	4.4	1.6
	Ethers	5	3.8	1.2
3.	Processes			
	Dehydration	22	8.8	13.2
	Dehydrogenation	15	7.1	7.9
	Catalytic conversion	12	3.3	8.7
	Condensation reactions	11	4.4	6.6
	Oxidation	9	7.0	2
	Synthesis	9	1.9	7.1
	Dewatering	9	1.4	7.6
	Condensation	8	3.4	4.6
	Adsorption	6	2.7	3.3
	Aldol condensation	5	2.7	2.3
	Catalytic dehydration	5	2.0	3
	Ethylene production	5	1.9	3.1
	Guerbet reaction	5		5
4	Materials			
	H-ZSM 5	15	3.1	11.9
	Zeolites	14	7.2	6.8
	Silica	12	5.1	6.9
	Magnesia	11	3.2	7.8
	Alumina	10	4.2	5.8
	Copper	9	3.7	5.3
	Zinc	7	1.4	5.6
	Gold	7	1.3	5.7
	Ruthenium	6	1.3	4.7

(Continued)

TABLE 108.8 (*Continued*)
The Most Prolific Keywords in Bioethanol Fuel-based Biochemical Production

No.	Keywords	Sample Papers (%)	Population Papers (%)	Surplus (%)
	Phosphorus	6	0.0	6
	Zirconium	5	5.2	–0.2
	Ketones	5	3.4	1.6
	Hydroxyapatite	5		5
	Silicate minerals	5		5
5.	Catalysis			
	Catalysts	39	20.6	18.4
	Catalyst selectivity	25	10.7	14.3
	Catalysis	25	8.6	16.4
	Catalyst activity	22	12.3	9.7
	Catalytic performance	11	6.4	4.6
	Heterogeneous catalysis	7	1.8	5.2
	Catalyst deactivation	6	2.1	3.9

TABLE 108.9
The Most Prolific Research Fronts in Bioethanol Fuel-based Biochemical Production

No.	Research Fronts	Sample Papers (%)
1	Bioethylenes	26
2	Biobutadienes	23
3	Biobutanols	21
4	Bioacetaldehydes	15
5	Bioethyl acetates	7
6	Biopropylenes	6
7	Bioacetic acids	5
8	Biodiethyl ethers	4
9	Bioisobutenes	3

Sample papers, the sample of the 100 most-cited papers.

108.3.10 The Most Prolific Research Fronts on Bioethanol-based Biochemical Production

Information about the most prolific research fronts for the sample papers in bioethanol fuel-based biochemical production is given in Table 108.9.

There are four primary research fronts in this field: Bioethylenes (Morschbacker, 2009; Takahara et al., 2005; Zhang and Yu, 2013), biobutadienes (Angelici et al., 2013; Makshina et al., 2012; Sushkevich et al., 2014), biobutanols (Fu et al., 2017; Ndou et al., 2003; Tsuchida et al., 2006), and bioacetaldehydes (Gong and Mullins, 2008; Jackson et al., 1991; Santacesaria et al., 2012) with 26, 23, 21, and 15 sample papers, respectively. The other research fronts are bioethyl acetates (Inui et al., 2002), biopropylenes (Goto et al., 2010), bioacetic acids (Christensen et al., 2006), biodiethyl ethers (Christiansen et al., 2013), and bioisobutenes (Liu et al., 2013), with at least three sample papers each.

108.4 DISCUSSION

108.4.1 Introduction

Crude oil-based gasoline fuels have been widely used in the transportation sector since the 1920s. However, there have been great public concerns over the adverse environmental and human impacts of these fuels. Hence, biomass-based bioethanol fuels have increasingly been used in blending gasoline fuels, in fuel cells, and in producing valuable biochemicals in a biorefinery context.

In the meantime, research in nanomaterials and nanotechnology has intensified in recent years. The research on the production of biochemicals from bioethanol fuels has primarily intensified in the research fronts of bioethylenes, biobutadienes, biobutanols, and bioacetaldehydes. The other research fronts have been bioethyl acetates, biopropylenes, bioacetic acids, biodiethyl ethers, and bioisobutenes. However, it is essential to develop efficient incentive structures for the primary stakeholders to enhance research in this field. The scientometric analysis has been used in this context to inform the primary stakeholders about the current state of research in a selected research field.

As there have been no scientometric studies on biochemical production in general and on biochemical production from bioethanol fuels, this chapter presents a scientometric study of the research in bioethanol fuel-based biochemical production. It examines the scientometric characteristics of both the sample and population data, presenting the scientometric characteristics of these both datasets in the order of documents, authors, publication years, institutions, funding bodies, source titles, countries, Scopus subject categories, keywords, and research fronts.

The search for this study was carried out using the Scopus database in September 2021. As a first step for the search of the relevant literature, the keywords were selected using the 200 most-cited papers. The selected keyword list was optimized to obtain a representative sample of papers for this research field. This keyword list was provided in the appendix for future replication studies. Additionally, the information about the most-used keywords was given in Section 108.3.9 to highlight the key research fronts in Section 108.3.10.

As the second step, two sets of data were used in this study. First, a population sample of over 1,100 papers was used to examine the scientometric characteristics of the population data. Second, a sample of 100 most-cited papers was used to examine the scientometric characteristics of these citation classics with over 46 citations each.

The scientometric characteristics of these both sample and population datasets were presented in the order of documents, authors, publication years, institutions, funding bodies, source titles, countries, Scopus subject categories, keywords, and research fronts. Lastly, the key scientometric findings for both datasets were discussed to highlight the research landscape for bioethanol fuel-based biochemical production. Additionally, a number of brief conclusions were drawn, and a number of relevant recommendations were made to enhance the future research landscape.

108.4.2 The Most Prolific Documents on Bioethanol-based Biochemical Production

The articles dominate both the sample and population datasets. The review papers and articles have a slight surplus and deficit, respectively.

Scopus differs from the Web of Science database in differentiating and showing articles and conference papers published in journals separately. Similarly, Scopus differs from the Web of Science database in introducing short surveys. However, there are no conference papers and short surveys in both samples.

It was observed during the search process that there has been inconsistency in the classification of documents in Scopus as well as in other databases such as Web of Science. This is especially relevant for the classification of papers as reviews or articles, as papers not involving a literature review may be erroneously classified as review papers. There is also a case of review papers being classified as articles. For example, although there are six review papers as classified by the Scopus database, eight of the sample papers are review papers.

In this context, it would be helpful to provide a classification note for published papers in books and journals at the first instance. It would also be helpful to use document types listed in Table 108.1 for this purpose. Book chapters may also be classified as articles or reviews as an additional classification to differentiate review chapters from experimental chapters, as it is done by the Web of Science. It would be further helpful to additionally classify the conference papers as articles or review papers, as well as it is done in the Web of Science database.

108.4.3 The Most Prolific Authors on Bioethanol-based Biochemical Production

There have been 32 most prolific authors with at least two sample papers and four population papers each, as given in Table 108.2. These authors have shaped the development of research in this field.

The most prolific authors are Tadahiro Fujitani, Atsushi Takahashi, Irina I. Ivanova, Vitaly L. Sushkevich, Carlo Angelici, Pieter C. A. Bruijnincx, and Bert M. Weckhuysen (Table 108.2).

It is important to note the inconsistencies in the indexing of author names in Scopus and other databases. It is especially an issue for names with more than two components, such as 'Judge Alex de Camp Sirous'. The probable outcomes are 'Sirous, J.A.D.C.', 'de Camp Sirous, J.A.', or 'Camp Sirous, J.A.D.'. The first choice is the gold standard of the publishing sector as the last word in the name is taken as the last name. In most of the academic databases, such as PubMed and EBSCO databases, this version is used predominantly. The second choice is a strong alternative, while the last choice is an undesired outcome as two last words are taken as the last name. It is good practice to combine the words of the last name with a hyphen: 'Camp-Sirous, J.A.D.'. It is notable that inconsistent indexing of author names may cause substantial inefficiencies in the search process for papers as well as in allocating credit to authors as there are different author entries for each outcome in the databases.

There is also a case of the shortening of Chinese names. For example, 'Yuoyang Wang' is often shortened as 'Wang, Y.', 'Wang, Y.-Y.', and 'Wang Y.Y.' as it is done in the Web of Science database as well. However, the gold standard in this case is 'Wang Y' where the last word is taken as the last name and the first word is taken as a single forename. In most of the academic databases, such as PubMed and EBSCO, this first version is used predominantly. Therefore, there have been difficulties locating papers for the Chinese authors. Nevertheless, it makes sense to use the third option to efficiently differentiate Chinese names: 'Wang Y.Y.'. In such cases, the use of the unique author codes provided for each author by the Scopus database has been helpful.

There is also a difficulty in allowing credit for the authors, especially for those with common names such as 'Wang, Y.', or 'Huang, Y.' or 'Zhu, Y.' in conducting scientometric studies. These difficulties strongly influence the efficiency of the scientometric studies as well as allocating credit to the authors, as there are the same author entries for different authors with the same name, e.g., 'Wang Y.' in the databases.

In this context, the coding of authors in the Scopus database is a welcome innovation compared to other databases such as Web of Science. In this process, Scopus allocates a unique number to each author in the database (Aman, 2018). However, there might still be substantial inefficiencies in this coding system, especially for common names. For example, some of the papers for a certain author maybe allocated to another researcher with a different author code. It is possible that Scopus uses a number of software programs to differentiate the author names, and the program may not be false-proof (Shin et al., 2014).

In this context, it does not help that author names are not given in full in some journals and books. This makes it difficult to differentiate authors with common names and makes scientometric studies further difficult in the author domain. Therefore, author names should be given in all books and journals at the first instance. There is also a cultural issue where some authors do not use their full names in their papers. Instead, they use initials for their forenames: 'Coutancy, A.P.', instead of 'Coutancy, Alas Padras'.

There are also inconsistencies in the naming of authors with more than two components by authors themselves in journal papers and book chapters. For example, 'Alaspanda, A.P.C.', 'Sakoura, C.E.', and 'Mentaslo, S.J.' might be given as 'Alaspanda, A.', 'Sakoura, C.', or 'Mentaslo, S.' in journals and books, respectively. This also makes scientometric studies difficult in the author's domain. Hence, contributing authors should use their names consistently in their publications.

Another critical issue regarding author names is the spelling of author names in the national spellings (e.g., Gonçalves, Übeiro) rather than in the English spellings (e.g., Goncalves, Ubeiro) in the Scopus database. Scopus differs from the Web of Science database and many other databases in this respect, where author names are given only in their English spelling. It is observed that the national spelling of author names does not help in conducting scientometric studies as well as in allocating credit to authors, as sometimes there are different author entries for English and National spellings in the Scopus database.

The most prolific institutions for the sample dataset are AIST, Hiroshima University, Nankai University, the National Academy of Sciences of Ukraine, the University of Naples, and Utrecht University. Similarly, the most prolific countries for the sample dataset are Japan, the Netherlands, China, Italy, the UK, Ukraine, and the USA. It is not surprising that authors from these countries dominate this prolific author list.

It is also notable that there is a significant gender deficit in the sample dataset surprisingly, nearly all of these top researchers are male. This finding is the most thought-provoking, with strong public policy implications. Hence, institutions, funding bodies, and policymakers should take efficient measures to reduce the gender deficit in this field as well as in other scientific fields with a strong gender deficit. In this context, it is worth noting that the level of representation of researchers from minority groups in science on the basis of race, sexuality, age, and disability, besides gender (Blankenship, 1993; Dirth and Branscombe, 2017; Konur, 2000, 2002a,b,c, 2006a,b, 2007a,b).

108.4.4 The Most Prolific Research Output by Years on Bioethanol-based Biochemical Production

The research output observed between 1970 and 2021 is illustrated in Figure 108.1. This figure clearly shows that the bulk of the research papers in the population dataset was published primarily in the 2010s. This was followed by the early 2020s and 2000s, while the publication rates for the pre-2000s were relatively negligible. Similarly, the bulk of the research papers in the sample dataset was published in the 2010s, followed by the 2000s.

However, there was a rising trend for the population papers between 2007 and 2017, and thereafter, it steadied around 7% each year after a sharp rise in 2017. This suggests that research in this field lost its momentum after 2017.

These data suggest that the most-cited sample and population papers were primarily published in the 2010s. These are thought-provoking findings, as there has been no significant research in this field in the pre-2000s, but there has been a significant research boom in the last two decades. In this context, the increasing public concerns about climate change (Change, 2007), greenhouse gas emissions (Carlson et al., 2017), and global warming (Kerr, 2007) have been certainly behind the boom in research in this field in the last two decades.

Based on these findings, the size of the population papers likely to more than double in the current decade, provided that the public concerns about climate change, greenhouse gas emissions, and global warming are efficiently translated into research funding in this field.

108.4.5 The Most Prolific Institutions on Bioethanol-based Biochemical Production

The 21 most prolific institutions publishing papers on bioethanol fuel-based biochemical production with at least two sample papers and 0.5% of the population papers each given in Table 108.3 have shaped the development of research in this field.

The most prolific institutions are the AIST, CNRS, Lomonosov Moscow State University, and Utrecht University. The top countries for these most prolific institutions are China, the USA, Japan, the Netherlands, and Russia.

On the other hand, the institutions with the highest citation impact are AIST, Utrecht University, and Lomonosov Moscow State University. Similarly, institutions with the lowest impact are Tianjin University and the National Academy of Sciences of Ukraine.

108.4.6 The Most Prolific Funding Bodies on Bioethanol-based Biochemical Production

The ten most prolific funding bodies funding at least two sample papers and 1% of the population papers each are given in Table 108.4.

The most prolific funding bodies are the National Natural Science Foundation of China, the US Department of Energy, and the National Science Foundation. It is notable that 39% and 40.1% of the sample and population papers are funded, respectively. The most prolific countries for these top funding bodies are China and the USA. In total, four countries and the European Union house these top funding bodies. The heavy funding by the Chinese and US funding bodies is notable, as these countries are all major producers of research in this field.

These findings on the funding of the research in this field suggest that the level of funding, mostly in the last two decades, has been largely instrumental in enhancing the research in this field (Ebadi and Schiffauerova, 2016) in the light of North's institutional framework (North, 1991). However, as the research output lost its momentum after 2017, there are question marks about the level of funding after 2017 suggesting that there is ample room to increase the funding rate for this field.

108.4.7 The Most Prolific Source Titles on Bioethanol-based Biochemical Production

The 17 most prolific source titles publishing at least two sample papers and 0.5% of the population papers each in bioethanol fuel-based biochemical production have shaped the development of research in this field (Table 108.5).

The most prolific source titles are 'ACS Catalysis', 'Applied Catalysis A General', 'Journal of Catalysis', 'Catalysis Letters', Journal of the American Chemical Society', and 'ChemSusChem'.

On the other hand, the source titles with the highest citation impact are ACS Catalysis', 'Journal of Catalysis', 'Applied Catalysis A General', and 'ChemSusChem'. Similarly, the source titles with the lowest impact are 'Chemical Engineering Journal', 'ACS Sustainable Chemistry and Engineering', and 'Catalysts'.

It is notable that these top source titles are related to catalysis and, to a lesser extent, chemistry. This finding suggests that the journals in these fields have significantly shaped the development of research in this field as they focus on the development of nanocatalysts and conventional catalysts for the production of biochemicals from bioethanol fuels.

108.4.8 The Most Prolific Countries on Bioethanol-based Biochemical Production

The 16 most prolific countries publishing at least two papers and 1% of the population papers each have significantly shaped the development of research in this field (Table 108.6).

The most prolific countries are the USA, China, Japan and, to a lesser extent, the Netherlands, Russia, and the UK. On the other hand, the countries with the highest citation impact are the USA, Japan, and the Netherlands, while the countries with the lowest citation impact are Spain, China, and Brazil.

It is notable that 11 European countries listed in Table 18.6 produce 43% and 28% of the sample and population papers, respectively, with a 13% surplus, making Europe the first largest producer of research in this field above the USA, China, and Japan, its key competitors.

A close examination of these findings suggests that Europe, the USA, China, and Japan are the major producers of research in this field. It is a fact that the USA has been a major player in science (Leydesdorff and Wagner, 2009; Leydesdorff et al., 2014). The USA has further developed a strong research infrastructure to support its corn and grass-based bioethanol industries (Vadas et al., 2008). The USA has also been very active in nanotechnology research (Dong et al., 2016) as well as biochemical research (Titov and Żiane, 2013).

However, China has been a rising star in scientific research in competition with the USA and Europe (Leydesdorff and Zhou, 2005). China is also a major player in this field as a major producer of bioethanol (Li and Chan-Halbrendt, 2009). China has also been very active in nanotechnology research (Dong et al., 2016) as well as biochemical research (Chen et al., 2018).

Next, Europe has been a persistent player in the scientific research in competition with both the USA and China (Leydesdorff, 2000). Europe has also been a persistent producer of bioethanol, along with the USA and Brazil (Gnansounou, 2010). Europe has also been very active in nanotechnology research (Scheufele et al., 2009) as well as biochemical research (Ballesteros et al., 2012).

Additionally, Japan has also been a persistent player in scientific research at a moderate level (Fukuzawa and Ida, 2016). Japan has also developed a strong research infrastructure to support its biomass-based bioethanol industry (Koga, 2008). Japan has also been very active in nanotechnology research (Katao, 2006) as well as biochemical research (Higashino et al., 2007).

108.4.9 The Most Prolific Scopus Subject Categories on Bioethanol-based Biochemical Production

The eight most prolific Scopus subject categories, indexing at least six and 10% of the sample and population papers, respectively, given in Table 108.7, have shaped the development of research in this field.

The most prolific Scopus subject categories in bioethanol fuel-based biochemical production are 'Chemical Engineering' and 'Chemistry'. On the other hand, the Scopus subject category with the highest citation impact is 'Chemical Engineering'. Similarly, the Scopus subject categories with the lowest citation impact are 'Physics and Astronomy', 'Materials Science', and 'Biochemistry'.

These findings are thought-provoking, suggesting that the primary subject categories are 'Chemistry' and 'Chemical Engineering'. The other key finding is that social sciences are not well represented in both the sample and population papers, unlike the field of evaluative studies in bioethanol fuels.

These findings are not surprising, as the key research fronts in this field are the development of catalysts for the production of biochemicals from bioethanol fuels. All these research fronts are related to the hard sciences and engineering.

108.4.10 The Most Prolific Keywords on Bioethanol-based Biochemical Production

A limited number of keywords have shaped the development of research in this field, as shown in Table 108.8 and the appendix.

These keywords are grouped under five headings: Bioethanol fuels, biochemicals, processes, materials, and catalysis related to the biochemical production from bioethanol fuels.

The most prolific keywords across all the subject headings are ethanol, catalysts, ethylene, catalyst selectivity, catalysis, dehydration, catalyst activity, acetaldehyde, alcohols, butadiene, ethanol dehydration, 1,3-butadiene, ethanol conversion, dehydrogenation, and H-ZSM 5.

These prolific keywords highlight the key research fronts in this field and reflect well the keywords used in the sample papers.

108.4.11 The Most Prolific Research Fronts on Bioethanol-based Biochemical Production

There are four primary research fronts in this field: Bioethylenes, biobutadienes, biobutanols, and bioacetaldehydes. The other research fronts are bioethyl acetates, biopropylenes, bioacetic acids, biodiethyl ethers, and bioisobutenes, with at least three sample papers each.

The focus of these 100 most-cited papers is the development of catalysts and catalyst supports based on nanomaterials and conventional materials for the production of biochemicals from bioethanol fuels.

Research in the field of biochemicals has intensified in recent years (Zhang et al., 2016; Nikolau et al., 2008). The production of biochemicals in the biorefinery context makes the production process cost-effective, reducing the production costs for the primary biofuels (Cherubini et al., 2009; Maity, 2015).

One innovative way to enhance biochemical production has been the use of nanomaterials for both the catalysts themselves and the catalyst supports (Bi et al., 2010; Gucbilmez et al., 2006). However, it is surprising that the application of nanotechnology in this sample has been scarce.

In the end, the efficiency of biochemical production from bioethanol fuels could be optimized using the structure, processing, and property relationships of the nanomaterials and conventional materials used as both catalyst materials and catalyst supports (Formela et al., 2016; Konur, 2018b, 2020b, 2021c,d,e,f; Konur and Matthews, 1989).

108.5 CONCLUSION AND FUTURE RESEARCH

The research on bioethanol fuel-based biochemical production has been mapped through a scientometric study of both sample and population datasets.

The critical issue in this study has been obtaining a representative sample of research, as in any other scientometric study. Therefore, the keyword set has been carefully devised and optimized after a number of runs in the Scopus database. It should be noted that the focus in this book chapter has been on the bioethylenes, biobutadienes, biobutanols, bioacetaldehydes, bioethyl acetates, biopropylenes, bioacetic acids, biodiethyl ethers, and bioisobutenes. It is a representative sample of the wider population studies, which include large numbers of biochemicals and biohydrocarbons in this field.

Another issue has been the selection of a multidisciplinary database to carry out the scientometric study of research in this field. For this purpose, the Scopus database has been selected. The journal coverage of this database has been wider than that of Web of Science.

The key scientometric properties of research in this field have been determined and discussed in this book chapter. It is evident that a limited number of documents, authors, institutions, publication periods, institutions, funding bodies, source titles, countries, Scopus subject categories, keywords, and research fronts have shaped the development of the research in this field.

There is ample scope to increase the efficiency of scientometric studies in this field in the author and document domains by developing consistent policies and practices in both domains across all academic databases. In this respect, authors, journals, and academic databases have a lot to do. Furthermore, the significant gender deficit, as in most scientific fields, emerges as a public policy issue. Potential deficits on the basis of age, race, disability, and sexuality also need to be explored in this field, as in other scientific fields.

Research in this field has boomed in the 2010s, possibly promoted by public concerns on global warming, greenhouse gas emissions, and climate change. However, it appears the research in this field lost its momentum after 2017.

The institutions from China, the USA, Japan, the Netherlands, and Russia have mostly shaped the research in this field.

The relatively low funding rates of 39% and 40% for sample and population papers, respectively, suggest that funding in this field significantly enhanced research in this field primarily in the 2010s, possibly more than doubling in the current decade. However, there is ample room for more funding, as these funding rates for this field are not much considering the loss of momentum after 2017.

The most prolific source titles have been mostly indexed by the subject categories of chemistry and chemical engineering, as the focus of the sample papers has been on the development of catalysts and catalyst supports to produce biochemicals from bioethanol fuels.

Europe, the USA, China, and Japan have been the major producers of research in this field as the major producers of bioethanol fuels from different types of biomass such as corn, sugarcane, and grass, as well as other types of biomass. These countries have well-developed research infrastructure in bioethanol fuels, bioethanol fuel-based biochemical production, nanomaterials, and conventional materials.

The primary subject categories have been 'Chemistry' and 'Chemical Engineering'. Due to the technological emphasis of this field on the development of catalysts and catalyst supports, the social sciences have not been fairly represented in both the sample and population papers, unlike the evaluative studies in bioethanol fuels.

Ethanol is more popular than bioethanol as a keyword, with strong implications for the search strategy. In other words, the search strategy using only the bioethanol keyword would not be much helpful.

These keywords are grouped under five headings: Bioethanol fuels, biochemicals, processes, materials, and catalysis related to the biochemical production from bioethanol fuels. These groups of keywords highlight the potential primary research fronts for these fields: the development of catalysts and catalyst supports to produce biochemicals from bioethanol fuels.

There are four primary research fronts in this field: Bioethylenes, biobutadienes, biobutanols, and bioacetaldehydes. The other research fronts are bioethyl acetates, biopropylenes, bioacetic acids, biodiethyl ethers, and bioisobutenes. The focus has been on the use of conventional materials for the development of catalysts and catalyst supports with 96% of the sample papers, suggesting that there is ample room of the nanotechnology applications in this field.

These findings are thought-provoking. The focus of these 100 most-cited papers is the development of catalysts and catalyst supports rather than the biochemical production processes themselves, such as dehydrogenation and dehydration. There are strong structure-processing-property relationships for all of the materials used for these research fronts. In the end, the efficiency of the materials science and nanotechnology applications could be optimized using the structure, processing, and property relationships of materials used in these applications.

Thus, scientometric analysis has a great potential to gain valuable insights into the evolution of research in this field as well as in other scientific fields.

It is recommended that further scientometric studies are carried out about the other aspects of both the production and utilization of bioethanol fuels. It is further recommended that reviews of the most-cited papers are carried out for each research front to complement these scientometric studies. Next, the scientometric studies of the hot papers in these primary fields are carried out.

ACKNOWLEDGMENTS

The contribution of the highly cited researchers in the field of bioethanol fuel-based biochemical production has been gratefully acknowledged.

APPENDIX: THE KEYWORD SET FOR BIOETHANOL-BASED BIOCHEMICAL PRODUCTION

(((TITLE (ethanol OR bioethanol) OR SRCTITLE (ethanol OR bioethanol)) AND (TITLE ("acetic acid*" OR "ethyl acetate" OR acetaldehyde OR "n-butanol" OR "1-butanol" OR *butadiene OR "diethyl ether" OR *ethylene OR bioethylene OR propylene OR isobutene OR isobutylene) OR SRCTITLE ("acetic acid*" OR "ethyl acetate" OR acetaldehyde OR "n-butanol" OR "1-butanol"

OR butadiene OR "1,3-butadiene" OR "diethyl ether" OR *ethylene OR bioethylene OR propylene OR isobutene OR isobutylene))) AND NOT (TITLE ("carbon dioxide" OR flames OR co2 OR combustion OR immuno* OR fruit* OR viscosity OR consumption OR emission* OR humidified OR glucose OR octane OR isopropanol OR decarbonylation OR carvedilol OR alkane* OR *cellular OR ptsn OR isobutanol OR "Carbon Monoxide" OR "fuel cells" OR sers OR ester* OR adsorp* OR separation OR ingestion OR sulfation OR adsorb* OR reform* OR cells OR lipid* OR soret OR dna OR hydrogenolysis OR hot OR burning OR membrane* OR hepatic OR sensing OR engine OR maleate OR distill* OR carcin* OR malono OR oxalate OR mutat* OR ferment* OR comicel* OR muscle OR solubility OR cardio* OR corrosion OR wine* OR isomer* OR zein OR styrne) OR SRCTITLE (biochem* OR thermo* OR carcin* OR spectrom* OR data OR combust* OR biol* OR entom* OR astro* OR biophys* OR mutat* OR biosensor* OR food* OR fluid* OR separation) OR SUBJAREA (medi OR phar OR immu OR neur OR psyc OR nurs OR vete OR dent OR heal))) OR ((TITLE (("catalytic conversion" OR guerbet OR coupling OR dimerization OR dimerization) AND (ethanol OR bioethanol))) AND TITLE (butanol)) AND (LIMIT-TO (SRCTYPE, "j") OR LIMIT-TO (SRCTYPE, "k") OR LIMIT-TO (SRCTYPE, "b")) AND (LIMIT-TO (DOCTYPE, "ar") OR LIMIT-TO (DOCTYPE, "re") OR LIMIT-TO (DOCTYPE, "no") OR LIMIT-TO (DOCTYPE, "le") OR LIMIT-TO (DOCTYPE, "ch") OR LIMIT-TO (DOCTYPE, "bk")) AND (LIMIT-TO (LANGUAGE, "English"))

REFERENCES

Aman, V. 2018. Does the Scopus author ID suffice to track scientific international mobility? A case study based on Leibniz laureates. *Scientometrics* 117:705–720.

Angelici, C., B. M. Weckhuysen and P. C. A. Bruijnincx. 2013. Chemocatalytic conversion of ethanol into butadiene and other bulk chemicals. *ChemSusChem* 6:1595–1614.

Antolini, E. 2007. Catalysts for direct ethanol fuel cells. *Journal of Power Sources* 170:1–12.

Ballesteros, R., J. J. Hernandez and J. Guillen-Flores. 2012. Carbonyls speciation in a typical European automotive diesel engine using bioethanol/butanol-diesel blends. *Fuel* 95:136–145.

Beaudry, C. and V. Lariviere. 2016. Which gender gap? Factors affecting researchers' scientific impact in science and medicine. *Research Policy* 45:1790–1817.

Bi, J., X. Guo, M. Liu and X. Wang. 2010. High effective dehydration of bio-ethanol into ethylene over nanoscale HZSM-5 zeolite catalysts. *Catalysis Today* 149:143–147.

Blankenship, K. M. 1993. Bringing gender and race in: US employment discrimination policy. *Gender & Society* 7:204–226.

Burnham, J. F. 2006. Scopus database: A review. *Biomedical Digital Libraries* 3:1–8.

Carlson, K. M., J. S. Gerber and D. Mueller, et al. 2017. Greenhouse gas emissions intensity of global croplands. *Nature Climate Change* 7:63–68.

Change, C. 2007. Climate change impacts, adaptation and vulnerability. *Science of the Total Environment* 326:95–112.

Chen, J. M., B. Yu and Y. M. Wei. 2018. Energy technology roadmap for ethylene industry in China. *Applied Energy* 224:160–174.

Cherubini, F., G. Jungmeier and M. Wellisch, et al. 2009. Toward a common classification approach for biorefinery systems. *Biofuels, Bioproducts and Biorefining* 3:534–546.

Christensen, C. H., B. Jorgensen and J. Rass-Hansen, et al. 2006. Formation of acetic acid by aqueous-phase oxidation of ethanol with air in the presence of a heterogeneous gold catalyst. *Angewandte Chemie International Edition* 45:4648–4651.

Christiansen, M. A., G. Mpourmpakis and D. G. Vlachos. 2013. Density functional theory-computed mechanisms of ethylene and diethyl ether formation from ethanol on γ-Al_2O_3(100). *ACS Catalysis* 3:1965–1975.

Dirth, T. P. and N. R. Branscombe. 2017. Disability models affect disability policy support through awareness of structural discrimination. *Journal of Social Issues* 73:413–442.

Dong, H., Y. Gao and P. J. Sinko, et al. 2016. The nanotechnology race between China and the United States. *Nano Today* 11:7–12.

Ebadi, A. and A. Schiffauerova. 2016. How to boost scientific production? A statistical analysis of research funding and other influencing factors. *Scientometrics* 106:1093–1116.

Formela, K., A. Hejna, L. Piszczyk, M. R. Saeb and X. Colom. 2016. Processing and structure-property relationships of natural rubber/wheat bran biocomposites. *Cellulose* 23:3157–3175.

Fu, S., Z. Shao, Y. Wang and Q. Liu. 2017. Manganese-catalyzed upgrading of ethanol into 1-butanol. *Journal of the American Chemical Society* 139:11941–11948.

Fukuzawa, N. and T. Ida. 2016. Science linkages between scientific articles and patents for leading scientists in the life and medical sciences field: The case of Japan. *Scientometrics* 106:629–644.

Garfield, E. 1955. Citation indexes for science. *Science* 122:108–111.

Geim, A. K. 2009. Graphene: Status and prospects. *Science* 324:1530–1534.

Geim, A. K. and K. S. Novoselov. 2007. The rise of graphene. *Nature Materials* 6:183–191.

Gnansounou, E. 2010. Production and use of lignocellulosic bioethanol in Europe: Current situation and perspectives. *Bioresource Technology* 101:4842–4850.

Gong, J. and C. B. Mullins. 2008. Selective oxidation of ethanol to acetaldehyde on gold. *Journal of the American Chemical Society* 130:16458–16459.

Goto, D., Y. Harada and Y. Furumoto, et al. 2010. Conversion of ethanol to propylene over HZSM-5 type zeolites containing alkaline earth metals. *Applied Catalysis A: General* 383:89–95.

Gucbilmez, Y., T. Dogu and S. Balci. 2006. Ethylene and acetaldehyde production by selective oxidation of ethanol using mesoporous V-MCM-41 catalysts. *Industrial and Engineering Chemistry Research* 45:3496–3502.

Higashino, H., K. Mita, H. Yoshikado, M. Iwata and J. Nakanishi. 2007. Exposure and risk assessment of 1, 3-butadiene in Japan. *Chemico-Biological Interactions* 166:52–62.

Hill, J., E. Nelson, D. Tilman, S. Polasky and D. Tiffany. 2006. Environmental, economic, and energetic costs and benefits of biodiesel and ethanol biofuels. *Proceedings of the National Academy of Sciences of the United States of America* 103:11206–11210.

Hill, J., S. Polasky and E. Nelson, et al. 2009. Climate change and health costs of air emissions from biofuels and gasoline. *Proceedings of the National Academy of Sciences of the United States of America* 106:2077–2082.

Hsieh, W. D., R. H. Chen, T. L. Wu and T. H. Lin. 2002. Engine performance and pollutant emission of an SI engine using ethanol-gasoline blended fuels. *Atmospheric Environment* 36:403–410.

Inui, K., T. Kurabayashi and S. Sato. 2002. Direct synthesis of ethyl acetate from ethanol carried out under pressure. *Journal of Catalysis* 212:207–215.

Jackson, N. B., C. M. Wang and Z. Luo, et al. 1991. Attachment of TiO_2 powders to hollow glass microbeads: Activity of the TiO_2-coated beads in the photoassisted oxidation of ethanol to acetaldehyde. *Journal of the Electrochemical Society* 138:3660–3664.

Katao, K. 2006. Nanomaterials may call for a reconsideration of the present Japanese chemical regulatory system. *Clean Technologies and Environmental Policy* 8:251–259.

Kerr, R. A. 2007. Global warming is changing the world. *Science* 316:188–190.

Koga, N. 2008. An energy balance under a conventional crop rotation system in northern Japan: Perspectives on fuel ethanol production from sugar beet. *Agriculture, Ecosystems & Environment* 125:101–110.

Konur, O. 2000. Creating enforceable civil rights for disabled students in higher education: An institutional theory perspective. *Disability & Society* 15:1041–1063.

Konur, O. 2002a. Access to nursing education by disabled students: Rights and duties of nursing programs. *Nurse Education Today* 22:364–374.

Konur, O. 2002b. Assessment of disabled students in higher education: Current public policy issues. *Assessment and Evaluation in Higher Education* 27:131–52.

Konur, O. 2002c. Access to employment by disabled people in the UK: Is the Disability Discrimination Act working? *International Journal of Discrimination and the Law* 5:247–279.

Konur, O. 2006a. Participation of children with dyslexia in compulsory education: Current public policy issues. *Dyslexia* 12:51–67.

Konur, O. 2006b. Teaching disabled students in higher education. *Teaching in Higher Education* 11:351–363.

Konur, O. 2007a. A judicial outcome analysis of the *Disability Discrimination Act*: A windfall for the employers? *Disability & Society* 22:187–204.

Konur, O. 2007b. Computer-assisted teaching and assessment of disabled students in higher education: The interface between academic standards and disability rights. *Journal of Computer Assisted Learning* 23:207–219.

Konur, O. 2011. The scientometric evaluation of the research on the algae and bio-energy. *Applied Energy* 88:3532–3540.

Konur, O. 2012a. Prof. Dr. Ayhan Demirbas' scientometric biography. *Energy Education Science and Technology Part A: Energy Science and Research* 28:727–738.

Konur, O. 2012b. The evaluation of the biogas research: A scientometric approach. *Energy Education Science and Technology Part A: Energy Science and Research* 29:1277–1292.

Konur, O. 2012c. The evaluation of the global energy and fuels research: A scientometric approach. *Energy Education Science and Technology Part A: Energy Science and Research* 30:613–628.

Konur, O. 2012d. The evaluation of the research on the biodiesel: A scientometric approach. *Energy Education Science and Technology Part A: Energy Science and Research* 28:1003–1014.

Konur, O. 2012e. The evaluation of the research on the bioethanol: A scientometric approach. *Energy Education Science and Technology Part A: Energy Science and Research* 28:1051–1064.

Konur, O. 2012f. The evaluation of the research on the biofuels: A scientometric approach. *Energy Education Science and Technology Part A: Energy Science and Research* 28:903–916.

Konur, O. 2012g. The evaluation of the research on the biohydrogen: A scientometric approach. *Energy Education Science and Technology Part A: Energy Science and Research* 29:323–338.

Konur, O. 2012h. The evaluation of the research on the microbial fuel cells: A scientometric approach. *Energy Education Science and Technology Part A: Energy Science and Research* 29:309–322.

Konur, O. 2012i. The scientometric evaluation of the research on the production of bioenergy from biomass. *Biomass and Bioenergy* 47:504–515.

Konur, O. 2015. Current state of research on algal bioethanol. In *Marine Bioenergy: Trends and Developments*, Ed. S. K. Kim and C. G. Lee, pp. 217–244. Boca Raton, FL: CRC Press.

Konur, O. 2016a. Scientometric overview in nanobiodrugs. In *Nanoarchitectonics for Smart Delivery and Drug Targeting*, Ed. A. M. Holban and A. M. Grumezescu, pp. 405–428. Amsterdam: Elsevier.

Konur, O. 2016b. Scientometric overview regarding nanoemulsions used in the food industry. In *Emulsions: Nanotechnology in the Agri-Food Industry*, Ed. A. M. Grumezescu, pp. 689–711. Amsterdam: Elsevier.

Konur, O. 2016c. Scientometric overview regarding the nanobiomaterials in antimicrobial therapy. In *Nanobiomaterials in Antimicrobial Therapy*, Ed. A. M. Grumezescu, pp. 511–535. Amsterdam: Elsevier.

Konur, O. 2016d. Scientometric overview regarding the nanobiomaterials in dentistry. In *Nanobiomaterials in Dentistry*, Ed. A. M. Grumezescu, pp. 425–453. Amsterdam: Elsevier.

Konur, O. 2016e. Scientometric overview regarding the surface chemistry of nanobiomaterials. In *Surface Chemistry of Nanobiomaterials*, Ed. A. M. Grumezescu, pp. 463–486. Amsterdam: Elsevier.

Konur, O. 2016f. The scientometric overview in cancer targeting. In *Nanoarchitectonics for Smart Delivery and Drug Targeting*, Ed. A. M. Holban and A. Grumezescu, pp. 871–895. Amsterdam: Elsevier.

Konur, O. 2017a. Recent citation classics in antimicrobial nanobiomaterials. In *Nanostructures for Antimicrobial Therapy*, Ed. A. Ficai and A. M. Grumezescu, pp. 669–685. Amsterdam: Elsevier.

Konur, O. 2017b. Scientometric overview in nanopesticides. In *New Pesticides and Soil Sensors*, Ed. A. M. Grumezescu, pp. 719–744. Amsterdam: Elsevier.

Konur, O. 2017c. Scientometric overview regarding oral cancer nanomedicine. In *Nanostructures for Oral Medicine*, Ed. E. Andronescu and A. M. Grumezescu, pp. 939–962. Amsterdam: Elsevier.

Konur, O. 2017d. Scientometric overview regarding water nanopurification. In *Water Purification*, Ed. A. M. Grumezescu, pp. 693–716. Amsterdam: Elsevier.

Konur, O. 2017e. Scientometric overview in food nanopreservation. In *Food Preservation*, Ed. A. M. Grumezescu, pp. 703–729. Amsterdam: Elsevier.

Konur, O. 2018a. Bioenergy and biofuels science and technology: Scientometric overview and citation classics. In *Bioenergy and Biofuels*, Ed. O. Konur, pp. 3–63. Boca Raton: CRC Press.

Konur, O., Ed. 2018b. *Bioenergy and Biofuels*. Boca Raton, FL: CRC Press.

Konur, O. 2019a. Cyanobacterial bioenergy and biofuels science and technology: A scientometric overview. In *Cyanobacteria: From Basic Science to Applications*, Ed. A. K. Mishra, D. N. Tiwari and A. N. Rai, pp. 419–442. Amsterdam: Elsevier.

Konur, O. 2019b. Nanotechnology applications in food: A scientometric overview. In *Nanoscience for Sustainable Agriculture*, Ed. R. N. Pudake, N. Chauhan and C. Kole, pp. 683–711. Cham: Springer.

Konur, O. 2020a. The scientometric analysis of the research on the bioethanol production from green macroalgae. In *Handbook of Algal Science, Technology and Medicine*, Ed. O. Konur, pp. 385–401. London: Academic Press.

Konur, O., Ed. 2020b. *Handbook of Algal Science, Technology and Medicine*. London: Academic Press.

Konur, O. 2021a. Nanotechnology applications in diesel fuels and the related research fields: A review of the research. In *Handbook of Biodiesel and Petrodiesel Fuels: Science, Technology, Health, and Environment. Volume 1. Biodiesel Fuels: Science, Technology, Health, and Environment*, Ed. O. Konur, pp. 89–110. Boca Raton, FL: CRC Press.

Konur, O. 2021b. Nanobiosensors in agriculture and foods: A scientometric review. In *Nanobiosensors in Agriculture and Food*, Ed. R. N. Pudake, pp. 365–384. Cham: Springer.

Konur, O., Ed. 2021c. *Handbook of Biodiesel and Petrodiesel Fuels: Science, Technology, Health, and Environment*. Boca Raton, FL: CRC Press.

Konur, O., Ed. 2021d. *Handbook of Biodiesel and Petrodiesel Fuels: Science, Technology, Health, and Environment. Volume 1. Biodiesel Fuels: Science, Technology, Health, and Environment.* Boca Raton, FL: CRC Press.

Konur, O., Ed. 2021e. *Handbook of Biodiesel and Petrodiesel Fuels: Science, Technology, Health, and Environment. Volume 2. Biodiesel Fuels based on the Edible and Nonedible Feedstocks, Wastes, and Algae: Science, Technology, Health, and Environment.* Boca Raton, FL: CRC Press.

Konur, O., Ed. 2021f. *Handbook of Biodiesel and Petrodiesel Fuels: Science, Technology, Health, and Environment. Volume 3. Petrodiesel Fuels: Science, Technology, Health, and Environment.* Boca Raton, FL: CRC Press.

Konur, O. and F. L. Matthews. 1989. Effect of the properties of the constituents on the fatigue performance of composites: A review. *Composites* 20:317–328.

Leydesdorff, L. 2000. Is the European Union becoming a single publication system? *Scientometrics* 47:265–280.

Leydesdorff, L. and C. Wagner. 2009. Is the United States losing ground in science? A global perspective on the world science system. *Scientometrics* 78:23–36.

Leydesdorff, L., C. S. Wagner and L. Bornmann. 2014. The European Union, China, and the United States in the top-1% and top-10% layers of most-frequently cited publications: Competition and collaborations. *Journal of Informetrics* 8:606–617.

Leydesdorff, L. and P. Zhou. 2005. Are the contributions of China and Korea upsetting the world system of science? *Scientometrics* 63:617–630.

Li, S. Z. and C. Chan-Halbrendt. 2009. Ethanol production in (the) People's Republic of China: Potential and technologies. *Applied Energy* 86:S162–S169.

Liu, C., J. Sun, C. Smith and Y. A. Wang. 2013. A study of $Zn_xZr_yO_z$ mixed oxides for direct conversion of ethanol to isobutene. *Applied Catalysis A: General* 467:91–97.

Ma, X., L. Sun and C. Song. 2002. A new approach to deep desulfurization of gasoline, diesel fuel and jet fuel by selective adsorption for ultra-clean fuels and for fuel cell applications. *Catalysis Today* 77:107–116.

Maity, S. K. 2015. Opportunities, recent trends and challenges of integrated biorefinery: Part I. *Renewable and Sustainable Energy Reviews* 43:1427–1445.

Makshina, E.V., W. Janssens, B. F. Sels and P. A. Jacobs. 2012. Catalytic study of the conversion of ethanol into 1,3-butadiene. *Catalysis Today* 198:338–344.

Morschbacker, A. 2009. Bio-ethanol based ethylene. *Polymer Reviews* 49:79–84.

Najafi, G., B. Ghobadian and T. Tavakoli, et al. 2009. Performance and exhaust emissions of a gasoline engine with ethanol blended gasoline fuels using artificial neural network. *Applied Energy* 86:630–639.

Ndou, A. S., N. Plint and N. J. Coville. 2003. Dimerisation of ethanol to butanol over solid-base catalysts. *Applied Catalysis A: General* 251:337–345.

Newman, P. W. G. and J. R. Kenworthy. 1989. Gasoline consumption and cities: A comparison of U.S. cities with a global survey. *Journal of the American Planning Association* 55:24–37.

Nikolau, B. J., M. A. D. Perera, L. Brachova and B. Shanks. 2008. Platform biochemicals for a biorenewable chemical industry. *Plant Journal* 54:536–545.

North, D. C. 1991. Institutions. *Journal of Economic Perspectives* 5:97–112.

Santacesaria, E., G. Carotenuto, R. Tesser and M. di Serio. 2012. Ethanol dehydrogenation to ethyl acetate by using copper and copper chromite catalysts. *Chemical Engineering Journal* 179:209–220.

Scheufele, D. A., E. A. Corley, T. J. Shih, K. E. Dalrymple and S. S. Ho. 2009. Religious beliefs and public attitudes toward nanotechnology in Europe and the United States. *Nature Nanotechnology* 4:91–94.

Shin, D., T. Kim, J. Choi and J. Kim. 2014. Author name disambiguation using a graph model with node splitting and merging based on bibliographic information. *Scientometrics* 100:15–50.

Sushkevich, V. L., I. I. Ivanova, V. V. Ordomsky and E. Taarning. 2014. Design of a metal-promoted oxide catalyst for the selective synthesis of butadiene from ethanol. *ChemSusChem* 7:2527–2536.

Takahara, I., M. Saito, M. Inaba and K. Murata. 2005. Dehydration of ethanol into ethylene over solid acid catalysts. *Catalysis Letters* 105:249–252.

Titov, M. and Y. Ziane. 2013. Price dynamics of propylene and ethylene in the United States. *Journal of Energy and Development* 39:207–217.

Tsuchida, T., S. Sakuma, T. Takeguchi and W. Ueda. 2006. Direct synthesis of n-butanol from ethanol over nonstoichiometric hydroxyapatite. *Industrial and Engineering Chemistry Research* 45:8634–8642.

Vadas, P. A., K. H. Barnett and D. J. Undersander 2008. Economics and energy of ethanol production from alfalfa, corn, and switchgrass in the Upper Midwest, USA. *Bioenergy Research* 1:44–55.

Zhang, K., Z. Pei and D. Wang. 2016. Organic solvent pretreatment of lignocellulosic biomass for biofuels and biochemicals: A review. *Bioresource Technology* 199:21–33.

Zhang, M. and Y. Yu. 2013. Dehydration of ethanol to ethylene. *Industrial and Engineering Chemistry Research* 52:9505–9514.

109 Bioethanol Fuel-based Biochemicals

Review

Ozcan Konur
(Formerly) Ankara Yildirim Beyazit University

109.1 INTRODUCTION

Crude oil-based gasoline fuels (Ma et al., 2002; Newman and Kenworthy, 1989) have been widely used in the transportation sector since the 1920s. However, there have been great public concerns over the adverse environmental and human impacts of these fuels (Hill et al., 2006, 2009). Hence, biomass-based bioethanol fuels (Hill et al., 2006; Konur, 2012, 2015, 2019a, 2020a) have increasingly been used in blending gasoline fuels (Hsieh et al., 2002; Najafi et al., 2009) and in fuel cells (Antolini, 2007, 2009).

In the meantime, research in nanomaterials and nanotechnology has intensified in recent years to become a major research field in scientific research, with over one and a half million published papers (Geim, 2009; Geim and Novoselov, 2007). In this context, large numbers of nanomaterials have been developed for nearly every research field. These materials offer an innovative way to increase efficiency in the production and utilization of bioethanol fuels and bioethanol fuel-based biochemical production, as in other scientific fields (Konur 2016a,b,c,d,e,f, 2017a,b,c,d,e, 2019b, 2021a,b).

Research in the field of chemicals has also intensified in recent years (Ames et al., 1990; Grandjean and Landrigan, 2006). Due to societal concerns on the sustainability of the environmental and human impacts of the chemicals, the biochemicals have become an alternative research field (Dodds and Gross, 2007; Nikolau et al., 2008). The primary focus of the research on biochemicals has been the optimization of the operating conditions and the development of catalysts to maximize biochemical production.

One way to produce biochemicals has been to produce them from bioethanol fuels (Angelici et al., 2013; Morschbacker, 2009; Zhang and Yu, 2013). The production of biochemicals from bioethanol fuels helps to reduce the production cost of bioethanol fuels (Aden and Foust, 2009; Hannon et al., 2020) in a biorefinery context (Fernando et al., 2006; Huang et al., 2008). There are four primary research fronts in this field: Bioethylenes (Morschbacker, 2009; Takahara et al., 2005; Zhang and Yu, 2013), biobutadienes (Angelici et al., 2013; Makshina et al., 2012; Sushkevich et al., 2014), biobutanols (Fu et al., 2017; Ndou et al., 2003; Tsuchida et al., 2006), and bioacetaldehydes (Gong and Mullins, 2008; Jackson et al., 1991; Santacesaria et al., 2012). The other research fronts are bioethyl acetates (Inui et al., 2002), biopropylenes (Goto et al., 2010), bioacetic acids (Christensen et al., 2006), biodiethyl ethers (Christiansen et al., 2013), and bioisobutenes (Liu et al., 2013).

However, it is essential to develop efficient incentive structures (North, 1991) for the primary stakeholders to enhance research in this field (Konur, 2000, 2002a,b,c, 2006a,b, 2007a,b).

Although there have been a number of review papers on the production of biochemicals from bioethanol fuels (Angelici et al., 2013; Morschbacker, 2009; Zhang and Yu, 2013), there has been no review of the 25 most-cited articles in this field.

DOI: 10.1201/9781003226574-145

This chapter presents a review of the 25 most-cited articles in the field of bioethanol-based production of biochemical production. Then, it discusses the key findings of these highly influential papers and comments on future research priorities in this field.

109.2 MATERIALS AND METHODS

The search for this study was carried out using the Scopus database (Burnham, 2006) in September 2021.

As the first step for the search of the relevant literature, keywords were selected using the 200 most-cited papers. The selected keyword list was optimized to obtain a representative sample of papers in this research field. This keyword list was provided in the appendix of Konur (2023) for future replication studies.

As the second step, a sample dataset was used for this study. The first 25 articles in the sample of 100 most-cited papers with at least 106 citations each were selected for the review study. Key findings from each paper were taken from the abstracts of these papers and discussed. Additionally, a number of brief conclusions were drawn and a number of relevant recommendations were made to enhance the future research landscape.

109.3 RESULTS

The brief information about the 25 most-cited papers with at least 123 citations each on the production of biobutanols from bioethanol fuels is given below (Table 109.1).

109.3.1 Bioethanol-based Biobutanols

A brief information about seven prolific studies with at least 123 citations each on the production of biobutanols from bioethanol fuels is given in Table 109.1. Furthermore, brief notes on the contents of these studies are also given.

Ndou et al. (2003) evaluated the conversion of bioethanol to 1-butanol in a paper with 166 citations. They used alkali earth metal oxides and modified magnesia (MgO) catalysts. They found that the MgO catalyst exhibited the highest reaction activity and 1-butanol selectivity among these catalysts, while the dimerization reaction did not proceed primarily through the aldol condensation reaction. They further noted that the C-H bond in the β-position in bioethanol was activated by the basic metal oxide and condensed with another molecule of bioethanol by dehydration to form 1-butanol.

TABLE 109.1
Bioethanol-based Biobutanols

No.	Papers	Catalysts	Catalyst Supports	Biochemicals	Process	Cits
1	Ndou et al. (2003)	Na	Earth metal oxides and modified MgO	1-Butanol	Dimerization	166
2	Fu et al. (2017)	Mn	Na	1-Butanol	Condensation	143
3	Ogo et al. (2011)	Ca and Sr	Hydroxyapatite	1-Butanol	Catalytic conversion	128
4	Ogo et al. (2012)	Sr	hydroxyapatite	1-Butanol	Catalytic conversion	106
5	Tsuchida et al. (2006)	Calcium phosphate	Na	n-Butanol	Catalytic conversion	165
6	Dowson et al. (2013)	Ru	Diphosphine	n-Butanol	Catalytic conversion	125
7	Carvalho et al. (2012)	Na	Mg-Al mixed oxides	n-Butanol	Catalytic conversion	123

Cits., the number of the citations received by each paper; Na, non available.

Tsuchida et al. (2006) evaluated the direct synthesis of n-butanol from bioethanol over hydroxyapatite (HPA) in a paper with 165 citations. They developed a highly active calcium phosphate compound. They found that this catalyst catalyzed the selective conversion of bioethanol to n-butanol in a single reaction at atmospheric pressure and low temperature, with a maximum selectivity of 76%, while higher alcohols were also formed and bioethanol was adsorbed and activated on HPA.

Fu et al. (2017) evaluated the production of 1-butanol from bioethanol in a paper with 143 citations. They explored the Guerbet-type condensation reaction of bioethanol. This process proceeded selectively in the presence of a well-defined manganese (Mn) pincer complex at the parts per million level. They observed that the developed reaction represented a sustainable synthesis of 1-butanol with excellent turnover number and turnover frequency. They further identified the essential role of the N–H moiety of the Mn catalysts and the major reaction intermediates related to the catalytic cycle.

Ogo et al. (2011) evaluated the synthesis of 1-butanol from bioethanol over strontium phosphate (Sr P) HPA catalysts in a paper with 128 citations. They noted that this catalyst exhibited the highest 1-butanol selectivity among the tested catalysts in the region of bioethanol conversions. The reaction mechanism of 1-butanol formation over this catalyst included the dehydrogenation of ethanol into acetaldehyde, the aldol condensation of acetaldehyde into crotonaldehyde, and the hydrogenations of crotonaldehyde, 2-buten-1-ol, and/or butyraldehyde into 1-butanol.

Dowson et al. (2013) evaluated the conversion of bioethanol to n-butanol in a paper with 125 citations. They noted that selectivity of over 94 % at good conversion with a ruthenium (Ru) diphosphine catalyst. They reasoned that the control over the uncontrolled acetaldehyde aldol condensation was critical for the high selectivity, and there was evidence for an on-metal condensation step.

Carvalho et al. (2012) evaluated the synthesis of n-butanol from bioethanol in a paper with 123 citations. They prepared Mg-Al mixed oxides by thermal decomposition of hydrotalcites employing two different Mg/Al ratios as well as MgO and alumina (Al_2O_3). They noted that Mg and aluminum (Al) mixed oxides promoted the synthesis of C_4 compounds from bioethanol. They reasoned that the adjacent acid and medium basic sites were needed in order to generate the intermediate compounds. As the higher the concentration of Mg was, the higher the hydrogenation capacity of the catalyst, they observed a greater selectivity to n-butanol.

Ogo et al. (2012) evaluated the 1-butanol synthesis from bioethanol over Sr P-HPA catalysts with various Sr/P molar ratios in a paper with 106 citations. They observed that these catalysts showed high 1-butanol selectivity in the gas-phase conversion of bioethanol, while these catalysts with higher Sr/P molar ratios showed higher catalytic activity and 1-butanol selectivity. The rate-determining step in bioethanol conversion was the dimerization process, while aldol condensation was mainly accelerated by base catalysis, and these catalysts with higher basic site density showed higher catalytic activity and 1-butanol selectivity.

109.3.2 Bioethanol-based Bioethylenes

The brief information about five prolific studies with at least 135 citations each on the production of bioethylenes from bioethanol fuels is given in Table 109.2. Furthermore, brief notes on the contents of these studies are also given.

Takahara et al. (2005) evaluated the dehydration of bioethanol into bioethylene over solid acid catalysts in a paper with 232 citations. They used zeolites and silica (SiO_2)-alumina at temperatures ranging 453–573 K under atmospheric pressure. Ethylene was produced via diethyl ether during the dehydration process. They observed that H-mordenites were the most active for dehydration. They correlated the catalyst activity with the number of strong Bronsted acid sites in the catalyst. Furthermore, H-mordenite was more stable with a SiO_2/Al_2O_3 ratio of 90 than with a SiO_2/Al_2O_3 ratio of 20.

Phillips and Datta (1997) evaluated the production of bioethylene from bioethanol on an HZSM-5 catalyst in a paper with 156 citations. They noted that the presence of water in the bioethanol feed

TABLE 109.2
Bioethanol-based Bioethylenes

No.	Papers	Catalysts	Biochemicals	Process	Cits
1	Takahara et al. (2005)	Zeolites	Ethylene	Dehydration	232
2	Phillips and Datta (1997)	HZSM-5	Ethylene	Catalytic conversion	156
3	Chen et al. (2007)	TiO_2/γ-Al_2O_3	Ethylene	Catalytic dehydration	153
4	Varisli et al. (2007)	Heteropolyacids	Ethylene and diethyl ether	Dehydration	147
5	Zhang et al. (2008)	γ-Al2O3, HZSM-5, SAPO-34, and NiAPSO-34	Ethylene	Catalytic dehydration	135

enhanced the steady-state activity and selectivity of these catalysts. They observed a sharp initial decline in catalyst activity within a few minutes on stream due to the formation of low-temperature coke from ethylene oligomerization. They reasoned that water in the bioethanol feed enhanced the steady-state catalytic activity and ethylene selectivity by moderating the acidity of the catalytic sites.

Chen et al. (2007) evaluated the catalytic dehydration of bioethanol to bioethylene over titania (TiO_2)/γ-Al_2O_3 catalysts in microchannel reactors in a paper with 153 citations. They noted that the catalysts doped with TiO_2 had a high ethanol conversion of 99.96% and an ethylene selectivity of 99.4%.

Varisli et al. (2007) evaluated the bioethylene and biodiethyl ether production by dehydration reaction of bioethanol over different heteropolyacid catalysts in a paper with 147 citations. They used tungstophosphoric acid (TPA), silicotungstic acid (STA), and molybdophosphoric acid (MPA). They obtained very high ethylene yields of over 0.75 at 250° C with TPA. At temperatures lower than 180° C the main product was diethyl ether. The presence of water vapor caused some decrease of catalyst activity. They noted that there were two parallel routes for the production of ethylene and diethyl ether, while the activity trend was STA> TPA> MPA.

Zhang et al. (2008) evaluated the catalytic dehydration of bioethanol to bioethylene in a paper with 135 citations. They compared the activity and stability of γ-Al_2O_3, H0–5, silicoaluminophosphate (SAPO-34) and Ni-substituted SAPO-34 (NiAPSO-34). They noted that the substitution of Ni and Al ions in the SAPO-34 framework led to an increase in weak and moderately strong acid strength and gave rise to weak acid sites. The conversion of bioethanol and selectivity to ethylene decreased in the order HZSM-5 > NiAPSO-34 > SAPO-34 > γ-Al_2O_3. They further noted that the stability NiAPSO-34 and SAPO-34 was better than that of the other two catalysts, while NiAPSO-34 was the suitable catalyst in the dehydration of bioethanol.

109.3.3 Bioethanol-based Acetaldehydes

A brief information about three prolific studies with at least 122 citations each on the production of bioacetaldeyhdes from bioethanol fuels is given in Table 109.3. Furthermore, brief notes on the contents of these studies are also given.

Liu and Hensen (2013) developed gold (Au) nanoparticles (NPs) on the $MgCuCr_2O_4$ spinel support for the aerobic oxidation of bioethanol to acetaldehyde in a paper with 284 citations. They observed that the conversion rate and yield were 100% and ~95%, respectively. They attributed this catalytic performance to the strong synergy between metallic Au NPs and surface Cu ions, while these Cu ions acted as sites for O_2 activation.

Wang et al. (1991) evaluated the photoassisted oxidation of bioethanol to acetaldehyde in a paper with 131 citations. They attached n-TiO_2 particles to hollow microbeads of 80–100 μm average

TABLE 109.3
Bioethanol-based Bioacetaldehydes

No.	Papers	Catalysts	Catalyst Supports	Process	Cits
1	Liu and Hensen (2013)	Au NPs	$MgCuCr_2O_4$	Aerobic oxidation	284
2	Wang et al. (1991)	Na	TiO_2 and SiO_2	Photoassisted oxidation	137
3	Gong and Mullins (2008)	Au	Na	Oxidation	122

diameter. These photoactive beads caused the oxidation of bioethanol molecules by dissolved dioxygen to acetaldehyde. The aluminosilicate glass beads produced 0.7 mol of acetaldehyde per Einstein. Acetaldehyde was produced at an efficiency of 0.5 mol/Einstein in the photoassisted oxidation of ethanol on TiO_2-activated borosilicate microbeads.

Gong and Mullins (2008) evaluated the oxidation of bioethanol to acetaldehyde in a paper with 122 citations. They employed Au(111) catalyst. They noted that bioethanol initially underwent O-H bond cleavage followed by selective β-C-H bond activation to form acetaldehyde and water on atomic oxygen precovered Au(111).

109.3.4 Bioethanol-based Acetic Acids

The brief information about three prolific studies with at least 161 citations each on the production of acetic acids from bioethanol fuels is given in Table 109.4. Furthermore, brief notes on the contents of these studies are also given.

Christensen et al. (2006) evaluated the production of acetic acid by aqueous-phase oxidation of bioethanol with air in the presence of a heterogeneous Au NP catalyst in a paper with 186 citations. They used this catalyst at temperatures of about 423 K and O_2 pressures of 0.6 MPa. They found that this reaction proceeded readily in aqueous acidic media with yields of up to 90% and CO_2 as the only major byproduct.

Jorgensen et al. (2007) evaluated the aerobic oxidation of aqueous bioethanol using heterogeneous Au catalysts to obtain acetic acid and ethyl acetate in a paper with 161 citations. They used $Au/MgAl_2O_4$ and Au/TiO_2. They noted that these catalysts exhibited similar performance in the reaction. By proper selection of the reaction conditions, they achieved a yield of 90%–95% of acetic acid at moderate temperatures and pressures. The rate-determining step was the dehydrogenation of ethanol to produce acetaldehyde. At low ethanol concentrations, the main product was acetic acid; at concentrations more than 60 wt %, it was ethyl acetate.

Tarnowski and Korzeniewski (1997) evaluated the electrochemical oxidation of bioethanol to acetic acid in a paper with 106 citations. They used Pt(111), Pt(557), and Pt(335) single crystal electrodes. They observed that the oxidation pathway leading to acetic acid showed a marked dependence on electrode surface structure, while acetic acid formation decreased as the surface step density increased. On the stepped surfaces, they reasoned that facile C-C bond cleavage and high surface poisoning accounted for the low acetic acid production. Ethanol oxidation to acetic acid was inhibited on Pt surfaces that contained high surface step densities.

109.3.5 Bioethanol-based Ethyl Acetates

The brief information about three prolific studies with at least 107 citations each on the production of ethyl acetates from bioethanol fuels is given in Table 109.5. Furthermore, brief notes on the contents of these studies are also given.

TABLE 109.4
Bioethanol-based Acetic Acids

No.	Papers	Catalysts	Catalyst Supports	Biochemicals	Process	Cits
1	Christensen et al. (2006)	Au NP	Na	Acetic acid	Aqueous-phase oxidation	186
2	Jorgensen et al. (2007)	Au	TiO_2 and $MgAl_2O_4$	Acetic acid and ethyl acetate	Aerobic oxidation	161
3	Tarnowski and Korzeniewski (1997)	Pt	Na	Acetic acid	Electrochemical oxidation	106

TABLE 109.5
Bioethanol-based Ethyl Acetates

No.	Papers	Catalysts	Catalyst Supports	Biochemicals	Process	Cits
1	Nielsen et al. (2012)	Ru	Na	Ethyl acetate	Acceptorless dehydrogenation	187
2	Jorgensen et al. (2007)	Au	TiO_2 and $MgAl_2O_4$	Acetic acid and ethyl acetate	Aerobic oxidation	161
3	Inui et al. (2002)	Cu-Zn-Zr-Al-O	Na	Ethyl acetate	Catalytic conversion	107

Nielsen et al. (2012) evaluated the synthesis of ethyl acetate through the acceptorless dehydrogenation of bioethanol in a paper with 186 citations. They observed that the reaction proceeded under mild reaction conditions in the presence of a Ru catalyst with the liberation of molecular hydrogen. They achieved high yields of ethyl acetate and excellent catalyst turnover numbers at low catalyst loading.

Inui et al. (2002) evaluated the direct synthesis of ethyl acetate from ethanol in a paper with 107 citations. They used Cu-Zn-Zr-Al-O catalyst under pressure conditions between 473 and 533 K. They noted that both the selectivity to ethyl acetate and the space-time yield of ethyl acetate increased with increasing reaction pressure, whereas bioethanol conversion decreased. During the process, ethanol was first dehydrogenated to acetaldehyde and was then coupled with another ethanol molecule to form hemiacetal, which was further dehydrogenated to ethyl acetate. The concentration of byproducts such as 1-butanol and butanone decreased with increasing reaction pressure. Since the equilibrium of the dehydrogenation of ethanol to acetaldehyde shifted to an ethanol-rich composition at high pressure, they reasoned that the decrease in the partial pressure of acetaldehyde explained the suppression of the byproducts formed through acetaldol.

109.3.6 Bioethanol-based Butadienes

The brief information about three prolific studies with at least 111 citations each on the production of bioacetaldeyhdes from bioethanol fuels is given in Table 109.6. Furthermore, brief notes on the contents of these studies are also given.

Makshina et al. (2012) evaluated the catalytic conversion of bioethanol into 1,3-butadiene (BD) in a paper with 144 citations. They used $MgO–SiO_2$ catalysts doped with transition metal oxides. They noted that the modification of the $MgO–SiO_2$ binary system using a consecutive impregnation step significantly increased the bioethanol conversion rate and BD yield. They obtained the

TABLE 109.6
Bioethanol-based Butadienes

No.	Papers	Catalysts	Catalyst Supports	Process	Cits
1	Makshina et al. (2012)	Na	MgO and SiO_2	Catalytic conversion	144
2	Sushkevich et al. (2014)	Ag, Cu, and Ni	MgO, ZrO_2, Nb_2O_5, TiO_2, and Al_2O_3	Catalytic conversion	117
3	Jones et al. (2011)	Bi- and trimetallic catalysts	SiO_2	Catalytic conversion	111

BD yield higher than 55 mol% at full bioethanol conversion for materials containing Cu and Ag modifiers. They obtained high BD productivity and high BD concentration in the product stream.

Sushkevich et al. (2014) evaluated the synthesis of 1,3-BD from bioethanol in a paper with 117 citations. They used metal-containing (M=Ag, Cu, and Ni) oxide catalysts (MO_x=MgO, ZrO_2, Nb_2O_5, TiO_2, and Al_2O_3) supported on silica. They noted that the key reaction steps of BD synthesis involved ethanol dehydrogenation, acetaldehyde condensation, and the reduction of crotonaldehyde with bioethanol into crotyl alcohol. They obtained the best catalytic performance over the Ag/ZrO_2/SiO_2 catalyst, which showed the highest selectivity toward BD (74 mol%).

Jones et al. (2011) evaluated the conversion of bioethanol to 1,3-BD in a paper with 111 citations. They used a variety of silica impregnated bi- and trimetallic catalysts. They noted that the highest selectivity was 67%, while there was a relationship between the pore diameter of the silica catalysts and the selectivity to 1,3-butadiene. When the same metals were impregnated on non-acidic supports the conversion was dramatically reduced.

109.3.7 Bioethanol-based Other Biochemicals

A brief information about three prolific studies with at least 122 citations each on the production of other biochemicals from bioethanol fuels is given in Table 109.3. Furthermore, brief notes on the contents of these studies are also given.

109.3.7.1 Bioethanol-based Biodiethyl Ethers

Varisli et al. (2007) evaluated the ethylene and diethyl ether production by the dehydration reaction of ethanol over different heteropolyacid catalysts in a paper with 147 citations. See Section 109.3.2 for the notes.

109.3.7.2 Bioethanol-based Biopropylenes

Song et al. (2010) evaluated the conversion of bioethanol to propylene over ZSM-5 zeolites in a paper with 113 citations. They used HZSM-5 (Si/Al_2=30, 80, and 280) and ZSM-5 (Si/Al_2=80) modified with various metals. They noted that HZSM-5 (Si/Al_2=80) afforded a high propylene yield, which indicated that a moderate surface acidity favored propylene production. Zr-modified ZSM-5(80) gave the highest yield (32%) of propylene at 773 K. Furthermore, the catalytic stability of the zeolite was improved by the modification of Zr. They reasoned that the surface acidity and the presence of metal ions played important roles on the production of propylene.

109.3.7.3 Bioethanol-based Bioisobutenes

Sun et al. (2011) evaluated the direct conversion of bioethanol to isobutene on nanosized $Zn_xZr_yO_z$ mixed oxides with balanced acid-base sites in a paper with 178 citations. They obtained a high-yield conversion. They added ZnO to ZrO_2 to selectively passivate Zr's strong Lewis acidic sites and weaken Bronsted acidic sites while simultaneously introducing basicity. As a result, the undesired

TABLE 109.7
Bioethanol-based Other Biochemicals

No.	Papers	Catalysts	Catalyst Supports	Biochemicals	Process	Cits
1	Varisli et al. (2007)	Na	Heteropolyacids	Ethylene and diethyl ether	Dehydration	147
2	Sun et al. (2011)	Na	$Zn_xZr_yO_z$	Isobutene	Direct conversion	178
3	Song et al. (2010)	Na	ZSM-5	Propylene	Catalytic conversion	113

reactions of bioethanol dehydration and acetone polymerization/coking were suppressed. Instead, a surface basic site-catalyzed ethanol dehydrogenation to acetaldehyde, acetaldehyde to acetone conversion via a complex pathway including aldol condensation/dehydrogenation, and a Bronsted acidic site-catalyzed acetone-to-isobutene reaction pathway dominated on this catalyst, leading to a highly selective process for direct conversion of bioethanol to isobutene (Table 109.7).

109.4 DISCUSSION

109.4.1 Introduction

Crude oil-based gasoline fuels have been widely used in the transportation sector since the 1920s. However, there have been great public concerns over the adverse environmental and human impacts of these fuels. Hence, biomass-based bioethanol fuels have increasingly been used in blending gasoline fuels and in fuel cells.

In the meantime, research in nanomaterials and nanotechnology has intensified in recent years to become a major research field in scientific research, with over one and a half million published papers. In this context, large numbers of nanomaterials have been developed nearly for every research field. These materials offer an innovative way to increase the efficiency in the production and utilization of bioethanol fuels and bioethanol fuel-based biochemical production, as in other scientific fields.

Research in the field of chemicals has also intensified in recent years. Due to societal concerns about the sustainability of the environmental and human impacts of the chemicals, the biochemicals have become an alternative research field. The primary focus of research on biochemicals has been the optimization of operating conditions and the development of catalysts to maximize the biochemical production.

Thus, research on the production of biochemicals from bioethanol fuels has primarily intensified in research fronts of bioethylenes, biobutadienes, biobutanols, and bioacetaldehydes. The other research fronts have been bioethyl acetates, biopropylenes, bioacetic acids, biodiethyl ethers, and bioisobutenes.

However, it is essential to develop efficient incentive structures for the primary stakeholders to enhance research in this field.

Although there has been a number of review papers on the production of biochemicals from bioethanol fuels, there has been no review of the 25 most-cited articles in this field.

This chapter presents a review of the 25 most-cited articles in the field of bioethanol-based production of biochemical production. Then, it discusses the key findings of these highly influential papers and comments on future research priorities in this field.

As the first step for the search of the relevant literature, the keywords were selected using the 200 most-cited papers. The selected keyword list was optimized to obtain a representative sample of papers in this research field. This keyword list was provided in the appendix of Konur (2023) for future replication studies.

TABLE 109.8
The Most Prolific Research Fronts in Bioethanol-based Biochemicals

No.	Research Fronts	Reviewed Papers (%)	Sample Papers (%)
1	Biobutanols	28	21
2	Bioethylenes	20	26
3	Bioacetaldehydes	12	15
4	Bioacetic acids	12	5
5	Biobutadienes	12	23
6	Bioethyl acetates	12	7
7	Others		
	Biodiethyl ethers	4	4
	Bioisobutenes	4	3
	Biopropylenes	4	6
	Sample	25	100

Reviewed papers, The sample of 25 reviewed papers; Sample papers, The sample of the 100 most-cited papers.

As the second step, a sample dataset was used for this study. The first 25 articles in the sample of 100 most-cited papers with at least 106 citations each were selected for the review study. Key findings from each paper were taken from the abstracts of these papers and discussed. Additionally, a number of brief conclusions were drawn, and a number of relevant recommendations were made to enhance the future research landscape (Table 109.8).

109.4.2 Bioethanol-based Biobutanols

A number of catalysts and catalyst supports were used in the production of biobutanols from bioethanol fuels: Mn (Fu et al., 2017), Ca (Ogo et al., 2011), Sr (Ogo et al., 2011, 2012), earth metal oxides and modified MgO (Ndou et al., 2003), HPA (Ogo et al., 2011, 2012), calcium phosphate (Tsuchida et al., 2006), and Mg-Al mixed oxides (Carvalho et al., 2012).

These studies used dimerization (Ndou et al., 2003), condensation (Fu et al., 2017), and catalytic conversion (Carvalho et al., 2012; Dowson et al., 2013; Ogo et al., 2012; Tsuchida et al., 2006) for the production of biobutanols.

These findings are thought-provoking in seeking ways to optimize bioethanol-based biobutanol production. The types and properties of the catalysts and the catalyst supports as well as the production processes emerge as the key variables impacting bioethanol-based biobutanol production.

109.4.3 Bioethanol-based Bioethylenes

A number of catalysts and catalyst supports were used in the production of bioethylenes from bioethanol fuels: HZSM-5 (Phillips and Datta, 1997), $TiO_2/\gamma\text{-}Al_2O_3$ (Chen et al., 2007), $\gamma\text{-}Al_2O_3$, HZSM-5, SAPO-34, and NiAPSO-34 (Zhang et al., 2008).

These studies used dehydration (Chen et al., 2007; Takahara et al., 2005; Varisli et al., 2007; Zhang et al., 2008) and catalytic conversion (Phillips and Datta, 1997) for the production of bioethylenes.

These findings are thought-provoking in seeking ways to optimize bioethanol-based bioethylene production. The types and properties of the catalysts and the catalyst supports as well as the production processes emerge as the key variables impacting bioethanol-based bioethylene production.

109.4.4 Bioethanol-based Bioacetaldehydes

A number of catalysts and catalyst supports were used in the production of acetaldehydes from bioethanol fuels: Au NPs and $MgCuCr_2O_4$ (Liu and Hensen, 2013), TiO_2 and SiO_2 (Wang et al., 1991), and Au (Gong and Mullins, 2008).

These studies used oxidation (Gong and Mullins, 2008), aerobic oxidation (Liu and Hensen, 2013), and photoassisted oxidation (Wang et al., 1991) for the production of acetaldehydes from bioethanol fuels.

These findings are thought-provoking in seeking ways to optimize bioethanol-based acetaldehyde production. The types and properties of the catalysts and the catalyst supports as well as the production processes emerge as the key variables impacting bioethanol-based acetaldehyde production.

109.4.5 Bioethanol-based Bioacetic Acids

A number of catalysts and catalyst supports were used in the production of bioacetic acids from bioethanol fuels: Au (Tarnowski and Korzeniewski, 1997), Ru (Christensen et al., 2006), Au, TiO2, and $MgAl_2O_4$ (Jorgensen et al., 2007).

These studies used electrochemical oxidation (Tarnowski and Korzeniewski, 1997), aqueous-phase oxidation (Christensen et al., 2006), and aerobic oxidation (Jorgensen et al., 2007) for the production of acetic acids.

These findings are thought-provoking in seeking ways to optimize bioethanol-based acetic acid production. The types and properties of the catalysts and the catalyst supports as well as the production processes emerge as the key variables impacting bioethanol-based acetic acid production.

109.4.6 Bioethanol-based Ethyl Bioacetates

A number of catalysts and catalyst supports were used in the production of bioethyl acetates from bioethanol fuels: Pt (Jorgensen et al., 2007), Ru (Nielsen et al., 2012), Cu-Zn-Zr-Al-O, (Inui et al., 2002), TiO_2, and $MgAl_2O_4$ (Jorgensen et al., 2007).

These studies used aerobic oxidation (Jorgensen et al., 2007), acceptorless dehydrogenation (Nielsen et al., 2012), and catalytic conversion (Inui et al., 2002) for the production of ethyl acetates.

These findings are thought-provoking in seeking ways to optimize bioethanol-based ethyl acetate production. The types and properties of the catalysts and the catalyst supports as well as the production processes emerge as the key variables impacting bioethanol-based ethyl acetate production.

109.4.7 Bioethanol-based Biobutadienes

A number of catalysts and catalyst supports were used in the production of biobutadienes from bioethanol fuels: MgO, SiO_2 (Makshina et al., 2012), Ag, Cu, Ni, MgO, ZrO_2, Nb_2O_5, TiO_2, Al_2O_3 (Sushkevich et al., 2014), and bi- and trimetallic catalysts, SiO_2 (Jones et al., 2011).

These studies used catalytic conversion for the production of butadienes (Jones et al., 2011; Makshina et al., 2012; Sushkevich et al., 2014).

These findings are thought-provoking in seeking ways to optimize bioethanol-based butadiene production. The types and properties of the catalysts and the catalyst supports as well as the production processes emerge as the key variables impacting bioethanol-based butadiene production.

109.4.8 Bioethanol-based Other Biochemicals

The other biochemicals were biodiethyl ethers (Varisli et al., 2007), bioisobutenes (Sun et al., 2011), and biopropylenes (Song et al., 2010).

A number of catalysts and catalyst supports were used in the production of other biochemicals from bioethanol fuels: heteropolyacids (Varisli et al., 2007), $Zn_xZr_yO_z$ (Sun et al., 2011), and ZSM-5 (Song et al., 2010).

These studies used dehydration (Varisli et al., 2007), direct conversion (Sun et al., 2011), and catalytic conversion (Song et al., 2010) for the production of other biochemicals.

These findings are thought-provoking in seeking ways to optimize bioethanol-based other biochemical production. The types and properties of the catalysts and the catalyst supports as well as the production processes emerge as the key variables impacting bioethanol-based other biochemical production.

109.4.9 The Overall Remarks

Research on the chemicals has focused on butanols (Groot et al., 1992), ethylenes (Flory, 1940), acetaldehydes (Jira, 2009), acetic acids (Yoneda et al., 2001), ethyl acetates (Kenig et al., 2001), butadienes (Liao et al., 2017), and other chemicals such as dietyl ethers, propylenes (Clerici et al., 1991), and isobutenes (Su et al., 2000).

A number of the prolific studies reviewed in the preceding sections hint that the primary catalysts used and researched for bioethanol-based biochemical production are Au (Chen and Goodman, 2006), Ca (Tsuchida et al., 2006), Mn (Aguilera et al., 2011), Sr (Tantirungrotechai, et al., 2013), Ru (Amendola et al., 2000), Pt (Mayrhofer et al., 2005), Ag (Bethke and Kung, 1997), Cu (Kuld et al., 2016), and Ni (Bengaard et al., 2002). As these studies suggest, these catalysts have a wide range of applications besides bioethanol-based biochemical production with the satisfactory results.

Similarly, the related catalyst supports are earth metal oxides (Mootabadi et al., 2010), modified MgO (Margitfalvi et al, 2002), hydroxyapatites (Ghantani et al., 2013), calcium phosphate (Stosic et al., 2012), Rh (Barnhart et al., 1994), Mg-Al mixed oxides (Chen et al., 2013), HZSM-5 (Lietz et al., 1995), TiO_2/γ-Al_2O_3 (Chen et al., 2007), γ-Al_2O_3 (Rynkowski et al., 1993), SAPO-34 (Liang et al., 1990), NiAPSO-34 (Kang and Inui, 1999), heteropolyacids (Atia et al., 2008), $MgCuCr_2O_4$ (Liu and Hensen, 2013), TiO_2 (Date and Haruta, 2001), SiO_2 (Brands et al., 1999), $MgAl_2O_4$ (Guo et al., 2007), ZrO_2 (Balducci et al., 1998), and Nb_2O_5 (Zhao et al., 2012). As these studies suggest, these catalyst supports have a wide range of applications besides bioethanol-based biochemical production with the satisfactory results.

The findings narrated in the previous sections are thought-provoking in seeking ways to optimize bioethanol-based biochemical production using a number of prolific catalysts and catalyst supports. The nanostructure, catalyst supports, catalysts and operating conditions of the biochemical production emerge as the key variables impacting bioethanol-based biochemical fuel production through a number of production processes.

One innovative way to enhance biochemical production has been the use of nanomaterials for both the catalysts themselves and the catalyst supports (Johanek et al., 2004). In parallel with the research on biochemicals, the research on nanomaterials and nanotechnology (Geim, 2009; Geim and Novoselov, 2007) has intensified in recent years with over one and a half million papers, enriching the material portfolio to be used for biochemical production. It is expected that this enriched portfolio of innovative nanomaterials would enhance bioethanol-based biochemical fuel production as both catalyst materials and catalyst supports in the future, benefiting from the superior properties of these nanomaterials.

In the end, these most-cited papers in this field hint that the efficiency of bioethanol-based biochemical fuel production could be optimized using the structure, processing, and property relationships of the nanomaterials and conventional materials used as both catalyst materials and catalyst supports (Formela et al., 2016; Konur, 2018, 2020b, 2021c,d,e,f; Konur and Matthews, 1989).

109.5 CONCLUSION AND FUTURE RESEARCH

The brief information about the key research fronts covered by the 25 most-cited papers with at least 106 citations each in the field of bioethanol fuel-based biochemical production is given under seven headings: the biobutanols, bioethylenes, bioacetaldehydes, bioacetic acids, bioethyl acetates, biobutadienes, and other biochemicals in Table 109.8. These findings hint that research on the biochemical production from bioethanol fuels has focused on a limited number of biochemicals.

It is notable that there is similarity of these research fronts in the sample of reviewed papers and the sample of the 100 most-cited papers (the first column data in Table 109.8) in this field. This finding suggests that the set of these 25 most prolific papers is a representative sample of the wider research on bioethanol fuel-based biochemical production. However, there is a slight overrepresentation and underrepresentation in the fields of bioacetic acids and biobutadienes, respectively.

The findings narrated in the previous sections are thought-provoking in seeking ways to optimize bioethanol-based biochemical production using a number of prolific catalysts and catalyst supports. The nanostructure, catalyst supports, catalysts, and operating conditions of the biochemical production emerge as the key variables impacting bioethanol-based biochemical production.

One innovative way to enhance biochemical production has been the use of nanomaterials for both the catalysts themselves and the catalyst supports. Research on nanomaterials and nanotechnology has intensified in recent years, enriching the material portfolio to be used for bioethanol-based biochemical production. It is expected that this enriched portfolio of innovative nanomaterials would enhance bioethanol-based biochemical production as both catalyst materials and catalyst supports in the future, benefiting from the superior properties of these nanomaterials as for the development of materials for catalysts and catalyst supports for biochemical production.

In the end, these most-cited papers in this field hint that the efficiency of bioethanol-based biochemical production could be optimized using the structure, processing, and property relationships of the nanomaterials and conventional materials used as both catalyst materials and catalyst supports.

It is recommended that such review studies should be performed for the other research fronts on both the production and utilization of bioethanol fuels, complementing the corresponding scientometric studies.

ACKNOWLEDGMENTS

The contribution of the highly cited researchers in the field of bioethanol fuel-based biochemical production has been gratefully acknowledged.

REFERENCES

Aden, A. and T. Foust. 2009. Technoeconomic analysis of the dilute sulfuric acid and enzymatic hydrolysis process for the conversion of corn stover to ethanol. *Cellulose* 16:535–545.

Aguilera, D. A., A. Perez, R. Molina and S. Moreno. 2011. Cu-Mn and Co-Mn catalysts synthesized from hydrotalcites and their use in the oxidation of VOCs. *Applied Catalysis B: Environmental* 104:144–150.

Amendola, S. C., S. L. Sharp-Goldman and M. S. Janjua, et al. 2000. An ultrasafe hydrogen generator: Aqueous, alkaline borohydride solutions and Ru catalyst. *Journal of Power Sources* 85:186–189.

Ames, B. N., M. Profet and L. S. Gold. 1990. Nature's chemicals and synthetic chemicals: Comparative toxicology. *Proceedings of the National Academy of Sciences* 87:7782–7786.

Angelici, C., B. M. Weckhuysen and P. C. A. Bruijnincx. 2013. Chemocatalytic conversion of ethanol into butadiene and other bulk chemicals. ChemSusChem 6:1595–1614.

Antolini, E. 2007. Catalysts for direct ethanol fuel cells. Journal of Power Sources 170:1–12.

Antolini, E. 2009. Carbon supports for low-temperature fuel cell catalysts. Applied Catalysis B: Environmental, 88:1–24.

Atia, H., U. Armbruster and A. Martin. 2008. Dehydration of glycerol in gas phase using heteropolyacid catalysts as active compounds. *Journal of Catalysis* 258:71–82.

Balducci, G., J. Kaspar, P. Fornasiero, M. Graziani and M. S. Islam. 1998. Surface and reduction energetics of the CeO_{2-ZrO2} catalysts. *Journal of Physical Chemistry B* 102:557–561.

Barnhart, R. W., X. Wang and P. Noheda, et al. 1994. Asymmetric catalysis. Asymmetric catalytic intramolecular hydroacylation of 4-pentenals using chiral rhodium diphosphine catalysts. *Journal of the American Chemical Society* 116:1821–1830.

Bengaard, H. S., J. K. Norskov and J. Sehested, et al. 2002. Steam reforming and graphite formation on Ni catalysts. *Journal of Catalysis* 209:365–384.

Bethke, K. A. and H. H. Kung. 1997. Supported Ag catalysts for the lean reduction of NO with C_3H_6. *Journal of Catalysis* 172:93–102.

Brands, D. S., E. K. Poels and A. Bliek. 1999. Ester hydrogenolysis over promoted Cu/SiO_2 catalysts. *Applied Catalysis A: General* 184:279–289.

Burnham, J. F. 2006. Scopus database: A review. *Biomedical Digital Libraries* 3:1–8.

Carvalho, D. L., R. R. de Avillez, M. T. Rodrigues, L. E. P. Borges and L. G. Appel. 2012. Mg and Al mixed oxides and the synthesis of n-butanol from ethanol. *Applied Catalysis A: General* 415-416:96–100.

Chen, F., G. Wang, W. Li and F. Yang. 2013. Glycolysis of poly (ethylene terephthalate) over Mg-Al mixed oxides catalysts derived from hydrotalcites. *Industrial & Engineering Chemistry Research* 52:565–571.

Chen, G., S. Li, F. Jiao and Q. Yuan. 2007. Catalytic dehydration of bioethanol to ethylene over TiO_2/γ-Al_2O_3 catalysts in microchannel reactors. *Catalysis Today* 125:111–119.

Chen, M. S. and D. W. Goodman. 2006. Structure-activity relationships in supported Au catalysts. *Catalysis Today* 111:22–33.

Chen, X. and S. S. Mao. 2007. Titanium dioxide nanomaterials: Synthesis, properties, modifications and applications. *Chemical Reviews* 107:2891–2959.

Christensen, C. H., B. Jorgensen and J. Rass-Hansen, et al. 2006. Formation of acetic acid by aqueous-phase oxidation of ethanol with air in the presence of a heterogeneous gold catalyst. *Angewandte Chemie International Edition* 45:4648–4651.

Christiansen, M. A., G. Mpourmpakis and D. G. Vlachos. 2013. Density functional theory-computed mechanisms of ethylene and diethyl ether formation from ethanol on γ-Al2O3(100). ACS Catalysis 3:1965–1975.

Clerici, M. G., G. Bellussi and U. Romano. 1991. Synthesis of propylene oxide from propylene and hydrogen peroxide catalyzed by titanium silicalite. *Journal of Catalysis* 129:159–167.

Date, M. and M. Haruta. 2001. Moisture effect on CO oxidation over Au/TiO_2 catalyst. *Journal of Catalysis* 201:221–224.

Dodds, D. R. and R. A. Gross. 2007. Chemicals from biomass. *Science* 318:1250–1251.

Dowson, G. R. M., M. F. Haddow, J. Lee, R. L. Wingad and D. F. Wass. 2013. Catalytic conversion of ethanol into an advanced biofuel: Unprecedented selectivity for n-butanol. *Angewandte Chemie International Edition* 52:9005–9008.

Fernando, S., S. Adhikari, C. Chandrapal and N. Murali. 2006. Biorefineries: Current status, challenges, and future direction. *Energy & Fuels* 20:1727–1737.

Flory, P. J. 1940. Molecular size distribution in ethylene oxide polymers. *Journal of the American Chemical Society* 62:1561–1565.

Formela, K., A. Hejna, L. Piszczyk, M. R. Saeb and X. Colom. 2016. Processing and structure-property relationships of natural rubber/wheat bran biocomposites. *Cellulose* 23:3157–3175.

Fu, S., Z. Shao, Y. Wang and Q. Liu. 2017. Manganese-catalyzed upgrading of ethanol into 1-butanol. *Journal of the American Chemical Society* 139:11941–11948.

Geim, A. K. 2009. Graphene: Status and prospects. *Science* 324:1530–1534.

Geim, A. K. and K. S. Novoselov. 2007. The rise of graphene. *Nature Materials* 6:183–191.

Ghantani, V. C., S. T. Lomate, M. K. Dongare and S. B. Umbarkar. 2013. Catalytic dehydration of lactic acid to acrylic acid using calcium hydroxyapatite catalysts. *Green Chemistry* 15:1211–1217.

Gong, J. and C. B. Mullins. 2008. Selective oxidation of ethanol to acetaldehyde on gold. *Journal of the American Chemical Society* 130:16458–16459.

Goto, D., Y. Harada and Y. Furumoto, et al. 2010. Conversion of ethanol to propylene over HZSM-5 type zeolites containing alkaline earth metals. Applied Catalysis A: General 383:89–95.

Grandjean, P. and P. J. Landrigan. 2006. Developmental neurotoxicity of industrial chemicals. *Lancet* 368:2167–2178.

Groot, W. J., R. G. J. M. van der Lans and K. C. A. Luyben. 1992. Technologies for butanol recovery integrated with fermentations. *Process Biochemistry* 27:61–75.

Guo, J., H. Lou and X. Zheng. 2007. The deposition of coke from methane on a $Ni/MgAl_2O_4$ catalyst. *Carbon* 45:1314–1321.

Hannon, J. R., L. R. Lynd and O. Andrade, et al. 2020. Technoeconomic and life-cycle analysis of single-step catalytic conversion of wet ethanol into fungible fuel blendstocks. *Proceedings of the National Academy of Sciences* 117:12576–12583.

Hill, J., E. Nelson, D. Tilman, S. Polasky and D. Tiffany. 2006. Environmental, economic, and energetic costs and benefits of biodiesel and ethanol biofuels. *Proceedings of the National Academy of Sciences of the United States of America* 103:11206–11210.

Hill, J., S. Polasky and E. Nelson, et al. 2009. Climate change and health costs of air emissions from biofuels and gasoline. *Proceedings of the National Academy of Sciences of the United States of America* 106:2077–2082.

Hsieh, W. D., R. H. Chen, T. L. Wu and T. H. Lin. 2002. Engine performance and pollutant emission of an SI engine using ethanol-gasoline blended fuels. *Atmospheric Environment* 36:403–410.

Huang, H. J., S. Ramaswamy, U. W. Tschirner and B. V. Ramarao. 2008. A review of separation technologies in current and future biorefineries. *Separation and Purification Technology* 62:1–21.

Inui, K., T. Kurabayashi and S. Sato. 2002. Direct synthesis of ethyl acetate from ethanol carried out under pressure. *Journal of Catalysis* 212:207–215.

Jackson, N. B., C. M. Wang and Z. Luo, et al. 1991. Attachment of TiO2 powders to hollow glass microbeads: Activity of the TiO2-coated beads in the photoassisted oxidation of ethanol to acetaldehyde. Journal of the Electrochemical Society 138:3660–3664.

Jira, R. 2009. Acetaldehyde from ethylene-A retrospective on the discovery of the Wacker process. *Angewandte Chemie International Edition* 48:9034–9037.

Johanek, V., M. Laurin and A. W. Grant, et al. 2004. Fluctuations and bistabilities on catalyst nanoparticles. *Science* 304:1639–1644.

Jones, M. D., C. G. Keir and C. D. Iulio, et al. 2011. Investigations into the conversion of ethanol into 1,3-butadiene. *Catalysis Science and Technology* 1:267–272.

Jorgensen, B., S. E. Christiansen, M. L. D. Thomsen and C. H. Christensen. 2007. Aerobic oxidation of aqueous ethanol using heterogeneous gold catalysts: Efficient routes to acetic acid and ethyl acetate. *Journal of Catalysis* 251:332–337.

Kang, M. and T. Inui. 1999. Dynamic reaction characteristics affected by water molecules during the methanol to olefin conversion on NiAPSO-34 catalysts. *Journal of Molecular Catalysis A: Chemical* 140:55–63.

Kenig, E. Y., H. A. Bader and H. A. Gorak, et al. 2001. Investigation of ethyl acetate reactive distillation process. *Chemical Engineering Science* 56:6185–6193.

Konur, O. 2000. Creating enforceable civil rights for disabled students in higher education: An institutional theory perspective. *Disability & Society* 15:1041–1063.

Konur, O. 2002a. Access to nursing education by disabled students: Rights and duties of nursing programs. *Nurse Education Today* 22:364–374.

Konur, O. 2002b. Assessment of disabled students in higher education: Current public policy issues. *Assessment and Evaluation in Higher Education* 27:131–152.

Konur, O. 2002c. Access to employment by disabled people in the UK: Is the Disability Discrimination Act working? *International Journal of Discrimination and the Law* 5:247–279.

Konur, O. 2006a. Participation of children with dyslexia in compulsory education: Current public policy issues. *Dyslexia* 12:51–67.

Konur, O. 2006b. Teaching disabled students in higher education. *Teaching in Higher Education* 11:351–363.

Konur, O. 2007a. A judicial outcome analysis of the *Disability Discrimination Act*: A windfall for the employers? *Disability & Society* 22:187–204.

Konur, O. 2007b. Computer-assisted teaching and assessment of disabled students in higher education: The interface between academic standards and disability rights. *Journal of Computer Assisted Learning* 23:207–219.

Konur, O. 2012. The evaluation of the research on the bioethanol: A scientometric approach. *Energy Education Science and Technology Part A: Energy Science and Research* 28:1051–1064.

Konur, O. 2015. Current state of research on algal bioethanol. In *Marine Bioenergy: Trends and Developments*, Ed. S. K. Kim and C. G. Lee, pp. 217–244. Boca Raton, FL: CRC Press.

Konur, O. 2016a. Scientometric overview in nanobiodrugs. In *Nanoarchitectonics for Smart Delivery and Drug Targeting*, Ed. A. M. Holban and A. M. Grumezescu, pp. 405–428. Amsterdam: Elsevier.

Konur, O. 2016b. Scientometric overview regarding nanoemulsions used in the food industry. In *Emulsions: Nanotechnology in the Agri-Food Industry*, Ed. A. M. Grumezescu, pp. 689–711. Amsterdam: Elsevier.

Konur, O. 2016c. Scientometric overview regarding the nanobiomaterials in antimicrobial therapy. In *Nanobiomaterials in Antimicrobial Therapy*, Ed. A. M. Grumezescu, pp. 511–535. Amsterdam: Elsevier.

Konur, O. 2016d. Scientometric overview regarding the nanobiomaterials in dentistry. In *Nanobiomaterials in Dentistry*, Ed. A. M. Grumezescu, pp. 425–453. Amsterdam: Elsevier.

Konur, O. 2016e. Scientometric overview regarding the surface chemistry of nanobiomaterials. In *Surface Chemistry of Nanobiomaterials*, Ed. A. M. Grumezescu, pp. 463–486. Amsterdam: Elsevier.

Konur, O. 2016f. The scientometric overview in cancer targeting. In *Nanoarchitectonics for Smart Delivery and Drug Targeting*, Ed. A. M. Holban and A. Grumezescu, pp. 871–895. Amsterdam; Elsevier.

Konur, O. 2017a. Recent citation classics in antimicrobial nanobiomaterials. In *Nanostructures for Antimicrobial Therapy*, Ed. A. Ficai and A. M. Grumezescu, pp. 669–685. Amsterdam: Elsevier.

Konur, O. 2017b. Scientometric overview in nanopesticides. In *New Pesticides and Soil Sensors*, Ed. A. M. Grumezescu, pp. 719–744. Amsterdam: Elsevier.

Konur, O. 2017c. Scientometric overview regarding oral cancer nanomedicine. In *Nanostructures for Oral Medicine*, Ed. E. Andronescu, A. M. Grumezescu, pp. 939–962. Amsterdam: Elsevier.

Konur, O. 2017d. Scientometric overview regarding water nanopurification. In *Water Purification*, Ed. A. M. Grumezescu, pp. 693–716. Amsterdam: Elsevier.

Konur, O. 2017e. Scientometric overview in food nanopreservation. In *Food Preservation*, Ed. A. M. Grumezescu, pp. 703–729. Amsterdam: Elsevier.

Konur, O., Ed. 2018. *Bioenergy and Biofuels*. Boca Raton, FL: CRC Press.

Konur, O. 2019a. Cyanobacterial bioenergy and biofuels science and technology: A scientometric overview. In *Cyanobacteria: From Basic Science to Applications*, Ed. A. K. Mishra, D. N. Tiwari and A. N. Rai, pp. 419–442. Amsterdam: Elsevier.

Konur, O. 2019b. Nanotechnology applications in food: A scientometric overview. In *Nanoscience for Sustainable Agriculture*, Ed. R. N. Pudake, N. Chauhan and C. Kole, pp. 683–711. Cham: Springer.

Konur, O. 2020a. The scientometric analysis of the research on the bioethanol production from green macroalgae. In *Handbook of Algal Science, Technology and Medicine*, Ed. O. Konur, pp. 385–401. London: Academic Press.

Konur, O., Ed. 2020b. *Handbook of Algal Science, Technology and Medicine*. London: Academic Press.

Konur, O. 2021a. Nanotechnology applications in diesel fuels and the related research fields: A review of the research. In *Handbook of Biodiesel and Petrodiesel Fuels: Science, Technology, Health, and Environment. Volume 1. Biodiesel Fuels: Science, Technology, Health, and Environment*, Ed. O. Konur, pp. 89–110. Boca Raton, FL: CRC Press.

Konur, O. 2021b. Nanobiosensors in agriculture and foods: A scientometric review. In *Nanobiosensors in Agriculture and Food*, Ed. R. N. Pudake, pp. 365–384. Cham: Springer.

Konur, O., Ed. 2021c. *Handbook of Biodiesel and Petrodiesel Fuels: Science, Technology, Health, and Environment*. Boca Raton, FL: CRC Press.

Konur, O., Ed. 2021d. *Handbook of Biodiesel and Petrodiesel Fuels: Science, Technology, Health, and Environment. Volume 1. Biodiesel Fuels: Science, Technology, Health, and Environment*. Boca Raton, FL: CRC Press.

Konur, O., Ed. 2021e. *Handbook of Biodiesel and Petrodiesel Fuels: Science, Technology, Health, and Environment. Volume 2. Biodiesel Fuels based on the Edible and Nonedible Feedstocks, Wastes, and Algae: Science, Technology, Health, and Environment*. Boca Raton, FL: CRC Press.

Konur, O., Ed. 2021f. *Handbook of Biodiesel and Petrodiesel Fuels: Science, Technology, Health, and Environment. Volume 3. Petrodiesel Fuels: Science, Technology, Health, and Environment*. Boca Raton, FL: CRC Press.

Konur, O. 2023. Bioethanol fuel-based biochemicals: Scientometric study. In *Evaluation and Utilization of Bioethanol Fuels. II.: Biohydrogen Fuels, Fuel Cells, Biochemicals, and Country Experiences. Handbook of Bioethanol Fuels Volume 6*, Ed. O. Konur, pp. 317–337. Boca Raton, FL: CRC Press.

Konur, O. and F. L. Matthews. 1989. Effect of the properties of the constituents on the fatigue performance of composites: A review. *Composites* 20:317–328.

Kuld, S., M. Thorhauge and H. Falsig, et al. 2016. Quantifying the promotion of Cu catalysts by ZnO for methanol synthesis. *Science* 352:969–974.

Liang, J., H. Li and S. Zhao, et al. 1990. Characteristics and performance of SAPO-34 catalyst for methanol-to-olefin conversion. *Applied Catalysis* 64:31–40.

Liao, P. Q., N. Y. Huang, W. X. Zhang, J.P. Zhang and X. M. Chen. 2017. Controlling guest conformation for efficient purification of butadiene. *Science* 356:1193–1196.

Lietz, G., K. H. Schnabel and C. Peuker, et al. 1994. Modifications of H-ZSM-5 catalysts by NaOH treatment. *Journal of Catalysis* 148:562–568.

Liu, P. and E. J. M. Hensen. 2013. Highly efficient and robust Au/$MgCuCr_2O_4$ catalyst for gas-phase oxidation of ethanol to acetaldehyde. *Journal of the American Chemical Society* 135:14032–14035.

Liu, C., J. Sun, C. Smith and Y. A. Wang. 2013. A study of $Zn_xZr_yO_z$ mixed oxides for direct conversion of ethanol to isobutene. Applied Catalysis A: General 467:91–97.

Ma, X., L. Sun and C. Song. 2002. A new approach to deep desulfurization of gasoline, diesel fuel and jet fuel by selective adsorption for ultra-clean fuels and for fuel cell applications. *Catalysis Today* 77:107–116.

Makshina, E. V., W. Janssens, B. F. Sels and P. A. Jacobs. 2012. Catalytic study of the conversion of ethanol into 1,3-butadiene. *Catalysis Today* 198:338–344.

Margitfalvi, J. L., A. Fasi and M. Hegedus, et al. 2002. Au/MgO catalysts modified with ascorbic acid for low temperature CO oxidation. *Catalysis Today* 72:157–169.

Mayrhofer, K. J. J., B. B. Blizanac and M. Arenz, et al. 2005. The impact of geometric and surface electronic properties of Pt-catalysts on the particle size effect in electrocatalysis. *Journal of Physical Chemistry B* 109:14433–14440.

Mootabadi, H., B. Salamatinia, S. Bhatia and A. Abdullah. 2010. Ultrasonic-assisted biodiesel production process from palm oil using alkaline earth metal oxides as the heterogeneous catalysts. *Fuel* 89:1818–1825.

Morschbacker, A. 2009. Bio-ethanol based ethylene. *Polymer Reviews* 49:79–84.

Najafi, G., B. Ghobadian and T. Tavakoli, et al. 2009. Performance and exhaust emissions of a gasoline engine with ethanol blended gasoline fuels using artificial neural network. *Applied Energy* 86:630–639.

Ndou, A. S., N. Plint and N. J. Coville. 2003. Dimerisation of ethanol to butanol over solid-base catalysts. *Applied Catalysis A: General* 251:337–345.

Newman, P. W. G. and J. R. Kenworthy. 1989. Gasoline consumption and cities: A comparison of U.S. cities with a global survey. *Journal of the American Planning Association* 55:24–37.

Nielsen, M., H. Junge, A. Kammer and M. Beller. 2012. Towards a green process for bulk-scale synthesis of ethyl acetate: Efficient acceptorless dehydrogenation of ethanol. *Angewandte Chemie International Edition* 51:5711–5713.

Nikolau, B. J., M. A. D. Perera, L. Brachova and B. Shanks. 2008. Platform biochemicals for a biorenewable chemical industry. *Plant Journal* 54:536–545.

North, D. C. 1991. Institutions. *Journal of Economic Perspectives* 5:97–112.

Ogo, S., A. Onda and Y. Iwasa, et al. 2012. 1-Butanol synthesis from ethanol over strontium phosphate hydroxyapatite catalysts with various Sr/P ratios. *Journal of Catalysis* 296:24–30.

Ogo, S., A. Onda and K. Yanagisawa. 2011. Selective synthesis of 1-butanol from ethanol over strontium phosphate hydroxyapatite catalysts. *Applied Catalysis A: General* 402:188–195.

Phillips, C. B. and R. Datta. 1997. Production of ethylene from hydrous ethanol on H-ZSM-5 under mild conditions. *Industrial and Engineering Chemistry Research* 36:4466–4475.

Rynkowski, J. M., T. Paryjczak and M. Lenik. 1993. On the nature of oxidic nickel phases in $NiO/\gamma\text{-}Al_2O_3$ catalysts. *Applied Catalysis A: General* 106:73–82.

Santacesaria, E., G. Carotenuto, R. Tesser and M. di Serio. 2012. Ethanol dehydrogenation to ethyl acetate by using copper and copper chromite catalysts. Chemical Engineering Journal 179:209–220.

Song, Z., A. Takahashi, I. Nakamura and T. Fujitani. 2010. Phosphorus-modified ZSM-5 for conversion of ethanol to propylene. *Applied Catalysis A: General* 384:201–205.

Stosic, D., S. Bennici and S. Sirotin, et al. 2012. Glycerol dehydration over calcium phosphate catalysts: Effect of acidic-basic features on catalytic performance. *Applied Catalysis A: General* 447:124–134.

Su, C., J. Li, D. He, Z. Cheng and Q. Zhu. 2000. Synthesis of isobutene from synthesis gas over nanosize zirconia catalysts. *Applied Catalysis A: General* 202:81–89.

Sun, J., K. Zhu and F. Gao, et al. 2011. Direct conversion of bio-ethanol to isobutene on nanosized $Zn_xZr_yO_z$ mixed oxides with balanced acid-base sites. *Journal of the American Chemical Society* 133:11096–11099.

Sushkevich, V. L., I. I. Ivanova, V. V. Ordomsky and E. Taarning. 2014. Design of a metal-promoted oxide catalyst for the selective synthesis of butadiene from ethanol. *ChemSusChem* 7:2527–2536.

Takahara, I., M. Saito, M. Inaba and K. Murata. 2005. Dehydration of ethanol into ethylene over solid acid catalysts. *Catalysis Letters* 105:249–252.

Tantirungrotechai, J., S. Thepwatee and B. Yoosuk. 2013. Biodiesel synthesis over Sr/MgO solid base catalyst. *Fuel* 106:279–284.

Tarnowski, D. J. and C. Korzeniewski. 1997. Effects of surface step density on the electrochemical oxidation of ethanol to acetic acid. *Journal of Physical Chemistry B* 101:253–258.

Tsuchida, T., S. Sakuma, T. Takeguchi and W. Ueda. 2006. Direct synthesis of n-butanol from ethanol over nonstoichiometric hydroxyapatite. *Industrial and Engineering Chemistry Research* 45:8634–8642.

Varisli, D., T. Dogu and G. Dogu. 2007. Ethylene and diethyl-ether production by dehydration reaction of ethanol over different heteropolyacid catalysts. *Chemical Engineering Science* 62:5349–5352.

Wang, C. M., Z. Luo and J. Schwitzgebel, et al. 1991. Attachment of TiO_2 powders to hollow glass microbeads: Activity of the TiO_2-coated beads in the photoassisted oxidation of ethanol to acetaldehyde. *Journal of the Electrochemical Society* 138:3660–3664.

Yoneda, N., S. Kusano, M. Yasui, P. Pujado and S. Wilcher. 2001. Recent advances in processes and catalysts for the production of acetic acid. *Applied Catalysis A: General* 221:253–265.

Zhang, M. and Y. Yu. 2013. Dehydration of ethanol to ethylene. *Industrial and Engineering Chemistry Research* 52:9505–9514.

Zhang, X., R. Wang, X. Yang and F. Zhang. 2008. Comparison of four catalysts in the catalytic dehydration of ethanol to ethylene. *Microporous and Mesoporous Materials* 116:210–215.

Zhao, Y., X. Zhou, L. Ye and S. C. E. Tsang. 2012. Nanostructured Nb_2O_5 catalysts. *Nano Reviews* 3:17631.

110 An Overview of Bioethanol Conversion to Hydrocarbons

Vannessa Caballero, Anthony William Savoy, Junming Sun, and Yong Wang
Washington State University
Institute for Integrated Catalysis

110.1 INTRODUCTION: OVERVIEW OF BIOETHANOL PRODUCTION AND ITS APPLICATIONS

The growing demand for energy and resources coupled with growing emission restrictions prompts the need to explore alternative sources in an effort to mitigate global dependence on fossil feedstocks. It is well known that biomass is vastly abundant material, and its renewable nature enables a CO_2 neutral pathway in industrial chemical processes. Production of ethanol from biomass and using it as a platform molecule for value-added chemicals is a promising alternative approach to replace fossil feedstocks.

Direct fermentation of biomass to ethanol from readily available biomass feedstocks, such as sugarcane and starch, is the most common route for ethanol production (Morschbacker 2009). Global ethanol production in 2020 was around 26 billion gallons (RFA, 2022), and 90% of the ethanol on the market is from biomass sources (Sun and Wang, 2014). However, the utilization of edible plants as a bioethanol feedstock has led to controversy about the competitiveness of food vs fuels production and its impact on society. Therefore, bioethanol production from non-edible biomass, such as lignocelluloses, hemicelluloses, and cellulose, has attracted much attention in recent years (Alvira et al., 2010; Gray et al., 2006; Groom et al., 2008; Hahn-Hagerdal et al., 2006).

Environmental regulations have mandated the blending of gasoline with bioethanol to supplement refined fossil fuels used in the transportation industry. In the USA, 98% of gasoline contains ethanol, usually in a mixture of 10:90 ethanol to gasoline (i.e., E10) (Dagle et al., 2020a). Additionally, higher blending ratios such as E15 for newer light-duty vehicles and even E85 mixtures for flexible fuel vehicles are also available in the market (Dagle et al., 2020a, USDA, 2022). However, these mixtures are not extensively used due to their limited infrastructure in addition to concerns about their fuel economy and possible side effects on conventional combustion engines.

In addition to fuel-based products, ethanol is capable of acting as a precursor molecule to a diverse array of industrially relevant chemicals (Angelici et al., 2013; Dagle et al., 2020a). A previous review by Dagle et al. (2020a) summarized the cost-to-profit ratio for a multitude of key ethanol-derived chemicals based on stoichiometric requirements, a large amount of which were predicted to be potentially profitable at industrial scale. Consequently, the vast availability of ethanol and its growing ubiquity through industry makes it a promising potential platform molecule to produce several high-value compounds for chemical industry.

The catalytic conversion of ethanol to produce hydrocarbons, such as 1,3-butadiene, isobutene, and other bulk chemicals, has been widely studied in the recent years (Angelici et al., 2013; Wang and Liu, 2012).

This chapter will discuss the details behind the reaction mechanism and active site requirements for ethanol conversion to value-added chemicals, particularly focusing on hydrocarbons such as

DOI: 10.1201/9781003226574-146

ethylene, C_3–C_4 olefins, and fuels. The presented knowledge implies the versatility of ethanol as a readily available feedstock in chemical industries and illustrates the complexities behind its catalytic conversion.

110.2 STRATEGIES FOR BIOETHANOL CONVERSION TO HYDROCARBONS

110.2.1 Ethanol to Ethylene via Dehydration

As the simplest olefin, ethylene is a platform molecule with a spectrum of applications, especially in polymerization processes to produce polyethylene (PE), polyethylene terephthalate (PET), polyvinyl chloride (PVC), and polystyrene (PS) (Angelici et al., 2013). With extensive demand, global ethylene consumption reached more than 150 million tons/year in 2017 (Gao et al., 2019). Currently, ethylene is primarily produced by steam cracking of light hydrocarbons such as ethane and naphtha (Morschbacker, 2009). However, the gradual shift from crude oil to sustainable feedstocks and the massive availability of bioethanol make its conversion into ethylene an attractive route, particularly from lignocellulosic biomass feedstocks. Equation (110.1) shows the overall ethanol-to-ethylene reaction.

$$CH_3CH_2OH \rightarrow C_2H_2 + H_2O\Delta H \cong 45.7\text{kj}\cdot\text{mol}^{-1} \tag{110.1}$$

110.2.1.1 Reaction Mechanism for Ethanol Dehydration to Ethylene

Depending on the catalyst and reactants used, El, ElcB, and E2 reaction pathways have been proposed as the alcohol dehydration mechanism (Figure 110.1) (Tanabe et al., 1990). In the El mechanism, a Brønsted acid site on zeolite catalyst can protonate the alcoholic oxygen, enabling C-O bond cleavage to form water and a carbocation intermediate bound to a basic site. From this intermediate, the H on the adjacent carbon is then abstracted by another basic lattice O to form the alkene product. On the other hand, an ElcB mechanism proceeds on strong basic catalysts through a carbanion intermediate by firstly undergoing C-H cleavage to a carbanion intermediate. This intermediate is then converted to ethylene via removal of the hydroxyl on a neighboring acid site. An E2 mechanism undergoes simultaneous C-H and C-O bond cleavage by surface acid sites and neighboring protons bound to base sites (Eagan et al., 2019; Zhang and Yu, 2013).

The El and E2 reaction pathways have been suggested to occur with secondary and tertiary alcohols, e.g., isopropanol and tert-butanol (Janik et al., 2009; Kwak et al., 2011). Generally, for primary alcohols such as ethanol, the E2 mechanism takes place because of the high energy barrier required to form the primary carbocation intermediate (Arnett and Hofelch, 1983; Eagan et al., 2019; Tanabe et al., 1990). Additionally, the El and E2 reactions are believed to behave irreversibly and will not generate ethanol while the opposite is true for the Eclb mechanism (Zhang and Yu,

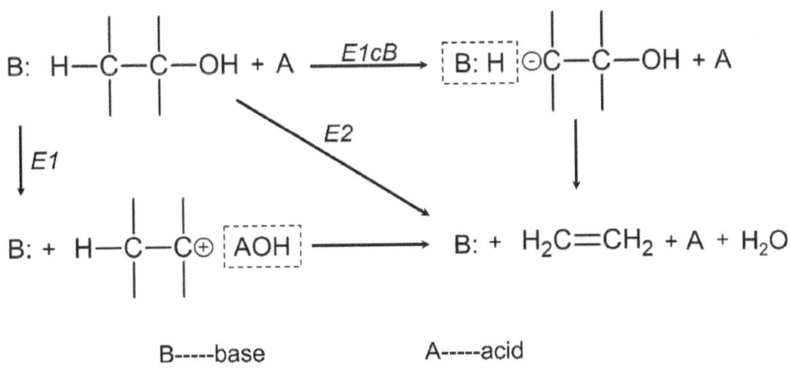

FIGURE 110.1 Reaction mechanism for alcohol dehydration. There are three mechanisms: R1, ElcB, and E2.

2013). Kinetics studies have also revealed another parallel reaction pathway for ethanol to ethylene, where the lone pairs electrons of the O atom can act as a weak base, adsorbing to an acid site, and further cleaving the C-O bond to possibly undergo the E1 mechanism to ethylene (Iwamoto, 2015; Morschbacker, 2009; Zhang and Yu, 2013). The endothermicity of the ethanol-to-ethylene pathway allows the mildly exothermic ethanol-to-diethyl ether reaction to take hold at low temperatures, thus requiring elevated temperatures typically above 300°C to ensure ethylene is selectively produced (Dagle et al., 2020b; Zhang and Yu, 2013). Such thermodynamic constraints cause industrial plants to typically utilize adiabatic reactors in parallel or multiparallel sequences with heating addition in between reactors to maintain high temperatures and ensure high productivity (Zhang and Yu, 2013).

Another less-mentioned parallel reaction pathway involves dehydrogenation of ethanol to acetaldehyde, a Tishchenko reaction to ethyl acetate, and subsequent cracking to acetic acid and ethylene; however, this reaction is expected to be slow and contribute minimal ethylene in parallel to ethanol dehydration (Iwamoto, 2015). Additionally, side reactions can also occur, such as dehydrogenation of ethanol to acetaldehyde, and further ethylene dimerization to butylenes (Kagyrmanova et al., 2011; Zhang and Yu, 2013). Concurrent experimental and simulation works have also revealed that production of ethylene can occur on metal-based catalysts (i.e., NiP) by a nondirected 'rake' mechanism. This can occur by an ethoxide surface species dehydrogenating to form acetaldehyde vinyl alkoxide via enolization followed by subsequent hydrodeoxygenation (Li et al., 2012).

110.2.1.2 Catalysts for Ethanol-to-Ethylene Reaction

Since 1745, thermal ethanol dehydration to ethylene has been carried out using sulfuric acid as a catalyst. However, this method achieves low yields and significant byproduct selectivity. Therefore, phosphoric acid, either in liquid form or dispersed on a solid support, was used to temporarily improve the process (Sun and Wang, 2014). Only until the early 1990s did gas-phase catalytic ethanol conversion to ethylene on heterogenous catalysis start to become attractive (Engelder, 1916; Pease and Yung, 1924). Current industrial practices for the ethanol-to-ethylene reaction typically utilize an activated alumina or a silica-alumina-based molecular sieve under isothermal or adiabatic conditions in a fluidized bed or fixed bed reactor (Morschbacker, 2009; Zhang and Yu, 2013). As a result, many researchers have focused on introducing additives or secondary catalytic species, altering pore size and structure, turning Si/Al, and experimenting with reaction conditions to optimize catalytic efficacy for this reaction (Eagan et al., 2019).

During the last few decades, conversion of ethanol into ethylene has been widely studied on a range of different solid catalysts such as alumina (Chen et al., 2007; Engelder, 1916; Kagyrmanova et al., 2011; Miciukiewicz et al., 1995; Tanabe et al., 1990; Wu and Marwil, 1980), zeolites (Diaz Alvarado and Gracia, 2010; Le van Mao et al., 1987, 1989; Phillips and Datta, 1997), transition metal oxides (Zaki, 2005), and heteropolyacids (Engelder, 1916; Janik et al., 2009; Varisli et al., 2017), where γ-Al_2O_3 and zeolites (e.g., ZSM-5) are the most studied due to their high activity and selectivity. Additionally, significant conversion was also found on transition metal oxides (Engelder, 1916).

In 1981, alumina-based catalysts were commercialized for ethanol-to-ethylene production and could reach conversions up to 99% and ethylene selectivity ~97% at 318°C with relevant stability for more than 8 months of operation. However, when bioethanol (~10 wt% ethanol in H_2O) is used as a feedstock for ethanol dehydration, water has a detrimental effect on alumina-based catalysts due to the suppression of ethanol conversion (Kochar and Padia, 1981). At low temperature, water content can directly affect the activity of the catalyst during the reaction. It was observed that water increment from ~5 wt% to ~ 90 wt% decreases the ethanol conversion from ~86% to ~ 65% coupled with increased selectivity toward diethyl ether over the TiO_2/γ-Al_2O_3 catalyst. Nonetheless, this water effect can be minimized at reaction temperatures above 420°C (Chen et al., 2007). Several works reported in literature (Chen et al., 2007; Kochar and Padia, 1981) have corroborated this with alumina-based catalytic systems where high temperatures (>400°C) can mitigate the impact of high-water content and prevent the production of diethyl ether to reach high ethylene selectivity

(Chen et al., 2007; Kagyrmanova et al., 2011). In addition, activated alumina-based catalysts also require a sufficient concentration of ethanol in the feed stream; otherwise, the reaction requires higher temperatures and lower space velocities to maintain ethylene selectivity, resulting in higher energy demand (Zhang and Yu, 2013).

Zeolites possess several appealing qualities in catalysis, including uniform pore structures, high surface area, and adjustable acidity, making them a commonly utilized catalyst in alcohol dehydration reactions. In the case of ethanol dehydration, reasonable activities have been observed on zeolites with reaction temperatures lower than 300°C (Bi et al., 2010; Le van Mao et al., 1987, 1989; Nguyen and Le van Mao, 1990; Phillips and Datta, 1997; Schulz and Bandermann, 1994; Takahara et al., 2005), and even lower than 200°C in some cases (Le van Mao et al., 1987). At high temperatures (>400°C), secondary reactions of ethylene produce longer chain hydrocarbons (e.g., from light olefins to gasoline-range hydrocarbons) which will be further discussed in a later section. Studies have also observed high catalytic activity of ZSM-5 or modified ZSM-5 catalysts in ethylene dehydration due to the hydrophobic nature of their surfaces (Hahn-Hagerdal et al., 2006; Le van Mao et al., 1987, 1989; Nguyen and Le Van Mao, 1990).

Similar to γ-Al_2O_3, zeolites have shown an ability to perform a simultaneous parallel-consecutive reaction mechanism to produce ethylene at lower temperatures (<270°C). Nguyen and Le Van Mao (1990) reported a systematic study of bioethanol-to-ethylene (BETE) conversion on H-ZSM-5 or modified H-ZSM-5 catalysts and found the reaction mechanisms of ethanol dehydration on steam-treated ZSM-5 or asbestos-derived ZSM-5 relied heavily on reaction temperatures (Le van Mao et al., 1987, 1999, 1990; Nguyen and Le van Mao, 1990). At low temperatures (<270°C), ethylene formation mainly takes place throughout a diethyl ether intermediate, while, at high temperatures (270°C–350°C), direct ethanol-to-ethylene formation occurs (Nguyen and Le van Mao, 1990).

Another important property in catalytic dehydration reactions is the surface acidity of the catalyst, which plays a crucial role in catalytic performances with regard to ethylene selectivity and stability (Moser et al., 1989). Concentrating strong Brønsted acidic sites on the surface of zeolites by lowering the Si/Al ratio has been observed to induce secondary reactions of ethylene (i.e., oligomerization, cracking, and coking) to longer chain hydrocarbons and carbonaceous deposits on ZSM-5 catalysts (Chaudhuri et al., 1990; Le van Mao et al., 1989).

Optimal Si/Al ratios for high ethylene selectivity, activity, and stability have been found to lie in the range of 35–55 (Nguyen and Le van Mao, 1990). The use of additives on a catalyst can also allow the adjustment of surface acidity to enhance ethylene selectivity and catalyst stability. For example, incorporation of Zn and Mg into ZSM-5 showed an enhancement of ethylene selectivity and a mitigation of diethyl ether and other light olefines formation (Le van Mao et al., 1987). Likewise, weak acid sites from P on ZSM-5 also showed high selectivity to ethylene with P loadings >3.4 wt% in a large range of reaction temperatures (300°C–440°C), whereas at P loadings <3.4 wt%, longer chain hydrocarbons were produced, especially at higher temperatures conditions (Zhang et al., 2008). Additives like trifluoromethanesulfonic acid (TFA) (Le van Mao et al., 1989) and Fe (Calsavara et al., 2008) on ZSM-5 generate catalytic materials that also display high activity and selectivity on a wide range of reaction temperatures (170–285°C) (Calsavara et al., 2008; Le van Mao et al., 1989). Studies with other kinds of zeolite catalysts, such as β-zeolite, have shown their moderate acidity is capable of high ethylene selectivity even at high temperatures (370°C) (Hutchings et al., 1994).

In contrast to γ-Al_2O_3, the addition of water to the feedstock on ZSM-5 enhances the selectivity of ethylene and catalyst stability (Nguyen and Le van Mao, 1990; Phillips and Datta, 1997). This is an important factor since zeolite catalysts generally exhibit better ethanol activity and require lower temperatures; however, unlike Al_2O_3, molecular sieves often suffer from instability and experience significant deactivation issues (Zhang and Yu, 2013). Phillips and Datta (1997) found in their kinetic study that water considerably improves the reaction rate of direct ethanol to ethylene and diethyl ether to ethylene, whereas the reaction rate for ethanol to diethyl is less affected. This improvement is due to a moderation of acidity on the catalyst surface by the incorporation of water, which mitigates secondary reactions of ethylene to longer chain hydrocarbons and coke formation. Furthermore, theoretical calculations have demonstrated that surface Brønsted acidity is modified by the presence of water (Krossner and Sauer, 1996):

It is worth mentioning that ZSM-5 exhibits uniform microporous structures and commonly have large primary particle sizes (a few micrometers), which can lead to mass transfer limitations that inhibit ethanol access to surface acid sites. Micropores can be promptly blocked by bulky molecules, promoting the deactivation of the catalyst (Choi et al., 2009; Zhu et al., 2011). Consequently, various approaches have been employed to overcome these diffusion limitations and provide access to the acid sites (Choi et al., 2006; Han et al., 2003; Su et al., 2003; Zhu et al., 2012). It was found that decreasing the ZSM-5 particle size increases the activity and stability of the catalyst for ethanol conversion and maintains high ethylene selectivity for over 630 h of time-on-stream (TOS) with a 95 vol% of ethanol feed (Bi et al., 2010).

110.2.2 Ethanol to C_3-C_4 Olefines

The production of C_3-C_4 olefins, such as propylene, 1- and 2-butene, isobutene, and butadiene, is in high demand in the production of several industrial compounds. They are the building blocks for the synthesis of bulk chemicals (e.g., ethylene oxide, acrolein), polymers (e.g., polyethylene, polypropylene), and fuels (e.g., diesel and gasoline) (Boyadjian et al., 2011). The importance of olefin production in industry further prompts the continual search for new sustainable feedstocks to replace or supplement petrochemical resources, making bioethanol to C_3-C_4 olefines conversion an attractive technology worth exploring. Ethanol conversion into C_3-C_4 olefins has been studied with many different catalytic materials like modified and unmodified zeolites (Aguayo et al., 2002; Gayubo et al., 2001, 2010, 2011; Song et al., 2009, 2010) and mixed metal oxides (Hayashi and Iwamoto, 2013; Iwamoto et al., 2011; Mizuno et al., 2012). Depending on the choice of catalytic system, different reaction mechanisms have been proposed.

110.2.2.1 Controlled Oligomerization/Cracking on Zeolites

The conversion of ethanol using zeolites can produce a variety of hydrocarbons including ethylene, C_3-C_4 light olefins, and C_{5+}. Formation of C_3-C_4 olefins on zeolites is mediated via an ethylene intermediate similar to the methanol-to-gasoline (MTG) conversion (Chang, 1991). First, ethanol is dehydrated to form ethylene, which can then convert into C_3-C_4 hydrocarbons through acidic catalyzed oligomerization-cracking and oligomerization-aromatization mechanisms (Boyadjian et al., 2011; Chang, 1991), though some works also suggest that C_3 olefin (propene) could be directly produced from ethylene on ZSM-F via carbene mechanism depending on the Si/Al ratio (Takahashi et al., 2013). This mechanism generally occurs via the protonation of the C=C bound to form an adsorbed carbenium intermediate on an acid catalyst which can then for a C-C bound with another olefin, thus leading to longer chain hydrocarbon formation (Eagan et al., 2019; Ghashghaee, 2018).

The carbenium intermediate is more stabilized with increasing carbon length. The electronic configuration on zeolites enable to stabilize secondary and tertiary carbeniums ions more easily than primary carbenium ions, and this is reflected in faster rates of oligomerization for C_{3+} olefins than in ethylene oligomerization (Eagan et al., 2019; Ghashghaee, 2018). Such cascade reaction can either overtake the standard zeolitic catalyst via coke formation or can diffuse out as smaller hydrocarbons by cracking into smaller olefins, aromatics, and paraffins (Eagan et al., 2019). The efficacy of ethanol as a feedstock for C_3–C_4 olefins is also controlled by thermodynamics behind oligomerization. Oligomerization reactions are generally exothermic but have a negative entropy chance, thus olefin production is less thermodynamically favorable with increasing molecular size, making light olefins the most favorable at relatively higher temperature (>300°C) and ambient pressure (Eagan et al., 2019).

Reaction conditions, including temperature and residence time, as well as catalytic acidity can affect the product distribution significantly. The acidity of the catalytic surface can be relatively controlled by parameters such as Si/Al ratio, additives in zeolites, and water content in the feed. For example, secondary reactions of ethylene to produce longer-chain hydrocarbons are favored by strong acidity and high temperatures. However, these conditions also favor the cracking and coking processes. Therefore, it is crucial to precisely control the acidity of the surface and the reaction conditions during the formation process of C_3–C_4 olefins. Song et al. studied the effect of H-ZSM-5 at different Si/Al ratios in ethanol conversion at 400°C. The optimum acid strength to produce C_3 olefins was obtained at Si/Al ratio of 80. Si/Al ratios >80 promote the C_{5+} hydrocarbon formation due to higher acid-site density and strength on the surface (Song et al., 2009). Gayubo et al. illustrated the effect of varying reaction conditions for ethanol-to-olefins conversion over ZSM-5 (Si/A_l=24). Using aqueous ethanol, an optimum residence time and a temperature of 450°C were essential to reach high selectivity to C_3–C_4 olefins.

Prolonged residence times promoted the secondary ethylene conversion to C_{5+} hydrocarbons, while short residence times produce mainly ethylene. Likewise, water had a beneficial effect on the process of forming light olefins due to the modification of acid-site strength and the inhibition of oligomerization and cracking reactions (Gayubo et al., 2001). In addition, water content in the reaction reduces the formation of coke, thereby improving catalytic stability. Nevertheless, at high water content and high temperatures, irreversible deactivation of the catalyst was observed due to the dealumination and collapse of the microporous structure (Aguayo et al., 2002; Masuda et al., 1998).

Additives can modify the surface acidity of zeolites in order to obtain high selectivity to C_3–C_4 olefins. In studies from Song et al., it was found that P and Zr additives on H-ZSM-5 passivate the strong acid sites, whereas moderate acid sites were maintained, inducing high selectivity toward propylene (31%–32%). Additionally, P and Zr have shown an inhibitive effect on the dealumination of the zeolite, improving its stability (Song et al., 2009, 2010). Likewise, Nickel ions supported on mesoporous silica MCM-41 (i.e., NiO/MCM-41) provide active sites for ethylene dimerization to propylene and subsequent metathesis to produce butenes from unreacted ethylene (Iwamoto et al., 2011). Furthermore, a more recent work found small particles sizes of ZSM-5 (Si/Al=~7.6) favor the formation of propylene at 500°C due to the absence of mass transfer limitations (Meng et al., 2012).

110.2.2.2 Aldol Pathway to Propylene, Isobutene, 1,3-BTD/N-Butenes on Metal Oxides

Mixed metal oxides with balanced acid-base sites have been developed for ethanol conversion into propylene (Hayashi and Iwamoto, 2013; Mizuno et al., 2012) and isobutene (Liu et al., 2013; Sun et al., 2011), where acetaldehyde and acetone are the primary intermediates, showing higher selectivity toward propylene and isobutene (>60%) in comparison with the well-studied zeolites (>30%). In addition, mixed metal oxides have been extensively studied to produce 1,3-butadiene from ethanol (Angelici et al., 2013). Albeit the mechanism for this reaction is still up for debate, researchers have broadly accepted the reaction proceeds through catalytic dehydrogenation on basic sites followed by aldol condensation reaction (Angelici et al., 2013; Jones et al., 2011; Makshina et al., 2012).

110.2.2.2.1 Production of Propylene

Supported metal oxides have been also studied as a catalytic material to produce propylene from ethanol. In contrast to zeolites, ethanol conversion on metal oxides such as NiO/MCM-41 can proceed in parallel two different reaction pathways to produce ethylene. In one pathway, ethanol is dehydrated through diethyl ether as intermediate, while the other pathway involves acetaldehyde and ethyl acetate as intermediates (Iwamoto et al. 2011). The ethylene formed is converted into propylene which subsequently follows at dimerization, isomerization, and metathesis reaction pathway (ethylene → butenes → isobutene → propylene). The experimental evidence suggests that layered nickel-silicate is the active phase for conversion of ethanol into propylene. Furthermore, in the presence of water (75 wt% ethanol), ethanol conversion and propylene selectivity are not altered significantly (Iwamoto, 2011).

Thermodynamic assessment of ethylene to propylene has also shed light on the obstacles surrounding the ethanol conversion to propylene process that are not strictly limited to the catalyst itself. It was found that propylene production is generally optimized with lower pressures and higher temperatures (Lehmann and Seidel-Morgenstern, 2014). A maximum yield of 42% was observed for ethanol to propylene with optimal reaction conditions at around 600°C with a pressure of 1 bar (Lehmann and Seidel-Morgenstern, 2014). Such thermodynamic bottlenecks imply the limitations behind achieving the conversion of ethanol to light olefins in addition to the kinetic factors at play in catalysis. Interestingly, the presence of water acting strictly as a dilutant enables similar propylene yields at lower temperatures (Lehmann and Seidel-Morgenstern, 2014).

The thermodynamic constraints mentioned above could be circumvented on other Lewis acid-base metal oxides with a different reaction mechanism. Iwamoto et al. reported propylene yields up to 60 mol% using a scandium-modified indium oxide catalyst (Sc/In_2O_3) at 550°C in presence of water and hydrogen. This high yield of propylene was mainly associated with a reaction pathway that proceeded via ethanol dehydrogenation to acetaldehyde, then acetaldehyde condensation/ketonization into acetone, and finally hydrogenation and dehydration of acetone to form propylene (ethanol → acetaldehyde → acetone → propylene). The addition of Sc and water vapor to the reaction system prevents deactivation of the catalyst, while the cofed hydrogen promotes the hydrogenation-dehydration of acetone (Mizuno et al., 2012). In subsequent work, a set of additives on CeO_2 was studied, where it was found that the catalyst consisting of Y/CeO_2 obtained 25% and 50% selectivity to propylene and ethylene, respectively. Additionally, the stability of these kind of catalysts was much higher than zeolites (Hayashi and Iwamoto, 2013). However, despite the advance in knowledge of this reaction, many reaction steps have not yet been well established. Note that although a higher propylene can be achieved by following the reaction pathway, 25% of C will be lost as CO_2 side product.

110.2.2.2.2 Production of Isobutene

In chemical industry, isobutene is an essential platform molecule to produce many other high-demanded products, such as butyl rubber, polyethylene terephthalate (PET), and tri-isobutene, which is a premium solvent and additive for jet fuel. The production of isobutene from ethanol opens the possibility to use renewable biomass as a feedstock. For this reaction, two processes have been reported which are commonly known as the two-step process and one-step process.

110.2.2.2.3 Two-Step Process

In this case, the ethanol-to-isobutene conversion requires two steps: (i) first, ethanol is converted into acetone on a basic metal oxide (Murthy et al., 1988; Nakajima et al., 1987, 1994; Nishiguchi et al., 2005) (equation 110.2) and (ii) then, acetone is converted into isobutene on an acidic zeolite (Dolejsek et al., 1991; Hutchings et al., 1994; Tago et al., 2011) (equation 110.3). During this process, H_2 is generated as a valuable byproduct while acetone is utilized as a valuable intermediate product capable of further producing a wide range of useful chemical compounds in addition to isobutene, like diacetone, alcohol, mesityl oxide, and methyl isobutyl ketone (Salvapati et al., 1989). Additionally, ~33% of C will be lost as CO_2 in this particular reaction pathway.

$$2CH_3CH_2OH + H_2O \rightarrow CH_3COCH_3 + CO_2 + 4H_2 \tag{110.2}$$

$$3CH_3COCH_3 \rightarrow 2i - C_4H_8 + CO_2 + H_2O \tag{110.3}$$

Ethanol to acetone reaction has been broadly studied and its mechanism is well understood. First, ethanol is hydrogenated on the base site of the catalyst to produce acetaldehyde, then can undergo either aldol addition and decarbonylation to form acetone (Murthy et al., 1988; Nishiguchi et al., 2005), or it can be oxidized to acetic acid followed by ketonization to form acetone (Nakajima et al., 1994). Mixed metal oxides such as ZnO-CaO (Nakajima et al., 1987, 1989), Fe-Zn mixed oxide (Nakajima et al., 1994), and $Cu\text{-}La_2Zr_2O_7$ (Bussi et al., 1998) have shown effective catalytic activity in this reaction, with almost complete ethanol conversion and carbon selectivity around 70% at reaction temperatures between 400°C and 440°C. Nishiguchi et al. studied aqueous ethanol (water/ethanol ratio=5) conversion on CuO/ CeO_2 and observed that at the reaction temperature of 380°C, the main product was acetone, whereas at 260°C, it was acetaldehyde. These results established that the oxygen supply from the CeO_2 surface plays a key role in the reaction (Nishiguchi et al., 2005).

During the acetone-to-isobutene process, multiple reactions are involved including condensation, dehydration, and decomposition reactions (Dolejsek et al., 1991; Xu et al., 1994). Both Brønsted (Hutchings et al., 1994; Xu et al., 1994) and Lewis acid-base pairs (Zaki et al., 2001) are effective active sites to catalyze this reaction. A comparative study of ZSM-5 and USY zeolites found that either Brønsted or Lewis acid sites are active sites, but they follow different reaction mechanisms (Panov and Fripiat, 1998). Essentially, the initial condensation reaction occurs with the interaction of one acetone molecule in gas phase and one acetone molecule absorbed on Lewis acid sites (i.e., Eley-Rideal mechanism), whereas the Brønsted acid-mediated reaction requires two absorbed acetone molecules (i.e., Langmuir-Hinsheldwood mechanism) (Panov and Fripiat, 1998).

The formation of isobutene has also been observed in secondary reactions of isobutylene or mesityl oxides intermediates, where depending on different parameters like surface acidity, reaction temperature, and residence time, it is possible to control the secondary reaction and hence the isobutene selectivity (Chang and Silvestri, 1977; Hutchings et al., 1994; Tago et al., 2011). A study with ZSM-5 at 329°C found acetone conversion was ~25% with a high isobutene selectivity of ~83%. But, when temperature was increased at 399°C the isobutene selectivity significantly dropped to ~4%, increasing the aromatic compounds selectivity as well as the acetone conversion to ~95% (Masuda et al., 1998). In the case of β-zeolite, which has a moderate acidity, it is possible to obtain an isobutene selectivity ~87% even at temperatures up to 400°C (Hutchings et al., 1994). On the other hand, with TOS, a rapid decline of acetone conversion is observed on ZSM-5 (Si/Al=80) at 400°C.

Therefore, isobutene selectivity increases at the expense of selectivity to aromatic compounds. Strong acid sites favor the secondary reaction of olefins to aromatics, but with TOS, these strong acid sites are blocked due to coke deposition which inhibits secondary reactions, leading to higher selectivity to isobutene. In addition, these strong acid sites can be passivated with alkali additives to suppress secondary reactions and enhance the isobutene selectivity (Tago et al., 2011). It is worth mentioning that the main challenge in the acetone-to-isobutene conversion on zeolites is the quick coke formation and subsequent deactivation of catalyst. Some studies have found that using nano-sized particles of ZSM-5 at 330°C mitigates the deactivation by avoiding the free access of large-sized molecules (Zhu et al., 2011).

110.2.2.2.4 One-Step Process

Contrary to the two-step process described above, the one-step conversion of ethanol to isobutene can be achieved by designing and synthesizing catalytic materials with proper acid-base balance properties. Hence, catalytic materials like $Zn_xZr_yO_z$ mixed oxide present well-balanced acid-base properties, enabling direct ethanol conversion to isobutene with yields greater than 80%. This catalytic cascade reaction involves dehydrogenation, condensation, dehydration, and decomposition reactions (ethanol → acetaldehyde → acetone → isobutene), where hydrogen is obtained as a valuable byproduct (equation 110.4) (Su et al., 2003; Zhu et al., 2012). The use of Zn as an additive on ZrO_2 passivates the strong acidic sites and generates basic sites on the catalytic surface. Therefore, the direct ethanol-to-acetone conversion is inhibited, leading to the enhancement of ethanol dehydration to acetaldehyde which can then convert into acetone via aldol addition/ketonization pathways on the basic sites of the catalyst.

A recent study (Li et al., 2021a) which combines surface characterization, kinetics measurements, and density functional theory (DFT) calculations reveals the essential role of water in the acetone-to-isobutene reaction catalyzed by Lewis acid-base pairs on $Zn_xZr_yO_z$. Water dissociation adsorbed on Lewis acid sites promotes the diacetone alcohol (DAA) decomposition to isobutene whereas hinders the mesityl oxide formation. It was stablished that reaction follows the acetone → DAA → isobutene mechanism, where the rate-limiting step is the decomposition of DAA to isobutene. The DFT calculations demonstrate that the vicinity of the OH on the surface inhibits the strong interactions with the carbonyl group from DAA, preventing its severe distortion which eventually lead dehydration reactions (e.g., polymerization, cooking), while a slight bend of the O of the hydroxyl group toward the C of the carbonyl group generates a rearrangement that facilitates the isobutene formation (Li et al., 2021a).

$$3CH_3CH_2OH + H_2O \rightarrow i - C_4H_8 + 2CO_2 + 6H_2 \quad (110.4)$$

110.2.2.2.5 Production of 1,3-Butadiene/N-Butenes

1,3-butadiene is an essential platform molecule in the petrochemical industry. This compound is widely use in the global production of polymers and its intermediates such as styrene-butadiene rubber and polybutadiene, which consume more than 50% of butadiene production (Angelici et al., 2013). The reaction mechanism of 1,3-butadiene from ethanol conversion is quite complex and is still debated; nevertheless, a general mechanism for this reaction is accepted. Typically, the transformation of ethanol into 1,3-butadiene requires: (i) ethanol dehydrogenation to acetaldehyde; (ii) aldol addition of acetaldehyde to acetaldol; (iii) dehydration of acetaldol to crotonalydehyde or Meerweing-Ponndorf-Verley (MPV) reaction between acetaldol and ethanol to produce 3-hydroxybutanol; iv) MPV reaction between crotonaldehyde and ethanol to form crotyl alcohol and acetaldehyde; and v) dehydration of crotyl alcohol or 3-hydroxybutanol to obtain 1,3-butadiene (Angelici et al., 2013; Jones et al., 2011; Makshina et al., 2012; Wang and Liu, 2012). From a thermodynamic perspective, the most unfavorable step in this reaction is the formation of 3-hydroxybutanol but is stabilized by the formation of crotonaldehyde due to its exergonic nature (Angelici et al., 2013). Similar to the one-step process of ethanol-to-isobutene (Sun et al., 2011) or ethanol-to-propylene reaction (Hayashi and Iwamoto, 2013), the proper acid and base strength is essential in order to maximize selectivity to the dehydration of crotyl alcohol to 1,3-butadiene pathway while also promoting the dehydrogenation and aldol-condensation reactions. (Hiroo et al. 1972).

Several metal oxide and mixed metal oxide catalysts have been studied on the ethanol-to-1,3-butadiene reaction and have been summarized in recent literature (Angelici et al., 2013; Wang and Liu, 2012), but only those with the appropriate acid-base balance displayed significant yield toward 1,3-butadiene in one-step ethanol conversion (Corson et al., 1950; Hiroo et al., 1972). Catalytic promoters of dehydrogenation, including Ag, Cu, and Pt, are commonly coupled with a Lewis acidic (e.g., TiO_2, ZrO_2, Y, etc.) or basic (e.g., MgO, ZnO, TaO_x, etc.) catalyst to facilitate the aldol condensation of acetaldehyde. For instance, MgO/SiO_2 catalysts have displayed high selectivity to 1,3-butadiene (Kitayama and Michishita, 1981; Kvisle et al., 1988; Makshina et al., 2012).

The proper reduction of acidity on this catalyst enhanced the 1,3-butadiene yield from 44% to 87% (Ohnishi et al., 1985). Other additives like Cu and Ag oxides have been also investigated on MgO/SiO_2, improving the catalytic activity in terms of 1,3-butadiene yield. Therefore, the ratio between acidic and basic metals and their redox nature plays a crucial role in the inhibition of ethanol dehydration to ethylene, thereby improving the 1,3-butadiene yield (Makshina et al., 2012). Combination of metal oxides like Zr-Zn supported on silica has attracted attention for the transformation of ethanol into 1,3-butadiene. The Lewis acidity from both Zn(II) and Zr(IV) is suggested to boost the catalytic activity, while the cofeeding ethanol/acetaldehyde helps the aldol-condensation reaction improve the selectivity to 1,3-butadiene (Jones et al., 2011). Tuning the interfacial bounds of Ag by ZrO_2, addition to a SiO_2 support was also found to enhance dehydrogenation capabilities via the Ag-O-Si bond, while simultaneous $Ag^{\delta+}$ interaction with the medium and weak Lewis acidic sites from the Zr-O-Si promoted H^+ transfer during MPV reduction of crotonaldehyde to crotyl alcohol (Li et al., 2021b). In general, product selectivity is the most-contended issue in this reaction and mitigating side product formation that inhibits selectivity to 1,3-butadiene and/or induces catalyst deactivation is essential (Dagle and Winkelman, 2020b).

Similar to 1,3-butadiene, n-butenes (1, and 2-butnene) are suitable light olefins as platform molecule for production of middle distillate fuels (i.e., diesel and jet blendstocks). A recent work demonstrated the mechanism for direct ethanol conversion into n-butenes over the Ag-ZrO_2/SBA-16 catalyst (ethanol → acetaldehyde → crotonaldehyde → 1,3-butadiene → n-butenes) (Dagle and Winkelman, 2020b) under a H_2 atmosphere, and with an optimal n-butene selectivity of 88% and almost complete ethanol conversion. The main reaction pathway for direct ethanol conversion involves the following reactions: (i) ethanol dehydrogenation to acetaldehyde over the metallic Ag active sites; (ii) aldol condensation of acetaldehyde to form crotonaldehyde over the Lewis acid sites from ZrO_2/SiO_2; (iii) MPV reduction of crotonaldehyde to crotyl alcohol followed by its dehydration to produce 1,3-butadiene; and iv) hydrogenation of 1,3-butadiene into n-butenes over the metallic Ag and ZrO_2 actives sites. Besides, a small fraction of n-butenes is also produced from the reduction of crotonaldehyde into butyraldehyde as intermediate (crotonaldehyde → butyraldehyde → n-butenes). For the later step, it was demonstrated by operando-NMR experiments and theoretical analysis (i.e., ab initio molecular dynamics (AIMD) simulations and DFT calculations) that ethanol is the source of hydrogen instead of H_2 for the crotonaldehyde hydrogenation to butyraldehyde (Dagle and Winkelman, 2020b).

Over the metal oxides, the one-step ethanol conversion to different products (i.e., propylene, isobutene, 1,3-butadiene, and n-butenes) all start from the ethanol dehydrogenation on basic sites of the catalyst to produce acetaldehyde, followed by acetone production in the ethanol-to-propylene (Hayashi and Iwamoto, 2013) and ethanol-to-isobutene (Sun et al., 2011) reaction mechanisms. In the case of ethanol-to-1,3-butadiene, it is proposed that acetaldehyde undergoes aldol addition to form acetaldol, which is dehydrated to produce crotonaldehyde (Jones et al., 2011). In general, H_2O and reaction temperature play significant roles in promoting the acetaldehyde reaction pathways. Cofeeding H_2O and operating at high temperatures (>400°C) promotes the acetaldehyde-to-acetone transformation via ketonization mechanism (Rahman et al., 2016), whereas low temperatures (<400°C) benefit the aldol-condensation of acetaldehyde (Jones et al., 2011). Additionally, H_2 is also essential in tuning the propylene and isobutene production from acetone (Baylon et al., 2016).

In presence of small quantities or no hydrogen during the reaction, isobutene production is favored, while at high hydrogen concentration, the production of propylene is dominant (Baylon et al., 2016; Hayashi and Iwamoto, 2013; Mizuno et al., 2012). For successful ethanol transformation into 1,3-butadiene, selective hydrogenation of crotonaldehyde to crotyl alcohol is required instead of C=C hydrogenation. Despite the increasing knowledge of the ethanol to 1,3-butadiene reaction, there is still continual debate surrounding the mechanism of the reaction, the rate-limiting step, and the chemistry required for optimizing catalytic performance (Angelici et al., 2013). Further studies are necessary to complete the comprehension of different mechanisms in order to control the reaction pathways and key intermediates toward the desire products.

110.2.3 Ethanol to Fuel-Range Hydrocarbons

The strong reliance of automotive transportation energy on petrochemical-based fuel production is gradually becoming more vulnerable as global society shifts toward more sustainable, environmentally conscious fuel sources. With a simultaneous push toward petroleum-free processes and an increase in energy demand, replacing gasoline and other commercial fuels with sustainable, green feedstocks is an appealing path to address these challenges. On this note, bioethanol has emerged as a promising platform molecule to achieve this endeavor since it is so readily available and is also already being incorporated into commercially available fuels. This section will dissect the mechanistic details and newly developed catalysts surrounding direct ethanol-to-gasoline (ETG) reaction and its potential to produce jet/diesel fuel in the future.

110.2.3.1 Ethanol-to-Gasoline (ETG)

Gasoline is the most commonly available commercial fuel and is composed of a well-blended mixture of hydrocarbons ranging from C_5 to C_{12} chain lengths. Due to its high energy density and readily temporary accessible nature, it's a very convenient source of energy for automotive combustion engines. The ethanol-to-gasoline (ETG) reaction consists of a complicated mechanism but is generally composed of ethanol dehydration to ethylene followed by further oligomerization to higher-order hydrocarbons that can then undergo aromatization or cracking to form various paraffins and aromatic products (Talukdar et al., 1997). It is generally accepted that the primary active sites for the ETG reaction are Brønsted acid sites; however, other potential sites have been proposed, such as radical sites generated during the reaction, while available Brønsted acid sites were blocked by carbonaceous species (Madeira et al., 2012; Trane et al., 2012). The dehydration of ethanol to ethylene is faster than ethylene oligomerization; therefore, in order to promote ethylene oligomerization, elevated temperatures above 300°C are typically employed (Eagan et al., 2019). At such temperatures, oligomerization of short-chain olefin products, such as linear butenes, takes hold since their secondary and tertiary carbenium intermediates are more stable than the primary carbenium from ethylene (Ghashghaee, 2018).

In addition to high temperature enabling ethylene dimerization, other factors including aromatization and cracking come into play as well. Thermodynamically, neither aromatization nor oligomerization is more favorable. Nevertheless, the H_2 from aromatization reaction hydrogenate other olefins (or transfer hydrogenation), promoting the aromatization pathway due to the high exothermicity of the hydrogenation of non-aromatic olefins. Therefore, operating at high temperatures (300°C–400°C) allows the ethylene oligomerization to produce a mixture of aromatic compounds and light paraffins (Eagan et al., 2019). Additionally, the efficacy of solid acid catalysts for ethanol to olefins are exemplified by reports of directly feeding 1-butene as reactant over molecular sieves catalysts. In these cases, >90% selectivity to C_{8+} products is readily obtained at temperatures lower than 200°C where cracking is mitigated (Eagan et al., 2019). Such products are too light for diesel fuels but can be readily distilled to obtain ideal fractions to makeup gasoline products (Eagan et al., 2019).

The most generally accepted mechanism of ethanol to C_{3+} hydrocarbons on zeolites typically involves the activation of ethanol at Brønsted acid sites in the zeolitic framework (Costa et al., 1985; Madeira et al., 2012). Due to the electrostatic components of zeolitic materials, carbenium ions can be readily stabilized, enabling further oligomerization to high-order carbon chain lengths (Costa et al., 1985). The general mechanism of ethanol conversion to liquid and gaseous hydrocarbons involves the reaction of gaseous olefins with carbenium ions absorbed to surface acid sites via oligomerization to form either absorbed liquid hydrocarbons or gaseous paraffins that can then desorb as products (Costa et al., 1985).

The most common catalyst employed for the ETG reaction is ZSM-5-type zeolites due to their unique structural configurations that can allow longer hydrocarbon chain formation from ethylene (Costa et al., 1985; Graschinsky et al., 2012; Madeira et al., 2009; Ramasamy et al., 2014a-b). As previously discussed in Section 110.2.1 and 110.2.2, surface acidity of the catalysts is an essential component it its activity. Tuning the acidity via Si/Al ratio adjustments highlights the impact of acidity, where low Si/Al ratios around 20 produce significantly more liquid aromatic hydrocarbons compared to Si/Al ratios of 100 which produce majorly light C_2-C_4 olefins (Talukdar et al., 1997). The acid density effect has been investigated further with exaggerated Si/Al ratios ranging from 16 to 500 where the catalyst lacked both selectivity and stability with a Si/Al ratio of 500, while a low Si/Al of 40 exhibited an optimal balance of Brønsted acidity and radical active sites capable of highly selective behavior (Madeira et al., 2012). Such findings have suggested the quantity of acidic sites in ZSM-5 is an essential component when obtaining C_{5+} hydrocarbons from ethanol.

Despite obtaining higher activity, highly acidic catalysts with low Si/Al ratios must also contend with several detriments, including deactivation via coking at strong acid sites as well as dealumination of the zeolite (Aguayo et al., 2002; Gayubo et al. 2010, 2011; Madeira et al., 2012). These factors can be mitigated by modifying the zeolite with an additional dopant on its surface, such as Ni, (Gayubo et al., 2010), Ga (Saha and Sivasanker, 1992), Zn (Saha and Sivasanker, 1992), or Fe (Calsavara et al., 2008). Such modifications can mitigate coking and dealumination by weakening the surface acid strength and therefore extending the lifetime of the catalyst significantly. This was especially highlighted with trimethyl-phosphite-modified ZSM-5, where only weak acid sites resided on the surface. The conversion of primary alcohols was mostly to ethers; however, the weak acidity was incapable of further converting the ethers to longer-chain hydrocarbons (Tynjala et al., 1998).

The importance of surface acid sites for ETG has been repeatedly demonstrated; however, the density and strength of acid sites play an important role in the product distribution and catalytic activity of zeolites for ETG. Alkenes are particularly known for their affinity to form coke on acidic surfaces (Huber et al., 2006). Since ethylene is primary intermediate in the ethanol to C_{3+} hydrocarbon reaction, coking is a primary obstacle cited for catalytic deficiencies (Farrell et al., 2006). As such, the main focus of many ETG mechanism studies look at the role of acidity, the optimal balance of surface acid density, and other possible dopants that can mitigate strong acid sites to prevent deactivation via coking (Gayubo et al., 2010; Madeira et al., 2012; Marcu et al., 2009; Saha and Sivasanker, 1992). However, more recent studies have suggested that radicals created during the initial periods of coke formation may act as active sited for C_{3+} hydrocarbon formation from ethanol (Farrell et al., 2006; Madeira et al., 2012). Such studies have observed the existence of radicals with EPR spectroscopy and a consumption of Brønsted acid sites in IR spectroscopy, while high activity for C_{3+} is simultaneously maintained (Farrell et al. 2006; Sanchez and Cardona, 2008). In this mechanism, it is proposed that the initial formation of highly energetic absorbed carbonaceous species early in the reaction form radicals that can stabilize ethylene intermediates, allowing it to undergo incremental condensation to longer chained carbons (Madeira et al., 2012). The radical-mediated pathway was exemplified by acid sites and microporosity but still maintained significant catalytic performance to C_{3+} hydrocarbons, implying the importance of free radicals during hydrocarbon production from ethanol on a coked catalyst (Madeira et al, 2010). However, catalytic activity via this radical-mediated pathway is only temporary and depreciates with increasing TOS, suggesting that the radical active sites begin to stabilize into more ordered, carbonaceous species less active for ETG (Madeira et al., 2012).

As with the previous Sections 110.2.1 and 110.2.2, reaction temperature and water also play an impactive role in the ETG reaction. Previous work has demonstrated that an increase from 4 to 15 wt% of water in an aqueous ethanol reactant stream reduced the lifetime of the catalyst by half (Costa et al., 1985). High ranges of water concentration up to 80 wt% have shown the presence of water suppressing the yield of liquid hydrocarbons significantly (Talukdar et al., 1997). Conversely, some studies have demonstrated the ETG reaction actually benefits from an optimal concentration of water in the feed since water was capable of mitigating coke formation that would otherwise prevent liquid hydrocarbon formation (Aguayo et al. 2002). Similar optimization effects have also been observed with reaction temperature, where temperatures between 350°C and 450°C provide maximal yields of liquid hydrocarbons from ethanol (Aguayo et al., 2002; Costa et al., 1985). Reaction temperatures above 450°C lead to cracking of longer-chain hydrocarbons into smaller olefins (Costa et al., 1985).

110.2.3.2 Ethanol to Diesel and Jet Fuels

Production of heavier fuel-range hydrocarbons such as jet and diesel fuels from renewable bioethanol is another attractive research topic. Jet fuels (i.e., kerosene) and diesel consist of hydrocarbons between C_8–C_{16} and C_{10}–C_{22} chain lengths, respectively, and are typically composed of isoparaffins and naphthenes.

Olefin oligomerization to heavy hydrocarbons is preferred at low temperatures (<300°C) in order to avoid aromatization, transfer hydrogenation, and cracking reactions. However, as mentioned above, ethylene activation requires high temperatures (>300°C) to enable the formation of primary carbenium intermediates (Eagan et al., 2019; Ghashghaee, 2018). Therefore, a two-step process is the best set up for ethanol conversion into heavy hydrocarbons. First, ethanol is converted into C_3–C_4 olefins at high temperatures (>400°C), which are then oligomerized to produce heavy olefins at low temperatures (<200°C). In the oligomerization step, high pressures shift the product distribution of the reaction to benefit the production of longer-chain olefins, avoiding side reactions (Eagan et al., 2019).

As previously mentioned, ethanol conversion into light olefins is carried out with short residence times to restrict secondary reactions. Additionally, cracking reactions of heavier olefins are promoted by high temperatures above 450°C. Therefore, butenes are the primary product from the ethylene oligomerization. Due to the formation of more stable secondary carbenium intermediate, butenes react faster than ethylene, and their yields do not exceed 15%. (Li et al., 2016; Takahashi et al., 2012). Propene is mostly obtained from cracking of C_{5+} olefins (Goguen et al., 1998; Ingram and Lancashire, 1995), which gives a yield of around 30% (Li et al., 2016). As a result, the maximum C_3-C_4 olefin yield from ethanol conversion is 45%, with ethylene being the major product (~40%) (Eagan et al., 2019). For the second stage, some studies have reported oligomerization of butenes on acidic zeolites that can achieve C_{8+} selectivity up to 90% with high conversions (>90%) at temperatures lower than 200°C and high pressure (63 bar) (Bond et al., 2015; Coelho et al., 2013; Kim et al., 2015). Furthermore, oligomerization of propene over modified zeolites at 200°C and 40 bar reached 60%–75% selectivity to diesel-range hydrocarbons (Corma et al., 2013). Likewise, propene conversion (200°C, 40 bar) over a solid phosphoric acid has demonstrated high selectivity to nonene (59.6%) and dodecane (19.1%) with 98% of conversion (Zhang et al., 2015). The high selectivity to C_{9+} hydrocarbons in part is because the propene oligomers enable to feature tertiary carbenium ions which are more reactive than propene. Therefore, formation of tertiary carbenium ions as intermediate facilitate the route to heavier oligomers (Eagan et al., 2019). In general, the optimization of proper reaction conditions and feed composition is crucial to achieve selective oligomerization to heavy hydrocarbons as well as employing sufficiently active catalysts that are capable of high conversions at low temperature.

110.3 CONCLUSIONS AND OUTLOOK

The utilization of carbon-neutral feedstocks in industrial chemical production is and will be a cornerstone of sustainable, green technologies in the fight against climate change. Successful valorization of bioethanol to chemicals that can supplement or replace fossil feedstocks entirely is an increasingly attractive route being pursued by industry and academic researchers. The facile production and availability of bioethanol makes it a versatile molecule that can readily be used as sustainable feedstock for the current societal demands; however, as this chapter highlighted, the technologies behind such processes must contend with the inherent complexities from upgrading such a simple molecule to chemicals of higher value.

As has been discussed in this chapter, researchers have already demonstrated the efficacy of using bioethanol as a platform molecule by converting it into C_3-C_4 olefins, n-butenes, 1,3-butadiene, isobutene, and even into a fuel simulant. The success of the various catalysts employed for these reactions was shown to stem from several key factors: reaction conditions, the overall acidity of the catalytic surface, and the presence of water (in the case of bioethanol). These factors have been studied in detail to overcome different challenges associated with the catalytic bioethanol conversion to hydrocarbons. Many common approaches employed involve tuning the acidity via the Si/Al ratio in zeolites, altering catalyst structure and composition, adding secondary metal dopants, or increasing/decreasing water content while simultaneously optimizing reaction conditions by varying reaction temperatures and residence time. By simply modifying the surface acidity and reaction conditions, ethanol can be selectively upgraded through a multitude of varying reaction pathways including dehydration/dehydrogenation, aldol condensation/self-deoxygenation, polymerization, cracking, etc.

Despite the plethora of knowledge obtained from ethanol valorization technologies in recent decades, there is still a lot of room for further innovation. The simplicity of ethanol is both a benefit and drawback since it requires relatively little effort to produce as a feedstock, but also demands complex catalytic designs and fundamental understandings to selectively control its conversion to higher-order hydrocarbons. Therefore, further research must be surround ethanol upgradation so that such technologies can be inevitably scaled to industrial capacities.

REFERENCES

Aguayo, A. T., A. G. Gayubo, A. Atutxa, M. Olazar and J. Bilbao. 2002. Catalyst deactivation by coke in the transformation of aqueous ethanol into hydrocarbons. Kinetic modeling and acidity deterioration of the catalyst. *Industrial & Engineering Chemistry Research* 41: 4216–4224.

Alvira, P., E. Tomas-Pejo, M. Ballesteros and M. J. Negro. 2010. Pretreatment technologies for an efficient bioethanol production process based on enzymatic hydrolysis: A review. *Bioresource Technology* 101:4851–4861.

Angelici, C., B. M. Weckhuysen and P. C. Bruijnincx. 2013. Chemocatalytic conversion of ethanol into butadiene and other bulk chemicals. *ChemSusChem* 6:1595–1614.

Arnett, E. M. and T. C. Hofelich 1983. Stabilities of carbocations in solution. 14. An extended thermochemical scale of carbocation stabilities in a common superacid. *Journal of the American Chemical Society* 105:2889–2895.

Baylon, R. A. L., J. Sun, K. J. Martin, P. Venkitasubramanian and Y. Wang. 2016. Beyond ketonization: Selective conversion of carboxylic acids to olefins over balanced Lewis acid-base pairs. *Chemical Communications* 52: 4975–4978.

Bi, J., X. Guo, M. Liu and X. Wang. 2010. High effective dehydration of bio-ethanol into ethylene over nanoscale HZSM-5 zeolite catalysts. *Catalysis Today* 149:143–147.

Bond, J. Q., D. M. Alonso, D. Wang, R. M. West and J. A. Dumesic. 2010. Integrated catalytic conversion of γ-valerolactone to liquid alkenes for transportation fuels. *Science* 327:1110–1114.

Boyadjian, C., K. Seshan, A. G. J. van der Ham and H. van den Berg. 2011. Production of C_3/C_4 olefins from n-hexane: Conceptual design of a catalytic oxidative cracking process and comparison to steam cracking. *Industrial & Engineering Chemistry Research* 50:342–351.

Bussi, J., S. Parodi, B. Irigaray and R. Keffer 1998. Catalytic transformation of ethanol into acetone using copper-pyrochlore catalysts. *Applied Catalysis A: General* 172(1): 117–129.

Calsavara, V., M. L. Baesso and N. R. C. Fernandes-Machado. 2008. Transformation of ethanol into hydrocarbons on ZSM-5 zeolites modified with iron in different ways. *Fuel* 87:1628–1636.

Chang, C. D. 1991. MTG Revisited. *Studies in Surface Science and Catalysis* 61:393–404.

Chang, C. D. and A. J. Silvestri. 1977. The conversion of methanol and other O-compounds to hydrocarbons over zeolite catalysts. *Journal of Catalysis* 47:249–259.

Chaudhuri, S. N., C. Halik and J. A. Lercher. 1990. Reactions of ethanol over HZSM-5. *Journal of Molecular Catalysis* 62:289–295.

Chen, G., S. Li, F. Jiao and Q. Yuan. 2007. Catalytic dehydration of bioethanol to ethylene over $TiO_2/\gamma\text{-}Al_2O_3$ catalysts in microchannel reactors. *Catalysis Today* 125:111–119.

Choi, M., H. S. Cho and R. Srivastava, et al. 2006. Amphiphilic organosilane-directed synthesis of crystalline zeolite with tunable mesoporosity. *Nature Materials* 5:718–723.

Choi, M., K. Na and J. Kim, et al. 2009. Stable single-unit-cell nanosheets of zeolite MFI as active and long-lived catalysts. *Nature* 461:246–249.

Coelho, A., G. Caeiro, M. A. N. D. A. Lemos, F. Lemos and J. R. Ribeiro. 2013. 1-Butene oligomerization over ZSM-5 zeolite: Part 1 - Effect of reaction conditions. *Fuel* 111:449–460.

Corma, A., C. Martinez and E. Doskocil. 2013. Designing MFI-based catalysts with improved catalyst life for C3=andC5= oligomerization to high-quality liquid fuels. *Journal of Catalysis* 300:183–196.

Corson, B. B., H. E. Jones, C. E. Welling, J. A. Hinckley and E. E. Stahly. 1950. Butadiene from ethyl alcohol. Catalysis in the one- and two-step processes. *Industrial Engineering Chemistry* 42:359–373.

Costa, E., A. Uguina, J. Aguado and P. J. Henrnandez. 1985. Ethanol to gasoline process - effect of variables, mechanism, and kinetics. *Industrial & Engineering Chemistry Process Design and Development* 24:239–244.

Dagle, R. A., A. D. Winkelman, K. K. Ramasamy, V. L. Dagle and R. S. Weber. 2020a. Ethanol as a renewable building block for fuels and chemicals. *Industrial & Engineering Chemistry Research* 59:4843–4853.

Dagle, V. L., A. D. Winkelman and N. R. Jaegers, et al. 2020b. Single-step conversion of ethanol to n-butene over Ag-ZrO2/SiO_2 catalysts. *ACS Catalysis* 10:10602–10613.

Diaz Alvarado, F. and F. Gracia 2010. Steam reforming of ethanol for hydrogen production: Thermodynamic analysis including different carbon deposits representation. *Chemical Engineering Journal* 165:649–657.

Dolejsek, Z., J. Novakova, V. Bosacek and L. Kubelkova. 1991. Reaction of ammonia with surface species formed from acetone on a HZSM-5 zeolite. *Zeolites* 11:244–247.

Eagan, N. M., M. D. Kumbhalkar, J. C. Buchanan, J. A. Dumesic and G.W. Huber. 2019. Chemistries and processes for the conversion of ethanol into middle-distillate fuels. *Nature Reviews Chemistry* 3:223–249.

Engelder, C. 1916. Decomposition of ethyl alcohol. *Journal of Physical Chemistry* 21:676–704.

Farrell, A. E., R. J. Plevin and B. T. Turner, et al. 2006. Ethanol can contribute to energy and environmental goals. *Science* 311:506–508.

Gao, Y., L. Neal and D. Ding, et al. 2019. Recent advances in intensified ethylene production-a review. *ACS Catalysis* 9:8592–8621.

Gayubo, A. G., A. Alonso and B. Valle, et al. 2010. Hydrothermal stability of HZSM-5 catalysts modified with Ni for the transformation of bioethanol into hydrocarbons. *Fuel* 89:3365–3372.

Gayubo, A. G., A. Alonso and B. Valle, et al. 2011. Kinetic modelling for the transformation of bioethanol into olefins on a hydrothermally stable Ni-HZSM-5 catalyst considering the deactivation by coke. *Chemical Engineering Journal* 167:262–277.

Gayubo, A. G., A. M. Tarrio, A. T. Aguayo, M. Olazar and J. Bilbao. 2001. Kinetic modelling of the transformation of aqueous ethanol into hydrocarbons on a HZSM-5 zeolite. *Industrial & Engineering Chemistry Research* 40:3467–3474.

Ghashghaee, M. 2018. Heterogeneous catalysts for gas-phase conversion of ethylene to higher olefins. *Reviews in Chemical Engineering* 34:595–655.

Goguen, P. W., T. Xu and D. H. Barich, et al. 1998. Pulse-quench catalytic reactor studies reveal a carbon-pool mechanism in methanol-to-gasoline chemistry on zeolite HZSM-5. *Journal of the American Chemical Society* 120:2650–2651.

Graschinsky, C., P. Giunta, N. Amadeo and M. Laborde. 2012. Thermodynamic analysis of hydrogen production by autothermal reforming of ethanol. *International Journal of Hydrogen Energy* 37:10118–10124.

Gray, K. A., L. Zhao and M. Emptage. 2006. Bioethanol. *Current Opinion in Chemical Biology* 10:141–146.

Groom, M. J., E. M. Gray and P. A. Townsend. 2008. Biofuels and biodiversity: Principles for creating better policies for biofuel production. *Conservation Biology* 22:602–609.

Hahn-Hagerdal, B., M. Galbe, M. F. Gorwa-Grauslund, G. Liden and G. Zacch. 2006. Bio-ethanol - the fuel of tomorrow from the residues of today. *Trends in Biotechnology* 24:549–556.

Han, Y., N. Li and L. Zhao, et al. 2003. Understanding of the high hydrothermal stability of the mesoporous materials prepared by the assembly of triblock copolymer with preformed zeolite precursors in acidic media. The *Journal of Physical Chemistry B* 107:7551–7556.

Hayashi, F. and M. Iwamoto 2013. Yttrium-modified ceria as a highly durable catalyst for the selective conversion of ethanol to propene and ethene. *ACS Catalysis* 3:14–17.

Hiroo, N., M. Saburo and E. Etsuro. 1972. Butadiene formation from ethanol over silica-magnesia catalysts. *Bulletin of the Chemical Society of Japan* 45:655–659.

Huber, G. W., S. Iborra and A. Corma. 2006. Synthesis of transportation fuels from biomass: Chemistry, catalysts, and engineering. *Chemical Reviews* 106:4044–4098.

Hutchings, G. J., P. Johnston and J. Silvestri. 1994. The conversion of methanol and other O-Compounds to hydrocarbons over zeolite β. *Journal of Catalysis* 147:177–185.

Ingram, C. W. and R. J. Lancashire. 1995. On the formation of C_3 hydrocarbons during the conversion of ethanol using H-ZSM-5 catalyst. *Catalysis Letters* 31:395–403.

Iwamoto, M. 2011. One step formation of propene from ethene or ethanol through metathesis on nickel ion-loaded silica. *Molecules* 16:7844–7863.

Iwamoto, M. 2015. Selective catalytic conversion of bio-ethanol to propene: A review of catalysts and reaction pathways. *Catalysis Today* 242: 243–248.

Iwamoto, M., K. Kasai and T. Haishi. 2011. Conversion of ethanol into polyolefin building blocks: Reaction pathways on nickel ion-loaded mesoporous silica. *ChemSusChem* 4:1055–1058.

Janik, M. J., J. Macht, E. Iglesia and M. Neurock. 2009. Correlating acid properties and catalytic function: A first-principles analysis of alcohol dehydration pathways on polyoxometalates. *Journal of Physical Chemistry C* 113:1872–1885.

Jones, M. D., C. G. Keir and C. di Iulio, et al. 2011. Investigations into the conversion of ethanol into 1,3-butadiene. *Catalysis Science & Technology* 1:267–272.

Kagyrmanova, A. P., V. A. Chumachenko, V. N. Korotkikh, V. N. Kashkin and A. S. Noskov. 2011. Catalytic dehydration of bioethanol to ethylene: Pilot-scale studies and process simulation. *Chemical Engineering Journal* 176-177:188–194.

Kim, Y. T., J. P. Chada and Z. Xu, et al. 2015. Low-temperature oligomerization of 1-butene with H-ferrierite. *Journal of Catalysis* 323: 33–44.

Kitayama, Y. and A. Michishita. 1981. Catalytic activity of fibrous clay mineral sepiolite for butadiene formation from ethanol. *Journal of the Chemical Society-Chemical Communications* 1981:401–402.

Kochar, N. K. and A. S. Padia 1981. Ethylene from ethanol. *Chemical Engineering Process* 77: 66–70.

Krossner, M. and J. Sauer 1996. Interaction of water with brønsted acidic sites of zeolite catalysts. Ab initio study of 1:1 and 2:1 surface complexes. *Journal of Physical Chemistry* 100:6199–6211.

Kvisle, S., A. Aguero and R. P. A. Sneeden. 1988. Transformation of ethanol into 1,3-butadiene over magnesium oxide/silica catalysts. *Applied Catalysis* 43:117–131.

Kwak, J. H., R. Rousseau, D. Mei, C. H. F. Peden and J. Szanyl. 2011. The origin of regioselectivity in 2-butanol dehydration on solid acid catalysts. *ChemCatChem* 3:1557–1561.

Le van Mao, R., P. Levesque, G. McLaughlin and L. H. Dao. 1987. Ethylene from ethanol over zeolite catalysts. *Applied Catalysis* 34:163–179.

Le van Mao, R., T. M. Nguyen and G. M. McLaughlin. 1989. The bioethanol-to-ethylene (B.E.T.E.) process. *Applied Catalysis* 48:265–277.

Le Van Mao, R., T. M. Nguyen and J. Yao. 1990. Conversion of ethanol in aqueous solution over ZSM-5 zeolites: Influence of reaction parameters and catalyst acidic properties as studied by ammonia TPD technique. *Applied Catalysis* 61:161–173.

Lehmann, T. and A. Seidel-Morgenstern 2014. Thermodynamic appraisal of the gas phase conversion of ethylene or ethanol to propylene. *Chemical Engineering Journal* 242:422–432.

Li, D., P. Bui and H. Y. Zhao, et al. 2012. Rake mechanism for the deoxygenation of ethanol over a supported Ni_2P/SiO_2 catalyst. *Journal of Catalysis* 290:1–12.

Li, H., D. Guo and L. Ulumuddin, et al. 2021a. Elucidating the cooperative roles of water and Lewis acid-base pairs in cascade C-C coupling and self-deoxygenation reactions. *JACS Au* 1:1471–1481.

Li, H., J. Pang and N. R. Jaegers, et al. 2021b. Conversion of ethanol to 1,3-butadiene over $Ag\text{-}ZrO_2/SiO_2$ catalysts: The role of surface interfaces. *Journal of Energy Chemistry* 54:7–15.

Li, X., A. Kant and Y. He, et al. 2016. Light olefins from renewable resources: Selective catalytic dehydration of bioethanol to propylene over zeolite and transition metal oxide catalysts. *Catalysis Today* 276:62–77.

Liu, C., J. Sun, C. Smith and Y. Wang. 2013. A study of ZnxZryOz mixed oxides for direct conversion of ethanol to isobutene. *Applied Catalysis A: General* 467: 91–97.

Madeira, F. F., K. Ben Tayeb and L. Pinard, 2012. Ethanol transformation into hydrocarbons on ZSM-5 zeolites: Influence of Si/Al ratio on catalytic performances and deactivation rate. Study of the radical species role. *Applied Catalysis A: General* 443–444: 171–180.

Madeira, F. F., N. S. Gnep and P. Magnoux, et al. 2010. Mechanistic insights on the ethanol transformation into hydrocarbons over HZSM-5 zeolite. *Chemical Engineering Journal* 161:403–408.

Madeira, F. F., N. S. Gnep, P. Magnoux, S. Maury and N. Cadran. 2009. Ethanol transformation over HFAU, HBEA and HMFI zeolites presenting similar Brønsted acidity. *Applied Catalysis A: General* 367:39–46.

Makshina, E. V., W. Janssens, B. F. Sels and P. A. Jacobs. 2012. Catalytic study of the conversion of ethanol into 1,3-butadiene. *Catalysis Today* 198:338–344.

Marcu, I.-C., D. Tichit, F. Fajula and N Tanchoux. 2009. Catalytic valorization of bioethanol over Cu-Mg-Al mixed oxide catalysts. *Catalysis Today* 147:231–238.

Masuda, T., Y. Fujikata, S. R. Mukai and K. Hashimoto. 1998. Changes in catalytic activity of MFI-type zeolites caused by dealumination in a steam atmosphere. *Applied Catalysis A: General* 172:73–83.

Meng, T., D. Mao, Q. Guo and G. Lu. 2012. The effect of crystal sizes of HZSM-5 zeolites in ethanol conversion to propylene. *Catalysis Communications* 21:52–57.

Miciukiewicz, J., T. Mang and H. Knozinger. 1995. Raman spectroscopy characterization of molybdena supported on titania-zirconia mixed oxide. *Applied Catalysis A: General* 122:151–159.

Mizuno, S., M. Kurosawa, M. Tanaka and M. Iwamoto. 2012. One-path and selective conversion of ethanol to propene on scandium-modified indium oxide catalysts. *Chemistry Letters* 41:892–894.

Morschbacker, A. 2009. Bio-ethanol based ethylene. *Polymer Reviews* 49:79–84.

Moser, W. R., R. W. Thompson, C.W. Chiang and H. Tong. 1989. Silicon-rich H-ZSM-5 catalyzed conversion of aqueous ethanol to ethylene. *Journal of Catalysis* 117:19–32.

Murthy, R. S., P. Patnaik, P. Sidheswaran and M. Jayamani. 1988. Conversion of ethanol to acetone over promoted iron oxide catalysis. *Journal of Catalysis* 109:298–302.

Nakajima, T., H. Nameta, S. Mishima and K. Tanabe. 1994. A highly active and highly selective oxide catalyst for the conversion of ethanol to acetone in the presence of water vapour. *Journal of Materials Chemistry* 4:853–858.

Nakajima, T., K. Tanabe, T. Yamaguchi, I. Matsuzaki and S. Mishima. 1989. Conversion of ethanol to acetone over zinc oxide-calcium oxide catalyst optimization of catalyst preparation and reaction conditions and deduction of reaction mechanism. *Applied Catalysis* 52:237–248.

Nakajima, T., T. Yamaguchi and K. Tanabe. 1987. Efficient synthesis of acetone from ethanol over Zno-CaO catalyst. *Journal of the Chemical Society-Chemical Communications* 1987:394–395.

Nguyen, T. M. and R. Le van Mao. 1990. Conversion of ethanol in aqueous solution over ZSM-5 zeolites: Study of the reaction network. *Applied Catalysis* 58:119–129.

Nishiguchi, T., T. Matsumoto and H. Kanai, et al. 2005. Catalytic steam reforming of ethanol to produce hydrogen and acetone. *Applied Catalysis A: General* 279:273–277.

Ohnishi, R., T. Akimoto and K. Tanabe. 1985. Pronounced catalytic activity and selectivity of MgO-SiO_2-Na_2O for synthesis of buta-1,3-diene from ethanol. *Journal of the Chemical Society, Chemical Communications* 1985:1613–1614.

Panov, A. G. and J. J. Fripiat. 1998. Acetone condensation reaction on acid catalysts. *Journal of Catalysis* 178:188–197.

Pease, R. N. and C. Y. Yung. 1924. The catalytic dehydration of ethyl alcohol and ether by alumina. *Journal of American Chemistry Society* 446:390–403.

Phillips, C. B. and R. Datta 1997. Production of ethylene from hydrous ethanol on H-ZSM-5 under mild conditions. *Industrial & Engineering Chemistry Research* 36: 4466–4475.

Rahman, M. M., S. D. Davidson, J. Sun and Y. Wang. 2016. Effect of water on ethanol conversion over ZnO. *Topics in Catalysis* 59:37–45.

Ramasamy, K. K., M. A. Gerber, M. Flake, H. Zhang and Y. Wang. 2014a. Conversion of biomass-derived small oxygenates over HZSM-5 and its deactivation mechanism. *Green Chemistry* 16:748–760.

Ramasamy, K. K., H. Zhang, J. Sun and Y. Wang. 2014b. Conversion of ethanol to hydrocarbons on hierarchical HZSM-5 zeolites. *Catalysis Today* 238:103–110.

RFA. 2022. *Annual Fuel Ethanol Production: U.S. and World Ethanol Production*. Renewable Fuels Association. https://ethanolrfa.org/markets-and-statistics/annual-ethanol-production

Saha, S. K. and S. Sivasanker 1992. Influence of Zn- and Ga-doping on the conversion of ethanol to hydrocarbons over ZSM-5. *Catalysis Letters* 15:413–418.

Salvapati, G. S., K. V. Ramanamurty and M. Janardanarao. 1989. Selective catalytic self-condensation of acetone. *Journal of Molecular Catalysis* 54:9–30.

Sanchez, O. J. and C. A. Cardona 2008. Trends in biotechnological production of fuel ethanol from different feedstocks. *Bioresource Technology* 99:5270–5295.

Schulz, J. and F. Bandermann 1994. Conversion of ethanol over zeolite H-ZSM-5. *Chemical Engineering & Technology* 17:179–186.

Song, Z., A. Takahashi, I. Nakamura and T. Fujitani. 2010. Phosphorus-modified ZSM-5 for conversion of ethanol to propylene. *Applied Catalysis A: General* 384:201–205.

Song, Z., A. Takahashi, N. Mimura and T. Fujitani. 2009. Production of propylene from ethanol over ZSM-5 zeolites. *Catalysis Letters* 131:364–369.

Su, L., L. Liu and J. Zhuang, et al. 2003. Creating mesopores in ZSM-5 zeolite by alkali treatment: A new way to enhance the catalytic performance of methane dehydroaromatization on Mo/HZSM-5 catalysts. *Catalysis Letters* 91:155–167.

Sun, J. and Y. Wang 2014. Recent advances in catalytic conversion of ethanol to chemicals. *ACS Catalysis* 4:1078–1090.

Sun, J., K. Zhu and F. Gao, et al. 2011. Direct conversion of bio-ethanol to isobutene on nanosized $Zn_xZr_yO_z$ mixed oxides with balanced acid-base sites. *Journal of the American Chemical Society* 133:11096–11099.

Tago, T., H. Konno, M. Sakamoto, Y. Nakasaga and T. Masuda. 2011. Selective synthesis for light olefins from acetone over ZSM-5 zeolites with nano- and macro-crystal sizes. *Applied Catalysis A: General* 403:183–191.

Takahara, I., M. Saito, M. Inaba and K. Murata. 2005. Dehydration of ethanol into ethylene over solid acid catalysts. *Catalysis Letters* 105:249–252.

Takahashi, A., W. Xia and I. Nakamura, H. Shimada and T. Fujitani. 2012. Effects of added phosphorus on conversion of ethanol to propylene over ZSM-5 catalysts. *Applied Catalysis A: General* 423–424:162–167.

Takahashi, A., W. Xia and Q. Wu, et al. 2013. Difference between the mechanisms of propylene production from methanol and ethanol over ZSM-5 catalysts. *Applied Catalysis A: General* 467:380–385.

Talukdar, A. K., K. G. Bhattacharyya and S. Sivasanker. 1997. HZSM-5 catalysed conversion of aqueous ethanol to hydrocarbons. *Applied Catalysis A: General* 148:357–371.

Tanabe, K., M. Misono, H. Hattori and Y. Ono. 1990. *New Solid Acids and Bases: Their Catalytic Properties*. Amsterdam: Elsevier Science.

Trane, R., S. Dahl, M. S. Skjoth-Rasmussen and A. D. Jensen. 2012. Catalytic steam reforming of bio-oil. *International Journal of Hydrogen Energy* 37:6447–6472.

Tynjala, P., T. T. Pakkanen and S. Mustamaki. 1998. Modification of ZSM-5 zeolite with trimethyl phosphite. 2. Catalytic properties in the conversion of C_{1-C4} alcohols. *Journal of Physical Chemistry B* 102:5280–5286.

USDA. 2022. *Ethanol Fuel Basics*. Washington, DC: U.S. Department of Energy Alternative Fuels Data Center. https://afdc.energy.gov/fuels/

Varisli, D., T. Dogu and G. Dogu. 2007. Ethylene and diethyl-ether production by dehydration reaction of ethanol over different heteropolyacid catalysts. *Chemical Engineering Science* 62:5349–5352.

Wang, Y. and S. Liu 2012. Butadiene production from ethanol. *Journal of Bioprocess Engineering and Biorefinery* 1:33–43.

Wu, Y. and S. J. Marwil. 1980. *Dehydration of Alcohols*. US Patent 4,234,752.

Xu, T., E. J. Munson and J. F. Haw. 1994. Toward a systematic chemistry of organic reactions in zeolites: *In situ* NMR studies of ketones. *Journal of the American Chemical Society* 116:1962–1972.

Zaki, M. I., M. A. Hasan and L. Pasupulety. 2001. *In Situ* FTIR spectroscopic study of 2-propanol adsorptive and catalytic interactions on metal-modified aluminas. *Langmuir* 17:4025–4034.

Zaki, T. 2005. Catalytic dehydration of ethanol using transition metal oxide catalysts. *Journal of Colloid and Interface Science* 284:606–613.

Zhang, D., R. Wang and X. Yang. 2008. Effect of P content on the catalytic performance of P-modified HZSM-5 catalysts in dehydration of ethanol to ethylene. *Catalysis Letters* 124:384–391.

Zhang, J., Y. Yan, Q. Chu and J. Feng. 2015. Solid phosphoric acid catalyst for propene oligomerization: Effect of silicon phosphate composition. *Fuel Processing Technology* 135:2–5.

Zhang, M. and Y. Yu. 2013. Dehydration of ethanol to ethylene. *Industrial & Engineering Chemistry Research* 52:9505–9514.

Zhu, K., J. Sun and J. Liu, et al. 2011. Solvent evaporation assisted preparation of oriented nanocrystalline mesoporous MFI zeolites. *ACS Catalysis* 1: 682–690.

Zhu, K., J. Sun, H. Zhang, J. Liu and Y. Wang. 2012. Carbon as a hard template for nano material catalysts. *Journal of Natural Gas Chemistry* 21:215–232.

Index